OBJECT DETECTION AND RECOGNITION IN DIGITAL IMAGES

OBJECT DETECTION AND RECOGNITION IN DIGITAL IMAGES

THEORY AND PRACTICE

Bogusław Cyganek

AGH University of Science and Technology, Poland

A John Wiley & Sons, Ltd., Publication

Library of Congress Cataloging-in-Publication Data

Cyganek, Boguslaw.
  Object detection and recognition in digital images : theory and practice / Boguslaw Cyganek.
      pages cm
  Includes bibliographical references and index.
  ISBN 978-0-470-97637-1 (cloth)
  1. Pattern recognition systems.   2. Image processing–Digital techniques.   3. Computer vision.   I. Title.
  TK7882.P3C94 2013
  621.39′94–dc23

                                                                                              2012050754

A catalogue record for this book is available from the British Library

ISBN: 978-0-470-97637-1

Typeset in 10/12pt Times by Aptara Inc., New Delhi, India

To my family with love

Contents

Preface

We live in an era of technological revolution in which developments in one domain frequently entail breakthroughs in another. Similar to the nineteenth century industrial revolution, the last decades can be termed an epoch of computer revolution. For years we have been witnessing the rapid development of microchip technologies which has resulted in a continuous growth of computational power at ever decreasing costs. This has been underpinned by the recent developments of parallel computational systems of graphics processing units and field programmable gate arrays. All these hardware achievements also open up new application areas and possibilities in the quest of making a computer see and understand what it sees – is a primary goal in the domain of computer vision. However, although fast computers are of great help in this respect, what really makes a difference are new and better processing methods and their implementations.

The book presents selected methods of object detection and recognition with special stress on statistical and – relatively new to this domain – tensor based approaches. However, the number of interesting and important methods is growing rapidly, making it difficult to offer a complete coverage of these methods in one book. Therefore the goal of this book is slightly different, namely the methods chosen here have been used by myself and my colleagues in many projects and proved to be useful in practice. Our main areas concern automotive applications in which we try to develop vision systems for road sign recognition or driver monitoring. When starting this book my main purpose was to not only give an overview of these methods, but also to provide the necessary, though concise, mathematical background. However, just as important are implementations of the discussed methods. I'm convinced that the connection of detailed theory and its implementation is a prerequisite for the in-depth understanding of the subject. In this respect the choice of the implementation platform is also not a surprise. The C++ programming language used throughout this book and in the attached software library is of worldwide industry standard. This does not mean that implementations cannot be done using different programming platforms, for which the provided code examples can be used as a guide or for direct porting. The book is accompanied by a companion website at www.wiley.com/go/cyganekobject which contains the code, color figures, as well as slides, errata and other useful links.

This book grew as a result of my fascination with modern computer vision methods and also after writing my previous book, co-authored with J. Paul Siebert and devoted mostly to the processing of 3D images. Thus, in some sense it can be seen as a continuation of our previous work, although both can be read as standalone texts.

Thus, the book can be used by all scientists and industry practitioners related to computer vision and machine pattern recognition, but can also be used as a tutorial for students interested in this rapidly developing area.

Bogusław Cyganek
Poland

Acknowledgements

Writing a book is a tremendous task. It would not be possible if not for the help of friends, colleagues, cooperators, and many other people, whose names I sometimes don't even know, but I know they did wonderful work to make this book happen.

I would particularly like to thank numerous colleagues from the AGH University of Science and Technology, as well as the Academic Computer Centre Cyfronet, in Kraków, Poland. Special thanks go to Professor Ryszard Tadeusiewicz and Professor Kazimierz Wiatr for their continuous encouragement and support.

I would also like to express my thanks to Professor Ralf Reulke from the Humboldt-Universität zu Berlin, and Deutsches Zentrum für Luft- und Raumfahrt, as well as to all the colleagues from his team, for our fruitful cooperation in interesting scientific endeavours.

I'm very grateful to the Wiley team who have helped to make this book possible. I'd like to express my special thanks to Richard Davies, Alex King, Nicky Skinner, Simone Taylor, Liz Wingett, as well as to Nur Wahidah Binte Abdul Wahid, Shubham Dixit, Caroline McPherson, and all the others whose names I don't know but I know they did a brilliant job to make this book happen. Once again – many thanks!

I'm also very grateful to many colleagues around the world, and especially readers of my previous book on 3D computer vision, for their e-mails, questions, suggestions, bug reports, and all the discussions we've had. All these helped me to develop better text and software. I also ask for your support now and in the future!

I would like to kindly express my gratitude to the National Science Centre NCN, Republic of Poland, for their financial support in scientific research projects conducted over the years 2007–2009, as well as 2011–2013 under the contract no. DEC-2011/01/B/ST6/01994, which greatly contributed to this book. I would also like to express my gratitude to the AGH University of Science and Technology Press for granting the rights to use parts of my previous publication.

Finally, I would like to thank my family: my wife Magda, my children Nadia and Kamil, as well as my mother, for their patience, support, and encouragement during all the days I worked on this book.

Notations and Abbreviations

B	Base matrix
C	Number of data classes
C	Coefficient matrix
$\mathbf{C}_x$	Correlation matrix of a data set $\{\mathbf{x}_i\}$
D	Data matrix
D	Distance function
E	Statistical expectation
i, j, k, m, n	Free coordinates, matrix indices
$\mathbf{1}_n$	Matrix of dimensions $n \times n$ with all elements set to 1
$\mathbf{I}_n$	Identity matrix of dimensions $n \times n$
I	Image; Intensity signal of an image
I_x, I_y	Spatial derivatives of an image I in the directions x, y
J	Number of components in a series
K	Kernel matrix
L	Number of components in a vector; Dimensionality of a space
M	Number of clusters; Number of image channels
N	Number of (data) points
P	Probability mass function
p	Probability density function
P, Q, C	Numbers of indices in tensors (tensor dimensions)
p, q	Covariant and contravariant degrees of a tensor
R	Number of principal components
$\mathfrak{R}$	Set of real numbers
$\mathcal{T}$	Tensor
$\mathbf{T}_{(k)}$	k-th flattening mode of a tensor $\mathcal{T}$
$\mathbf{T}_C$	Compact structural tensor
$\mathbf{T}_E$	Extended structural tensor
t	Time coordinate
W	Vector space
W^*	Dual vector space
$\mathbf{X}$	Matrix
$\mathbf{X}^T$	Transposed matrix $\mathbf{X}$
$\mathbf{X}_i$	i-th matrix (from a series of matrices)
x, y	Spatial coordinates

$\mathbf{x}$	Column vector
$\mathbf{x}_i$	i-th vector (from a series of vectors)
$\{\mathbf{x}_i\}$	Set of vectors $\mathbf{x}_i$ for a given range of indices i
$\mathbf{x}_i^{(k)}$	k-th column vector from a matrix $\mathbf{X}_i$
$\hat{\mathbf{x}}$	Normalized column vector
$\bar{\mathbf{x}}$	Mean vector
$\tilde{\mathbf{x}}$	Orthogonal residual vector
x_i	i-th component of the vector $\mathbf{x}$
Σ_x	Covariance matrix of a data set $\{\mathbf{x}_i\}$
ρ	Number of bins in the histogram
Δ	Width of a bin in the histogram
Ω	Set of class labels
$\odot$	Khatri–Rao product
$\otimes$	Kronecker product
$\circledast$	Elementwise multiplication (Hadamard product)
$\oslash$	Elementwise division
$\circ$	Outer product of vectors
$\vee$	Max product
$\wedge$	Min product
$\times$	Morphological outer product
$\forall$	For all
AD	Anisotropic Diffusion
ALS	Alternating Least-Squares
AMI	Affine Moment Invariants
AWG	Adaptive Window Growing
CANDECOMP	CANonical DECOMPosition (of tensors)
CID	Color Image Discriminant
CNMF	Constrained NMF
CP	CANDECOMP / PARAFAC
CST	Compact Structural Tensor
CST	Convolution Standardized Transform
CV	Computer Vision
CVS	Computer Vision System
DAS	Driver Assisting System
DFFS	Distance From Feature Space
DIFS	Distance In Feature Space
DSP	Digital Signal Processing
DT	Distance Transform
EMML	Expectation Maximization Maximum Likelihood
EMD	Earth Mover's Distance
EST	Extended Structural Tensor
FCM	Fuzzy c-Means
FIR	Far Infra-Red
FN	False Negative
FP	False Positive

GHT	Generalized Hough Transform
GLOH	Gradient Location and Orientation Histogram (image descriptor)
GP	Gaussian Processes / Genetic Programming
GPU	Graphics Processing Unit (graphics card)
HDR	High Dynamic Range (image)
HI	Hyperspectral Image
HNN	Hamming Neural Network
HOG	Histogram of Gradients
HOOI	Higher-Order Orthogonal Iteration
HOSVD	Higher-Order Singular Value Decomposition
ICA	Independent Component Analysis
IMED	Image Euclidean Distance
IP	Image Processing
ISM	Implicit Shape Model
ISRA	Image Space Reconstruction Algorithms
KFCM	Kernel Fuzzy c-Means
k-NN	k-Nearest-Neighbor
KPCA	Kernel Principal Component Analysis
K–R	Khatri–Rao product
LP	Log-Polar
LN	Local Neighborhood of pixels
LSH	Locality-Sensitive Hashing
LSQE	Least-Squares Problem with a Quadratic Equality Constraint
LQE	Linear Quadratic Estimator
MAP	Maximum A Posteriori classification
MICA	Multilinear ICA
ML	Maximum Likelihood
MNN	Morphological Neural Network
MoG	Mixture of Gaussians
MPI	Message Passing Interface
MRI	Magnetic Resonance Imaging
MSE	Mean Square Error
NIR	Near Infra-Red
NMF	Nonnegative Matrix Factorization
NTF	Nonnegative Tensor Factorization
NUMA	Non-Uniform Memory Access
OC-SVM	One-Class Support Vector Machine
PARAFAC	PARAllel FACtors (of tensors)
PCA	Principal Component Analysis
PDE	Partial Differential Equation
PDF	Probability Density Function
PERCLOSE	Percentage of Eye Closure
PR	Pattern Recognition
PSNR	Peak Signal to Noise Ratio
R1NTF	Rank-1 Nonnegative Tensor Factorization
RANSAC	RANdom SAmple Consensus

RBF	Radial Basis Function
RLA	Richardson–Lucy algorithm
RMSE	Root Mean Square Error
ROC	Receiver Operating Characteristics
ROI	Region of Interest
RRE	Relative Reconstruction Error
SAD	Sum of Absolute Differences
SIMCA	Soft Independent Modeling of Class Analogies
SIFT	Scale-Invariant Feature Transform (image descriptor)
SLAM	Simultaneous Localization And Mapping
SMO	Sequential Minimal Optimization
SNR	Signal to Noise Ratio
SOM	Self-Organizing Maps
SPD	Salient Point Detector
SSD	Sum of Squared Differences
ST	Structural Tensor
SURF	Speeded Up Robust Feature (image descriptor)
SVM	Support Vector Machine
TDCS	Tensor Discriminant Color Space
TIR	Thermal Infra-Red
TN	True Negative
TP	True Positive
WOC-SVM	Weighted One-Class Support Vector Machine
WTA	Winner-Takes-All

1

Introduction

Look in, let not either the proper quality, or the true worth of anything pass thee, before thou hast fully apprehended it.

—Marcus Aurelius *Meditations*, 170–180 AD
(Translated by Meric Casaubon, 1634)

This book presents selected object detection and recognition methods in computer vision, joining theory, implementation as well as applications. The majority of the selected methods were used in real automotive vision systems. However, two groups of methods were distinguished. The first group contains methods which are based on tensors, which in the last decade have opened new frontiers in image processing and pattern analysis. The second group of methods builds on mathematical statistics. In many cases, object detection and recognition methods draw from these two groups. As indicated in the title, equally important is the explanation of the main concepts of the methods and presentation of their mathematical derivations, as their implementations and usage in real applications. Although object detection and recognition are strictly connected, to some extent both domains can be seen as pattern classification and frequently detection precedes recognition, we make a distinction between the two. Object detection in our definition mostly concerns answering a question about whether a given type of object is present in images. Sometimes, their current appearance and position are also important. On the other hand, the goal of object recognition is to tell its particular type. For instance, we can detect a face, or after that identify a concrete person. Similarly, in the road sign recognition system for some signs, their detection unanimously reveals their category, such as "Yield." However, for the majority of them, we first detect their characteristic shapes, then we identify their particular type, such as "40km/h speed limit," and so forth.

Detection and recognition of objects in the observed scenes is a natural biological ability. People and animals perform this effortlessly in daily life to move without collisions, to find food, avoid threats, and so on. However, similar computer methods and algorithms for scene analysis are not so straightforward, despite their unprecedented development. Nevertheless, biological systems after close observations and analysis provide some hints for their machine realizations. A good example here are artificial neural networks which in their diversity resemble biological systems of neurons and which – in their software realization – are frequently used by computers to recognize objects. This is how the branch of computer science, called

---

computer vision (CV), developed. Its main objective is to make computers see as humans, or even better. Sometimes it becomes possible.

Due to technological breakthroughs, domains of object detection and recognition have changed so dynamically that preparation of even a multivolume publication on the majority of important subjects in this area seems impossible. Each month hundreds of new papers are published with new ideas, theorems, algorithms, etc. On the other hand, the fastest and most ample source of information is Internet. One can easily look up almost all subjects on a myriad of webpages, such as Wikipedia. So, nowadays the purpose of writing a book on computer vision has to be stated somewhat differently than even few years ago. The difference between an ample set of information versus knowledge and experience starts to become especially important when we face a new technological problem and our task is to solve it or design a system which will do this for us. In this case we need a way of thinking, which helps us to understand the state of nature, as well as a methodology which takes us closer to a potential solution. This book grew up in just this way, alongside my work on different projects related to object recognition in images. To be able to apply a given method we need first to understand it. At this stage not just a final formula summarizing a method, but also its detailed mathematical background, are of great use. On the other hand, bare formulas don't yet solve the problem. We need their implementations. This is the second stage, sometimes requiring more time and work than the former. One of the main goals of this book is to join the two domains on a selected set of useful methods of object detection and recognition. In this respect I hope this book will be of practical use, both for self study and also as a reference when working on a concrete problem. Nevertheless, we are not able to go through all stages of all the methods, but I hope the book will provide at least a solid start for further study and development in this fascinating and dynamically changing area.

As indicated in the title, one of my goals was to join theory and practice. My experience is that such composition leads to an in-depth understanding of the subject. This is further underpinned by case studies of mostly automotive applications of object detection and recognition. Thus, sections of this book can be grouped as follows:

- Presentations of methods, their main concepts, and mathematical background.
- Method implementations which contain C++ code listings (sections of this type are indicated with word IMPLEMENTATION).
- Analysis of special applications (their names start with CASE STUDY).

Apart from this we have some special entries which contain brief explanations of some mathematical concepts with examples which aim is to help in understanding the mathematical derivation in the surrounding sections.

A comment on code examples. I have always been convinced that in a book like this we should not spoil pages with an introduction to C, C++ or other basic principles of computer science, as sometimes is the case. The reasons are at least twofold: the first is that for computer science there are a lot of good books available, for which I provide the references. The second reason, is so to not divert a Reader from the main purpose of this book, which is an in-depth presentation of the modern computer vision methods and algorithms. On the other hand, Readers who are not familiar with C++ can skip detailed code explanations and focus on implementation in other platforms. However, there is no better way of learning the method than through practical testing and usage in applications.

This book is based on my experience gathered while working on many scientific projects. Results of these were published in a number of conference and journal articles. In this respect, two previous books are special. The first, *An Introduction to 3D Computer Vision Techniques and Applications*, written together with J. Paul Siebert, was published by Wiley in 2009 [1]. The second is my habilitation thesis [2], also issued in 2009 by the AGH University of Science and Technology Press in Kraków, Poland. Extended parts of the latter are contained in different sections of this book, permission for which was granted by the AGH University Press.

Most of all, I have always found being involved in scientific and industry projects real fun and an adventure leading to self-development. I wish the same to you.

1.1 A Sample of Computer Vision

In this section let us briefly take a look at some applications of computer vision in the systems of driver monitoring, as well as scene analysis. Both belong to the on-car Driver Assisting System aimed at facilitating driving, for example by notifying drivers of incoming road signs, and most of all by preventing car accidents, for example due to the driver falling asleep.

Figure 1.1 depicts a system of cameras mounted in a test car. The cameras can observe the driver and allow the system to monitor his or her state. Cameras can also observe the front of the car for pedestrian detection or road sign recognition, in which case they can send an image like the one presented in Figure 1.2.

What type of information can we draw from such an image? This depends on our goal, certainly. In the real traffic situation depicted we are mainly interested in driving the car safely, avoiding pedestrians and other vehicles in motion or parked, as well as spotting and reacting to

Figure 1.1 System of cameras mounted in a car. The cameras can observe a driver to monitor his/her state. Cameras can also observe the front of a car for pedestrian detection or road sign recognition. Such vision modules will probably soon become standard equipment, being a part of the on-board Driver Assisting System of a car.

Figure 1.2 A traffic scene. A car-mounted computer with cameras can provide information on the road scene to help safe driving. But computer vision can also help you identify where the picture was taken.

traffic signals and signs. However, in a situation where someone sent us this image we might be interested in finding out the name of that street, for instance. What can computer vision do for us? To some extent all of the above, and soon driving a car, at least in special conditions. Let us look at some stages of processing by computer vision methods, details of which are discussed in the next chapters.

Let us first observe that even a single color image has three dimensions, as shown in Figure 1.3(a). In the case of multiple images or a video stream, dimensions grow. Thus, we need tools to analyze such structures. As we will see, tensors offer new possibilities in

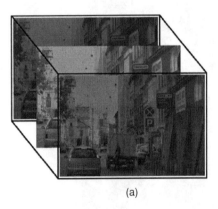

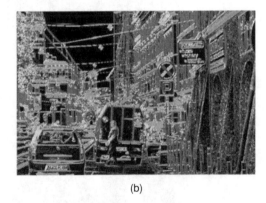

(a)   (b)

Figure 1.3 A color image can be seen as a 3D structure (a). Internal properties of such multidimensional signals can be analyzed with tensors. Local structures can be detected with the structural tensor (b). Here different colors encode orientations of areas with strong signal variations, such as edges. Areas with weak texture are in black. These features can be used to detect pedestrians, cars, road signs and other objects. (For a color version of this figure, please see the color plate section.)

this respect. Also, their recently developed decompositions allow insight into information contained in such multidimensional structures, as well as their compression or extraction of features for further classification. Much research into computer vision and pattern recognition is on feature detection and their properties. In this respect such transformations are investigated which change the original intensity or color pixels into some new representation which provides some knowledge about image contents or is more appropriate for finding specific objects. An example of an application of the structural tensor to image in Figure 1.2 for detection of areas with strong local structures is shown in Figure 1.3(b). Found structures are encoded with color – their orientation is represented by different colors, whereas strength is by color saturation. Let us observe that areas with no prominent structures show no response of this filter – in Figure 1.3(b) they are simply black. As will be shown, such representation proves very useful in finding specific figures in images, such as pedestrians, cars, or road signs, and so forth.

Let us now briefly show the possible steps that lead to detection of road signs in the image in Figure 1.2. In this method signs are first detected with fast segmentation by specific colors characteristic to different groups of expected signs. For instance, red color segmentation is used to spot all-red objects, among which could also be the red rims of the prohibitive signs, and so on for all colors of interest.

Figure 1.4 shows binary maps obtained of the image in Figure 1.2 after red and blue segmentations, respectively. There are many segmentation methods which are discussed in this book. In this case we used manually gathered color samples which were used to train the support vector classifiers.

From the maps in Figure 1.4 we need to find a way of selecting objects whose shape and size potentially correspond to the road signs we are looking for. This is done by specific methods which rely on detection of salient points, as well as on fuzzy logic rules which define the potential shape and size of the candidate objects.

Figure 1.5 shows the detected areas of the signs. These now need to be fed to the next classifier which will provide a final response, first if we are really observing a sign and not for instance a traffic light, and then what the type of particular sign it is. However, observed signs can be of any size and can also be rotated. Classifiers which can cope with such patterns are for instance the cooperating groups of neural networks or the decomposition of tensors of deformed prototypes. Both of the aforementioned classifiers respond with the correct type of signs visible in Figure 1.2. These, as well as many other methods of object detection and recognition, are discussed in this book.

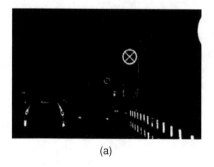
(a)
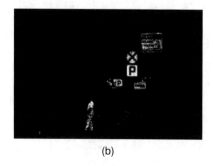
(b)

Figure 1.4   Segmentation of image in Figure 1.2. Red (a), blue color segmentation (b).

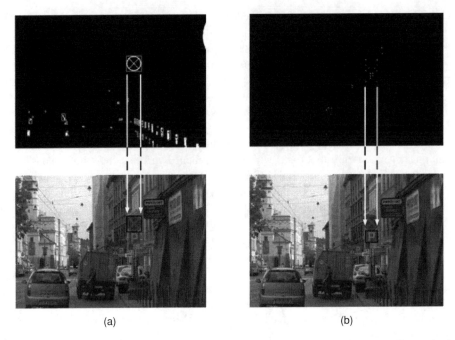

(a)   (b)

Figure 1.5 Circular signs are found by outlining all red objects detected in the scene. Then only those which fulfill the definition and relative size expected for a sign are left (a). Triangular and rectangular shapes are found based on their corner points. The points are checked for all possible rectangles and again only those which comply with fuzzy rules defining sought figures are left (b). (For a color version of this figure, please see the color plate section.)

1.2 Overview of Book Contents

Organizing a book is not straightforward due to many the interrelations between the topics discussed. Such relations are not linear, and in this respect electronic texts with inner links show many benefits. The printed version has its own features. On the one hand, the book can be read linearly, from the beginning to its end. On the other, selected topics can be read independently, especially when looking for a specific method or its implementation. The book is organized into six chapters, starting with the Introduction.

Chapter 2 is entirely devoted to different aspects of tensor methods applied to numerous tasks of computer vision and pattern recognition. We start with basic explanations of what tensors are, as well as their different definitions. Then basic properties of tensors, and especially their distances, are discussed. The next section provides some information on filtering of tensor data. Then structural tensor is discussed, which proves very useful in many different tasks and different types of images. A further important topic is tensor of inertia, as well as statistical moments, which can be used at different stages of object detection and recognition. Eigendecomposition of tensors, as well as their invariants, are discussed next. A separate topic are multi-focal tensors which are used to represent relations among corresponding points in multiple views of the same scene.

The second part of Chapter 2 is devoted to multilinear methods. First the most important concepts are discussed, such as k-mode product, tensor flattening, as well as different ranks

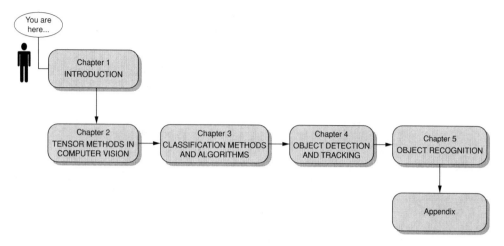

Figure 1.6   Organization of the book.

of tensors. These are followed by the three main important tensor decompositions, namely Higher Order Singular Value Decomposition, best rank-1, as well as best rank-$(R_1, \ldots, R_P)$ where R_1 to R_P represent desired ranks of each of the P dimension of the tensor. The chapter ends with a discussion of subspace data representation, as well as nonnegative decompositions of tensors.

Chapter 3 presents an overview of classification methods. We start with a presentation of subspace methods with one of the most important data representation methods – Principal Component Analysis. The majority of the methods have their roots in mathematical statistics, so the next chapters present a concise introduction to the statistical framework of object recognition. Not surprisingly the key concept here is the Bayes theorem. Then we discuss the parametric methods as well as the Kalman filter, frequently used in tracking systems but whose applications reach far beyond this. A discussion on the nonparametric follows, starting with simple, but surprisingly useful, histogram methods. Then the Parzen approach is discussed with its connections to nearest-neighbor methods. Mean shift methods are discussed in the consecutive parts of Chapter 3. Then the probabilistic, Hamming, as well as morphological neural networks are presented.

A separate topic within Chapter 3 concerns kernel processing. These are important novel classification methods which rely on smart data transformation into a higher dimensional space in which linear classification is possible. From this group come Support Vector Machines, one of the most important types of data classifier.

The last part of Chapter 3 is devoted to the family of k-means data clustering methods which find broad application in many areas of data processing. They are used in many of the discussed applications, for which special attention to ensembles of classifiers is deserved, such as the one discussed at the end of Chapter 3.

Chapter 4 deals with object detection and tracking. It starts with a discussion on the various methods of direct pixel classification, used mostly for fast image segmentation, as shown with the help of two applications. Methods of detection of basic shapes and figures follow. These are discussed mostly in the context of automotive applications. Chapter 4 ends with a brief overview of the recent methods of pedestrian detection.

Object recognition is discussed in Chapter 5. We start with recognition methods that are based on analysis of phase histograms of objects which come from the structural tensor. Discussion on scale-space template matching in the log-polar domain follows. This technique has found many applications in CV. From these, two are discussed. Two very important topics are discussed next. The first is the idea of object recognition in the domain of deformable prototypes. The second concerns ensembles of classifiers. As was shown, these show superior results even compared to very sophisticated but single classifiers.

Chapter 5 concludes with a presentation of the road sign classification systems based on ensembles of classifiers and deformable patterns, but realized in two different ways. The first employs Hamming neural networks. The second is based on decomposition of a tensor of deformable prototype patterns. The latter is also shown in the context of handwritten digit recognition.

A very specific topic discussed at the end of Chapter 5 is eye recognition, used for monitoring the driver's state to prevent dangerous situations arising from the driver falling asleep. Chapter 5 concludes with a discussion on the recent methods of object category recognition.

Appendix A discusses a number of auxiliary topics. It starts with a presentation of the morphological scale-space. Then a domain of morphological tensors operators is briefly discussed. Next, the geometry of quadratic forms is provided. Then the problem of testing classifiers is discussed. This section gathers different approaches to classifier testing, as well as containing a list of frequent parameters and measures used to assess classifiers. The rest of Appendix A briefly presents the OpenMP library used to convert serial codes into functionally corresponding but concurrent versions. In the last section some useful MATLAB® functions for matrix and tensor processing are presented.

As already mentioned, the majority of the presented topics are accompanied by their full C++ implementations. Their main parts are also discussed in the book. The full implementation in the form of a software library can be downloaded from the book webpage [3]. This webpage also contains some additional materials, such as the manual to the software platform, color images, and other useful links.

Last but not least, I will be very grateful to hear your opinion of the book.

References

[1] Cyganek B., Siebert J.P.: An Introduction to 3D Computer Vision Techniques and Algorithms, Wiley, 2009.
[2] Cyganek B.: Methods and Algorithms of Object Recognition in Digital Images. Habilitation Thesis. AGH University of Science and Technology Press, 2009.
[3] http://www.wiley.com/go/cyganekobject

2

Tensor Methods in Computer Vision

2.1 Abstract

This chapter gathers different computer vision techniques which make use of tensors, as well as their decomposition and analysis. As will be shown, the discussed methods have found application in many methods for object detection and recognition in images. Although tensors have been known in mathematics for over a hundred years, their application in computer vision (CV) and pattern recognition (PR) has been a matter of the last two decades. The real power of tensor processing in these areas comes from their natural ability to represent the multidimensional nature of processed data well.

Based on the fundamental sampling theorem, continuous signals when sampled with sufficient frequency can be unambiguously represented by their discrete samples [1, 2]. This fundamental property transforms physical measurements with the world of computer processing, since digital signals are just data in computer memory. As will be shown, tensors are the right tools for processing a variety of digital signals, such as sound, vision, seismic, medical electroencephalogram (EEG), as well as magnetic resonance imaging (MRI), which opens vast possibilities in medical diagnosis. In MRI, for instance, it is assumed that the motion of water molecules in tissues can be approximated by a Brownian motion in the voxels of the image. However, the Brownian motion is entirely described by a symmetric and positive definite matrix, called the *diffusion tensor*. Processing and visualization of diffusion tensors is one of the most rapidly growing domains, joining mathematics, physics, medicine, and computer vision.

The goal of this chapter is to present different areas of CV and PR which can be well represented and analyzed with tensors. We start with definitions of tensors, as well as basic properties of tensors. The two most pronounced characteristics of tensors are their transformation rules with respect to changes of the coordinate systems. The other is their multidimensionality, which makes them the right tool to process data which depend on many factors, as will be discussed. We present the structural tensor and its variants, as well as the tensor of inertia.

The former is based on signal differentiation, whereas the latter is related to the statistical moments computed from the signal. Both are useful to represent local areas, as well as whole objects, in the images. We also discuss methods of filtering of tensor data, as well as their eigendecomposition and invariants. Tensors are also the right tool to represent mutual relations between features of real objects imaged in multiple views. The next part of this chapter is devoted to the second aspect of tensors – their ability to represent and analyze multidimensional data. Presented are the most important tensor decompositions, the Higher-Order Singular Value Decomposition (HOSVD), best rank-1, as well as best rank-$(R_1, \ldots, R_P)$, where R_1 to R_P are the desired ranks of each of the P dimensions of the tensor. Finally, the nonnegative matrix and tensor factorizations, as well as the subspace data representation, are discussed.

Apart from a presentation of the mathematical background of the methods, their object-oriented implementations are also presented and discussed. Since implementations are generic, that is they can be used with user specified data types, they can be directly used in other projects. Also, some applications are provided which aim to exemplify the most important features of the methods. The subsequent chapters of the book contain further examples of applications of the presented tensor methods in real computer vision systems (CVS).

2.2   Tensor – A Mathematical Object

The tensor was developed to facilitate mathematical representation of physical laws in changing coordinate systems. One of the most famous application of tensors is the theory of relativity originally provided by A. Einstein at the beginning of the 20th century [3]. In the following sections we outline their basic properties which constitute a foundation for further concepts presented in this book. However, many valuable readings on tensors are available, among which the classical text by Bishop and Goldberg is especially interesting [4], as well as a dissertation on the frontiers of physics, relativity and mathematical physics, including tensor analysis and manifolds, by Penrose [5]. A unique treatment on tensors and their nonlinear functions is discussed in the book by Dimitrienko [6]. Regarding computer vision, tensors have found broad applications in such tasks as multiple view analysis [7, 8], structural tensor [11, 12], as well as multidimensional pattern recognition and view synthesis [11, 12], to name a few. These are further analyzed in the subsequent sections. An introduction to the domain of tensor analysis for the purpose of CV can be found, for example, in the papers by Triggs [13, 14], or in the book by Cyganek and Siebert [15].

2.2.1   Main Properties of Linear Spaces

A characteristic feature of tensor notation is the existence of upper and lower indices, which do not denote powers as in the case of polynomials. The number and position of the indices determine a type (valence) of a tensor. The other characteristic property of tensors is a *summation convention*, originally proposed by Einstein, to shorten mathematical formulas [3]. It simply assumes elimination of the summation symbol when there are two opposite indices which uniquely indicate summation. Hence, instead of $\sum_{i=1}^{n} a_i x_i$ simply $a^i x_i$ is written, assuming that summation spans through the same index i which for a is in the *contravariant* (upper) position, whereas for x it is in the *covariant* (lower) position.

A vector $\mathbf{x}$ in the L dimensional space with basis $\mathbf{b}_i$ can be expressed as

$$\mathbf{x} = \sum_{i=1}^{L} x^i \mathbf{b}_i, \tag{2.1}$$

which in Einstein's notation takes the following form

$$\mathbf{x} = x^i \mathbf{b}_i. \tag{2.2}$$

Hence, knowing a base, we can write $\mathbf{x} = (x^1, x^2, \ldots, x^L)$. The linear form (2.2), as well as its dual, constitute the base of tensor algebra which is especially useful if the base of a coordinate system changes. The tensor algebra is mainly constructed on two mathematical concepts:

- Linear operators.
- Change of the coordinate system (the Jacobian).

Explanation of the above concepts can be found in the majority of the books dealing with concepts of vector spaces, e.g. [4, 15, 16, 17].

2.2.2 Concept of a Tensor

A definition of a tensor is based on the concept of a vector space and its dual space, as follows [4]:

Definition 2.1 Let W and W^* be the vector space and its dual, respectively. A *tensor* over the base W is a multilinear scalar function f with arguments from W and W^*

$$f : \underbrace{W \times \ldots \times W}_{p \ \ times} \times \underbrace{W^* \times \ldots \times W^*}_{q \ \ times} \to \Re. \tag{2.3}$$

The number of arguments p and q denote a covariant and a contravariant degrees (a valence) of a tensor. □

From the above definition we easily notice that a tensor of valence (0,0) is a scalar, whereas tensors (1,0) and (0,1) are contra- and co-variant vectors, respectively. Moreover, tensors defined in the above manner create a vector space by themselves. It is called a tensor space over W.

Definition 2.1 provides a very convenient explanation of a tensor as a multilinear function. Thanks to this, the concept of a tensor is a very versatile one. However, in some respects such an ample definition is cumbersome for interpretation and other equivalent representations become more common. For example, knowing that all linear functions can be uniquely determined exclusively based on their values at base vectors, an analogous property for tensors is expressed as

Theorem 2.1   A tensor is uniquely determined by its values on the basis and its dual basis. The values are products of the tensor and the elements of the base and the dual base.   ☐

Based on the above tensors, being multilinear functions, are represented by the indexed values, such as $t_{i..j}{}^{m..n}$, in the basis and its dual. In accordance with Equation (2.3), there are p lower indices (covariant) and q upper indices (contravariant). Such a representation is frequently used in physics. However, the most important thing is the way these values change on change of the base of a coordinate's space. This is a key property of tensors which is sometimes used as a second means of definition. That is, a tensor is defined by its values $t_{i..j}{}^{m..n}$ which *change in a way characteristic of the type of a tensor*, i.e. its degree and valence (1.1). This also paves the way to the method of checking whether a physical value is a tensor or not, and if the answer is positive, what its valence is [18]. We would like to point up that this is the proper way to determine if a given mathematical object is, or is not, a tensor.

Let us now specify dimension of a tensor, based on values used in Equation (2.3). It is given by:

Theorem 2.2   A dimension of a tensor of p covariant and q contravariant indices is L^{p+q}, where $L = dim(W)$ denotes dimension of the vector space W.   ☐

The above two theorems allow us to treat a tensor as an P-dimensional "cube" of data, where $P = p + q$ denotes its common dimension. In computer science terms, this is an P-dimensional array in which to access an element P indices need to be provided. In this interpretation all indices are treated in the same way, i.e. we do not distinguish between contra- and co-variant ones since the transformation laws are not considered in this case, but only the number of independent indices. This way the tensor analysis can be used in such domains which require representation and manipulation of multidimensional data which we encounter frequently in physics, chemistry, data mining, as well as image processing, to name a few.

Summarizing, we encounter three definitions of tensors:

1. A multilinear function which takes the p-dimensional vector space and q-dimensional its dual, into the space of real values $\Re$, as defined in Equation (2.3);
2. A mathematical object described by the indexed values $t_{i..j}{}^{m..n}$, in which there is p covariant and q contravariant indices. These values transform in accordance with the transformation laws with a change of the base of the spaces [4, 15, 18];
3. A *multidimensional array* of data with $P = p + q$ independent indices.

As already indicated, the first interpretation is rather mathematical but allows a coherent introduction of the tensor algebra, discussed in the next section. The second interpretation is used frequently in physics, especially in mechanics, relativity, and so forth. The third simplifies the definition given in point (2) and is mostly used in multidimensional data representation and analysis. In CV the second and third interpretations are the most common, as will be discussed. However, once again we would like to stress that the proper way to check if a given mathematical object is, or is not, a tensor is to check its behavior on the change of the basis of the space. Only if this follows the tensor transformation laws for tensors of a given valence, is such an object a tensor. On the other hand, having a multidimensional data controlled by P independent indices does not mean that this is a tensor in the sense of definitions (1) or (2). Nevertheless, in data mining a multidimensional array is used to be named as a tensor due to many similarities with definition (2).

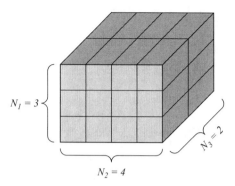

Figure 2.1 An example of a 3D tensor with the total of $P = 3$ dimensions $N_1 = 3$, $N_2 = 4$, and $N_3 = 2$.

2.3   Tensor – A Data Object

As alluded to previously, tensors can be relaxed from their strict transformation laws, and can be regarded as *multidimensional arrays of data*, in which each dimension separately corresponds to a different feature of described objects or phenomena. Following this idea, Figure 2.1 depicts a 3D tensor of dimensions $3 \times 4 \times 2$. This point of view is "parallel" to the previous one, rather than "instead" of, since even with explicitly provided n covariant and m contravariant indices, a tensor in a given base is unambiguously described by $n \cdot m$ real (or complex) values. This way we obtain a multilinear approach which constitutes an extension to the classical two-dimensional (matrix) analysis. Such an approach was recently undertaken to tackle problems of data mining, pattern recognition, neuroscience and psychometrics, chemistry, signal processing, computer vision, and many more [11, 19, 20, 121, 22].

A simple color image together with its three color channels red, green, and blue is shown in Figure 2.2. Since each element of an image, a pixel, has exactly three independent coordinates,

Figure 2.2 An image and its three color channels as an example of a 3D tensor. (For a color version of this figure, please see the color plate section.)

which are columns, rows, and channels, it can be interpreted as a three-dimensional array of data or just a 3D tensor. For a video sequence there is a fourth free index, time (frame number).

This interpretation is a fundamental one if the internal structure of an image has to be analyzed, e.g. for object recognition. On the other hand, an approach in which images are vectorized can lead to loss of information, frequently resulting in limited robustness of such methods.

However, we can give yet another interpretation of an image, such as the one in Figure 2.2. It can be seen as *a single point* in a multidimensional space of dimensions being a product of all allowable values of columns, rows, and channels [23]. Thus, we can suspect that similar images in such a space will tend to be places close together or, in other words, they will be contained in subspaces. But how can we measure a distance between points in this space? This requires a geometrical notion of a distance. However, to set some geometric properties such as a distance, an angle, etc. we need to join the space with a coordinate system. We can do this locally, obtaining so called manifolds which in some neighborhoods of the points from the space provide real coordinate functions. This way it becomes possible to determine the position of a point in that space and topological properties of its neighborhood. In other words, a space becomes locally Cartesian. These issues are further discussed in Sections 2.5 and 5.5 devoted to deformable models.

Let us analyze another type of image – hyperspectral imaging (HI) is the technology of acquiring a series of images, each at different electromagnetic spectral bands, starting from the ultraviolent up to the long-infrared spectrum. The main reason is that different objects reveal different properties at different wavelengths. Thanks to this property, hyperspectral images can show what is inaccessible in just the human visible spectrum. Different objects have unique characteristics in hyperspectra, which is called an object's *spectral signature*. Interestingly, many life creatures possess the ability to perceive wide wavebands. For instance, mantis shrimps are sensitive to the spectrum from infrared up to ultraviolet, which allows them to detect their prey [24]. Also humans at a certain stage of evolution, thanks to the acquired ability to detect colors, became able to distinguish ripe fruits from others which obviously affected the diet. Hyperspectral images are used in many areas, such as medical imaging for tumor detections, dentistry to detect tooth decay, remote sensing for Earth surface monitoring, agriculture for monitoring of a crop's health, the food industry for food quality assessment, pharmacy for detection of chemical components, and forensic science for ink examination on checks, to name a few [24, 25]. Acquisition of hyperspectral images requires special devices, such as push-broom cameras. One of the most famous is the Airborne Visible InfraRed Imaging Spectrometer (AVIRIS) which is an optical sensor developed by NASA [26]. It delivers calibrated images of the upwelling spectral radiance in 224 contiguous spectral channels in the range from 400 to 2500 nm. Another example is the eye monitoring system operating in the near infrared (NIR) spectrum, discussed in Section 5.9. NIR images, being invisible to a driver, allow eye monitoring for detection of a driver's fatigue, sleepiness or inattention in order to react on time.

From the above discussion we easily conclude that HI naturally leads to 3D tensors, with two spatial and one spectral dimension. An example is depicted in Figure 2.3. However, such 3D tensors can be acquired at different time steps, which adds yet another dimension leading to 4D tensors. The most common signal processing tasks related to the domain of HI are dimensionality reduction, object detection, change detection, classification, as well as spectral unmixing. Of special importance is the last one, that is spectral unmixing of the

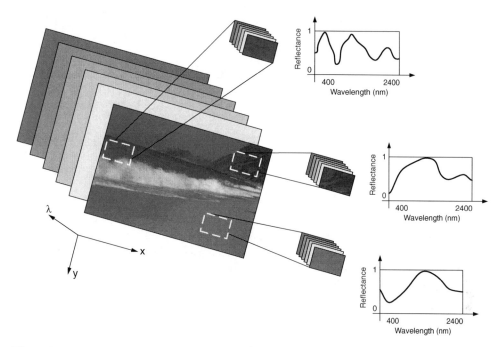

Figure 2.3   An example of a hyperspectral image[1] with selected regions of different reflectance properties. Series of hyperspectral images naturally form 3D tensors with two spatial dimensions (x,y) and one spectral λ.

hyperspectral data. In this case the linear model is assumed which allows the collected spectra to be expressed as a linear combination of spectrally pure signatures (so called endmembers) and their corresponding abundance factors. Discussion on these and their parallel implementations are discussed in the paper by Plaza *et al.* [27].

2.4   Basic Properties of Tensors

As alluded to previously, tensors transform in a way specific to a change of the base of the vector space. This transformation depends on the type of a tensor. To show this property for the simplest one-dimensional case, let us assume that the base **b** linearly transforms into **b**' as follows

$$\mathbf{b}'_j = A^i_j \mathbf{b}_i. \tag{2.4}$$

In the above, the elements $A^i_j \in \mathfrak{R}$ form a square matrix **A** which we assume is invertible, so the change of base is possible in two directions, i.e.

$$A'^j_i \mathbf{b}'_j = \mathbf{b}_i, \tag{2.5}$$

---

[1]Color versions of the images can be found on the accompanying web page [28].

where $A'^i_{\ j}$ are elements of $\mathbf{A}^{-1}$. Let us now find coefficients of a vector $\mathbf{x}$, given in Equation (2.2), in this new base $\mathbf{b}'$. Substituting Equation (2.4) into (2.2) yields vector $\mathbf{x}$ in a new (primed) base

$$\mathbf{x} = x^i \mathbf{b}_i = x^i A'^j_{\ i} \mathbf{b}'_j = \underbrace{A'^i_{\ j} x^j}_{x'^i} \mathbf{b}'_i, \qquad (2.6)$$

thus

$$x'^i = A'^i_{\ j} x^j. \qquad (2.7)$$

In other words, when the new base (primed) is obtained from the original one by multiplication with the matrix $\mathbf{A}$, as in Equation (2.4), coordinates of a vector $\mathbf{x}'$ in this new (primed) space are related to the original vector $\mathbf{x}$ by its inverse, i.e. $\mathbf{A}^{-1}$, seen in Equation (2.7).Such transformations are characteristic of all tensors and depend exclusively on their type. Continuing with the simplest one-dimensional case, when going from one space U into U', the covariant components of a tensor $\mathcal{T}$ transform as

$$T'_i = \alpha^k_{i'} T_k, \qquad (2.8)$$

whereas the contravariant in accordance with

$$T'^i = \alpha^{i'}_k T^k. \qquad (2.9)$$

In the above the values $\alpha^k_{i'}$ and $\alpha^{i'}_k$ denote coefficients of the direct and indirect transformation of the coordinate spaces, respectively. These are elements of the matrix $\mathbf{A}$ in Equation (2.4) for a linear transformation of the coordinate systems, however. In special cases, if these are Cartesian (rectangular), then $\alpha^k_{i'}$ are identical to the cosines between corresponding axes of these systems, respectively. In cases of higher order tensors, such as the mixed ones, the transformation rules get more complicated. However, notation for higher order tensors can be confusing. For this purpose Penrose devised a very original system of pictograms denoting tensor operations which facilitate their manipulations [5].

2.4.1   Notation of Tensor Indices and Components

In this and following sections we will treat an P-th order tensor as a *multidimensional matrix* with P indices (an P-dimensional array, or P-mode matrix), which define the P-dimensional *multilinear mapping* over the vector spaces. This is justified by the linear properties of tensors, discussed in Section 2.2.2, which make possible their unanimous representation providing simply their values on the current base. Thus, the most important is total dimensionality of a tensor regardless of its co- and contra-variant partitioning of components. This way, a zero-order tensor is a scalar, first-order is a vector, second-order a matrix, third-order a "cube," and so on. Obviously visualization of tensors of order higher than three is cumbersome.

Before we follow with further definitions, let us introduce the ways of representing tensors and their elements. Figure 2.4a depicts a $3 \times 3 \times 3$ tensor with exemplary values. For

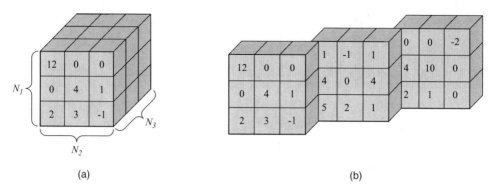

(a)   (b)

Figure 2.4   An example of a 3 × 3 × 3 tensor (a) and possible representation with three frontal 2D slices (b). Other slicing schemes are also possible.

visualization its values are presented on separate fronto-parallel slices of the tensor, as shown in Figure 2.4b. The latter obviously are 2D structures, i.e. matrices.

Thus, a tensor $\mathcal{T}$ of P dimensions of values $N_1, N_2, \ldots, N_P$, is denoted as follows

$$\mathcal{T} \in \mathfrak{R}^{N_1 \times N_2 \times \ldots \times N_P}. \tag{2.10}$$

In practice, we need to provide concrete values for the indices, such as $N_1 = 5$, $N_2 = 3$ for a $P = 2$ two-dimensional tensor (i.e. a matrix). Analogies with the well established matrix notations simplify the way we think about tensors. Similarly, we set $N_1 = 5$, $N_2 = 3$, and $N_3 = 12$, for a $P = 3$ three-dimensional tensor, and so on.

On the other hand, a single element t of $\mathcal{T}$ is addressed providing its precise position by a series of indices $n_1, n_2, \ldots, n_P$, such that

$$t_{n_1 n_2 \ldots n_P} \equiv \mathcal{T}_{n_1 n_2 \ldots n_P}, \tag{2.11}$$

where

$$1 \leq n_1 \leq N_1, \quad 1 \leq n_2 \leq N_2, \quad \ldots, \quad 1 \leq n_P \leq N_P. \tag{2.12}$$

An equivalent function-like notation is sometimes useful, i.e.

$$t_{n_1 n_2 \ldots n_P} \equiv \mathcal{T}(n_1, n_2, \ldots, n_P). \tag{2.13}$$

Notation (2.11) is encountered frequently in mathematical papers, while (2.13) is used in programming platforms, such as MATLAB® [29]. The latter is shown to be more legible in representing range of indices. As an example, let us start from a simple matrix $\mathbf{A}$ for which by convention the first index pertains to rows, whereas the second pertains to the columns. Hence, to express, for instance, its second column we need to specify a range of all possible indices for the rows and specifically the second column. This is denoted as $\mathbf{A}(:,2)$ in which a colon expresses a notion of all indices at this index position. This notation can be easily extended to the tensor notation. For instance, in the case of a 3D tensor to express its foremost "slice," which is two-dimensional, we write $\mathcal{T}(:,:,1)$.

Last but not least, let us also note that in many computer languages, such as C/C++, Python, etc., indexing starts from 0, rather than from 1 as in (2.12). This needs to be taken into consideration when translating mathematical notations to the code.

Let us exemplify component indexing with help of the 3D tensor of dimensions $3 \times 3 \times 3$, that is $P = 3$ and $N_1 = 3$, $N_2 = 3$, $N_3 = 3$, as shown in Figure 2.4. However, instead of a drawing, like the one in Figure 2.4, we can explicitly write the elements in a somewhat structured form, in which gray arrows denote indexing order, as follows

$$\mathcal{T} = \left[\begin{array}{ccc|ccc|ccc} 12 & 0 & 0 & 1 & -1 & 1 & 0 & 0 & -2 \\ 0 & 4 & 1 & 4 & 0 & 4 & 4 & 10 & 0 \\ 2 & 3 & -1 & 5 & 2 & 1 & 2 & 1 & 0 \end{array} \right]$$

Now, we can write some of its components and their values, as follows

$$t_{123} = \mathcal{T}(1, 2, 3) = 0, \quad t_{133} = \mathcal{T}(1, 3, 3) = -2, \quad t_{312} = \mathcal{T}(3, 1, 2) = 5. \quad (2.14)$$

A middle slice of $\mathcal{T}$ can be written using the colon notation, as follows

$$\mathcal{T}(:, :, 2) = \begin{bmatrix} 1 & -1 & 1 \\ 4 & 0 & 4 \\ 5 & 2 & 1 \end{bmatrix}.$$

It is interesting to observe that $\mathcal{T}(:, 2, 3)$ denotes a specific vector cut out from $\mathcal{T}$ (known also as a fiber). In this case we have

$$\mathcal{T}(:, 2, 3) = \begin{bmatrix} 0 & 10 & 1 \end{bmatrix},$$

since only the first index N_1 traverses all possible values (see Figure 2.4a), while the other two are fixed.

Finally, it holds that

$$\mathcal{T}(:, :, :) = \mathcal{T}.$$

2.4.2   Tensor Products

Tensors contain many interesting properties, can be transformed in specific ways, and are also endowed with tensor algebra [4, 6, 15, 18, 30]. Thus, tensors can be added, multiplied, and so forth. In this section we provide definitions and discuss the main properties of the three tensor products: the inner, the contraction, and the outer tensor products, respectively.

However, although these are general definitions, there are also special products of a tensor and a matrix, as well as a tensor and a series of matrices. These are discussed later in Section 2.12.1.

Definition 2.2   Outer product of tensors

The outer product of two tensors $\mathcal{S} \in \Re^{N_1 \times \ldots \times N_P}$ and $\mathcal{T} \in \Re^{M_1 \times \ldots \times M_Q}$ is defined as $\mathcal{W} \in \Re^{N_1 \times \ldots \times N_P \times M_1 \times \ldots \times M_Q}$

$$(\mathcal{S} \circ \mathcal{T})_{n_1 n_2 \ldots n_P m_1 m_2 \ldots m_Q} = \mathcal{W}_{n_1 n_2 \ldots n_P m_1 m_2 \ldots m_Q} = s_{n_1 n_2 \ldots n_P} \cdot t_{m_1 m_2 \ldots m_Q}. \; \Box \qquad (2.15)$$

In other words, the outer product results in a tensor of dimensions being a sum of indices of the multiplied tensors.

Definition 2.3   Contracted product of tensors

The contracted product of two tensors $\mathcal{S} \in \Re^{K_1 \times \ldots \times K_C \times N_1 \times \ldots \times N_P}$ and $\mathcal{T} \in \Re^{K_1 \times \ldots \times K_C \times M_1 \times \ldots \times M_Q}$ is defined as $\mathcal{U} \in \Re^{N_1 \times \ldots \times N_P \times M_1 \times \ldots \times M_Q}$

$$\mathcal{U}_{n_1 \ldots n_P m_1 \ldots m_Q} = \langle \mathcal{S}, \mathcal{T} \rangle_{\{K_1 \times \ldots \times K_C\}} (n_1, \ldots, n_P, m_1, \ldots, m_Q)$$

$$= \sum_{k_1=1}^{K_1} \sum_{k_2=1}^{K_2} \ldots \sum_{k_C=1}^{K_C} s_{k_1 \ldots k_C n_1 \ldots n_P} t_{k_1 \ldots k_C m_1 \ldots m_Q}. \quad \Box \qquad (2.16)$$

In other words, the contracted product allows reduction of the *first C* dimensions which need to be the same in both of the multiplied tensors $\mathcal{S}$ and $\mathcal{T}$. The remaining indices retain their original order, i.e. $N_1, \ldots, N_P$ come before $M_1, \ldots, M_Q$. However, let us indicate in this place that this is a slightly different scheme than in the case of a tensor and matrix product, which will be discussed in Section 2.12.1.3.

From these special attention deserves to be paid to a tensor contraction with respect to *all* of the indices. This is called *an inner product* of tensors of any, but the same, dimensions.

Definition 2.4   Inner (scalar) product of tensors

The inner (scalar) product of two tensors $\mathcal{S}, \mathcal{T} \in \Re^{K_1 \times \ldots \times K_C}$ is defined as

$$\langle \mathcal{S}, \mathcal{T} \rangle = \mathcal{S} \cdot \mathcal{T} = \sum_{k_1=1}^{K_1} \sum_{k_2=1}^{K_2} \ldots \sum_{k_C=1}^{K_C} s_{k_1 k_2 \ldots k_C} t_{k_1 k_2 \ldots k_C}. \quad \Box \qquad (2.17)$$

In other words, the inner product is a sum of all possible products of pairs of components with the same indices and the result is a single real scalar value.

Definition 2.5   Orthogonal tensors

If the inner product of the two tensors $\mathcal{S}, \mathcal{T}$ fulfills

$$\mathcal{S} \cdot \mathcal{T} = 0, \qquad (2.18)$$

then tensors $\mathcal{S}, \mathcal{T}$ are *orthogonal*. $\Box$

Definition 2.6   Frobenius norm of a tensor

The Frobenius norm of a tensor is induced by its scalar product

$$D_F\left(\mathcal{S}\right) = \|\mathcal{S}\|_F = \sqrt{\mathcal{S}\cdot\mathcal{S}} = \sqrt{\sum_{k_1=1}^{K_1}\sum_{k_2=1}^{K_2}\cdots\sum_{k_C=1}^{K_C} s_{k_1 k_2\dots k_C}^2}\cdot\quad\Box \qquad (2.19)$$

The above definitions constitute extensions of the well known inner product and the norm of vectors and matrices [31, 32].

For the two following tensors of dimensions $2 \times 2 \times 3$

$$\mathcal{S} = \begin{bmatrix} 1 & -2 & 3 & 0 & 0 & -8 \\ 0 & 1 & 7 & -2 & 11 & -11 \end{bmatrix} \quad\text{and}\quad \mathcal{T} = \begin{bmatrix} -2 & 1 & -1 & 2 & 0 & 4 \\ 0 & 2 & 4 & 1 & 1 & 0 \end{bmatrix},$$

their scalar product (2.18) is 0, what means that $\mathcal{S}$, $\mathcal{T}$ are orthogonal. On the other hand, their squared Frobenius norms (2.19) are 374 and 48, respectively.

A contraction of $\mathcal{S}$ and $\mathcal{T}$ with respect to their two first common indices $K_1 = 2$ and $K_2 = 2$, yields the following 3×3 matrix

$$\mathcal{U}_{3\times 3} = \langle\mathcal{S},\mathcal{T}\rangle_{\{K_1\times K_2\}} = \begin{bmatrix} -2 & -4 & -8 \\ -10 & 23 & 7 \\ -30 & 17 & -21 \end{bmatrix},$$

to see this, from Equation (2.16) let us compute $\mathcal{U}\left(3, 2\right)$

$$\mathcal{U}\left(3, 2\right) = \sum_{k_1=1}^{2}\sum_{k_2=1}^{2} s_{k_1 k_2 3}t_{k_1 k_2 2}$$

$$= \sum_{k_1=1}^{2}\sum_{k_2=1}^{2}\mathcal{S}\left(k_1, k_2, 3\right)\mathcal{T}\left(k_1, k_2, 2\right) = 0 + (-8)\cdot 2 + 11\cdot 4 - 11\cdot 2 = 17.$$

As already mentioned, a definition and discussion of the multiplication of a tensor and a matrix, the so called k-mode product, as well as with a series of matrices, is postponed to Section 2.12.1.

2.5   Tensor Distance Measures

A notion of a distance between data points is a fundamental scientific concept. A common "human like" measure in a 3D space is the Euclidean distance. However, even in daily life we know that such a measure is not sufficient, e.g. if looking for distance between two cities on a map. For problems of daily life we have evolved useful measures of taste, similarity or dissimilarity of architecture styles, etc. However, for computer vision and pattern recognition we need measures which are precisely defined and which as near as possible allow us to distinguish between objects of different categories. Thus, we expect a different measure, or measures, to compare colors, and a different measure, or measures, to compare shapes, for instance. Also tensors require definitions of their specific distances to convey characteristics of the physical quantities they represent. Hence, different measures are considered to compare multi-spectral images, and are different for interpolation of the MRI tensors [33], for instance.

In this section we summarize different measures especially devoted to tensor comparisons, based on recent publications [34, 35, 36]. Of special interest are the invariant properties of each measure, as well as the group of the symmetric positive definite tensors. However, before we provide these, let us recall the basic definitions and properties of measures [16].

Definition 2.7   A semi-metric

A nonnegative function D for two tensors $\mathcal{S}$ and $\mathcal{T}$ is a *semi-metric* if it satisfies the following two conditions

$$\mathcal{S} = \mathcal{T} \Leftrightarrow D\left(\mathcal{S}, \mathcal{T}\right) = 0, \tag{2.20}$$

$$D\left(\mathcal{S}, \mathcal{T}\right) = D\left(\mathcal{T}, \mathcal{S}\right) \quad \square \tag{2.21}$$

The first of the above conditions allows the distinguishing of equal tensors, whereas the second makes the measure independent of the order of its arguments. Additionally, if D has to be a metric, then it needs to fulfill the triangle condition, i.e.

Definition 2.8   A metric

A function D for two tensors $\mathcal{S}$ and $\mathcal{T}$ is a *metric* if it fulfills Equation (2.20) and the following triangle condition

$$D\left(\mathcal{S}, \mathcal{T}\right) \leq D\left(\mathcal{S}, \mathcal{U}\right) + D\left(\mathcal{U}, \mathcal{T}\right) \tag{2.22}$$

where $\mathcal{U}$ is a tensor of the same dimensions as $\mathcal{S}$ and $\mathcal{T}$. $\square$

The triangle condition (2.22) plays important role in methods which need to compute means or interpolate tensors. Also, the triangle condition implies nonnegativity and symmetry conditions on D, which had to be assumed in Definition 2.7.

In the case of tensor processing it is important to know invariant properties of a given measure D to transformations of its arguments. In this respect let us consider such features as size, orientation, as well as shape which are specific to many types of tensors, such as the diffusion tensor [34].

Definition 2.9   Size invariant measure

A measure D is *size invariant* if the following is fulfilled

$$D\left(a\mathcal{S}, b\mathcal{T}\right) = D\left(\mathcal{S}, \mathcal{T}\right), \tag{2.23}$$

where a and b are scalar values. $\square$

Definition 2.10   Rotation invariant measure

A measure D is *rotation invariant* if the following is fulfilled

$$D\left(\mathbf{R}^T \boldsymbol{\mathcal{S}} \mathbf{R}, \mathbf{P}^T \boldsymbol{\mathcal{T}} \mathbf{P}\right) = D\left(\boldsymbol{\mathcal{S}}, \boldsymbol{\mathcal{T}}\right), \qquad (2.24)$$

where $\mathbf{R}$ and $\mathbf{P}$ are rotation matrices.   □

Definition 2.11   Shape invariant measure

A measure D is *shape invariant* if it does not change when changing ratio between eigenvalues of $\boldsymbol{\mathcal{S}}, \boldsymbol{\mathcal{T}}$, or both.   □

Additionally, the robustness of a measure can be considered which means that small perturbations of tensor components result in changes of D constrained to some predefined threshold value ε.

Depending on an application, the above invariance properties can be useful especially if generalization properties are expected. However, these need to be thoroughly considered since invariance to a given feature makes the measure transparent to that feature.

Finally, let us recall the notion of the positive definite and semidefinite matrices.

Definition 2.12   Positive definite (semidefinite) matrix

A matrix $\mathbf{M}$ is *positive definite* if

$$\mathbf{x}^T \mathbf{M} \mathbf{x} > 0, \qquad (2.25)$$

for all $\mathbf{x} \neq 0$. A matrix $\mathbf{M}$ is *positive semidefinite* if

$$\mathbf{x}^T \mathbf{M} \mathbf{x} \geq 0. \quad □ \qquad (2.26)$$

Property (2.25) is sometimes shortened to $\mathbf{M} > 0$ for positive definite matrices. Analogously, instead of (2.26) we usually write $\mathbf{M} \geq 0$. Also a useful property is that positive definite matrices are *always invertible*.

2.5.1   Overview of Tensor Distances

The L^n norm, defined as follows

$$D_{Ln}\left(\boldsymbol{\mathcal{S}}, \boldsymbol{\mathcal{T}}\right) = n \sqrt{\sum_{k_1=1}^{K_1} \sum_{k_2=1}^{K_2} \cdots \sum_{k_C=1}^{K_C} \left(s_{k_1 k_2 \ldots k_C} - t_{k_1 k_2 \ldots k_C}\right)^n}, \qquad (2.27)$$

can be used to measure the distance of two tensors $\mathcal{S}$ and $\mathcal{T}$, assuming they are of the same dimensions. A special case for $n = 2$ is known as the Frobenius distance of tensors,

$$D_F(\mathcal{S}, \mathcal{T}) = D_{L2}(\mathcal{S}, \mathcal{T}) = \sqrt{\sum_{k_1=1}^{K_1} \sum_{k_2=1}^{K_2} \cdots \sum_{k_C=1}^{K_C} \left(s_{k_1 k_2 \ldots k_C} - t_{k_1 k_2 \ldots k_C}\right)^2}, \qquad (2.28)$$

which follows from Equation (2.19), after inserting squared differences of both tensor components. The second popular measure of tensor distances is the already discussed scalar product (2.17). However, its value is close to 0 if at least one of the tensors has components close to 0, and the two tensors are not necessarily similar in some sense.

Nevertheless, the above distances ignore an intrinsic structure of the compared objects, i.e. the number and order of the internal dimensions do not have any impact on the result. Therefore other measures can be more appropriate for tensors, as discussed. However, a special case is a group of positive definite and symmetric tensors which rise in many applications such as the structural tensor or the MRI diffusion tensor [33]. Therefore, these deserve special attention and the rest of this section will be devoted exclusively to the distances of the *symmetric and positive-definite tensors*.

A recent review of mathematical models and computational methods for the processing of diffusion MRI is presented in the paper by Lenglet *et al.* [37]. In this paper reconstruction of diffusion models, an analysis of the cerebral white matter connectivity, as well as segmentation techniques are discussed. This paper is also a good source of further references to the subject.

One approach to representing the aforementioned diffusion tensor is to use the covariance matrix of the Gaussian distribution, which describes a diffusion process of mainly water molecules in tissues. For these reasons, an appropriate measure can be obtained with the statistical approach, i.e. a statistical divergence of the two distributions can be applied. In this case tensor comparison follows the Kullback–Leibler distance between probability distributions. Its symmetric version, called J-divergence, was proposed for tensor distances by Wang and Vemuri, as follows [36]

$$D_J(\mathbf{S}, \mathbf{T}) = \frac{1}{2} \sqrt{\operatorname{Tr}\left(\mathbf{S}^{-1}\mathbf{T} - \mathbf{T}^{-1}\mathbf{S}\right) - 2K_C} \qquad (2.29)$$

where K_C stands for C-th dimension of the tensor (for the MRI tensors $C = 2$ and $K_C = 3$).

Another class of measures of the positive definite tensors come from the field of the *Riemannian geometry*. In this framework a tensor distance follows *geodesics* in the manifold of symmetric positive defined matrices. From this group a measure for diffusion tensors arising in MRI was proposed by Bachelor *et al.* [38] and by Pennec *et al.* [35]. It is defined as follows

$$D_R(\mathbf{S}, \mathbf{T}) = \sqrt{\operatorname{Tr}\left[\log^2\left(\mathbf{S}^{-\frac{1}{2}}\mathbf{T}\mathbf{S}^{-\frac{1}{2}}\right)\right]}, \qquad (2.30)$$

in which for symmetric tensors $\mathbf{T}$ it holds that $\mathbf{T}^2 = \mathbf{T}\mathbf{T} = \mathbf{T}\mathbf{T}^T$. This measure is invariant to linear transformations.

However, computation of Equation (2.30) can be time demanding. Therefore Arsigny *et al.* proposed a simplified Log-Euclidean distance, as follows [39]

$$D_{LF}\left(\mathbf{S}, \mathbf{T}\right) = D_F\left(\log\left(\mathbf{S}\right), \log\left(\mathbf{T}\right)\right) = \sqrt{\operatorname{Tr}\left[\left(\log\left(\mathbf{S}\right) - \log\left(\mathbf{T}\right)\right)^2\right]}. \qquad (2.31)$$

For the sake of computational complexity the most appropriate measures are D_F and D_{LF}. They have also been verified experimentally by Rittner *et al.* as those which show the best performance for comparisons of color images [40]. Moreover, D_{LF} can be seen as a simple Frobenius norm on *log preprocessed* input signals. The latter property also simplifies implementations since none of the classifiers operating with the Frobenius norm need to be changed – only their input signals need to be preprocessed with the logarithm function. Such an approach also has the very useful property pointed out by Arsigny *et al.* [39]. Namely, contrary to the affine invariant distance (2.30) the Log-Euclidean version (2.31), being simply Euclidean distance in the logarithmic domain, allows the transformation of tensors first into the vector space of symmetric matrices. These matrices, in turn, can be even further simplified into 6D vectors, in accordance with the following scheme

$$\mathbf{S} \in Sym_+ \rightarrow \log\left(\mathbf{S}\right) \rightarrow \mathbf{s}, \qquad (2.32)$$

where

$$\mathbf{s} = \left[\left[\log\left(\mathbf{S}\right)\right]_{11} \ \left[\log\left(\mathbf{S}\right)\right]_{22} \ \left[\log\left(\mathbf{S}\right)\right]_{33} \ \sqrt{2}\cdot\left[\log\left(\mathbf{S}\right)\right]_{12} \ \sqrt{2}\cdot\left[\log\left(\mathbf{S}\right)\right]_{13} \ \sqrt{2}\cdot\left[\log\left(\mathbf{S}\right)\right]_{23}\right]^T.$$
$$(2.33)$$

This procedure can be simply extended on tensors of higher dimensions. In this approach, the "classical" Euclidean distances of such 6D vectors become equivalent to the Log-Euclidean similarity invariant distance between tensors they represent [39].

Finally, let us mention a very interesting paper by Tuzel *et al.* on pedestrian detection obtained by classifiers operating on Riemannian manifold used to represent set of nonsingular covariance matrices describing objects of interest (e.g. pedestrians) [41]. However, such matrices do not form a vector space, so classification with such features needs to be done on the Riemannian manifold.

2.5.1.1 Computation of Matrix Exponent and Logarithm Functions

After presenting the measures (2.30) and (2.31) that employ the logarithms of the tensors let us briefly discuss implementation issues of matrix exponential and logarithmic functions. These two functions are defined as follows.

Definition 2.13   Exponential and logarithm of a matrix

The exponential of a matrix S is given by the following formula

$$\exp\left(\mathbf{S}\right) = \sum_{i=1}^{\infty} \frac{\mathbf{S}^i}{i!} = \mathbf{I} + \mathbf{S} + \frac{\mathbf{S}^2}{2!} + \frac{\mathbf{S}^3}{3!} + \dots. \qquad (2.34)$$

If there exists a matrix $\mathbf{S}$ for which $\mathbf{T} = \exp(\mathbf{S})$, then $\mathbf{S} = \log(\mathbf{T})$ is a natural logarithm of $\mathbf{T}$.   □

The topics of matrix exponential and its computation have found broad applications in science and techniques. However, in the general case the logarithm of a real invertible matrix may *not even exist*, and if it exists it may *not be unique*. Therefore development of efficient algorithms for computation of the matrix exponent and logarithm for broad range of different types of matrices does not belong with trivial computational tasks. Fortunately, the group of the *symmetric positive-definite* tensors has nice properties that make computations of their exponent and logarithm always possible and unique, as will be discussed. Further in this section we provide a brief overview of the simplest computation methods, as well as providing further references to the literature on the subject.

One of the most commonly used group of methods for computation of *exp* and *log* functions derives from the algorithm based on argument scaling and squaring. It scales the input matrix by a power of 2 in order to bring its norm closer to 1. Then the Padé approximation to the matrix exponential is computed. In the last stage the result is repeatedly squared to undo the effect of matrix scaling. In this respect the paper by Moller and Van Loan is highly recommended reading [42]. Also, a detailed overview is given in the paper by Higham [43].

In the case of the natural logarithm function, which we are mostly interested in due to its computation in distances (2.30) and (2.31), the strategy of scaling and squaring can be also applied. When the matrix $\mathbf{T}$ is sufficiently close to the identity matrix $\mathbf{I}$, then the direct method which is based on a simple truncated Taylor series

$$\log\left(\mathbf{T}\right) = -\left(\mathbf{I} - \mathbf{T}\right) - \frac{(\mathbf{I} - \mathbf{T})^2}{2} - \frac{(\mathbf{I} - \mathbf{T})^3}{3} - \frac{(\mathbf{I} - \mathbf{T})^4}{4} \cdots, \qquad (2.35)$$

can be used. However, if $\mathbf{T}$ differs from $\mathbf{I}$ then the above cannot be directly used. Instead, $\mathbf{T}$ is scaled as follows

$$\log\left(\mathbf{T}\right) = 2^k \log\left(\mathbf{T}^{1/2^k}\right). \qquad (2.36)$$

Then, increasing k successively makes $\mathbf{T}^{1/2^k} \rightarrow \mathbf{I}$, so the direct series computation can be applied to the scaled matrix [44]. In other words, the scaling consists of computation of the square roots of the matrix. Finally, the reverse scaling needs to be applied to obtain the logarithm of the input matrix $\mathbf{T}$. Therefore this approach is called the scaling and squaring method. In this respect, a discussion on efficient implementation methods for approximation of the logarithm of a matrix to a specified accuracy is presented in the paper by Cheng *et al.* [44]. The proposed algorithm allows relatively simple computations since only the basic matrix operations of multiplication, LU decomposition, and inversion are involved. A good source of information on functions of matrices with a strong implementation background is the book by Higham [45].

It can be shown that any matrix $\mathcal{S}$ of dimensions $N \times N$ which eigenvectors are linearly independent can be diagonalized to the following representation[2] [17]

$$\mathbf{S} = \mathbf{E}\boldsymbol{\Lambda}\mathbf{E}^{-1}. \qquad (2.37)$$

---

[2] See also Sections 2.9 and 3.3.1.

In the above

$$E = \begin{bmatrix} e_1 & e_2 & \cdots & e_N \end{bmatrix} \tag{2.38}$$

is a matrix of stacked eigenvectors of $\mathcal{S}$, whereas

$$\Lambda = \begin{bmatrix} \lambda_1 & & & \\ & \lambda_2 & & \\ & & \ddots & \\ & & & \lambda_N \end{bmatrix} \tag{2.39}$$

is a diagonal matrix containing eigenvalues of $\mathcal{S}$. However, in the general case, the eigenvalues can be of any sign, i.e. they are positive or negative, which can be important in further computation of the matrix functions.

It is now easy to show that thanks to the decomposition (2.37) any power of a matrix can be simply computed as follows

$$S^n = E\Lambda^n E^{-1.} \tag{2.40}$$

This, in turn, makes computation of any matrix function expressed as a power series as restricted only to the diagonal matrix Λ. Thus, Equation (2.34) yields

$$\exp(S) = E \exp(\Lambda) E^{-1}, \tag{2.41}$$

where $\exp(\Lambda)$ simply denotes computation of the exponent on each of its diagonal elements (scalars), that is

$$\exp(\Lambda) = \begin{bmatrix} \exp(\lambda_1) & & & \\ & \exp(\lambda_2) & & \\ & & \ddots & \\ & & & \exp(\lambda_N) \end{bmatrix}. \tag{2.42}$$

Finally, in analogy to Equation (2.41) the natural logarithm can be computed as follows

$$\log(S) = E \log(\Lambda) E^{-1} \tag{2.43}$$

assuming that $\lambda_i > 0$ for all $1 \leq i \leq n$. As already mentioned, the latter requirement is easily fulfilled for the group of symmetric and positive-definite matrices. In this case it holds also that the set of eigenvectors is orthogonal, that is

$$S \in Sym_+ : E^{-1} = E^T, \quad EE^T = E^T E = I. \tag{2.44}$$

Thanks to these properties, the matrix exponent realizes a *one-to-one mapping* between the space of symmetric matrices to the space of tensors [35, 39]. Hence, the aforementioned discussion on computation of the logarithm of a symmetric positive-definite matrix can be summarized in Algorithm 2.1.

1. Perform eigendecomposition of **S**:

$$\mathbf{S} = \mathbf{E}\mathbf{\Lambda}\mathbf{E}^T; \tag{2.45}$$

2. Compute the natural logarithm of each of the elements $\lambda_i > 0$ for all $1 \leq i \leq n$ of the matrix $\mathbf{\Lambda}$;
3. Multiply the matrices to obtain the natural logarithm of **S**:

$$\log(\mathbf{S}) = \mathbf{E}\log(\mathbf{\Lambda})\mathbf{E}^T. \tag{2.46}$$

Algorithm 2.1 Computation of the natural logarithm of the symmetric positive-definite tensor **S**.

Algorithm 2.1 can be easily used for computation of the matrix exponent function, substituting simply *log* with *exp*. In the same fashion other functions defined as a power series can be computed (e.g. square root).

For the two following tensors of dimensions 2×2

$$\mathbf{S} = \begin{bmatrix} 2 & -1 \\ -1 & 10 \end{bmatrix} \quad \text{and} \quad \mathbf{T} = \begin{bmatrix} 5 & 3 \\ 3 & 6 \end{bmatrix},$$

let us compute their Euclidean and Log-Euclidean distances, in accordance with Equations (2.28) and (2.31). Eigendecomposition of **S** leads to

$$\mathbf{E} = \begin{bmatrix} -0.9925 & -0.1222 \\ -0.1222 & 0.9925 \end{bmatrix}, \quad \mathbf{\Lambda} = \begin{bmatrix} 1.8769 & 0 \\ 0 & 10.1231 \end{bmatrix}$$

and also $\log(\mathbf{\Lambda}) = \begin{bmatrix} 0.6296 & 0 \\ 0 & 2.3148 \end{bmatrix}.$

This shows that **S** is symmetric and positive-definite, so that Algorithm 2.1 can be applied. Thus

$$\log(\mathbf{S}) = \begin{bmatrix} 0.6548 & -0.2044 \\ -0.2044 & 2.2897 \end{bmatrix}.$$

The same computations for **T** result in the following

$$\mathbf{E} = \begin{bmatrix} -0.7630 & 0.6464 \\ 0.6464 & 0.7630 \end{bmatrix}, \quad \mathbf{\Lambda} = \begin{bmatrix} 2.4586 & 0 \\ 0 & 8.5414 \end{bmatrix},$$

and also $\log(\mathbf{\Lambda}) = \begin{bmatrix} 0.8996 & 0 \\ 0 & 2.1449 \end{bmatrix}.$

T also is symmetric and positive-definite. Thus, from the above matrices we obtain

$$\log(\mathbf{T}) = \begin{bmatrix} 1.4199 & 0.6142 \\ 0.6142 & 1.6246 \end{bmatrix}.$$

Finally, the two distances, D_F in (2.28) and D_{LF} (2.31), between S and **T** are as follows

$$D_F(\mathbf{S}, \mathbf{T}) = 7.5498, \quad D_{LF}(\mathbf{S}, \mathbf{T}) = 1.5387.$$

We see that the two distances are quite different.

Algorithm 2.2 lists code of the *Orphan_Log_Of_Matrix* function for computation of a natural logarithm of a symmetric positive matrix. It implements the procedure outlined in Algorithm 2.1. For this purpose the *Compute_SVD* object is employed to decompose the input matrix in accordance with Equation (2.45). Then the natural logarithm is computed from the diagonal matrix of eigenvalues. Finally, the output matrix is assembled as in Equation (2.46).

```
///////////////////////////////////////////////////////////////
// This function computes the natural logarithm of
// a symmetric and positive-definite matrix.
///////////////////////////////////////////////////////////////
//
// INPUT:
//          inImage - input matrix
//
// OUTPUT:
//          log(inImage) - output matrix of the same
//               dimensions as inImage
//
// REMARKS:
//          Symmetry of a matrix is an absolute requirement
//          for this functions since eigendecomposition is
//          computed with the SVD.
//          Positive definite requirement is imposed by log.
//
RIAP Orphan_Log_Of_Matrix( const TRealImage & inImage )

{
    REQUIRE( inImage.GetCol() == inImage.GetRow() );
    // a matrix should be square

    // At first perform the eigendecomposition
    Compute_SVD      svd_calculator;
    // with SVD this is ok ONLY for SYMMETRIC matrices !!

    svd_calculator( inImage );

    // Then compute the logarithm on the eigenvalue matrix

    TRealImage * eigenvalueMatrix = svd_calculator.Get_V();
```

```
      int eigen_dim = eigenvalueMatrix->GetCol();
      REQUIRE( eigenvalueMatrix->GetRow() == 1 );

      // Compose everything back - here we reuse some of the matrices
      REQUIRE( eigen_dim == inImage.GetCol() );
      RIAP outImage( new TRealImage( inImage.GetCol(), inImage.GetRow(), 0.0 ) );

      for( int i = 0; i < eigen_dim; ++ i )
      {
              double eigen_value = eigenvalueMatrix->GetPixel( i, 0 );
              if( eigen_value <= 0.0 )
                      return RIAP( 0 );

              outImage->SetPixel( i, i, log( eigen_value ) );
      }

      // Compose everything back - here we reuse some of the matrices
      bool op_status = Mult_Matrix< double, double >( * svd_calculator.Get_D(),
                                                       * svd_calculator.Get_S(),
                                                       * outImage );

      op_status = op_status && Mult_Matrix_With_SECOND_Transposed< double, double >(
                      * outImage, * svd_calculator.Get_D(), * svd_calculator.Get_S() );

      return outImage;
}
```

Algorithm 2.2 Implementation of the *Orphan_Log_Of_Matrix* function for computation of a natural logarithm of a symmetric positive matrix.

It is worth noticing that with only small modification the function can be used to compute other matrix functions, such as exponent. A prefix "orphan" in the name of the above function means that this function allocates memory for the return object which is then abandoned to a caller of this function. This means that the responsibility of deallocating memory is left to the caller as well. A matrix is represented by the TRealImage object. Software framework is discussed in Section A.8.

2.5.2   *Euclidean Image Distance and Standardizing Transform*

As we saw in the previous section tensors are compared by their elements. However, frequently the position of an element within a structure, as well as the values of its neighbors, can also provide important information. This is a situation which we encounter when processing visual signals, such as monochrome and color images and video streams. Nevertheless, the second aspect is not easily accounted for due to geometrical transformation. On the other hand, image recognition relies heavily on comparison of images for which the Euclidean metric is the most frequently used, mostly due to its popularity and simplicity in computations. However, Wang *et al.* proposed a modification to the Euclidean metric which also takes into account the spatial relationship among pixels [46]. It is called *IMage Euclidean Distance* (IMED) and shows many useful properties, among which the most important is its insensitivity to small geometrical deformations of compared images.

Let us start with the Frobenius distance (2.28), however for two 2D images $\mathbf{S}, \mathbf{T} \in \mathfrak{R}^{N \times M}$ which can be expressed as follows

$$D_F (\mathbf{s}, \mathbf{t}) = \sum_{k=1}^{MN} \left(s^k - t^k \right)^2 = (\mathbf{s} - \mathbf{t})^T (\mathbf{s} - \mathbf{t}), \tag{2.47}$$

where $\mathbf{s}$ and $\mathbf{t}$ are column vectors which are vectorized versions of the images $\mathbf{S}$ and $\mathbf{T}$, respectively. However, the above is a special case of a more general distance defined by a metric matrix, as follows

$$D_G (\mathbf{s}, \mathbf{t}) = \sum_{k,l=1}^{MN} g_{kl} \left(s^k - t^k \right) \left(s^l - t^l \right) = (\mathbf{s} - \mathbf{t})^T \, \mathbf{G} \, (\mathbf{s} - \mathbf{t}), \tag{2.48}$$

where g_{kl} are elements of the symmetric nonnegative matrix $\mathbf{G}$ of dimensions $MN \times MN$ which defines the metric properties of the space. The metric coefficients g_{kl} are defined by the inner product of the corresponding base vectors $\mathbf{b}_k$ and $\mathbf{b}_l$, as follows

$$g_{kl} = \langle \mathbf{b}_k, \mathbf{b}_l \rangle = \|\mathbf{b}_k\| \, \|\mathbf{b}_l\| \cos \beta_{kl}, \tag{2.49}$$

where β_{kl} denotes an angle between $\mathbf{b}_k$ and $\mathbf{b}_l$. For the case of the base vectors of equal lengths, g_{kl} depend solely on the angle β_{kl} between corresponding vectors.

We easily see that the Euclidean distance (2.47) constitutes a special case of Equation (2.48) in which $\mathbf{G} = \mathbf{I}$. However, the identity matrix $\mathbf{I}$ implies an orthogonal basis which cannot convey information on relations of pixels on slightly different positions than directly corresponding ones. In other words, the orthogonal base vectors don't provide information on spatial relationships (distances) between pixels. Therefore to account for the fact that slightly deformed patterns should be similar to each other, the used metric must consider also pixel distances. This in turn implies that the metric coefficients g_{ij} of the metric matrix $\mathbf{G}$ have to be related to the pixel distances. In this respect in Equation (2.48) we observe two different distances: The first concerns pixel values $s_{k,l}$ and $t_{k,l}$. However, the second accounts for pixel distance and through the values of g_{kl} indicates for which pixels k and l their corresponding pixel values need to be compared and taken into the sum in Equation (2.48). Obviously, this happens for all k,l for which their corresponding metric coefficients g_{kl} are different from 0. Let us recall that for the "traditional" Euclidean distance this happens only for $k = l$, i.e. $g_{kl} = 1$ iff $k = l$, otherwise $g_{kl} = 0$. This means that the basis of the new (transformed) space should be *nonorthogonal* to better convey distances between pixels. Thanks to this property we can embed information on the spatial position of pixels into g_{kl}. In other words, the closer the pixels are the higher the value of g_{kl} should be, reaching its maximum for $k = l$. The distance between pixel positions (not values) is defined on an integer image lattice simply as the following "traditional" Euclidean distance

$$\|\mathbf{P}_k - \mathbf{P}_l\| = \sqrt{\left(p_k^1 - p_l^1 \right)^2 + \left(p_k^2 - p_l^2 \right)^2}, \tag{2.50}$$

where each pixel point $\mathbf{P}_i = \left[p_i^1, p_i^2 \right]^T$. Thus, it follows that g_{kl} need to be defined as functions of the distances $\|\mathbf{P}_k - \mathbf{P}_l\|$, which can be done as follows

$$g_{kl} = f\left(\|\mathbf{P}_k - \mathbf{P}_l\|\right) = \frac{1}{2\pi\sigma} e^{-\frac{\|\mathbf{P}_k - \mathbf{P}_l\|^2}{2\sigma^2}}, \qquad (2.51)$$

where σ is the width parameter, usually set to 1 [46]. Finally, incorporating Equations (2.50) and (2.51) into Equation (2.48) the IMED distance between images S and T and their vectorized versions $\mathbf{s}$ and $\mathbf{t}$, is obtained in the following form

$$D_{IMED}\left(\mathbf{s}, \mathbf{t}\right) = \frac{1}{2\pi\sigma} \sum_{k,l=1}^{MN} e^{-\frac{\left(p_k^1 - p_l^1\right)^2 + \left(p_k^2 - p_l^2\right)^2}{2\sigma^2}} \left(s^k - t^k\right)\left(s^l - t^l\right). \qquad (2.52)$$

It is worth noticing that, thanks to the aforementioned properties of D_{IMED}, the images are treated in a two-fold way. That is, pixel values of images $\mathbf{S}$ and $\mathbf{T}$ can be stacked in any possible way into the vectors $\mathbf{s}, \mathbf{t}$. If alone, this process leads to loss of information on the spatial neighborhood of pixels. However, D_{IMED} remedies this drawback by encoding information on the mutual positions of pixels within each of the input images into the metrics g_{kl}.

The D_{IMED} image metric in Equation (2.52) can be used for a direct comparison of images, such as in the case of the k-nearest-neighbor method, but it can be also incorporated into other classification algorithms, such as HOSVD classifier discussed in Section 5.8. This can be achieved by substituting D_{IMED} into all the places in which D_E has been used. However, for large databases of images direct computation of Equation (2.52) can be expensive. To overcome this problem, Wang *et al.* propose to represent Equation (2.52) as follows

$$D_{IMED}\left(\mathbf{s}, \mathbf{t}\right) = (\mathbf{s} - \mathbf{t})^T \mathbf{G}\left(\mathbf{s} - \mathbf{t}\right) = (\mathbf{s} - \mathbf{t})^T \mathbf{A}^T \mathbf{A}\left(\mathbf{s} - \mathbf{t}\right) = (\mathbf{u} - \mathbf{v})^T\left(\mathbf{u} - \mathbf{v}\right), \quad (2.53)$$

where

$$\mathbf{u} = \mathbf{As}, \quad \mathbf{v} = \mathbf{At}. \qquad (2.54)$$

In consequence, we can avoid repetitive computations of D_{IMED} transforming the input images $\mathbf{S}$ and $\mathbf{T}$ into their new representations $\mathbf{U}$ and $\mathbf{V}$ in accordance with Equation (2.54), and then using the new ones in "old" recognition algorithms that rely on the simple Euclidean metric without any further modifications. This feature makes D_{IMED} very attractive since it does not involve any changes to the existing algorithms and requires *only transformation of the input signals*. To find out $\mathbf{A}$ from $\mathbf{G}$ Wang *et al.* propose the following transformation

$$\mathbf{G} = \mathbf{G}^{\frac{1}{2}}\mathbf{G}^{\frac{1}{2}} = \mathbf{A}^T \mathbf{A}. \qquad (2.55)$$

Then we have

$$\mathbf{G}^{\frac{1}{2}} = \Gamma \Lambda^{\frac{1}{2}} \Gamma^T, \qquad (2.56)$$

Figure 2.5 Visualization of the Standardizing Transform applied to the road sign pictograms. Original pictograms (upper row) and their IMED versions (lower row).

where $\mathbf{\Lambda}$ is a diagonal matrix with eigenvalues of $\mathbf{G}$, and $\mathbf{\Gamma}$ is an orthogonal matrix which columns are eigenvectors of $\mathbf{G}$. Thus, for $\mathbf{A}$ in Equation (2.54) one can use $\mathbf{G}^{\frac{1}{2}}$ from Equation (2.56) which is called the Standardizing Transform.

Computation of Equation (2.56) can be done by the methods discussed in Section 2.5.1.1. Nevertheless, in practice such computations can be time consuming since the number of elements is $NM \times NM$, so for instance a HDTV standard makes $\sim 4 \cdot 10^{12}$ elements in $\mathbf{G}$, which may be prohibitive. A practical solution to overcome this computational problem was proposed by Sun and Feng [47] after observing that computation of D_{IMED} can be equivalently stated as transform domain smoothing. They developed the *Convolution Standardized Transform* (CST) which approximates well the D_{IMED} . For this purpose the following separable filter was used

$$\mathbf{H} = \mathbf{h} \otimes \mathbf{h}^T = \mathbf{h}\mathbf{h}^T, \tag{2.57}$$

where $\otimes$ denotes the Kronecker product of two vectors $\mathbf{h}$ and $\mathbf{h}^T$.[3] For the latter, Sun and Feng propose a filter with the following components

$$\mathbf{h} = \begin{bmatrix} 0.0053 & 0.2171 & 0.5519 & 0.2171 & 0.0053 \end{bmatrix}^T. \tag{2.58}$$

The filter $\mathbf{h}$ given by Equation (2.58) was also used in our computations since it offers much faster computations than direct application of Equation (2.56).

Figure 2.5 shows examples of application of the Standardizing Transformation for selected pictograms of road signs. The filter with coefficients $\mathbf{h}$ in (2.58) was used which results in low-pass smoothing of the pictograms (lower row).

D_{IMED} was employed by Liu *et al.* [48] in the tensor-based method for dimensionality reduction. Examples of D_{IMED} connected with the tensor based classifiers, as well as a discussion of its influence on classification accuracy, can also be found in the paper by Cyganek [49].

---

[3]The Kronecker product and its properties are discussed on p. 115.

2.6   Filtering of Tensor Fields

Filtering of nonscalar data, such as vector, matrices, or tensors in general, is a challenge. The simplest approach is to apply a scalar filtering in each channel, or component, separately. However, such a solution does not usually provide satisfactory results. On the other hand, the nonlinear filters outperform the linear ones since they allow smoothing which obeys edge discontinuities. One category within the nonlinear group constitute the order statistic filters [50]. A version of this method adapted to the filtering of tensor data was proposed in the paper [51]. The second interesting type of filter is the anisotropic diffusion filter which allows preservation of important details in images. Both types are related, however. Both also rely on the chosen comparison measure. These and related issues are discussed in the following sections.

2.6.1   Order Statistic Filtering of Tensor Data

Compared to the linear methods, the order statistic filters show many desirable features, such as edge preservation, removal of impulsive noise, etc. Usually they are assumed to operate within a fixed window in a 2D image, for which *the algebraic order* of samples is used to find the output value. However, there is a significant difference in finding order for scalar and nonscalar data, since for the latter there is no unique ordering relation.

The most popular example of the order statistic filters is *the median filter*. For the scalar valued images its operation is very simple. The image intensity values found in a window are sorted and the middle sample is taken as an output. However, such simple sorting is not possible even for color images, not to mention tensor data fields.

The concept of median filtering of vector data was further extended to the matrix and tensor valued data by Welk *et al.* [52]. Filtering of tensor data is important for processing of the diffusion tensors of the MRI, structural tensor (2.7), etc. Welk *et al.* propose a median filter of tensors based on the Frobenius norm used to measure their distance. This norm has the advantage of being rotation invariant which is a desirable property in the case of tensors. The minimization problem was solved with a modified gradient descent method. The median value in this case does not necessarily belong to the set of input data, however.

In contrast to the above, the method discussed in this section provides a nonnumerical way of tensor data processing for *different weighted* order statistics (not only median), as well as for different norms of tensors. The order statistic for scalars, vectors, and tensor data differ in the ordering scheme. For nonscalars it should provide equal importance to all components and use them all in data ordering. The usual way is to define a distance measure between each pair of data $\mathbf{x}_i$ and $\mathbf{x}_j$ [53]

$$D\left(\mathbf{x}_i, \mathbf{x}_j\right),\tag{2.59}$$

where D fulfills the distance properties, as discussed in Section 2.5. The aggregated weighted distance $D_{(*)}\left(\mathbf{x}_i, \{\mathbf{x}\}_K, \{c\}_N\right)$ of a sample $\mathbf{x}_i$ (for $i = 1,2,\ldots,N$) with respect to the set $\{\mathbf{x}\}_N$ of N other samples can be defined as follows

$$D_{(*)}\left(\mathbf{x}_i, \{\mathbf{x}\}_N, \{c\}_N\right) = \sum_{j=1}^{N} c_j D\left(\mathbf{x}_i, \mathbf{x}_j\right),\tag{2.60}$$

where $\{c\}_N$ is a set of N scalar weights.

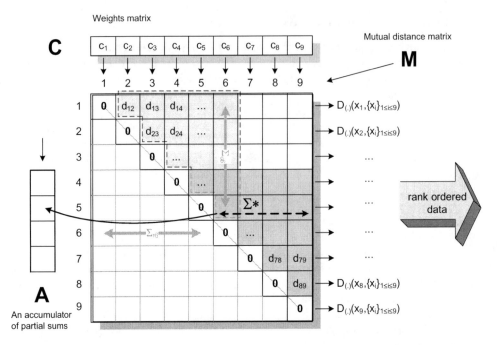

Figure 2.6 Data structures for computation of the aggregated distances $D_{(*)}$ (based on [77]).

The order statistic filters assume operation over a set of data $\{\mathbf{x}\}_N$. For each $\mathbf{x}_i \in \{\mathbf{x}\}_N$ the scalar $D_{(*)}(\mathbf{x}_i)$ is computed from Equation (2.60). These scalars are then sorted which leads to a new order among them. This new order is finally used to set the $\mathbf{x}_i$ values associated with each $D_{(*)}(\mathbf{x}_i)$. This can be denoted as follows

$$D^{(1)} \leq D^{(2)} \leq \cdots \leq D^{(p)} \leq \cdots \leq D^{(N)},$$

$$\mathbf{x}^{(1)} \rightarrow \mathbf{x}^{(2)} \rightarrow \cdots \rightarrow \mathbf{x}^{(p)} \rightarrow \cdots \rightarrow \mathbf{x}^{(N)},$$

(2.61)

where $D^{(p)} \in \{D(\mathbf{x}_1), D(\mathbf{x}_2), \ldots, D(\mathbf{x}_N)\}$ and $\mathbf{x}^{(p)} \in \{\mathbf{x}_1, \mathbf{x}_2, \ldots, \mathbf{x}_N\}$ for $1 \leq p \leq N$. Finally one data $\mathbf{x}^{(p)}$ is chosen as an output of the order statistic filter, depending on the desired characteristics of that filter. The special and very important case is the weighted median filter, defined as follows

$$\mathbf{x}_m = \underset{\mathbf{x}_i \in \{\mathbf{x}\}_N}{\arg \min} D_{(*)}\left(\mathbf{x}_i, \{\mathbf{x}\}_N, \{c\}_N\right),$$

(2.62)

where $\mathbf{x}$ are tensors (i.e. matrices, vectors, scalars, etc.) from a set $\{\mathbf{x}\}_N$ of data with a chosen aggregated distance measure $D_{(*)}$, $\{c\}_N$ is a set of scalar weights, and $\mathbf{x}_m$ is the median.

If for all i in Equation (2.60) $c_i = 1$, then Equation (2.62) denotes the classical median filter. The characteristic feature of the proposed method is that the result is a data from the filtered set. This is in contrast to the already mentioned approach by Welk *et al.* [37].

Table 2.1 Execution times of the median filtering of the $512 \times 512 \times 3$ tensor data in C++ implementation.

Filter size	3×3	5×5	7×7	9×9
Number of d_i	36	300	1176	3240
Run time – simple implementation [s]	0.6	2.8	9.9	26.7
Run time – optimized with matrix M [s]	**0.3**	**1.1**	**3.6**	**9.5**

Implementation of the method follows Equation (2.60) with a specific distance chosen for Equation (2.59). To speed up computations a special data structure, shown in Figure 2.6, was developed. It is a triangular matrix of dimensions $N^2 \times N^2$ for a filtering window of size $N \times N$. In Figure 2.6 a 3×3 filtering window is assumed. The matrix stores the mutual distances D for all data values. Due to their metric properties the matrix is triangular with a zero diagonal. In general, for N elements in the window ($\binom{N}{2}$) different pairs of distances need to be determined. The weights c_i are stored in a separate matrix **C**. The aggregated distances $D_{(*)}$ are weighted sums of the corresponding rows, e.g. $D(\mathbf{x}_7, \{\mathbf{x}_i\}_{1 \le i \le 9}) = \Sigma_{1 \le i \le 9}\, c_i \mathbf{m}[7,i]$. $D_{(*)}$ need to be computed only once, then they are rank ordered. For the median filter the output is x_p which corresponds to a minimal $D_{(*)}$.

The proposed data structure proves useful when moving filtering from point to point in an image. In this case, in the previously constructed matrix **M** only one column of data needs to be updated. Thanks to this strategy computation time was shortened by about three times compared to the not optimized implementation. Some execution time values for the C++ implementation are presented in Table 2.1.

Figure 2.7(a) depicts a HSV visualization of a region of the structural tensor, which will be discussed in Section (2.7), computed from the image in Figure 2.2. The same region filtered

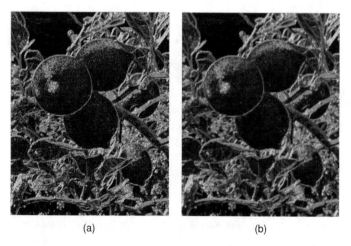

(a)   (b)

Figure 2.7 Color visualization of the structural tensor (2.7) computed from a region in Figure 2.2 in the HSV color encoding (a). Its 5×5 median filtered version (b). (For a color version of this figure, please see the color plate section.)

with a 5×5 median filter with the Frobenius measure D_F in Equation (2.62) is shown in Figure 2.7(b). The structures were smoothed while details still remain present.

Although in the case of 2D symmetrical tensors there are three components, exactly as in the case of color image filtering, the presented framework is general and can be applied to any dimensional data for which a suitable metric can be established.

Finally, let us notice that the described median filter belongs to a wider group of order statistic filters, such as morphological operators. These, however, encounter definition problems when going to higher dimensional signals. Nevertheless, it is possible to define an order relation endowed with a notion of maximum and minimum values, which in turn allow construction of the morphological operators for tensors. Such methods are further discussed in Appendix A.3. The other approach to multidimensional data ordering can be achieved with the help of the data reduction method, called principal component analysis, as will be discussed in Section 3.3.1.2. In this case the lexicographical order of vector data is assumed, however, on data already transformed into the principal component space, i.e. data are already ordered in accordance with their statistical variance.

In the context of tensor filtering an interesting read is the paper by Zhang and Hancock [54]. It deals with Riemannian median filtering in the broader context of techniques for directional and tensorial image data processing.

2.6.2 Anisotropic Diffusion Filtering

Many physical processes are well modeled by the diffusion equation, such as heat conduction [55]. This appears possible to be used for image processing as well. For a certain function f the diffusion equation is defined as follows [56, 57]

$$\frac{\partial f(x, y, t)}{\partial t} = c\nabla^2 f(x, y, t), \tag{2.63}$$

where ρ is a diffusion constant. For the input image $I(x,y)$ – assumed to be a continuous function for the further discussion – the border condition for the above are stated as follows

$$f(x, y, 0) = I(x, y). \tag{2.64}$$

Equation (2.63) can be written as

$$\partial_t f(x, y, t) = div(c \cdot \nabla f(x, y, t)). \tag{2.65}$$

The diffusion equation for a constant value c leads to a low pass filtering of an image [58]. Thus, after exchanging t to scale parameter σ, it leads to the linear image scale-space [15].

However, linear filtering of images leads to blurring of important details, such as edges [50]. To overcome this limitation Perona and Malik propose using anisotropic diffusion (AD) filtering, in which filtering is selective depending on the contents of an image [59]. Their idea was to make the parameter c in Equation (2.65) *dependent* on the signal gradient. In other words, filtering proceeds in areas bounded by strong signal variations which denote edges. Thus, such features as edges are preserved and sharp, whereas image areas characteristic of low frequency are filtered. Their idea can be incorporated into the basic equation (2.65) introducing a controlling function g to the inner gradient instead of c, as follows

$$\partial_t f(x, y, t) = div\left(g\left(\|\nabla f(x, y, t)\|\right) \cdot \nabla f(x, y, t)\right). \tag{2.66}$$

The argument of this control function g is *a module of the gradient* of a signal. Thus, it is not a surprise that g for a large argument should approach zero to stop smoothing in this direction. For this purpose Perona and Malik propose the following two functions [59]

$$g_1(x) = \left(1 + \frac{x^2}{\sigma^2}\right)^{-1}, \tag{2.67}$$

$$g_2(x) = \exp\left(-\frac{x^2}{(2\sigma)^2}\right), \tag{2.68}$$

where σ is a parameter which depends on the application (usually chosen empirically). A third function g_3 was proposed by Weickert and Brox [60]

$$g_3(x) = \frac{1}{x^r + \varepsilon}, \tag{2.69}$$

where r is a positive integer value, ε is a constant that assures numerical stability of the function if the signal gradient approaches zero. However, as shown by Sapiro, the Tukey biweight function, given as follows [61]

$$g_4(x) = \begin{cases} \frac{1}{2}\left(1 - \frac{x^2}{\sigma^2}\right)^2, & |x| \leq \sigma \\ 0, & otherwise \end{cases} \tag{2.70}$$

shows its superiority in leaving untouched strong signals. For the parameter σ in (2.70) the so called *robust scale*

$$\sigma_r = 1.4826 \cdot med\left(\|\nabla I - med\left(\|\nabla I\|\right)\|\right) \tag{2.71}$$

is used, that is $\sigma = \sigma_r$. It is computed from the gradient ∇I of the monochrome version of the original image, whereas *med* denotes a median [61].

Let us recall the basic relations among *gradient* and *divergence* operators [16, 17]:

$$\nabla \equiv i\frac{\partial}{\partial x} + j\frac{\partial}{\partial y} + k\frac{\partial}{\partial z}, \tag{2.72}$$

$$\nabla f(x, y, z) = grad\, f(x, y, z) = i\frac{\partial f}{\partial x} + j\frac{\partial f}{\partial y} + k\frac{\partial f}{\partial z}, \tag{2.73}$$

$$\nabla \cdot f(x, y, z) = div\, f(x, y, z) = \frac{\partial f}{\partial x} + \frac{\partial f}{\partial y} + \frac{\partial f}{\partial z}, \qquad (2.74)$$

$$\nabla^2 \equiv \nabla \cdot \nabla = \Delta = \frac{\partial^2}{\partial x^2} + \frac{\partial^2}{\partial y^2} + \frac{\partial^2}{\partial z^2}, \qquad (2.75)$$

$$\Delta f(x, y, z) = div\, grad\, f(x, y, z) = \nabla \cdot [\nabla f(x, y, z)] = \nabla^2 f(x, y, z), \qquad (2.76)$$

where $\mathbf{i}$, $\mathbf{j}$, $\mathbf{k}$ are Euclidean base vectors, while operator $(\cdot)$ denotes a scalar product of two vectors. In the above Equations (2.72) and (2.75) denote operators, i.e. they act on the supplied function, such as f.

For example, if $f(x, y, z) = 2x^2 + 3y^3 + 4z^4$, then we have

$$\frac{\partial f}{\partial x} = 4x, \quad \frac{\partial f}{\partial y} = 9y^2, \quad \frac{\partial f}{\partial z} = 16z^3.$$

From these the gradient is obtained as the following vector

$$\nabla f(x, y, z) = 4x\mathbf{i} + 9y^2\mathbf{j} + 16z^3\mathbf{k} = \begin{bmatrix} 4x & 9y^2 & 16z^3 \end{bmatrix}^T,$$

which squared norm is

$$\|\nabla f(x, y, z)\|^2 = (4x)^2 + (9y^2)^2 + (16z^3)^2.$$

Divergence of f is computed from Equation (2.74) as the following scalar

$$div\, f(x, y, z) = 4x + 9y^2 + 16z^3.$$

Finally, after computation of the second partial derivatives, from Equation (2.76) we easily compute the nabla operator $\Delta f(x, y, z)$, as follows

$$\Delta f(x, y, z) = 4 + 18y + 48z^2.$$

The above filters can be used exclusively to the scalar fields. However, Gerig et al. [62] proposed a diffusion for images of M channels, as follows

$$\underset{k}{\forall} : \partial_t f_k = div \left[g \left(\sum_{m=1}^{M} \|\nabla f_m\|^2 \right) \nabla f_k \right]. \qquad (2.77)$$

The above equation is nonlinear but still isotropic[4] which is an effect of application of the scalar function g. However, a qualitative change can be obtained replacing g with a tensor, as follows

$$\forall_k : \partial_t f_k = div \left[\mathbf{T} \left(\sum_{m=1}^{M} \nabla f_m \nabla f_m^T \right) \nabla f_k \right] \tag{2.78}$$

where $\mathbf{T}$ is a positive definite tensor. Thanks to this, diffusion is adapted locally to both, the data and to the direction of smoothing [63]. In other words, in this scheme it is possible to smooth along the edges while inhibiting smoothing across the edges. This results in smoothing while preserving important details. In practice $\mathbf{T}$ is connected with the structural tensor (ST), discussed in Section 2.7. It has the same eigenvalues which depend on the gradient of the intensity signal. This way, one of the eigenvectors is tangent to an edge in a local neighborhood, whereas the other is normal since the two are orthogonal, as discussed in Section 2.7.1. On the other hand, eigenvalues of $\mathbf{T}$ are usually chosen as functions of Equations (2.67)–(2.69) with arguments being eigenvalues of the ST [64]. Based on the above discussion the control tensor $\mathbf{T}$ in (2.78) can be stated as [60]

$$\mathbf{T} = g_3 \left(\lambda_1 \right) w_1 w_1^T + g_3 \left(\lambda_2 \right) w_2 w_2^T, \tag{2.79}$$

where $\lambda_{1,2}$ are eigenvalues, w_1 and w_2 are eigenvectors of the ST (2.90), for 2D images given by Equation (2.93). Function g_3 is given in Equation (2.69). Tensor driven diffusion shows very promising results, especially in the field of image filtering, precise detection of image features, as well as in the segmentation. Further details can be found in the cited literature.

As alluded to previously, the main benefit of using AD instead of linear low-pass filtering is preservation of local structures in images. In this respect AD has much in common with the nonlinear median filter, discussed in Section 2.6.1. For this reason, in many applications that operate in scale-spaces it is better to use AD rather than a simple Gaussian filter, since such space will preserve important features. For instance such modification was applied in the system for dental implant recognition, details of which are discussed in Section 5.4.3. In this case, the Gaussian filter smears details which are important for implant detection. On the other hand, AD preserves sharp object edges and allows proper detection.

As we shall see in Section 2.7.4, AD is also used in computations of the extended structural tensor. Thanks to this sharp object edges and other small details are not smeared.

Let us also notice that AD can be used to filter other types of signals, such as probability fields [61]. For instance, AD based filtering of the probability fields from road scenes was investigated in the paper by Cyganek [65]. In this proposition a slight modification to the process of anisotropic diffusion was applied, i.e. although the probability field is filtered, the control function is guided by the gradient of the "original" intensity signal.

---

[4]This is different from the formulation of Perona and Malik who call this process anisotropic [59]. However, we follow the observations by Weickert [57], in which anisotropy is attributed exclusively to the case of the tensor diffusivity factor.

2.6.3   *IMPLEMENTATION of Diffusion Processes*

The base for implementation constitutes the discrete versions of Equations (2.66) and (2.77), for the single and multiple channel images, respectively. In all cases computation of partial directional derivatives I_x and I_y is essential. For this purpose the derivative filters for discrete signals are used. However, instead of the finite difference method [56, 66], the directional filters by Farid and Simoncelli are used [67]. In this approach a discrete signal to be differentiated is first approximated with the continuous polynomial, for which the derivatives are computed, which are finally back sampled to the discrete domain. Implementation of this approach is discussed in [15]. In effect, computation of PDE in Equations (2.66) and (2.77) is stable and efficient since the filters are also separable.

Figure 2.8 depicts the hierarchy of classes responsible for implementation of the anisotropic diffusion processes for images. The central is the *AnisotropicDiffusion* class which implements anisotropic diffusion for a single as well as a series of images. As already discussed, the anisotropic process is an iterative algorithm. For this reason, there are two pairs of connected

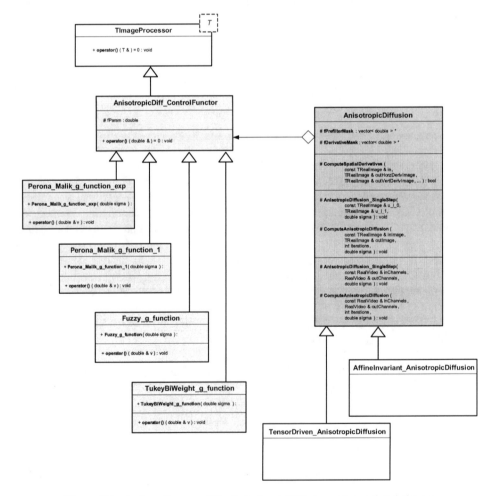

Figure 2.8   A class diagram of the *AnisotropicDiffusion* and associated classes.

members: *AnisotropicDiffusion_SingleStep*, which realizes a single step of the anisotropic diffusion, and the *ComputeAnisotropicDiffusion* which does a number of iterations which result in final output signal.

The mixin class *AnisotropicDiff_ControlFunction*, derived from the *TImageProcessor* template, starts a hierarchy of functor classes which implement control functions (2.67)–(2.70) for the AD, respectively. A functor object with a chosen control parameter is supplied to the constructor of the *AnisotropicDiffusion* class. However, the parameter σ can be changed in run-time. Algorithm 2.3 presents definitions of the most important members of the *AnisotropicDiffusion* class. The derived classes *AffineInvariant_AnisotropicDiffusion* and *TensorDriven_AnisotropicDiffusion* implement the ideas of the affine invariant [61], as well as the already outlined tensor-driven AD.

```
/////////////////////////////////////////////////////////////
// This class does anisotropic diffusion in images.
/////////////////////////////////////////////////////////////
class AnisotropicDiffusion
{
    public:

        // ====================================================
        AnisotropicDiffusion( AnisotropicDiff_ControlFunctor * controlFunctorObj= 0);
        virtual ~AnisotropicDiffusion();      // class virtual destructor
        // ====================================================

    protected:

        /////////////////////////////////////////////////////////////
        // These functions compute spatial derivative for a frame
        /////////////////////////////////////////////////////////////
        //
        // INPUT:
        //              theVideo - the video stream
        //              frame_no - frame number in the video
        //              theFrame - explicitly provided frame
        //
        // OUTPUT:
        //              true if ok,
        //              false otherwise
        //
        // REMARKS:
        //
        //
        bool ComputeSpatialDerivatives( const MonochromeVideo & theVideo,
            int frame_no, TRealImage & outHorzDerivImage,
            TRealImage & outVertDerivImage,
            vector< double > * fPrefilterMask, vector< double > * fDerivativeMask );
        bool ComputeSpatialDerivatives( const MonochromeImage & theFrame,
            TRealImage & outHorzDerivImage, TRealImage & outVertDerivImage,
            vector< double > * fPrefilterMask, vector< double > * fDerivativeMask );
        bool ComputeSpatialDerivatives( const TRealImage & theFrame,
            TRealImage & outHorzDerivImage, TRealImage & outVertDerivImage,
            vector< double > * fPrefilterMask, vector< double > * fDerivativeMask );
```

```
/////////////////////////////////////////////////////////
// Single diffusion step
/////////////////////////////////////////////////////////
//
// INPUT:
//              u_i_0 - input frame at step 0
//              u_i_1 - output frame at step 1
//              sigma - control function parameter
//
// OUTPUT:
//              none
//
// REMARKS:
//
//
virtual void AnisotropicDiffusion_SingleStep( const TRealImage & u_i_0,
             TRealImage & u_i_1, double sigma );
// Processes each frame of the inChannels video stream
virtual void AnisotropicDiffusion_SingleStep( const RealVideo & inChannels,
             RealVideo & outChannels, double sigma );


public:

/////////////////////////////////////////////////////////
// Does the anisotropic diffusion process.
/////////////////////////////////////////////////////////
//
// INPUT:
//              inImage - input image
//              outImage - output (filtered) images
//              iterations - number of iterations
//              sigma - control function parameter
//
// OUTPUT:
//              none
//
// REMARKS:
//
//
virtual void ComputeAnisotropicDiffusion( const TRealImage & inImage,
             TRealImage & outImage, int iterations, double sigma );
virtual void ComputeAnisotropicDiffusion( const MonochromeImage & inImage,
             MonochromeImage & outImage, int iterations, double sigma );
virtual void ComputeAnisotropicDiffusion( const ColorImage & inImage,
             ColorImage * & outImage, int iterations, double sigma );
virtual void ComputeAnisotropicDiffusion( const RealVideo & inChannels,
             RealVideo * & outChannels, int iterations, double sigma );

};
```

Algorithm 2.3 Definition of the *AnisotropicDiffusion* class (the most important members shown).

```
//                                                           in              out
void AnisotropicDiffusion::
    AnisotropicDiffusion_SingleStep( const TRealImage & u_i_0, TRealImage & u_i_1,
                                     double sigma )
{
1        int frame_cols = u_i_0.GetCol();
2        int frame_rows = u_i_0.GetRow();
3
4
5        // Create local images for derivatives in three directions
6        TRealImage & Dx = u_i_1;
7        TRealImage Dy( frame_cols, frame_rows, 0.0 );
8
9
10       bool retVal = ComputeSpatialDerivatives( u_i_0, Dx, Dy,
11                                                 fPrefilterMask, fDerivativeMask );
12       REQUIRE( retVal == true );
13
14
15       // In Dx and Dy we have a gradient field
16       // Now compute the squared module of the gradient field
17
18       TRealImage Dx2( Dx ), Dy2( Dy );
19
20       Dx2 *= Dx;
21       Dy2 *= Dy;
22
23       TRealImage squareModOfGradient( Dx2 );
24       squareModOfGradient += Dy2;
25
26       // Now we apply the control function on the squared module of the gradient
27       GenerateStagField( squareModOfGradient, sigma );
28
29
30       // The final stage is to combine the fields and to compute a divergence
31       Dx *= squareModOfGradient;
32       Dy *= squareModOfGradient;
33
34       // Now we compute the divergence
35       ComputeSpatial_x_Derivative( Dx, Dx, fPrefilterMask, fDerivativeMask );
36       // this can be done in situ
37       ComputeSpatial_y_Derivative( Dy, Dy, fPrefilterMask, fDerivativeMask );
38
39       Dx += Dy;
40
41       //u_i_1 = Dx;          Dx is already u_i_1 for simplicity
}
```

Algorithm 2.4 Implementation of the *AnisotropicDiffusion_SingleStep* function.

Apart from the constructor/destructor pair, in Algorithm 2.3 we see three basic groups of members: The first one (named *ComputeSpatialDerivatives*) is responsible for computation of the spatial derivatives. The second set of members (*AnisotropicDiffusion_SingleStep*) implement a single step of Equations (2.66) and (2.77), while the complete iterative anisotropic diffusion is realized by the third group of members (*ComputeAnisotropicDiffusion*) from the interface of the class. Steps of the two last groups are explained in code listings in Algorithm 2.4 and Algorithm 2.5, respectively. These implement the diffusion process for single-channel (scalar) images, described in Equation (2.66).

The base data structures come from the HIL library [68, 69]. In line <10> the first partial directional derivatives are computed with help of optimized filters which components are stored in the internal vectors *fPrefilterMask* and *fDerivativeMask*. Their values can be looked up in the attached source code, whereas details of their computation are discussed in [15]. Based on these, in lines <18–24> the squared module of the gradient vector is computed. Then in line <27> the control function is computed which as an argument takes $\|\nabla f\|$. In this step, the supplied σ parameter is also used. As already mentioned, the control function is represented by the functor object supplied to the *AnisotropicDiffusion* constructor. Finally, in steps <31–39> the divergence (2.66) is computed.

```
void AnisotropicDiffusion::ComputeAnisotropicDiffusion( const TRealImage & inImage,
                          TRealImage & outImage, int iterations, double sigma )
{
        TRealImage u0( inImage );

        TRealImage div( inImage.GetCol(), inImage.GetRow(), 0.0 );

        for( int i = 0; i < iterations; ++ i )
        {
                AnisotropicDiffusion_SingleStep( u0, div, sigma ); // u0 ==> div
                u0 += div;
        }

        // Make just a copy
        outImage = u0;
}
```

Algorithm 2.5   Implementation of the *ComputeAnisotropicDiffusion* function.

The interface function *ComputeAnisotropicDiffusion*, shown in Algorithm 2.5, simply calls the *AnisotropicDiffusion_SingleStep* the requested number of times. The number of iterations, as well as the parameter for one of the chosen control functions, belong to the externally provided parameters for the AD implementation. Usually these are chosen experimentally, as will be discussed.

2.7   Looking into Images with the Structural Tensor

The idea of the *structural tensor* (ST) in an *L*-dimensional space is to find regular structures in local neighborhoods of each point of this space [9]. If such a regularity around a

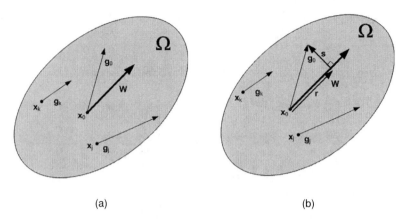

(a)   (b)

Figure 2.9 The structural tensor can represent a local neighborhood Ω with a single object vector **w**, which is accurate if most of the gradients in Ω coincide with **w** (a). It is computed by minimizing a cumulative sum of the residual vectors **s** (b).

point indeed exists, then it can be associated to a vector pointing toward the direction of the maximal change together with a measure of coherency of such a representation. In the $L = 2$ dimensional continuous space, a local neighborhood around a point x_0 is depicted in Figure 2.9(a) (a discrete case will be discussed in the next section). In this section we follow a more general approach presented in the work by Brox *et al.* [70]. A different and simpler derivation can be accessed from the books by Jähne [58, 71]. A discussion on properties and computational aspects of ST can be found in the book by Cyganek and Siebert [15].

An attempt is made to describe each local neighborhood (LN) Ω of pixels with a single object, denoted by a vector **w**, as presented in Figure 2.9(a). This is possible if the LN shows some regularity of its intensity signal, which can be measured by gradient vectors g_i, computed for each point of the LN. If these gradients are close in orientation to **w**, then we can say that **w** represents well given the LN.

In other words, **w** can represent all g_i. However, we need a measure of its fit to gradients, such as an average of the squared modules of vectors **s** which are perpendicular projections of g_i onto **w**. In each point the residual vector **s** in Figure 2.9(b) can be expressed as follows

$$s = g - r = g - \frac{w}{\|w\|} \|r\| . \tag{2.80}$$

In the above the subscript was omitted. Substituting

$$g \cdot w = \|g\| \|w\| \cos \alpha = \|g\| \|w\| \|r\| / \|g\| = \|w\| \|r\| , \tag{2.81}$$

into Equation (2.80) yields

$$s = g - \frac{w}{\|w\|} \frac{g \cdot w}{\|w\|} = g - \frac{wg^T w}{w^T w} . \tag{2.82}$$

Assuming that

$$\mathbf{w}^T \mathbf{w} = 1, \tag{2.83}$$

the error function can be now defined as follows [70]

$$e(\mathbf{x}, \mathbf{x}_0) = \|\mathbf{s}\| = \left\| \mathbf{g}(\mathbf{x}) - \mathbf{w}(\mathbf{x}_0) \mathbf{g}^T(\mathbf{x}) \mathbf{w}(\mathbf{x}_0) \right\|. \tag{2.84}$$

The total error E is obtained by integrating the square of $e(\mathbf{x}, \mathbf{x}_0)$ over all possible locations $\mathbf{x}$ in the neighborhood Ω, using a Gaussian soft averaging filter G_σ,

$$E(\mathbf{x}_0) = \int_\Omega e^2(\mathbf{x}, \mathbf{x}_0) G_\sigma(\mathbf{x}, \mathbf{x}_0) d\mathbf{x}. \tag{2.85}$$

In a robust approach, the G_σ is replaced with *a robust function* [70] which makes filtering *dependant* on a local structure. This makes the orientation vector $\mathbf{w}$ well adapted to local structures. However, this implies an iterative solution which usually requires more computations than a closed form solution. Inserting Equation (2.84) into Equation (2.85) and expanding the following is obtained

$$\begin{aligned}
E(\mathbf{w}) &= \int_\Omega \left\| \mathbf{g}(\mathbf{x}) - \mathbf{w}(\mathbf{x}_0) \mathbf{g}^T(\mathbf{x}) \mathbf{w}(\mathbf{x}_0) \right\|^2 G_\sigma(\mathbf{x}, \mathbf{x}_0) d\mathbf{x} \\
&= \int_\Omega \left(\mathbf{g}^T \mathbf{g} - 2\mathbf{g}^T \mathbf{g} + \mathbf{w}^T \mathbf{g} \mathbf{g}^T \mathbf{w} \right) G_\sigma(\mathbf{x}, \mathbf{x}_0) d\mathbf{x} = \\
&= - \int_\Omega \mathbf{g}^T \mathbf{g} G_\sigma(\mathbf{x}, \mathbf{x}_0) d\mathbf{x} + \mathbf{w}^T \left(\int_\Omega \mathbf{g} \mathbf{g}^T G_\sigma(\mathbf{x}, \mathbf{x}_0) d\mathbf{x} \right) \mathbf{w},
\end{aligned} \tag{2.86}$$

where we have used Equation (2.83) and the equality $\mathbf{w}^T \mathbf{g} = \mathbf{g}^T \mathbf{w}$ since this is a scalar value. To find $\mathbf{w}$ we have to solve the following optimization problem

$$\arg \min_{\mathbf{w}} \|E(\mathbf{w})\|, \tag{2.87}$$

subject to the constraint (2.83). Substituting Equation (2.86) into Equation (2.87) yields

$$\min_{\mathbf{w}} \|E\| = \min_{\mathbf{w}} \left\| \int_\Omega \mathbf{g}^T \mathbf{g} G_\sigma(\mathbf{x}, \mathbf{x}_0) d\mathbf{x} - \mathbf{w}^T \left(\int_\Omega \mathbf{g} \mathbf{g}^T G_\sigma(\mathbf{x}, \mathbf{x}_0) d\mathbf{x} \right) \mathbf{w} \right\|. \tag{2.88}$$

Since it holds that $\mathbf{g}^T\mathbf{g} \geq 0$, then Equation (2.88) is equivalent to the maximization problem

$$
\min_{\mathbf{w}} \|E\| = \max_{\mathbf{w}} \left\{ \mathbf{w}^T \left(\underbrace{\int_{\Omega} \mathbf{g}\mathbf{g}^T G_{\sigma}\left(\mathbf{x}, \mathbf{x}_0\right) d\mathbf{x}}_{\mathbf{T}} \right) \mathbf{w} \right\} \Bigg|_{\mathbf{w}^T\mathbf{w}=1} , \tag{2.89}
$$

where the expression

$$
\mathbf{T} = \int_{\Omega} \mathbf{g}\mathbf{g}^T G_{\sigma}\left(\mathbf{x}, \mathbf{x}_0\right) d\mathbf{x}, \tag{2.90}
$$

is called a structural tensor.

2.7.1   Structural Tensor in Two-Dimensional Image Space

In the case of images, which are two-dimensional discrete signals, the corresponding ST can be written as follows

$$
\mathbf{T} = F\left(\begin{bmatrix} I_x \\ I_y \end{bmatrix} \begin{bmatrix} I_x & I_y \end{bmatrix} \right) = F\left(\mathbf{U}\mathbf{U}^T\right) = F\left(\begin{bmatrix} I_x I_x & I_x I_y \\ I_y I_x & I_y I_y \end{bmatrix} \right) = \begin{bmatrix} T_{xx} & T_{xy} \\ T_{yx} & T_{yy} \end{bmatrix}, \tag{2.91}
$$

where I_x, I_y are discrete spatial derivatives in x and y direction, respectively, and F denotes a discrete averaging operator. However, although there are four components in Equation (2.91), actually only three count, since $T_{xy} = T_{yx}$. Thus, $\mathbf{T}$ is a positive symmetric matrix. Both derivative and averaging operators can be defined in many ways. For example, derivatives can be computed with the Sobel filter, whereas averaging can be done with the binomial filter [58, 72, 73, 74]. However, in many real applications more subtle filters can provide better results. For instance, the spatial derivatives can be better approximated with the separable filters which first approximate a discrete signal with a polynomial, compute derivative on the polynomial, and then sample the values back to the discrete domain, as proposed by Farid and Simoncelli [67]. On the other hand, averaging can be done with the already introduced a robust filter G_{σ} or the anisotropic diffusion filter, discussed in Section 2.6.1. A discussion on ST, as well as its implementation, can be also found in the book by Cyganek and Siebert [15].

In the 2D space eigenvalues of $\mathbf{T}$ can be found analytically. They are simply roots of the square equation and can be computed as follows [15, 58]

$$
\lambda_{1,2} = \frac{1}{2}\left[\left(T_{xx} + T_{yy}\right) \pm \sqrt{\left(T_{xx} - T_{yy}\right)^2 + 4T_{xy}^2} \right], \tag{2.92}
$$

where components of $\mathbf{T}$ are defined in Equation (2.91).

Using the method of Lagrange multipliers it can be shown that the directional vector $\mathbf{w}$, shown in Figure 2.9, is an eigenvector of $\mathbf{T}$ corresponding to its largest eigenvalue. Thus, it is given as follows

$$\mathbf{w} = \left[\, T_{xx} - T_{yy} \quad 2T_{xy} \,\right]^T . \tag{2.93}$$

Finally, the phase φ of $\mathbf{w}$ is

$$\varphi(\mathbf{w}) = \text{atan2}\left(\, T_{xx} - T_{yy}, \quad 2T_{xy} \,\right), \tag{2.94}$$

where function *atan2* is defined as follows [75]

$$\text{atan2}(x, y) = \begin{cases} \text{atan}\left(\frac{y}{x}\right) & \text{if} \quad x > 0 \\ \text{atan}\left(\frac{y}{x}\right) + \pi\,\text{sign}(y) & \text{if} \quad x < 0 \\ \frac{\pi}{2}\,\text{sign}(y) & \text{if} \quad x = 0,\, y \neq 0 \\ \text{undefined} & \text{if} \quad x = y = 0, \end{cases} \tag{2.95}$$

where *atan* denotes a "standard" arctangent function returning an angle in the range $\pm\pi$. The sign function is defined as follows

$$\text{sign}(x) = \begin{cases} +1 & \text{if} \quad x \geq 0 \\ -1 & \text{if} \quad x < 0. \end{cases} \tag{2.96}$$

Let us notice that *sign* in Equation (2.96) does not return 0, i.e. it differs from the *sgn* function since the latter returns 0 for $x = 0$. Thus, *atan2* differs from the arctangent function by taking into account the signs and magnitudes of both its components, as well as placing the returned angle value in the correct quadrant of the coordinate system. This function is not defined only if both its arguments are 0.

The orientation vector $\mathbf{w}$ is well defined only in the areas with *distinctive structure*. Such places are indicated by the coherence component c which is dependent on the eigenvalues $\lambda_{1,2}$ of $\mathbf{T}$. It is defined as follows [58]

$$c = \frac{(\lambda_1 - \lambda_2)^2}{(\lambda_1 + \lambda_2)^2} = \frac{\|\mathbf{w}\|^2}{\left(T_{xx} + T_{yy}\right)^2}, \quad \text{for} \quad \lambda_1 + \lambda_2 \neq 0. \tag{2.97}$$

Values of c which are close to 1 are characteristic of the places with very strong local orientation, whereas values near zero indicate either a constant signal or an isotropic structure. In the case of $T_{xx} + T_{xy} = 0$ we assume that also $c = 0$. The denominator in Equation (2.97) denotes a cumulative magnitude

$$m = \text{Tr}(\mathbf{T}) = T_{xx} + T_{yy}, \tag{2.98}$$

 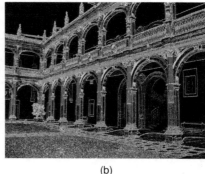

(a)   (b)

Figure 2.10   Exemplary color image (a). Three components of a 2D structural tensor encoded into the HSV color. Phase of the orientation vector **w** represented by the hue channel (H), coherence by saturation (S), and trace with value (V). (For a color version of this figure, please see the color plate section.)

which is also a trace of **T** and which is independent of the used coordinate system (i.e. it is an invariant). Thus, the structural tensor can be represented either by the components T_{xx}, T_{xy}, and T_{yy} in Equation (2.91), or by the vector

$$\mathbf{s} = \begin{bmatrix} m & \varphi & c \end{bmatrix}^{T}. \tag{2.99}$$

In practice, the latter allows easier interpretation of local structures in images [15, 71].

Figure 2.10(a) presents an exemplary color image, whereas in Figure 2.10(b) three components of the structural tensor are computed from this image. To show this in a single view, these are encoded into three channels of the HSV color space. Hue (H) denotes phase φ (2.94) of the vector of local orientation **w**, saturation (S) is responsible for the coherence (2.97), and value (V) encodes the trace m of **T** given in Equation (2.98), respectively. It is easily seen that local structures in this image which have similar orientations have also similar colors. For example the vertical edges are red, whereas the horizontal ones are blue. Also the uniform areas, such as the walls or the sky, result in black since the corresponding directional derivatives in Equation (2.91) are zero. Let us also note that the ST can be computed from monochrome but also from the multichannel image, as the color image in this example. These features of the ST were used in the matching of the stereo images [76]. The other application of ST is direct computation of the characteristic features in the 2D images, such as corners or specifically oriented edges.

It is interesting to analyze Equation (2.91) from the computational point of view. Computation of a discrete ST requires three convolutions and one multiplication. Each convolution, depending on the type of mask used, requires a number of multiplications and additions. This number depends on the size of the filter mask but important improvement can be obtained by employing so called separable filters [15].

In order to compare ST a suitable distance measure needs to be chosen. The most natural choice is the Euclidean distance, defined in Equation (2.28). However, other measures which account for specifics of tensor processing can be of interest. In this respect Fillard *et al.* analyzed a Riemannian framework for the processing of different tensor-valued images [77]. In this group measures for the structural tensors were also investigated. Their results indicate that in the low signal-to-noise ratio (SNR) better results were obtained for the simple Euclidean metric. On the other hand, in the case of well conditioned signals, i.e. high SNR, the Riemannian based metrics allow desirable magnification of low contrast structures in images, discussed in Section 2.5.1.

Here it is also interesting to check some properties of matrices which are simply outer products of a vector $\mathbf{m}$, such as in Equation (2.91). Thus, a matrix $\mathbf{M}$ can be expressed simply as

$$\mathbf{M} = \mathbf{m}\,\mathbf{m}^T. \tag{2.100}$$

Thus $\mathbf{M}$ is obviously symmetric. Substituting Equation (2.100) into Equation (2.26) we easily obtain

$$\mathbf{x}^T \mathbf{M} \mathbf{x} = \mathbf{x}^T \left(\mathbf{m}\,\mathbf{m}^T \right) \mathbf{x} = \left(\mathbf{x}^T \mathbf{m} \right) \left(\mathbf{m}^T \mathbf{x} \right) = \underbrace{\left(\mathbf{x}^T \mathbf{m} \right)}_{s} \underbrace{\left(\mathbf{x}^T \mathbf{m} \right)}_{s}^T = s^2 \geq 0, \tag{2.101}$$

where $s \in \Re$ denotes a scalar value. The above means that a matrix which is an outer product of vectors is always positive semidefinite. This makes possible application of metrics that require positive definite matrices, such as D_{LF} in Equation (2.31), to comparison of outer product tensors, such as the structural tensor or its extended version discussed in Section 2.7.4. However, we need to specifically treat the cases for which the value is exactly 0, as will be discussed.

2.7.2   Spatio-Temporal Structural Tensor

The benefit of the tensor calculus is revealed when moving into higher dimensions. For instance, color images can be analyzed as 3D tensors, as shown in Figure 2.2. However, a monochrome video sequence also forms a 3D tensor with two spatial and one time coordinates. In this case the structural tensor becomes a 3×3 matrix. Further extensions are discussed in previous sections. In higher dimensions the eigenvalues are usually found numerically. However, this is not necessary always the case since an analysis of the trace of the ST can provide sufficient information on the 3D structures, as pointed out in [71, 73].

Optical flow is a process which allows measurement of movements of objects in the observed sequence. There are many methods of computation of the optical flow. In this context it is very interesting to show how ST can be directly used to compute the optical flow. The basic equation of optical flow is the following [15, 78]

$$I_x u + I_y v + I_t = 0, \tag{2.102}$$

where I_x, I_y, and I_t are the partial derivatives of the intensity signal of the images with respect to the spatial coordinates x, y and time t, respectively. The values of u and v are horizontal and vertical velocities in a sequence, respectively. The above equation is called a brightness

Figure 2.11   Frames from the "Hamburg taxi" sequence (provided for tests by the Universität Karlsruhe, Fakultät für Informatik, Institut für Algorithmen und Kognitive Systeme, Group Prof. Dr. H.-H. Nagel [79]).

constancy constraint [80]. Equation (2.102) can be transformed into the minimization task which, in turn, leads to the following set of equations [15]

$$
\begin{bmatrix}
\displaystyle\int_{\Omega(x_0,y_0)} I_x^2 dxdy & \displaystyle\int_{\Omega(x_0,y_0)} I_x I_y dxdy \\[2em]
\displaystyle\int_{\Omega(x_0,y_0)} I_x I_y dxdy & \displaystyle\int_{\Omega(x_0,y_0)} I_y^2 dxdy
\end{bmatrix}
\begin{bmatrix} u \\ v \end{bmatrix}
= -
\begin{bmatrix}
\displaystyle\int_{\Omega(x_0,y_0)} I_x I_t dxdy \\[2em]
\displaystyle\int_{\Omega(x_0,y_0)} I_y I_t dxdy
\end{bmatrix},
\quad (2.103)
$$

which after considering Equation (2.90) can be written in terms of the ST, as follows

$$
\begin{bmatrix} T_{xx} & T_{xy} \\ T_{yx} & T_{yy} \end{bmatrix}
\begin{bmatrix} u \\ v \end{bmatrix}
= -
\begin{bmatrix} T_{xt} \\ T_{yt} \end{bmatrix}.
\quad (2.104)
$$

In the above it is convenient to explicitly write subscripts of $\mathbf{T}$ with letters x, y for the spatial, and t for the time domains, respectively. From Equation (2.104) it follows that in the places where the ST is not singular the local velocities $[u,v]^T$ at a point (x_0,y_0) can be computed as [15]

$$
u = \frac{T_{yt}T_{xy} - T_{xt}T_{yy}}{T_{xx}T_{yy} - T_{xy}^2} \quad \text{and} \quad v = \frac{T_{xt}T_{xy} - T_{yt}T_{xx}}{T_{xx}T_{yy} - T_{xy}^2},
\quad (2.105)
$$

assuming that the denominators in Equation (2.105) are different from 0. Alternatively, $\mathbf{T}$ can be inverted, first applying the SVD decomposition, i.e. $\mathbf{T} = \mathbf{SVD}^T$, and then inverting the non-zero diagonal elements of the matrix $\mathbf{V}$ (elements close or equal to 0 are set to 0). Finally, $\mathbf{T}^{-1} = \mathbf{DV}^{-1}\mathbf{S}^T$.

Figure 2.11 depicts three monochrome frames of the "Hamburg taxi" sequence commonly used in testing optical flow algorithms. From these, the five components of the ST were computed which were used in Equation (2.105). These are visualized in Figure 2.12.

There are three spatial and two temporal components. Whiter places in Figure 2.12 correspond to the objects t had changed their positions in the sequence in Figure 2.11 .

Figure 2.13 depicts the motion field computed in accordance with Equation (2.105). The horizontal and vertical displacement u and v are shown in Figure 2.13(a) and Figure 2.13(b), respectively. The displacement arrows are in Figure 2.13c. In the last case only those that were not shorter than 12 pixels were displayed for a clear view.

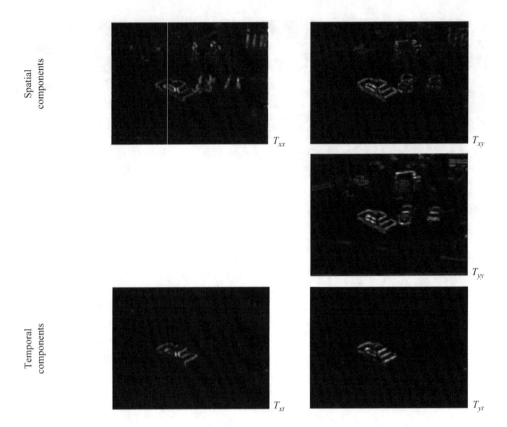

Figure 2.12 Components of the structural tensor of the taxi sequence in Figure 2.11.

2.7.3 Multichannel and Scale-Space Structural Tensor

Computation of the gradient vector **g** in Equation (2.90) for multichannel images can be done by a method proposed by Di Zenzo [81]. Such a method is common in other CV tasks when dealing with multichannel images. For example it was undertaken in the vision framework proposed by Sochen *et al.* [82] and in the segmentation method proposed by Brox *et al.* [83],

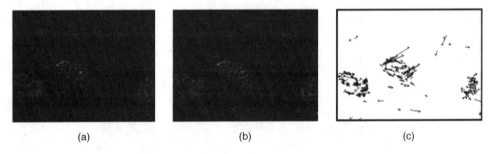

         (a)   (b)   (c)

Figure 2.13 Optical flow of the sequence in Figure 2.11 computed with the ST. Motion fields **u** (a), **v** (b), combined (c).

to name a few. In this approach partial gradients are computed independently in each channel and then these are summed up. With this assumption, Equation (2.90) for an image with M channels becomes

$$\mathbf{T}(\mathbf{x}_0) = \int_{\Omega} \left(\sum_{k=1}^{M} \mathbf{g}_k\,(\mathbf{x}_0)\,\mathbf{g}_k^T\,(\mathbf{x}_0) \right) d\mathbf{x} = \sum_{k=1}^{M} \left(\int_{\Omega} \mathbf{g}_k\,(\mathbf{x}_0)\,\mathbf{g}_k^T\,(\mathbf{x}_0)\,d\mathbf{x} \right) = \sum_{k=1}^{M} \mathbf{T}_k(\mathbf{x}_0), \quad (2.106)$$

in which a square window was used for G_σ. Thus, computation of the ST for a multichannel image boils down to independent computation of ST in each channel separately and then summing up the results. Moreover, Equation (2.106) is quite general, i.e. it can be used for each type of multichannel signal, not just color or video images. Thanks to this a characteristic local structure can be analyzed in the seismic images, for instance.

Further extensions of the multichannel structural tensor (2.106) are possible [84], such as a linear combination of the component tensors $\mathbf{T}_k$

$$\mathbf{T}(\mathbf{x}_0) = \sum_{k=1}^{M} c_k \mathbf{T}_k(\mathbf{x}_0), \quad (2.107)$$

where c_k are constants. Thanks to this it is possible to control the influence of each channel separately.

It is interesting to note that there are also two different space dimensions involved in the ST. The first is a dimension of $\mathbf{T}$ which comes directly from the dimension of the gradient vector, i.e. the most common are 2D and 3D, as already discussed. The second dimension follows the number of image channels, given by M in Equation (2.106) and (2.107). Similarly, there are two scale-spaces involved in the ST:

1. The scale associated with the input images, that is in the domain of computation of the gradients $\mathbf{g}_i$.
2. The scale imposed by the averaging, that is size of the filter F in Equation (2.91).

The two scales can be explicitly modeled into Equation (2.91), as follows

$$\hat{T}_{ij}(\rho, \xi) = F_\rho\big(R_i^{(\xi)} R_j^{(\xi)}\big), \quad (2.108)$$

where $R_i^{(\xi)}$ is a ξ-tap discrete directional operator, F_ρ is a smoothing kernel of scale ρ [15].

Figure 2.14 depicts three channels of the ST encoded into the HSV color space, as was already presented in Figure 2.10. However, different scales were used. These are the directional filters proposed by Farid and Simoncelli [67] and the smoothing Gaussian filter with the parameters in Equation (2.108) set to $\xi = 5$ and $\rho = 5$ in Figure 2.14(a), and $\xi = 5$ and $\rho = 9$ in Figure 2.14b, respectively. Detailed discussion of these filters, as well as values of their components are discussed in the book [15]. Their values can be also be looked up in the attached code [28].

(a)   (b)

Figure 2.14 Structural tensor computed in different scales from the fragment of Figure 2.10(a). Directional filters of $\xi = 5$ and $\rho = 5$ (a), $\xi = 5$ and $\rho = 9$ (b). (For a color version of this figure, please see the color plate section.)

2.7.4 Extended Structural Tensor

As alluded to previously, the structural tensor provides valuable information on the structure of local regions in the input image (Section 2.7.1). However, in many applications it is desirable to use intensity *and* the structural tensor together. For instance, such compound information is used in the stereo matching system, proposed by Cyganek [76]. Luis-García *et al.* propose such an extension for image segmentation [85]. Their idea consists of creating mixed products of gradients and intensity signal. This way the nonlinear *extended structural tensor* (EST) is obtained, as follows

$$\mathbf{T}_E = F\left(\begin{bmatrix} I_x \\ I_y \\ I \end{bmatrix} \begin{bmatrix} I_x & I_y & I \end{bmatrix} \right) = F\left(\begin{bmatrix} I_x^2 & I_x I_y & I_x I \\ I_y I_x & I_y^2 & I_y I \\ I_x I & I_y I & I^2 \end{bmatrix} \right) = F\left(\mathbf{U}_E \mathbf{U}_E^T \right), \quad (2.109)$$

where I_x and I_y denote directional derivatives of the intensity signal I in the x and y directions, respectively. For EST, the anisotropic diffusion filter is usually chosen as the averaging operator F, discussed in Section 2.6.2, although other filters are also possible especially if fast computations are required. In the case of color images $\mathbf{U}_E$ in the above it is further extended to

$$\mathbf{U}_E = \begin{bmatrix} I'_x & I'_y & I_R & I_G & I_B \end{bmatrix}^T, \quad (2.110)$$

where

$$I' = \frac{1}{3}\left(I_R + I_G + I_B \right) \quad (2.111)$$

is simply an averaged intensity signal (i.e. a monochrome intensity signal). The obtained tensors are symmetric, thus if a tensor $\mathbf{U}_E$ is of dimension L, then there is

$$k = \frac{1}{2}L(L+1) \tag{2.112}$$

of independent components. For example, for $L = 2, 3, 4, 5, \ldots$ we obtain $k = 3, 6, 10, 15, \ldots$ and so on of independent components.

Although EST conveys very useful information on local structures in images, in the case of color images the number of independent components is $k = 15$ which for some applications can be prohibitive due to long computation times. For this reason Luis-García et al. propose applying the principal component analysis (PCA) transformation and retain only the two most important components (see Section 3.3.1). Thus, from Equation (2.110) the following is obtained

$$\mathbf{U}_C = \text{PCA}(\mathbf{U}_E) = \begin{bmatrix} U_1 & U_2 \end{bmatrix}^T, \tag{2.113}$$

where U_1 and U_2 are first two principal components of $\mathbf{U}_E$. This leads to the nonlinear compact structural tensor $\mathbf{T}_C$ (CST) which counts only three independent values, i.e. the same as the ST. It is given as follows

$$\mathbf{T}_C = F\left(\mathbf{U}_C \mathbf{U}_C^T\right). \tag{2.114}$$

However, conversion of EST to CST with help of PCA reduces the number of independent components from 15 to only 3 and therefore sometimes leads to a severe loss of important information. One solution to this is to check the eigenvalues and set a threshold on the percentage of the total variance. This way the *adaptive* compact structure tensor can be obtained [85].

It is also interesting to notice that the extended tensors, $\mathbf{T}_E$ in Equation (2.109) and $\mathbf{T}_C$ in Equation (2.114), due to multiplications of internal components, show high variability of their values. Therefore, to assure well balanced comparisons, they should be normalized. However, as indicated by Luis-García et al. in their paper [85], a simple normalization is not recommended since it amplifies noise in the channels containing no information. Instead, a replacement of the values by their square roots is proposed. However, in this respexct the D_{LF} measure, defined in Equation (2.31), has interesting properties due to the applied logarithm. However, we need to remember that D_{LF} is defined for positive definite matrices.

Figure 2.15 depicts $k = 15$ independent components of the extended structural tensor computed for the color image shown in the upper left pane of Figure 2.15. The anisotropic diffusion filter was chosen for the signal averaging operator F in (2.109). The number of iterations $i = 25$ and the robust scale parameter in Equation (2.71) is $\sigma_r = 10.14$.

Figure 2.16 shows visualization of the $k = 3$ independent components of the CST computed for the color image shown in the upper left pane of Figure 2.15. In this case the number of principal components was set to 2. These convey 97.7% of the total variation in the image data.

The idea of EST can be extended to video sequences as well, adding an additional component I_t' to Equation (2.110) responsible for time, as denoted in Equation (2.103).

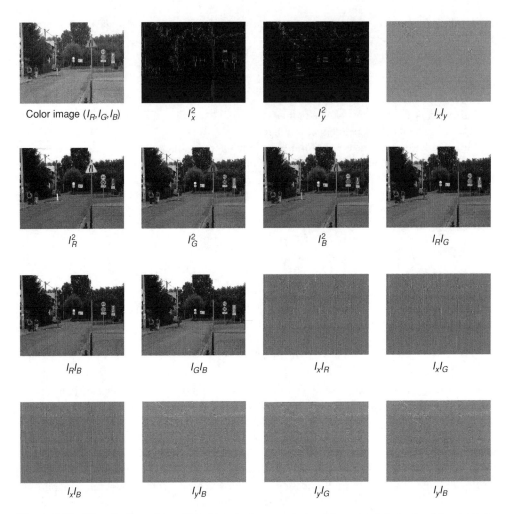

Figure 2.15 Visualization of $k = 15$ independent components of the extended structural tensor computed for the color image shown in the upper left pane. (For a color version of this figure, please see the color plate section.)

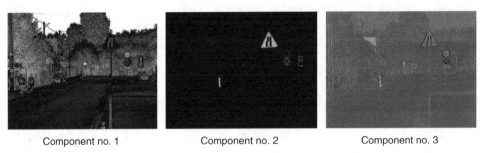

Figure 2.16 Visualization of the $k = 3$ independent components of the compact structural tensor with two principal components computed for the color image shown in the upper left pane of Figure 2.15.

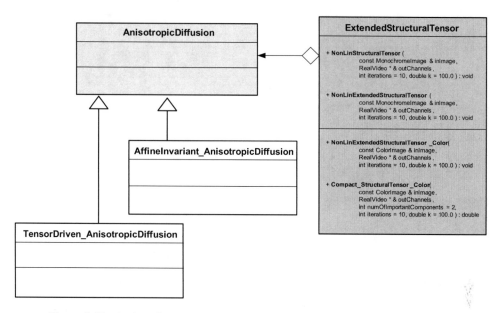

Figure 2.17   A class diagram of the *ExtendedStructuralTensor* and associated classes.

2.7.4.1   IMPLEMENTATION of the Linear and Nonlinear Structural Tensor

Figure 2.17 presents a class diagram of the *ExtendedStructuralTensor* and the *AnisotropicDiffusion* hierarchy used in the filtering stage, which was discussed in Section 2.6.3.

Algorithm 2.6 shows the implementation of the *ExtendedStructuralTensor* class for computation of the extended and compact structural tensors.

```
///////////////////////////////////////////////////////////
// This class implements the extended and compact
// structural tensors for monochrome and color images
///////////////////////////////////////////////////////////
class ExtendedStructuralTensor
{
  protected:

          AnisotropicDiffusion *        fAnisotropicDiffusionObj;


  protected:



      ///////////////////////////////////////////////////////////
      // These functions compute vectors which outer product
      // forms the tensor
      ///////////////////////////////////////////////////////////
      //
```

```
// INPUT:
//                    inImage - input image
//                    theChannels - output (orphaned) object
//                            with the computed 3 components:
//                            Ix2, Iy2, IxIy
//
// OUTPUT:
//                    none
//
// REMARKS:
//                    theChannels is orphaned
//
void PrepareChannelsForNonlinStructuralTensor( const MonochromeImage & inImage,
                                               RealVideo * & theChannels );



/////////////////////////////////////////////////////////////////
// The function computes 6 channels of the extended nonlinear
// structural tensor
/////////////////////////////////////////////////////////////////
//
// INPUT:
//                    inImage - input monochrome image
//                    theChannels - output (orphaned) object
//                            with the computed 6 components:
//                            Ix2, Iy2, IxIy, IxI, IyI, II
//
// OUTPUT:
//                    none
//
// REMARKS:
//                    Channels are indexed as follows:
//                    kDx2 = 0, kDy2, kDxDy, kDx_I, kDy_I, kIntensity2
//
//                    theChannels is orphaned
//
void PrepareChannelsForExtendedNonlinStructuralTensor(
            const MonochromeImage & inImage, RealVideo * & theChannels );


/////////////////////////////////////////////////////////////////
// This function computes 5 specific feature channels from
// the input color image.
/////////////////////////////////////////////////////////////////
//
// INPUT:
//                    inColorImage - the input color image
//                    theChannels - upon success contains 5 channels:
//                            Ix, Iy, Ir, Ig, Ib
// OUTPUT:
//                    if failure then theChannels is 0
//
// REMARKS:
//                    theChannels is orphaned
//
```

```
      void Prepare_Basic_ChannelsForExtendedNonlinStructuralTensor_Color(
                  const ColorImage & inColorImage, RealVideo * & theChannels );



      ///////////////////////////////////////////////////////////
      // This function computes 15 specific feature channels from
      // the input color image.
      ///////////////////////////////////////////////////////////
      //
      // INPUT:
      //                 inColorImage - the input color image
      //                 theChannels - upon success contains 15 channels:
      //                          Ix2,  Iy2,  Ir2,  Ig2,  Ib2,
      //                          IxIy, IyIr, IrIg, IgIb,
      //                          IxIr, IyIg, IrIb,
      //                          IxIg, IyIb,
      //                          IxIb
      //
      // OUTPUT:
      //                 if failure then theChannels is 0
      //
      // REMARKS:
      //                 theChannels is orphaned
      //
      void PrepareChannelsForExtendedNonlinStructuralTensor_Color(
                  const ColorImage & inColorImage, RealVideo * & theChannels );


public:


      ///////////////////////////////////////////////////////////
      //     Dx^2,    Dy^2, Dx*Dy,  Dx*I  Dy*I  intensity^2
      enum { kDx2 = 0, kDy2, kDxDy, kDx_I, kDy_I, kIntensity2, kNumOfInChannels };
      ///////////////////////////////////////////////////////////



      ///////////////////////////////////////////////////////////
      // This function computes the nonlinear structural tensor
      // from a monochrome image.
      ///////////////////////////////////////////////////////////
      //
      // INPUT:
      //                 inImage - input monochrome image
      //                 outChannels - an output series of real outputs
      //                         of the nonlinearly diffused channels
      //                         Ix2, Iy2, IxIy
      //                 iterations - number of iterations for diffusion
      //                 k - parameter for the control function
      //                         for the anisotropic filter
      //
      // OUTPUT:
      //                 orphaned (!) outChannels
      //
```

```
// REMARKS:
//                        outChannels - is an orphaned object;
//                        ONLY DERIVATIVE CHANNELS
//
virtual void NonLinStructuralTensor( const MonochromeImage & inImage,
      RealVideo * & outChannels, int iterations = 10, double k = 100.0 );



///////////////////////////////////////////////////////////////
// This function computes the extended nonlinear
// structural tensor from a monochrome image.
// Extended means that additionally the intensity channel
// is also added.
///////////////////////////////////////////////////////////////
//
// INPUT:
//                        inImage - input monochrome image
//                        outChannels - an output series of real outputs
//                                of the nonlinearly diffused channels
//                                Ix2, Iy2, IxIy, IxI, IyI, II
//                        iterations - number of iterations for diffusion
//                        k - parameter for the control function
//                                for the anisotropic filter
//
// OUTPUT:
//                        orphaned (!) outChannels
//
// REMARKS:
//                        outChannels - is an orphaned object
//                        ADDITIONALLY INTENSITY CHANNEL
//
virtual void NonLinExtendedStructuralTensor( const MonochromeImage & inImage,
      RealVideo * & outChannels, int iterations = 10, double k = 100.0 );




///////////////////////////////////////////////////////////////
// This function computes the extended nonlinear
// structural tensor from a color image.
// Extended means that the color channels
// are also added.
///////////////////////////////////////////////////////////////
//
// INPUT:
//                        inImage - input color image
//                        outChannels - an output series of real outputs
//                                of the nonlinearly diffused channels;
//                                Upon success contains 15 channels:
//                                Ix2,  Iy2,  Ir2,  Ig2,  Ib2,
//                                IxIy, IyIr, IrIg, IgIb,
//                                IxIr, IyIg, IrIb,
//                                IxIg, IyIb,
//                                IxIb
```

```
//                          iterations - number of iterations for diffusion
//                          k - parameter for the control function
//                                  for the anisotropic filter
//
// OUTPUT:
//                          if failure then outChannels is 0
//                          orphaned (!) outChannels
//
// REMARKS:
//                          outChannels - is an orphaned object
//                          ADDITIONALLY INTENSITY CHANNEL
//
virtual void NonLinExtendedStructuralTensor_Color( const ColorImage
    & inImage,  RealVideo * & outChannels, int iterations = 10,
    double k = 100.0 );



////////////////////////////////////////////////////////////////////
// This function computes the compact nonlinear
// structural tensor from a color image,
// This does PCA on extended structural tensor before
// doing the anisotropic diffusion
////////////////////////////////////////////////////////////////////
//
// INPUT:
//                          inImage - input color image
//                          outChannels - an output series of real outputs
//                                  of the nonlinearly diffused channels
//                          numOfImportantComponents - the number of the most
//                                  important channels (size of outChannels)
//                          iterations - number of iterations for diffusion
//                          k - parameter for the control function
//                                  for the anisotropic filter
//
// OUTPUT:
//                          orphaned (!) outChannels
//
// REMARKS:
//                          outChannels - is an orphaned object
//                          ADDITIONALLY INTENSITY CHANNEL
//
virtual double Compact_StructuralTensor_Color( const ColorImage
              & inImage, RealVideo * & outChannels,
              int numOfImportantComponents = 2,  int iterations = 10,
              double k = 100.0 );


};
```

Algorithm 2.6 Implementation of the *ExtendedStructuralTensor* class for computation of the extended and compact structural tensors.

Among the members of the *ExtendedStructuralTensor* class there are two groups: The first one, internal to the class, contains a number of overloaded *PrepareChannelsForNonlin-StructuralTensor* functions. Their role is computation of the vectors, such as $\mathbf{U}_E$ in Equation (2.110). In the second group we have members to compute the extended and compact structural tensors for monochrome and color images. These are *NonLinExtendedStructuralTensor*, *Non-LinExtendedStructuralTensor_Color*, and *Compact_StructuralTensor_Color*. Implementation details can be looked up in the attached code [28].

2.8   Object Representation with Tensor of Inertia and Moments

Natural objects encountered in digital images are views of real 3D objects, therefore some methods of modeling them used in classical mechanics can be adapted to the CV domain. In this respect *the inertia tensor* is very useful, which is a two-dimensional 3×3 tensor describing rotation of an object around a center of its mass.

Definition 2.14   Inertia tensor of a 3D rigid body

Elements of the tensor of inertia for a 3D rigid body are defined as follows [86]

$$T_{ij} = \int_V \left(x^k x^k \delta_{ij} - x^i x^j \right) \rho \, d\tau, \qquad (2.115)$$

where $d\tau = dxdydz$ denotes a volume element, and scalar function $\rho(\mathbf{x})$ is the density (a mass element in a volume element), δ_{ij} is the Kronecker delta (i.e. it is 1 iff $i = j$, 0 otherwise), V is the volume of an object.   □

In the above it is assumed that the origin of the coordinate system is located at the gravity center of the object. The indices i, j change from 1 to 3 and although this makes possible nine components (2.115), the inertia tensor is symmetrical, i.e. $T_{ij} = T_{ji}$, as will be seen very soon from the next derivations.

In the above definition k is the so called "free" index which can take any value which is related to all possible index values in this context, i.e. to the values of the i and j indices. Since in 3D there are three free coordinates x^1, x^2 and x^3, sometimes denoted as x, y, z, respectively, then k can take also 1, 2, and 3. Thus, in accordance with the summation convention (Section 2.2.1), the inner expression $x^k x^k$ stands for $x^1 x^1 + x^2 x^2 + x^3 x^3 = (x^1)^2 + (x^2)^2 + (x^3)^2$. From this we see that in tensor notation we should not confuse variable indices with their exponents. However, returning to the last expression we notice that it is further modulated by the δ_{ij} which value is 1 for all identical indices, i.e. for $i = j$, and 0 otherwise. Thus, $x^k x^k \delta_{ij}$ is not 0 only if $i = j$. To see how this operates, let us compute T_{11} and T_{12} from Equation (2.115), respectively.

$$T_{11} = \int_V \left(x^k x^k \delta_{11} - x^1 x^1 \right) \rho \, d\tau = \int_V \left(x^1 x^1 + x^2 x^2 + x^3 x^3 - x^1 x^1 \right) \rho \, d\tau$$
$$= \int_V \left[(x^2)^2 + (x^3)^2 \right] \rho \, d\tau,$$

$$T_{12} = \int_V \left(x^k x^k \delta_{12} - x^1 x^2 \right) \rho \, d\tau = \int_V \left(0 - x^1 x^2 \right) \rho \, d\tau = - \int_V x^1 x^2 \rho \, d\tau.$$

In usual coordinate notation these can be written in a slightly more legible form

$$T_{xx} = \int_V \left(y^2 + z^2\right) \rho \, d\tau, \ T_{xy} = -\int_V xy\rho \, d\tau. \tag{2.116}$$

Having defined the tensor of inertia, the scalar called the *moment of inertia* in a direction given by the vector **r**, is defined as follows [55]

$$M_{\mathbf{r}} = \mathbf{u}_{\mathbf{r}}^T \mathbf{T} \mathbf{u}_{\mathbf{r}}, \tag{2.117}$$

where $\mathbf{u}_{\mathbf{r}}$ is the unit vector in direction **r**. With this, to characterize the "nature" of an object, what is interesting are the *directions* in which $M_{\mathbf{r}}$ takes on its minimal M_{min} and maximal M_{max} values. Moments of inertia can be determined from the eigendecomposition of **T**, as for a general case as described in Section 2.9. The value of M_{min}/M_{max} is called *the elongation ratio*. The orientation φ of an object is an angle between the vector **r** which corresponds to M_{min} and the **x** axis. However, these were already computed for all 2D symmetrical tensors in (2.92)–(2.94). The main difference, however, is how components of these tensors are computed, assuming that in the case of images intensity, signal is substituted for the density ρ. For the ST these are based on the first derivatives of the intensity signal, whereas inertia tensors are based directly on the intensity signal. This has some consequences, though. The derivative based techniques are more prone to noise whereas intensity based methods suffer from outliers. Finally, a very interesting thing to note is that the coherence component (2.97) in the case of the inertia tensor becomes a measure of eccentricity, which for circular objects takes on zero, and becomes one for a line.

It is interesting to observe that Equations (2.116) are identical to the expressions for central statistical moments of second order for two variables, which are written as follows [87, 88]

$$c_{ab} = \int_{-\infty}^{+\infty} \int_{-\infty}^{+\infty} (x - \bar{x})^a \, (y - \bar{y})^b \, \rho \, dx dy, \tag{2.118}$$

where the coefficients $a,b \in \{1,2\}$ are exponents and define the order of the moments, while $(\bar{x}, \bar{y})$ denotes a point of center of mass. It is also interesting to notice that the three-dimensional integral from Equation (2.216) is now changed to a two-dimensional integration. In a similar way for noncentered data the "ordinary" statistical moments are defined

$$m_{ab} = \int_{-\infty}^{+\infty} \int_{-\infty}^{+\infty} x^a y^b \rho \, dx dy. \tag{2.119}$$

Intuitively, the tensor of inertia conveys information on mass distribution within a rigid body. This idea can be extended into the domain of pattern recognition, since each object to some extent can be defined by its "mass" distribution, where the "mass" is defined specifically to the pattern domain. In the case of objects contained in digital 2D and 3D images, they are characterized by pixel values. However, "mass" can also be any other measure, such as probability or fuzzy membership values, as we shall see in the next sections.

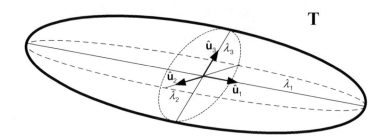

Figure 2.18   Tensor decomposition in the space spanned by its eigenvectors.

To model a real object, usually *an equivalent ellipsoid* (such as the one presented in Figure 2.18) is chosen, i.e. the one that rotates with the minimal inertia (2.117) exactly as the modeled object. For instance, such an approach was used by Orkisz *et al.* to model cylindrical shapes in the framework of vascular image processing [89]. It is also a common representation of objects in the continuous mean shift tracking method, discussed in Section 3.8.2.

However, when going into a discrete domain of digital images we need to substitute intensity for "mass" function and change integration to summation. This way, in the simplest approach the following discrete approximations are obtained

$$m_{ab} = \sum_{r=1}^{R} \sum_{s=1}^{S} I_{rs} r^a s^b, \tag{2.120}$$

$$c_{ab} = \sum_{r=1}^{R} \sum_{s=1}^{S} I_{rs} (r - \bar{x})^a (s - \bar{y})^b, \tag{2.121}$$

where I_{rs} denotes scalar image intensity values at a discrete position (r,s). However, a more precise computation can be obtained with properly integrated polynomials [90, 91]. This way the following is obtained

$$
\begin{aligned}
\tilde{m}_{ab} &= \sum_{r=1}^{R} \sum_{s=1}^{S} \left(I_{rs} \cdot \int_{P_{rs}} \int x^a y^b dx dy \right) \\
&= \frac{1}{a+1} \frac{1}{b+1} \sum_{r=1}^{R} \sum_{s=1}^{S} \left\{ I_{rs} \cdot \left[\left(r + \frac{1}{2} \right)^{a+1} - \left(r - \frac{1}{2} \right)^{a+1} \right] \left[\left(s + \frac{1}{2} \right)^{b+1} - \left(s - \frac{1}{2} \right)^{b+1} \right] \right\},
\end{aligned}
\tag{2.122}
$$

where P_{rs} denotes the area of a pixel at position (r,s).

The central point $(\bar{x}, \bar{y})$ can be computed from three moments (2.120), as follows

$$\bar{x} = \frac{m_{10}}{m_{00}}, \qquad \bar{y} = \frac{m_{01}}{m_{00}}, \qquad (m_{00} \neq 0). \tag{2.123}$$

It is also called a center of gravity or a centroid. Based on the above the inertia tensor can be expressed in terms of the central moments, as follows

$$\mathbf{T} = \begin{bmatrix} c_{20} & -c_{11} \\ -c_{11} & c_{02} \end{bmatrix}. \tag{2.124}$$

As alluded to previously, with the inertia tensor (2.124) an object can be approximated by an ellipse (called a blob) which is characteristic of its length l, width w, as well as an angle φ (see Appendix A.4). These are as follows

$$l = 2\sqrt{\lambda_1/m_{00}}, \ w = 2\sqrt{\lambda_2/m_{00}}, \quad \text{and} \quad \varphi = \frac{1}{2}\text{atan2}\,(2c_{11}, c_{20} - c_{02}), \tag{2.125}$$

where function *atan2* is defined in Equation (2.95), and $\lambda_1 \geq \lambda_2$ are eigenvalues of the inertia tensor $\mathbf{T}$. For their computation we can reuse Equation (2.92)

$$\lambda_{1,2} = \frac{1}{2}\left[(c_{20} + c_{02}) \pm \sqrt{4c_{11}^2 + (c_{20} - c_{02})^2}\right]. \tag{2.126}$$

An interesting extension to the central are the scale-invariant moments [58]

$$\hat{c}_{ab} = \frac{c_{ab}}{m_{00}^{(a+b+2)/2}}. \tag{2.127}$$

However, what is really interesting in pattern recognition are features which do not change with a change of a viewpoint or illumination. These are called invariants. Probably the most common are the Hu invariants, developed 50 years ago [92]. Also well known are the affine moment invariants proposed by Flusser and Suk [93, 94]. Their application for practical object recognition is discussed in Section 5.3.1.

An interesting approach, with the locally weighted moments computed at multiple scales, was proposed by Sühling *et al.* [95]. Their method allows more precise detection of elongated structures, multiscale Savitzky–Golay filters as well as multiscale optical flow when compared to the methods utilizing one-scale moments (2.121).

Finally, it is interesting to note that tensors allow representation of higher-order statistics, in which the most important concept is the cumulants since these are tensors and are blind to the additive Gaussian noise [96]. This feature is used to construct filters with high signal-to-noise-ratio (SNR) properties, as well as being used in independent component analysis (ICA). For the latter the fourth order cumulant tensor is employed. Its eigenvalue analysis leads directly to the so called FastICA algorithm [97].

2.8.1   *IMPLEMENTATION of Moments and their Invariants*

Statistical moments are computed in the software framework with the help of the two template functions *Moment* and *CentralMoment* for ordinary and central moments, respectively. Algorithm 2.7 and Algorithm 2.8 present declarations of the two functions, respectively. These are defined as template functions and therefore they can accept images with pixels of any scalar type. However, the moments are computed from the input images which are of type *TMasked-ImageFor* which allow definition of any object since they actually contain two images [15, 69]. The first with pixels and the second which is a bitmap that defines which pixels belong to the object(s) of interest.

```
///////////////////////////////////////////////////////////////
// This function computes a 2D moment of the rank:
// x_moment_rank, y_moment_rank
//
// The formula for 2D moments is as follows:
//
//              m_ab( I ) = SUM{ x^a * y^b * I(x,y) }
//
// where: a == x_moment_rank, b == y_moment_rank
// The "object" is defined by a mask associated
// with the original image (used TMaskedImageFor<>
// class)
//
///////////////////////////////////////////////////////////////
//
// INPUT:
//                    in - reference to the image
//                    x_moment_rank - the first rank of the moment
//                    y_moment_rank - the second rank of the moment
//
// OUTPUT:
//
//
// REMARKS:
//                    See e.g. Klette: "Digital Geometry" p. 541.
//
template< class T >
double Moment( const TMaskedImageFor< T > & in, const double x_moment_rank,
                        const double y_moment_rank );
```

Algorithm 2.7 Declaration of the *Moment* template function for computation of statistical moments of objects defined in the masked image object.

Implementations of the presented functions follow the formulas (2.120)–(2.123). Implementation details can be analyzed from the source code.

```
///////////////////////////////////////////////////////////////
// This function computes a 2D central moment of the rank:
// x_moment_rank, y_moment_rank
//
// This function assumes that the center of gravity
// is already given (e.g. from the CenterOfGravity(...))
//
```

```
// The formula for 2D moments is as follows:
//
//                 m_ab( I ) = SUM{ (x-ox)^a * (y-oy)^b * I(x,y) }
//
// where: a == x_moment_rank, b == y_moment_rank
// and a point (ox, oy) is a center of gravity (centroid)
//
// The "object" is defined by a mask associated
// with the original image (thus we use TMaskedImageFor<>
// classes).
//
// The central moments are computed the same way as moments
// if we use a coordinate system that has its origin
// at the center of gravity.
//
////////////////////////////////////////////////////////////////
//
// INPUT:
//                     in - reference to the image
//                     theCentroid - the center of gravity point,
//                             if already known
//                     x_moment_rank - the first rank of the moment
//                     y_moment_rank - the second rank of the moment
//
// OUTPUT:
//                     the central moment of order (x_moment_rank, y_moment_rank)
//
// REMARKS:
//
template< class T >
double CentralMoment( const TMaskedImageFor< T > & in, const Real_2D_Point
                      the Centroid, const double x_moment_rank, const
                      double y_moment_rank );
```

Algorithm 2.8 Declaration of the *CentralMoment* template function for computation of statistical central moments of objects defined in the masked image object.

Affine moment invariants are computed calling one template function *AffineMomentInvariants* which declaration is listed in Algorithm 2.9.

```
////////////////////////////////////////////////////////////////
// This function computes the set of the four affine
// moment invariants as defined by Flusser & Suk
//
// This function assumes that the center of gravity
// is already given (e.g. from the CenterOfGravity(...))
//
// The "object" is defined by a mask associated
// with the original image (thus we use TMaskedImageFor<>
// classes).
```

```
//
/////////////////////////////////////////////////////////////
//
// INPUT:
//                          in - reference to the image
//                          theCentroid - the center of gravity point,
//                                  if already known
//                          theInvariants - the output data structure,
//                                  a vector that, if returned true, contains
//                                  the four affine invariants.
//
// OUTPUT:
//                          true - if operation successful
//                          false - otherwise
//
// REMARKS:
//
template< class T >
bool AffineMomentInvariants( const TMaskedImageFor< T > & in, const Real_2D_Point
                                theCentroid, vector< double > & theInvariants );
```

Algorithm 2.9 Declaration of the *AffineMomentInvariants* template function for computation of affine moment invariants (AMI) of objects defined in the masked image object.

The affine moment invariants, which are discussed in Section 5.3.1, are computed in accordance with the formula (5.9). These are returned in a vector of floating-point values.

2.9 Eigendecomposition and Representation of Tensors

An analysis of tensors expressed in general form is not always intuitive and does not allow us to find characteristic behavior or structure expressed with these quantities. Of special interest, therefore, is the eigendecomposition of tensors, i.e. decomposition of tensors to the principal axes.

The eigenproblem for a linear transformation denotes a notion of a vector that's phase is not changed by this transformation. Many technical and scientific problems reduce to an analysis of this invariant property. The problem can be stated in the language of tensors as follows. For a given second order tensor $\mathbf{T}$ an eigenvector $\mathbf{u}$ is a nonzero complex tensor which fulfills the following equation [5]

$$\mathbf{Tu} = \lambda\mathbf{u}, \tag{2.128}$$

where λ is an eigenvalue which, in general case, can be a complex number. In tensor notation (Section 2.4.1) the above takes the following form

$$T_i^k u^i = \lambda u^k. \tag{2.129}$$

It is also easy to see that the second order tensors can be analyzed as matrices since both have two free indices. Therefore in the following we constrain our discussion to this type of tensors in which we can rely on the results of matrix analysis [32].

After a simple decomposition of Equation (2.128)

$$(\mathbf{T} - \lambda \mathbf{I})\mathbf{u} = 0, \tag{2.130}$$

it is easy to see that the above can have a nontrivial solution only if

$$\det(\mathbf{T} - \lambda \mathbf{I}) = 0. \tag{2.131}$$

This is the so called characteristic equation of $\mathbf{T}$ [32, 98]. Solution to this equation provides eigenvalues of $\mathbf{T}$.

For a tensor of an L-dimensional space the characteristic equation is a polynomial of order L, that is

$$\lambda^L + c_1 \lambda^{L-1} + c_2 \lambda^{L-2} + \cdots + c_{L-1} \lambda + c_L = 0, \tag{2.132}$$

where $c_M = (-1)^M \Sigma(all\ M \times M\ principal\ minors)$. A principal minor of order M is obtained from $\mathbf{T}$ after deleting L-M rows and columns with the same index. It holds also that

$$Tr(\mathbf{T}) = \lambda_1 + \lambda_2 + \ldots + \lambda_L = -c_1, \tag{2.133}$$

and

$$\det(\mathbf{T}) = \lambda_1 \lambda_2 \ldots \lambda_L = (-1)^L c_L. \tag{2.134}$$

Let us also observe that if a vector $\mathbf{u}$ is an eigenvector of $\mathbf{T}$ then also vector $\mathbf{u}$ which is a normalized version of $\mathbf{u}$ fulfills the eigenvalue equation (2.128), since from Equation (2.128) we obtain

$$\mathbf{T}\frac{\mathbf{u}}{\|\mathbf{u}\|} = \lambda \frac{\mathbf{u}}{\|\mathbf{u}\|},$$

thus

$$\mathbf{T}\hat{\mathbf{u}} = \lambda \hat{\mathbf{u}}, \quad \text{where} \quad \hat{\mathbf{u}} \equiv \frac{\mathbf{u}}{\|\mathbf{u}\|}. \tag{2.135}$$

In CV, of special interest are 3D *symmetric second order* tensors. These are, for instance, the motion ST or the diffusion tensor. Thanks to being symmetric, these have some specific and very important properties, summarized in the following list [32]:

- all eigenvalues are real positive, i.e. $\forall i: \lambda_i \in \Re_+$;
- all eigenvectors are orthogonal, i.e. $\forall i,j: \mathbf{u}^T_i \mathbf{u}_j = 0\ if\ i \neq j$;
- there exists a unitary matrix $\mathbf{U}$, i.e. $\mathbf{U}^T \mathbf{U} = \mathbf{1}$, such that

$$\mathbf{U}^T \mathbf{T} \mathbf{U} = diag\left[\lambda_1, \quad \lambda_2, \quad \ldots, \quad \lambda_L\right]_{L \times L}, \quad \text{where} \quad \mathbf{U}^T \mathbf{U} = \mathbf{U}\mathbf{U}^T = \mathbf{1}. \tag{2.136}$$

Moreover, columns $\mathbf{u}_i$ of the matrix $\mathbf{U}$ are orthonormal eigenvectors corresponding to the eigenvalues $\lambda_1, \lambda_2, \ldots, \lambda_n$, respectively. It holds also that

$$\mathbf{T} = \mathbf{U} diag\left[\lambda_1, \quad \lambda_2, \quad \ldots, \quad \lambda_L\right]\mathbf{U}^T = \sum_{p=1}^{L}\lambda_p\mathbf{u}_p\mathbf{u}_p^T. \tag{2.137}$$

For the considered case of a 3D symmetric second order tensor, the characteristic equation in (2.132) can be expressed as follows:

$$\lambda^3 - c_1\lambda^2 + c_2\lambda - c_3 = 0, \tag{2.138}$$

where coefficients c_i are

$$c_1 = T_{11} + T_{22} + T_{33} = Tr\left(\mathbf{T}\right), \tag{2.139}$$

$$c_2 = \begin{vmatrix} T_{11} & T_{12} \\ T_{21} & T_{22} \end{vmatrix} + \begin{vmatrix} T_{11} & T_{13} \\ T_{31} & T_{33} \end{vmatrix} + \begin{vmatrix} T_{22} & T_{23} \\ T_{32} & T_{33} \end{vmatrix}, \tag{2.140}$$

$$c_3 = \begin{vmatrix} T_{11} & T_{12} & T_{13} \\ T_{21} & T_{22} & T_{23} \\ T_{31} & T_{32} & T_{33} \end{vmatrix} = |\mathbf{T}| = \det\left(\mathbf{T}\right). \tag{2.141}$$

All c_i are *invariant* with respect to the chosen coordinate system, i.e. they depend solely on $\mathbf{T}$ and not on a coordinate system.

Equation (2.138) indicates that the 3D symmetrical tensor contains three eigenvalues, say λ_1, λ_2 and λ_3. They can be ordered, so

$$\lambda_1 \geq \lambda_2 \geq \lambda_3 \geq 0. \tag{2.142}$$

Now Equation (2.135) can be rewritten for these three eigenvalues, as follows

$$\begin{cases} \mathbf{T}\hat{\mathbf{u}}_1 = \lambda_1\hat{\mathbf{u}}_1 \\ \mathbf{T}\hat{\mathbf{u}}_2 = \lambda_2\hat{\mathbf{u}}_2. \\ \mathbf{T}\hat{\mathbf{u}}_3 = \lambda_3\hat{\mathbf{u}}_3 \end{cases} \tag{2.143}$$

Each *i-th* equation in the above is then multiplied by $\hat{\mathbf{u}}_i^T$, after which all are summed up. As a result, because $\hat{\mathbf{u}}_i$ are orthonormal, the following is obtained

$$\mathbf{T}\underbrace{\left(\hat{\mathbf{u}}_1\hat{\mathbf{u}}_1^T + \hat{\mathbf{u}}_2\hat{\mathbf{u}}_2^T + \hat{\mathbf{u}}_3\hat{\mathbf{u}}_3^T\right)}_{1} = \lambda_1\hat{\mathbf{u}}_1\hat{\mathbf{u}}_1^T + \lambda_2\hat{\mathbf{u}}_2\hat{\mathbf{u}}_2^T + \lambda_3\hat{\mathbf{u}}_3\hat{\mathbf{u}}_3^T,$$

$$\mathbf{T} = \lambda_1\hat{\mathbf{u}}_1\hat{\mathbf{u}}_1^T + \lambda_2\hat{\mathbf{u}}_2\hat{\mathbf{u}}_2^T + \lambda_3\hat{\mathbf{u}}_3\hat{\mathbf{u}}_3^T. \tag{2.144}$$

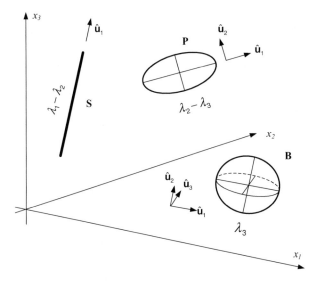

Figure 2.19   The three components **S** stick, **P** plate, and **B** ball, for the 3D symmetrical tensors of second order.

This is a very interesting result which indicates a possible *decomposition* of the tensor **T** in the space spanned by its eigenvectors (*eigendecomposition*) – visualized in Figure 2.18. This is also directly expressed by the theorem (2.137).

Assuming Equation (2.142) and after simple decomposition of the last equation, another important relation is obtained

$$\mathbf{T} = (\lambda_1 - \lambda_2)\underbrace{\hat{\mathbf{u}}_1\hat{\mathbf{u}}_1^T}_{\mathbf{S}} + (\lambda_2 - \lambda_3)\underbrace{\left(\hat{\mathbf{u}}_1\hat{\mathbf{u}}_1^T + \hat{\mathbf{u}}_2\hat{\mathbf{u}}_2^T\right)}_{\mathbf{P}} + \lambda_3\underbrace{\left(\hat{\mathbf{u}}_1\hat{\mathbf{u}}_1^T + \hat{\mathbf{u}}_2\hat{\mathbf{u}}_2^T + \hat{\mathbf{u}}_3\hat{\mathbf{u}}_3^T\right)}_{\mathbf{B}} \quad (2.145)$$

where **S**, **P** and **B** are called *stick, plate*, and *ball* (sphere) tensors, respectively. Their names come from the fact that **S** corresponds to the linear 1D structures, **P** to the 2D structures and finally **B** to the 3D structure components of the tensor **T**, as shown in Figure 2.19.

Tensor decomposition (2.145) allows classification of tensors in accordance with their type in terms of influence of the stick, plate, or ball components. Their influence is given directly by the relation of the eigenvalues. For the purpose of tensor classification it is thus convenient to define the following parameters [99]

$$d_S = 1 - \frac{\lambda_2}{\lambda_1}, \quad d_P = \frac{\lambda_2 - \lambda_3}{\lambda_1}, \quad d_B = \frac{\lambda_3}{\lambda_1}. \quad (2.146)$$

From the above it is evident that

$$d_S + d_P + d_B = 1. \quad (2.147)$$

The aforementioned tensor decomposition is a basic mechanism in the tensor voting framework proposed by Mordohai and Medioni [100], which aims at providing the methods of perceptual organization postulated in the famous gestalt theory [101]. The main postulate of this theory is that whenever points in an image have something common, such as shape, mutual distance, color, continuity, direction, etc., then they are grouped by the human visual system to create a new visual object which is higher in the semantics hierarchy. The important property of the laws of gestalt theory is that they reflect the common experience of the physical laws observed in daily life. The best validation of gestalt laws comes from observation of so called "impossible objects," i.e. drawings which do not comply with the observation of any physical object.

Finally, let us mention that the outlined method of eigendecomposition of tensors finds application to object recognition from the phase histograms, as presented in Section 5.2.

2.10 Tensor Invariants

In the previous section it was shown that tensors can be described in terms of inherent *shapes* and *orientations*. It appears that the inherent shapes – given by the eigenvalues in the decomposition (2.144) – can be quantified by tensor invariants which are independent of the coordinate system [102]. The other already discussed invariants are expressed in (2.139)–(2.141). But tensor eigenvalues are also invariant by themselves.

From the aforementioned basic invariants the other invariants can be created in the form of, usually nolinear, functions of the basic invariants. The main purpose of this operation is to provide a more intuitive means of assessment of tensor type, as e.g. in the analysis of the magnetic resonance imaging (MRI) signals [103]. For instance the following central moments

$$\mu_1 = \langle \lambda_i \rangle = \frac{1}{3}(\lambda_1 + \lambda_2 + \lambda_3), \ \mu_2 = \langle (\lambda_i - \mu_1)^2 \rangle, \quad \text{and} \quad \mu_3 = \langle (\lambda_i - \mu_1)^3 \rangle, \quad (2.148)$$

where $\langle . \rangle$ is an averaging operator, as well as the following values

$$\sigma = \sqrt{\mu_2}, \ \alpha_3 = \frac{\mu_3}{\sigma^3}, \quad (2.149)$$

allow analysis of medical MRI signals since they are invariants of the MRI tensor [102, 104, 105, 106]. In the same way they can be used in analysis of other images for recognition of objects. The most common invariant for 2D images is the tensor trace m in (2.98).

2.11 Geometry of Multiple Views: The Multifocal Tensor

Multiple views of a scene provide not only more information but also a qualitatively new type of information. Object recognition can be successive even in a single view. However, if multiple views of a scene are available then more information on objects is provided since the objects are imaged from different viewpoints. This can facilitate such tasks as object tracking up to detection of a scene geometry and multiple motions [8, 7, 15, 107]. In the task of object

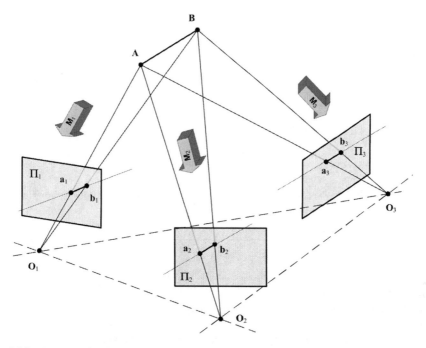

Figure 2.20   An example of a an object imaged into three views simultaneously. The pin-hole model of a camera is assumed. $\mathbf{O}_i$ is a projective centre point of the i-th camera. $\mathbf{\Pi}_i$ denotes an i-th camera plane. An image of the $\mathbf{O}_i$ point on a j-th camera plane is an epipolar point.

recognition this stage can serve scene segmentation which greatly simplifies detection and subsequent recognition of objects.

A single pin-hole camera does projective transformation of a 3D point $\mathbf{A}$ into a 2D point $\mathbf{a}_i$ on an i-th camera plane $\mathbf{\Pi}_i$, as follows

$$\mathbf{M}_i \mathbf{A} = \gamma_i \mathbf{a}_i, \tag{2.150}$$

where $\mathbf{M}_i$ is a 3×4 matrix defining the projective transformation of the camera [7, 15]. It is important to notice that in Equation (2.150) the points are expressed in the so called homogeneous coordinates, in which 2D and 3D points in the Euclidean space contain 3 and 4 coordinates in the homogeneous space, respectively. In systems acquiring many images at the same time, the pin-hole cameras are observing the same 3D point $\mathbf{P}$ at the same time step. This is different from the video sequence in which each image is acquired at different times. Dependencies among multiple views of the 3D points can be concisely written with the help of tensor notation, as will be shown.

Observation of an object with multiple cameras can be written in a compact way. For example, for the three cameras shown in Figure 2.20, observing the point $\mathbf{A}$ at the same time, the following is obtained [15, 108]

$$
\underbrace{\begin{bmatrix} \mathbf{M}_1 & \mathbf{a}_1 & 0 & 0 \\ \mathbf{M}_2 & 0 & \mathbf{a}_2 & 0 \\ \mathbf{M}_3 & 0 & 0 & \mathbf{a}_3 \end{bmatrix}}_{\mathbf{H}_3}{}_{9 \times 7} \begin{bmatrix} \mathbf{A} \\ -\gamma_1 \\ -\gamma_2 \\ -\gamma_3 \end{bmatrix}_{7 \times 1} = 0, \tag{2.151}
$$

where $\mathbf{M}_i$ denotes a projective transformation of the *i-th* camera, $\mathbf{a}_i$ is the image of $\mathbf{A}$ created on the *i-th* camera plane, γ_i is a scaling parameter. In this case, $\mathbf{H}_3$ is a 9×7 matrix composed of the three matrices $\mathbf{M}_i$ and vectors $\mathbf{a}_i$. Its rank has to be up to six to assure a nontrivial null space. In other words, Equation (2.151) denotes a set of homogeneous equations of seven unknowns, and for a nontrivial solution $det(\mathbf{H}_3)$ has to be 0 [16]. In the general case of m cameras, $\mathbf{H}_m$ is of rank at most $m+3$. The same type of equation can be written for the point $\mathbf{B}$ and its images $\mathbf{b}_1$, $\mathbf{b}_2$, and $\mathbf{b}_3$ in Figure 2.20, as well as for other tuples containing a point belonging to the 3D space and its camera images. However, an interesting issue is the relation between the scene parameters and positions of the points.

The simplest case involves two images. Hence, from Equation (2.151) a matrix $\mathbf{H}_2$ can be built taking two images indexed, say by 1 and 2 in Figure 2.20. From $\mathbf{M}_1$ and $\mathbf{M}_2$ their three rows are considered separately (upper index denoting a row number)

$$
\mathbf{H}_2 = \begin{bmatrix} \mathbf{m}_1^1 & a_1^1 & 0 \\ \mathbf{m}_1^2 & a_1^2 & 0 \\ \mathbf{m}_1^3 & 1 & 0 \\ \mathbf{m}_2^1 & 0 & a_2^1 \\ \mathbf{m}_2^2 & 0 & a_2^2 \\ \mathbf{m}_2^3 & 0 & 1 \end{bmatrix}_{6 \times 6}. \tag{2.152}
$$

In the above, $\mathbf{m}_j^i$ stands for the *i-th* raw of the *j-th* camera matrix, a_j^i is the *i-th* component of the *j-th* image point. A rank of $\mathbf{H}_2$ is at most five which yields

$$
det\,(\mathbf{H}_2) = 0, \tag{2.153}
$$

since $\mathbf{H}_2$ is 6×6. The above equation can be written in a tensor notation as follows [7]

$$
det\,(\mathbf{H}_2) = F_{ij}a_1^i a_2^j = 0, \tag{2.154}
$$

where a_k^i denotes *i-th* component of the *k-th* point $\mathbf{a}_k$, while F_{ij} are elements of the so called *fundamental matrix*. This matrix can be viewed as a *bifocal tensor* in accordance with the following

Definition 2.15   The bifocal tensor (the fundamental matrix)
Let ε_{ijk} be a permutation symbol and let $\mathbf{m}^i{}_j$ denote *i-th* row of the *j-th* camera matrix $\mathbf{M}_j$. The bifocal tensor is given as follows

$$F_{ij} = \varepsilon_{ief}\varepsilon_{jgh} \det \begin{bmatrix} \mathbf{m}_1^e \\ \mathbf{m}_1^f \\ \mathbf{m}_2^g \\ \mathbf{m}_2^h \end{bmatrix} . \quad \square \tag{2.155}$$

The elements F_{ij} are covariant components of the second degree tensor. Hence, due to the tensor transformation law, a change of the coordinate systems associated with images 1 and 2 induces a concordant change of values F_{ij}, analogously to Section 2.8.

The further extensions of the presented analysis of minors of the matrix (2.151) lead to tensors of higher degree [108, 109]. For instance the trifocal tensor can be formulated in an analogous way, taking now the matrix $\mathbf{H}_3$. Then again the rank condition $det(\mathbf{H}_3) = 0$ leads to the following trifocal constraint

$$\varepsilon_{jge}\varepsilon_{khf}\mathbf{a}_1^i\mathbf{a}_2^j\mathbf{a}_3^k T_i^{gh} = \mathbf{0}_{ef}, \tag{2.156}$$

where $\mathbf{0}_{ef}$ denotes a two-dimensional tensor with zero entries, and

$$T_i^{gh} = \varepsilon_{ief} \det \begin{bmatrix} \mathbf{m}_1^e \\ \mathbf{m}_i^f \\ \mathbf{m}_2^g \\ \mathbf{m}_3^h \end{bmatrix} \tag{2.157}$$

are components of the trifocal tensor. It is a third order *mixed tensor* in which the order of images is important since the first image with the *i-th* index is treated differently.

2.12   Multilinear Tensor Methods

As alluded to previously, tensors are well adapted to multidimensional data which are frequently encountered in different measurements (Section 2.3). This property was observed by many researchers applying tensors to data representation and analysis. Almost half a century ago Tucker proposed tensors for multifactor analysis of neuroscience and psychometric problems [110]. Further applications relate to various domains of science and technology, such as the food industry (Bro [111]), seismology (Le Bihan [112]), chemistry (Montoto [112]), data mining (Kolda and Bader [20]), as well as signal processing (Cichocki *et al.* [19], Muti *et al.* [113]) and particularly computer vision (Vasilescu and Terzopoulos [11, 12, 114]), and so forth. Other noticeable works in this field are by de Lathauwer [96], de Lathauwer *et al.*

[115], Kroonenberg [21], and Kolda *et al.* [116], to name a few. Finally, for a good historical overview of the matrix and multilinear algebra see the thesis by Franc [117].

For multidimensional arrays of data the essential question is that of the methods of their analysis. In this respect, special importance is found in the multidimensional extension of the Singular Value Decomposition (SVD), which is one of the most important and information-providing decompositions in the case of matrices. SVD allows comprehension of an influence from different factors of data (or "dominating directions") and their interaction. As shown by many researchers, such a decomposition exists also for multidimensional matrices although with some restrictions [96].

At the beginning it is interesting to compare the SVD decomposition of a 2D matrix $\mathbf{T}$ with a decomposition of a 3D tensor $\mathcal{T}$. In the case of matrices it is given as follows

$$\mathbf{T} = \mathbf{SVD}^T = \sum_{r=1}^{R} v_r \left(\mathbf{s}^{(r)} \circ \mathbf{d}^{(r)} \right), \tag{2.158}$$

where v_r are singular values which lie on diagonal the matrix $\mathbf{V}$, $\mathbf{s}^{(r)}$ and $\mathbf{d}^{(r)}$ are the *r-th* columns of the *orthogonal* matrices $\mathbf{S}$ and $\mathbf{D}$, respectively. R is a rank of the matrix $\mathbf{T}$. In the above $\circ$ denotes an outer product of the vectors. Let us now assume that a 3D tensor, of dimensions spanning $N_1 \times N_2 \times N_3$, can be expressed with a sum of rank-1 tensors, as follows

$$\mathcal{T} = \sum_{p=1}^{N_1} \sum_{q=1}^{N_2} \sum_{r=1}^{N_3} v_{pqr} \left(\mathbf{a}^{(p)} \circ \mathbf{b}^{(q)} \circ \mathbf{c}^{(r)} \right), \tag{2.159}$$

where $\mathbf{a}^{(p)}$, $\mathbf{b}^{(q)}$, and $\mathbf{c}^{(r)}$ are real vectors of dimensions N_1, N_2, and N_3, respectively, with *no* assumption of their potential orthogonality, however. Equation (2.159) is also known as the Tucker decomposition, for which an orthogonality constraint is not required [110]. Nevertheless, if $\mathbf{a}^{(p)}$, $\mathbf{b}^{(q)}$, and $\mathbf{c}^{(r)}$ in the above constitute columns of the orthogonal matrices, then Equation (2.159) is referred to as the higher-order SVD (HOSVD) [96] or the Tucker decomposition[5] [110]. Finally, $\mathcal{V} = [v_{pqr}]$ is known as a *core tensor* which usually is not diagonal (in contrast to the "classical" SVD for matrices).

Although the two decompositions (2.158) and (2.159) look similar up to this point, definition of rank for tensors of an order higher than two is more complicated than for the matrices. For the matrices the upper bound on their rank can be inferred directly from their dimensions, i.e. it is less or equal to the minimum of the two dimensions. However, this property does not hold for the higher dimensional tensors. That is, in the general case, neither a rank of a tensor, nor an upper bound of its rank, are given directly by its dimensions [118].

---

[5]There are also other names for this type of tensor decomposition, such as *N-mode* Principal Component Analysis or *N-mode* Singular Value Decomposition.

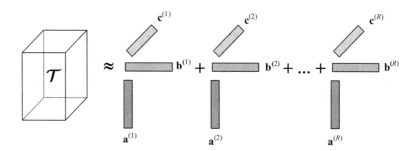

Figure 2.21   Visualization of the rank-1 decomposition of 3D tensors.

Let us now consider a special case, in which a tensor $\mathcal{T}$ with dimensions $N_1 \times N_2 \times N_3$ can be expressed as a pure sum of rank-1 tensors[6] (i.e. vectors). In this case, representation (2.159) reduces to the following

$$\mathcal{T} = \sum_{r=1}^{R} \mathbf{a}^{(r)} \circ \mathbf{b}^{(r)} \circ \mathbf{c}^{(r)}. \qquad (2.160)$$

In the above R is called a rank of the tensor $\mathcal{T}$ if R is minimal, i.e. it is the smallest number of rank-1 tensors necessary to produce a given tensor by their outer product. Figure 2.21 depicts this rank-1 decomposition for a 3D tensor.

Based on Equation (2.160) it is easy to show that an element t of a tensor $\mathcal{T}$ at a position indexed by i_1, i_2, i_3 is simply a product of three components of the vectors $\mathbf{a}$, $\mathbf{b}$ and $\mathbf{c}$, with corresponding indices, i.e.

$$t_{i_1 i_2 i_3} = \mathcal{T}(i_1, i_2, i_3) = a_{i_1} b_{i_2} c_{i_3}. \qquad (2.161)$$

This can be easily extended to higher dimensions.

For example if $\mathbf{a} = [1, 1, 0]^T$, $\mathbf{b} = [2, -1, 1]^T$, and $\mathbf{c} = [10, 2, 2]^T$, then $\mathcal{T}$ obtained from Equation (2.161), contains the following components

$$\mathcal{T} = \begin{bmatrix} 20 & -10 & 10 & 4 & -2 & 2 & 4 & -2 & 2 \\ 20 & -10 & 10 & 4 & -2 & 2 & 4 & -2 & 2 \\ 0 & 0 & 0 & 0 & 0 & 0 & 0 & 0 & 0 \end{bmatrix}, \qquad (2.162)$$

where each pane corresponds to the third index of $\mathcal{T}$, as in the scheme presented in Figure 2.4.

---

[6]The rank-1 decomposition of tensors is also known as canonical decomposition – CANDECOMP – as introduced in the 1970s in the psychometrics community, and also as parallel factor analysis – PARAFAC – as coined by Harshman [119]. Thus, it is often called the CANDECOMP/PARAFAC or CP decomposition, as proposed by Kiers [120]. The historical background as well as a systematic approach to these and other tensor decompositions can be found in the paper by Kolda and Bader [20].

A more detailed presentation of properties as well as methods of computation of the HOSVD follow in Section 2.12.2. Further, HOSVD appears to be a very useful technique for object classification – these issues are discussed in Section 5.8. On the other hand, the CP decomposition outlined here, with an algorithm for its computation, is discussed in Section 2.12.6. However, before we go further into the subject of tensor decompositions we need to provide basic definitions of multilinear algebra.

2.12.1   Basic Concepts of Multilinear Algebra

Although tensors are multidimensional, in many contexts it is convenient to represent them as two-dimensional structures, i.e. matrices, which are well established mathematical objects. However, this does not mean that tensors are just matrices, since as already mentioned in some contexts a matrix can be seen as a two-dimensional tensor. What is important is the inner structure of a tensor, expressed by its numbers of dimensions. If this information is retained, then data representation can be as convenient as necessary. This will be easily seen when discussing practical methods of computer representations of tensors, as provided in Section 1.1. So, we start providing methods of converting tensors into equivalent matrix formats. This conversion is called tensor flattening, unfolding or matricization. Then the methods of tensor multiplication are discussed. Finally, definitions of different ranks of tensors are outlined.

2.12.1.1   Tensor Flattening

Tensor flattening is the process of representing tensors in a matrix-like form. This does not mean that we lose information on a number of indices of a tensor, nor that tensors are equivalent to matrices. Such "flattened" representation of tensors is just convenient in many contexts, as will be discussed in this and the next sections. We start from the definition of fibers in a tensor, called k-mode vectors of a tensor.

Definition 2.16   The k-*mode* vector of a tensor

Given an P-*th* order tensor[7] $\mathcal{T} \in \mathfrak{R}^{N_1 \times N_2 \times \dots N_P}$ the k-*mode vector* (or *a fiber*) of $\mathcal{T}$ is a vector obtained from the elements of $\mathbf{T}$ by varying only one index n_k while keeping all other indices fixed.   □

Figure 2.22 depicts three types of fibers of a three-dimensional tensor. The 1-mode vector is called a column fiber, 2-mode – a row fiber, and finally the 3-mode – a tube fiber. In literature a 4-mode vector is sometimes called a pipe fiber.

Definition 2.17   Tensor flattening (unfolding[8], matricization)

For a given P-*th* order tensor $\mathcal{T} \in \mathfrak{R}^{N_1 \times N_2 \times \dots N_P}$ and for the chosen index N_k, the k-mode flattening of a tensor is a matrix $\mathbf{T}_{(k)}$ of dimensions $N_k \times (N_1 \ N_2 \ \dots \ N_{k-1} \ N_{k+1} \ \dots \ N_P)$. The position of an element $t_{n_1 n_2 \dots nP}$ is one-to-one mapping with a function Q into row index $r = n_k$, and column index $c = Q(n_1, n_2, \dots \ n_{k-1}, n_{k+1} \ \dots \ ,nP, N_1, N_2, \dots \ ,N_P)$ of the matrix $\mathbf{T}_{(k)}$.   □

---

[7]Although the majority of the presented definitions, theorems and results hold also for complex valued tensors, in image processing we are mostly interested in real values.
[8]The term "unfolding" has other special meaning in psychometrics, and therefore some authors suggest avoiding this term in the context of tensor representation [121].

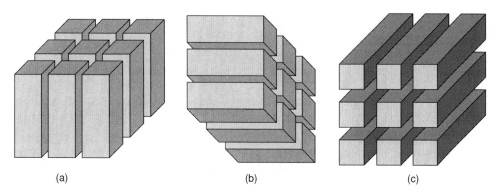

Figure 2.22 Visualization of three types of k-mode vectors (fibers) for a three dimensional tensors: 1-mode (column) fibers (a), 2-mode (row) fibers (b), and 3-mode (tube) fibers (c)

Let us observe that for a chosen k-mode flattening of a tensor, the number of the rows is uniquely given by the k-th dimension of the tensor, and a row index of a tensor element is the same as its k-th row index in the corresponding matrix representation. On the other hand, the number of columns is a product of all the remaining dimensions of the tensor. However, the column index of an element is determined by a function Q which can be any that takes a set of indices $n_1, n_2, \ldots n_{k-1}, n_{k+1} \ldots, n_P$ into the integer range $1 \ldots (N_1 N_2 \ldots N_{k-1} N_{k+1} \ldots N_P)$. From many possible choices of Q few have practical applications, as we shall see.

The k-mode flattenings of tensors can be easily obtained from the k-mode vectors, given in Definition 2.16. Figure 2.23 illustrates this process for 3D tensors, chosen for obvious reasons.

If we take the value of a tensor from Figure 2.4 and apply index ordering as shown in Figure 2.23, we obtain the following flattenings of the tensor

$$
\mathbf{T}_{(1)} = \begin{bmatrix} 12 & 1 & 0 & 0 & -1 & 0 & 0 & 1 & -2 \\ 0 & 4 & 4 & 4 & 0 & 10 & 1 & 4 & 0 \\ 2 & 5 & 2 & 3 & 2 & 1 & -1 & 1 & 0 \end{bmatrix},
\tag{2.163}
$$

$$
\mathbf{T}_{(2)} = \begin{bmatrix} 12 & 0 & 2 & 1 & 4 & 5 & 0 & 4 & 2 \\ 0 & 4 & 3 & -1 & 0 & 2 & 0 & 10 & 1 \\ 0 & 1 & -1 & 1 & 4 & 1 & -2 & 0 & 0 \end{bmatrix},
\tag{2.164}
$$

$$
\mathbf{T}_{(3)} = \begin{bmatrix} 12 & 0 & 0 & 0 & 4 & 1 & 2 & 3 & -1 \\ 1 & -1 & 1 & 4 & 0 & 4 & 5 & 2 & 1 \\ 0 & 0 & -2 & 4 & 10 & 0 & 2 & 1 & 0 \end{bmatrix}.
\tag{2.165}
$$

This type of index ordering is called a *backward* flattening mode.

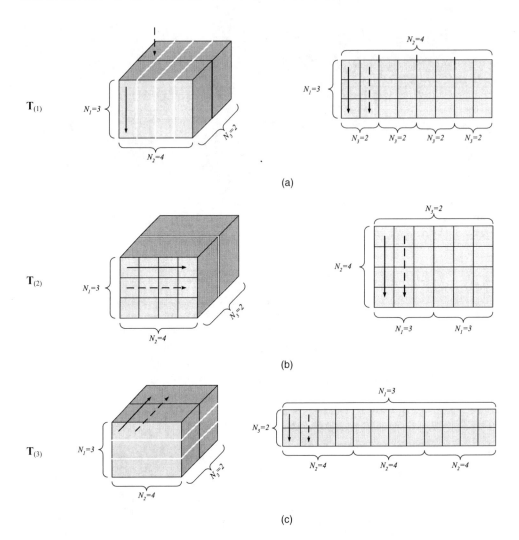

Figure 2.23   Flattening of a $N_1 \times N_2 \times N_3$ tensor $\mathcal{T}$ leads to the matrices $\mathbf{T}_{(1)}$ of dimensions $N_1 \times N_3N_2$, $\mathbf{T}_{(2)} - N_2 \times N_1N_3$, and $\mathbf{T}_{(3)} - N_3 \times N_2N_1$. Assumed backward order of index permutations.

From one flattening it is possible to obtain all other flattenings by the exchange of elements in accordance with the change of their coordinates. It is also possible to change the ordering (perturbation) of the indices in the already presented scheme of tensor flattening. This way we obtain a different, but equivalent, ordering of elements. Such a scheme was proposed by Kiers [120]. These topics are discussed also in the paper by Bader and Kolda [122]. Summarizing, when considering a flattening mode for objects with, say, three indices N_1, N_2 and N_3, the first thing to do is to select a single index, and then an order of the remaining indices. For the considered three indices we have $3! = 6$ options in general, but not all of them are equally convenient with respect to data analysis and interpretation, as will be discussed.

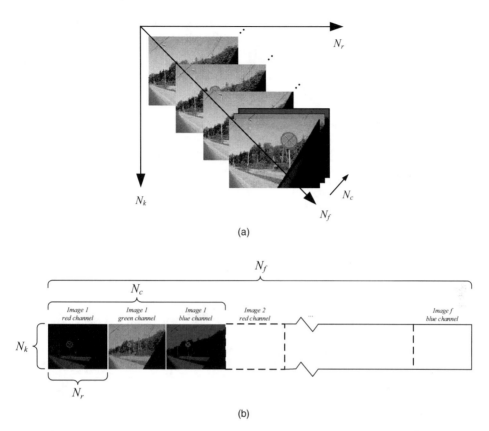

Figure 2.24   An example of a 4D tensor representing a traffic video scene with color frames[9] (a). A practical way of tensor flattening – the forward mode – which maps images to the memory in the scanning order (b). (For a color version of this figure, please see the color plate section.)

A color video stream can be represented as a four-dimensional tensor. An example of such a tensor of dimensions $320 \times 240 \times 4 \times 3$ (*columns* N_k × *rows* N_r × *frames* N_f × *color-channels* N_c) is depicted in Figure 2.24(a).

Figure 2.24(b) shows the chosen flattening mode for the video tensor which is called the *forward* flattening mode. This type of flattening is shown to be practical in video representation due to the orientation of axes which follow row by row, frame by frame video transmissions. Thus, in the forward flattening all rows and columns of each frame are kept together, then the index for the three color channels, red, green, blue, respectively, while the last index corresponds to the frame number in a video sequence.

Let us now take a closer look at the two types of index permutations which for the case of three indices are depicted in Figure 2.25.

In the *forward* permutations the process of generating new values in a permutation starts from a chosen index from which the next ones are obtained incrementing by 1. After reaching the largest value, the counter wraps up and starts from the lowest value in the series, as

---

[9]Color versions of the images are accessible from the book website: see [28].

Forward permutations (clockwise)   Backward permutations (anti-clockwise)

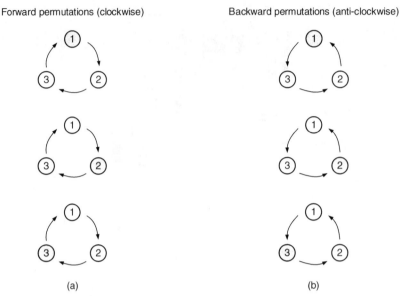

(a)   (b)

Figure 2.25 Two types of permutations of the three indices 1,2, and 3. Grayed circle denotes a starting index. Forward permutations (adding 1) are 1-2-3, 2-3-1 and 3-1-2 (a). Backward permutations (subtracting 1) from the top to the bottom rows are 1-3-2, 2-1-3 and 3-2-1 (b).

depicted in Figure 2.25(a). For example, starting from 2, the next value of a counter is 3, which after wrapping up drops to 1, and the process continues. In this case the permutation 2-3-1 is obtained, and so on. Figure 2.26 depicts a forward ordering complementary to the backward scheme already presented in Figure 2.23.

The starting value always denotes the first index, i.e. the number of rows in the flattened matrix. Next generated values index exclusively columns of this matrix. The first of them denotes the index which changes the fastests, and so on.

In the *backward* permutations scheme the process of generating new values in a permutation is reversed. That is, having a sequence of integer indices, their *backward* permutation starts from a chosen value and proceeds generating consecutive values by subtractions of 1 [120]. However, when the first value of the series is reached, to avoid negative values the counter wraps up, i.e. it is reset to the largest value from the series, and the process continues just to bounce the starting index, as depicted in Figure 2.25(b). For instance, starting from 2, the next index is 1, then after wrapping up the last index is 3. This gives a permutation 2-1-3. Starting from 3 we obtain 3-2-1, whereas from 1 we get 1-2-3.

Using the *forward* flattening scheme to the tensor in Figure 2.4 (p. 17) results in the following matrix

$$
\mathbf{T}_{(1)} = \begin{bmatrix} 12 & 0 & 0 & 1 & -1 & 1 & 0 & 0 & -2 \\ 0 & 4 & 1 & 4 & 0 & 4 & 4 & 10 & 0 \\ 2 & 3 & -1 & 5 & 2 & 1 & 2 & 1 & 0 \end{bmatrix}. \tag{2.166}
$$

Comparing the above $\mathbf{T}_{(1)}$ with $\mathbf{T}_{(1)}$ from (2.163) we notice that only some columns changed their positions. This is not so important as long as we keep the same order of permutations in all formulas.

Finally, let us note the following observations regarding tensor unfolding. If we unfold a tensor $\mathbf{T}$ to a corresponding matrix

$$\mathbf{T}_{(k)} \in \mathfrak{R}^{N_k \times (N_1 N_2 \dots N_{k-1} N_{k+1} \dots N_P)} \tag{2.167}$$

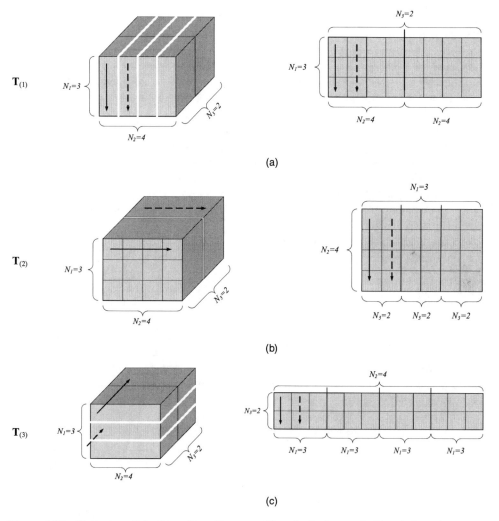

(a)

(b)

(c)

Figure 2.26 Flattening of the $N_1 \times N_2 \times N_3$ tensor $\mathcal{T}$ to obtain the matrix $\mathbf{T}_{(1)}$ of dimensions $N_1 \times N_2 N_3$, $\mathbf{T}_{(2)} - N_2 \times N_3 N_1$, and $\mathbf{T}_{(3)} - N_3 \times N_1 N_2$. Forward order of index permutations.

then, in accordance with Definition 2.16, the *columns* of $\mathbf{T}_{(k)}$ are the *k-mode* vectors of $\mathbf{T}$. The second useful observation is that the inner (scalar) product (2.17) of tensors, both of dimensions P, can be expressed as an inner product of their common *k*-mode flattened versions, i.e.

$$\langle \mathcal{S}, \mathcal{T} \rangle = \langle \mathbf{S}_{(k)}, \mathbf{T}_{(k)} \rangle \text{ for any } 1 \leq k \leq P. \tag{2.168}$$

These are shown to be useful in many derivations which follow, as well as simplifying implementation, as will be discussed in Section 2.12.1.5.

2.12.1.2 IMPLEMENTATION Tensor Representation

As already mentioned, a tensor can be seen as a multidimensional array of data. However, due to computer architectures all data structures despite their inner complexities must be transformed to fit into a linear memory space. Thus, when designing computer representation for tensors, the first question is how to efficiently store such amounts of data, as well as information on tensor dimensionality. Efficiency in this context does not mean only that data has to fit into the memory, but also that its access and manipulations have to be optimized for speed. In this respect an observation that a tensor can be always stored in one of its flattened, i.e. matrix like, representations with a chosen mode became very influential. Also many tensor operations are conveniently defined for flattened versions of tensors, as discussed in previous section. Although this does not follow the usual mathematical way in which a matrix can be seen as a special two-dimensional tensor, in our computer framework a tensor will be always represented in its flattened version, i.e. as a matrix endowed with additional information on tensor dimensions. This observation influenced design of the classes for tensor representations which hierarchy is shown in Figure 2.27.

The next question is on a relation between tensor and matrix representations. The first idea was to take advantage of existing data structures for two-dimensional objects and to impose a "has a" relation between the tensor and matrix objects. However, under closer examination it appears that a more universal idea would be to extend the concept of a matrix with additional data on the inner dimensionality of the tensor it represents, as well as with the chosen flattening mode, as discussed in Section 2.12.1.1. Thus, in the proposed representation, tensor "is a" kind of a matrix, i.e. the tensor class is derived from the matrix class, as shown in Figure 2.27.

The base template class *TImageFor* comes from the HIL library, which was designed especially for image operations and therefore the base objects are pixels and images. However, an image can be seen as a matrix, so in HIL we did not create a separate representation for matrices and for 2D data structures the *TImageFor* is used. Moreover, all the classes have the possibility of defining a type of basic elements, giving additional flexibility in choosing optimal data representation and arithmetic. This is especially useful when processing large data sets since the fixed-point representation (such as *FixedFor* in HIL) can be easily substituted for the ubiquitous floating-point type (i.e. *float* or *double* in C++), which for many applications can provide sufficient precision consuming a much lower number of bits, however.

The main class for tensor representation is the template class *TFlatTensorFor*. Due to its twofold properties, the *TFlatTensorFor* has two sets of methods for accessing its elements: The first pair *Get/SetElement* takes as an argument the *TensorIndex*, which is a vector of indices with length equal to the dimension of the tensor. The second set comprises the *Get/SetPixel*

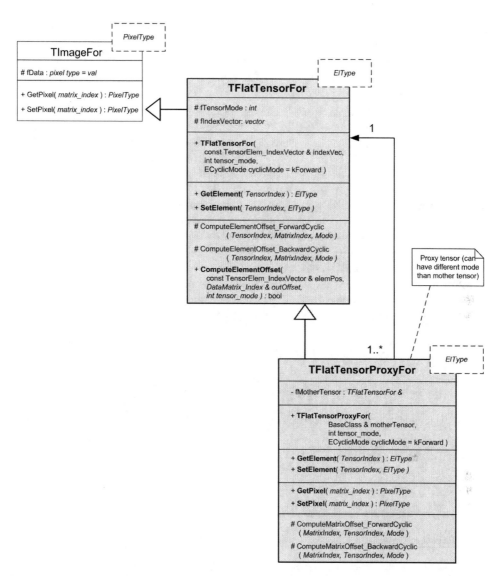

Figure 2.27 A class hierarchy for representation of tensors and basic operations on tensors. The central class for tensor representation is *TFlatTensorFor*, which is derived from the *TImageFor* class which can be used to represent all 2D data structures (such as images or matrices). Tensors are stored in flattened form with a chosen flattening mode. To allow efficient change of the flattening mode and other operations without duplicating tensors the *TFlatTensorProxyFor* class is created. All are template classes which can be instantiated with different types of basic elements.

methods inherited from its base *TImageFor*, which allow access to the matrix data provided just row and column (*r*,*c*) indices. The only, but substantial, difference when comparing matrix and tensor representations is an additional abstraction superimposed on the latter in the form of a number and the way of permutations of tensor indices. These are stored in each tensor object in the *fTensorMode* and *fIndexVector* private data members, respectively. *fTensorMode* stores the tensor flatness mode which is just one of the chosen indices in *fIndexVector*. Thus, it can reach values from 0 up to size of *fIndexVector* minus 1, in accordance with the indexing rule in the C++ language (i.e. C++ index is one less than a corresponding index in the mathematical formulas such as (2.189), for instance). However, a third parameter of each tensor object is the cyclic mode which reveals the chosen permutation order of its indices. This can be either *kBackward* or *kForward* mode, respectively, with the latter being a default for the reasons already discussed in Section 2.12.1. It is chosen once during construction of an object and then stored indirectly by setting a pointer *fCyclicFunction*, shown in Algorithm 2.10, to one of the two member functions, the *ComputeElementOffset_BackwardCyclic* and *ComputeElementOffset_ForwardCyclic*, respectively. Their role is to convert a multidimensional index of an element in a tensor, conveyed by the *fIndexVector* member, into its corresponding two-dimensional index in the low-level matrix data storage. However, since a cyclic mode is set once for an object's lifetime, then there can be only one conversion function *ComputeElementOffset* which indirectly calls either one of the backward or forward conversions.

```
/////////////////////////////////////////////////////////////
// This class implements a tensor in the matricized format.
/////////////////////////////////////////////////////////////
template < typename T = double >
class TFlatTensorFor : public TImageFor< T >
{
public:

        typedef TImageFor< T >                          BaseClass;

        typedef T                                       TensorElem_Type;


        // element access by value, put TensorElem_Type & to set by reference
        typedef TensorElem_Type                         ElemAccessType;

        typedef const ElemAccessType                    ConstElemAccessType;

        typedef                                         BaseClass DataMatrix;

        typedef int                                     TensorElem_Index;

        typedef vector< TensorElem_Index >              TensorElem_IndexVector;

        typedef Index                                   DataMatrix_Index;


protected:

        // selected index (in fIndexVector)
        int                                             fTensorMode;
```

```
        // the vector of indices of the tensor
        TensorElem_IndexVector               fIndexVector;

public:

        int         GetDataMatrix_Row( void ) const { return BaseClass::fRow; }
        int         GetDataMatrix_Col( void ) const { return BaseClass::fCol; }

        int         GetTensorMode( void ) const { return fTensorMode; }

        const TensorElem_IndexVector & GetIndexVector( void ) const { return
                                                            fIndexVector; }

        int         GetTensorDimension( void ) const { return fIndexVector.size(); }

protected:

        // A pointer to the index cycling function
        bool ( TFlatTensorFor ::* const fCyclicFunction )(
                    const TensorElem_IndexVector & elementPosition,
                    DataMatrix_Index & outOffset, int tensor_mode ) const;

        ///////////////////////////////////////////////////////////
        // This function converts a multidimensional tensor
        // index (given by a vector) into an offset to the
        // "flat" matricized representation. Used to access
        // elements of a tensor.
        ///////////////////////////////////////////////////////////
        //
        // INPUT:
        //                  elementPosition - tensor index of an element
        //                      (its size must be the same as fIndexVector,
        //                      and each value less than a corresponding
        //                      in fIndexVector)
        //                  outOffset - upon success contains an offset
        //                      of an element from the beginning of the
        //                      data buffer in the fDataMatrix
        //
        // OUTPUT:
        //                  true on success
        //                  false otherwise
        //
        // REMARKS:
        //                  Backward cycle starts from the selected index
        //                      (given by fTensorMode) and reaches indices in
        //                      a LOWERING direction but cyclically
        //
        bool ComputeElementOffset_BackwardCyclic(
                    const TensorElem_IndexVector & elementPosition,
                    DataMatrix_Index & outOffset, int tensor_mode ) const;
```

```
        // This is a forward version of the above - from fTensorMode this
        // goes in a direction of INCREASING indices, also cyclically
        bool ComputeElementOffset_ForwardCyclic(
                    const TensorElem_IndexVector & elementPosition,
                    DataMatrix_Index & outOffset, int tensor_mode ) const;

public:

        // Does either ComputeElementOffset_BackwardCyclic() or
        // ComputeElementOffset_ForwardCyclic() depending on which one was
        // chosen during
        // construction
        bool ComputeElementOffset( const TensorElem_IndexVector & elementPosition,
                DataMatrix_Index & outOffset, int tensor_mode ) const;


public:

        enum ECyclicMode { kBackward, kForward };

        ECyclicMode GetCyclicMode( void );


protected:

        // ====================================================

        /////////////////////////////////////////////////////////////
        // PROTECTED class constructor
        /////////////////////////////////////////////////////////////
        //
        // INPUT:
        //           tensor_mode - mode of the tensor (0 by default)
        //                    Tensor mode describes selected index of the
        //                    tensor which becomes row counter of the
        //                    data matrix.
        //           cyclicMode - determins cyclic mode of the tensor
        //                    which can be forward of backward (default)
        //
        // OUTPUT:
        //
        //
        // REMARKS:
        //
        //
        TFlatTensorFor( int tensor_mode = 0, ECyclicMode cyclicMode = kForward );
public:

        /////////////////////////////////////////////////////////////
        // Class constructor
        /////////////////////////////////////////////////////////////
        //
        // INPUT:
        //                   indexVec - a vector with tensor dimensions
```

```
//                                Its size (i.e. number of entries) determins
//                                dimensionality of a tensor. Eg. if indexVec
//                                is 4, 4, 5, 10 then this will be 4D tensor
//                                with each dimension 4, 4, 5, 10,
//                                respectively.
//                     tensor_mode - mode of the tensor (type of its "flatness")
//                                Tensor mode describes selected index of the
//                                tensor which becomes row counter of the
//                                data matrix.
//                                Cannot exceed number of dimensions
//                                of the tensor,
//                                i.e. cannot excedd size of indexVec;
//                     cyclicMode - determines cyclic mode of the tensor
//                                which can be forward (default) of backward
//
// OUTPUT:
//
//
// REMARKS:
//                     IMPORTANT - n_mode spans from 0 up to indexVec.size()-1
//                     This is lowered by 1 as denoted in the math books !!
//
//                     In case of wrong parameters or low memory it
//                     can throw T_Standard_HIL_Exception with a code:
//                     T_Standard_HIL_Exception::kCannotCreateAnObject
//
TFlatTensorFor( const TensorElem_IndexVector & indexVec, int tensor_mode,
                     ECyclicMode cyclicMode = kForward );


TFlatTensorFor( const TFlatTensorFor & );          // class copy constructor
TFlatTensorFor & operator = ( const TFlatTensorFor & );//
                                      class assignement operator


virtual ~TFlatTensorFor();                         // class virtual destructor
// =====================================================

public:



//////////////////////////////////////////////////////////
// This function returns a tensor's element at given index.
//////////////////////////////////////////////////////////
//
// INPUT:
//          elemIdx - a vector with an index of an element
//                to read. Length of this vector must be
//                the same as a length of the indexVec provided
//                during construction of the tensor. Moreover,
//                each corresponding value should be positive
//                but less than that in the indexVec.
//
// OUTPUT:
//          tensor element at position
//
```

```
        // REMARKS:
        //
        //
        ElemAccessType          GetElement( const TensorElem_IndexVector & elemIdx )
const;


        /////////////////////////////////////////////////////////////////
        // This function sets a value of an element of the tensor
        /////////////////////////////////////////////////////////////////
        //
        // INPUT:
        //          elemIdx - a vector with an index of an element
        //                  to read. Length of this vector must be
        //                  the same as a length of the indexVec provided
        //                  during construction of the tensor. Moreover,
        //                  each corresponding value should be positive
        //                  but less than that in the indexVec.
        //          elemVal - value to be set
        //
        //
        // OUTPUT:
        //          none
        //
        // REMARKS:
        //
        //
        void    SetElement( const TensorElem_IndexVector & elemIdx,
                                ElemAccessType elemVal );


public:

        // Returns true if the two tensors are the same (i.e. have
        // the same dimensions and the same elements)
        bool operator == ( const TFlatTensorFor< T > & refObj );

};
```

Algorithm 2.10   Definition of the *TFlatTensorFor* class for tensor representation in the matricized (flattened) form (the most important members shown). A tensor is defined providing its series of indices, flattening mode, and index permutation (cyclic) mode.

A tensor represented by the *TFlatTensorFor* object assumes a specific flattening mode, as well as a cyclic mode and a number of indices. However, in many tensor processing algorithms, other flattening modes are required. This can be done easily by duplicating a given tensor with a different flattening mode. However, such an operation is usually prohibited due to the huge amounts of data contained in tensors. Thus, the same functionality needs to be obtained by a simple alternative representation of the same tensor without the necessity of data copying. For this purpose the proxy object is created, which follows the proxy design pattern, implemented in our framework with the *TFlatTensorProxyFor* template class [123]. It is directly derived from the *TFlatTensorFor*, so it can be treated as any other tensor object. However, during its construction an existing tensor object needs to be provided, as shown in the class diagram depicted in Figure 2.27. That one is called a mother tensor, i.e. it is a tensor which data are indirectly accessed through the proxy. At the same time, the proxy object can be constructed with a different flattening mode, so the tensor it mimics can have different mode than its mother object. Last but not least, it can have also a different index permutation mode. Algorithm 2.11 contains a full listing of major members of the *TFlatTensorProxyFor* with additional comments on their usage and the meaning of the parameters.

```
// This class allows different modes of the base tensor
// without data copying, however.
template < typename T >
class TFlatTensorProxyFor : public TFlatTensorFor< T >
{
protected:

        typedef typename TFlatTensorFor< T > BaseClass;

private:

        // this must be a reference since a proxy cannot live without an
           external object
        BaseClass & fMotherTensor;

public:

        // =====================================================


        /////////////////////////////////////////////////////////////
        // Class constructor
        /////////////////////////////////////////////////////////////
        //
        // INPUT:
        //                  motherTensor - a main object to which this one is
        //                      a proxy.
        //                  tensor_mode - mode of the proxy-tensor
        //                      The supplied tensor_mode can be DIFFERENT than
        //                      in the mother tensor !!
        //                  cyclicMode - determins cyclic mode of the tensor
        //                      which can be forward of backward (default)
        //
        // OUTPUT:
        //
        //
        // REMARKS:
        //
        //
        TFlatTensorProxyFor( BaseClass & motherTensor, int tensor_mode,
                             ECyclicMode cyclicMode = kForward );

        virtual ~TFlatTensorProxyFor;
```

```
protected:

        TFlatTensorProxyFor( const TFlatTensorProxyFor< T > & r );
        TFlatTensorProxyFor & operator = ( const TFlatTensorProxyFor< T > & r );


        // =======================================================

public:

        ///////////////////////////////////////////////////////////
        // This function returns a tensor's element at given index.
        ///////////////////////////////////////////////////////////
        //
        // INPUT:
        //                elemIdx - a vector with an index of an element
        //                          to read. Length of this vector must be
        //                          the same as a length of the indexVec provided
        //                          during construction of the tensor. Moreover,
        //                          each corresponding value should be positive
        //                          but less than that in the indexVec.
        //
        // OUTPUT:
        //                tensor element at position
        //
        // REMARKS:
        //
        //
        ElemAccessType    GetElement( const TensorElem_IndexVector & elemIdx ) const;

        ///////////////////////////////////////////////////////////
        // This function sets a value of an element of the tensor
        ///////////////////////////////////////////////////////////
        //
        // INPUT:
        //                elemIdx - a vector with an index of an element
        //                          to read. Length of this vector must be
        //                          the same as a length of the indexVec provided
        //                          during construction of the tensor. Moreover,
        //                          each corresponding value should be positive
        //                          but less than that in the indexVec.
        //                elemVal - value to be set
        //
        //
        // OUTPUT:
        //                none
        //
        // REMARKS:
        //
        //
        void SetElement( const TensorElem_IndexVector & elemIdx,
                         ElemAccessType elemVal );

public:
```

```
/////////////////////////////////////////////////////////
// This function sets a pixel at position (x,y)or (col,row)
// of this image.
/////////////////////////////////////////////////////////
//
// INPUT:
//                  xPixPosition - the horizontal (or column)
//                                 position of a pixel
//                  yPixPosition - the vertical (or row) position of a pixel
//                  value - a value to be set at pixel position
//
// OUTPUT:
//                  none
//
// REMARKS:
//                  From the OOP point of view this function should be.
//                  However, to avoid run-time panalty it is not,
//                  and the "value" parameter is passed by value. In derived
//                  classes this strategy can be overloaded.
//
void SetPixel( Dimension xPixPosition, Dimension yPixPosition,
               const T value ) const;


/////////////////////////////////////////////////////////
// This function gets a VALUE of a pixel at position
// (x,y) or (col,row) of this image.
/////////////////////////////////////////////////////////
//
// INPUT:
//                  xPixPosition - the horizontal (or column)
//                                 position of a pixel
//                  yPixPosition - the vertical (or row) position of a pixel
//
// OUTPUT:
//                  a copy of a pixel, of type T, from the given position
//
// REMARKS:
//                  The xPixPosition should span from 0 to max_columns-1, while
//                  the yPixPosition from 0 to max_rows-1.
//
//                  From the OOP point of view this function should be.
//                  However, to avoid run-time panalty it is not.
//
T GetPixel( Dimension xPixPosition, Dimension yPixPosition ) const;


protected:

    // Vice versa...
    bool ComputeMatrixOffset_BackwardCyclic( const DataMatrix_Index & inOffset,
               TensorElem_IndexVector & outElementPosition,
                                    int tensorMode ) const;
```

```
       bool ComputeMatrixOffset_ForwardCyclic( const DataMatrix_Index & inOffset,
                        TensorElem_IndexVector & outElementPosition,
                                          int tensor_mode ) const;



       ///////////////////////////////////////////////////////////////
       // This function converts a given matrix index into
       // a corresponding one if the tensor were in a different
       // mode.
       ///////////////////////////////////////////////////////////////
       //
       // INPUT:
       //
       //
       // OUTPUT:
       //
       //
       // REMARKS:
       //
       //
       void ConvertMatrixIndices( const DataMatrix_Index
                               & sourceTensorMatrixOffset,
                               const int sourceTensorMode,
                               DataMatrix_Index & destTensorMatrixOffset,
                               const int destTensorMode ) const;

};
```

Algorithm 2.11 Definition of the *TFlatTensorProxyFor* class (the most important members are shown).

All these twofold index transformations in the *TFlatTensorProxyFor* are possible thanks to the pair *ComputeMatrixOffset_BackwardCyclic* and *ComputeMatrixOffset_ForwardCyclic* metods which are able to recompute tensor-matrix indices in two ways and in two cyclic modes (backward and forward), as well as for different flattening modes. An index of an element of a tensor $\mathcal{T}$ of dimension P is given by a tuple $(i_1, i_2, \ldots, i_P)$ of P indices. This transforms into an offset

$$\textit{off} = (((i_1 n_2 + i_2) n_3) + \ldots) n_P + i_P, \tag{2.169}$$

of a linear data structure, where a tuple $(n_1, n_2, \ldots, n_P)$ defines dimensions of $\mathcal{T}$. As already mentioned, matrix representation needs two selected dimensions, say r for rows and c for columns, which can be computed from the tuple $(n_1, n_2, \ldots, n_P)$ as follows

$$(r, c) = \left(n_m, \prod_{z=1, z \neq m}^{P} n_z \right). \tag{2.170}$$

In the above, m denotes a flattening mode of the tensor. In tensor processing an element with a given index (a P-th dimensional tuple) needs to fit into such a matrix. In the tensor proxy object the problem is inversed, i.e. given a matrix offset *off* a corresponding tensor index tuple needs to be determined due to different modes of the tensors. This is done by successive division of the *off* in Equation (2.169) by n_p, starting from $p = P$ and up to $p = 1$, since for all $p \leq P$ it holds that $i_p < n_p$. As a result a series of indices i_p is obtained which are simply the residua of the successive divisions.

2.12.1.3 The *k*-mode Product of a Tensor and a Matrix

Definition 2.18   The k-*mode product* $\times_k$ of a tensor and a matrix

Given a tensor $\mathcal{T} \in \mathfrak{R}^{N_1 \times N_2 \times \dots N_P}$ and a matrix $\mathbf{M} \in \mathfrak{R}^{Q \times N_k}$, the k-mode product $\mathcal{T} \times_k \mathbf{M}$ is a tensor $\mathcal{S} \in \mathfrak{R}^{N_1 \times N_2 \times \dots N_{k-1} \times Q \times N_{k+1} \times \dots N_P}$ which elements are given as follows

$$\mathcal{S}_{n_1 n_2 \dots n_{k-1} q n_{k+1} \dots n_P} = (\mathcal{T} \times_k \mathbf{M})_{n_1 n_2 \dots n_{k-1} q n_{k+1} \dots n_P} = \sum_{n_k=1}^{N_k} t_{n_1 n_2 \dots n_{k-1} n_k n_{k+1} \dots n_P} m_{q n_k}. \qquad \Box \quad (2.171)$$

Thus, in the k-mode product $\mathcal{S} = \mathcal{T} \times_k \mathbf{M}$ an index n_k of $\mathcal{T}$ is actually exchanged by the (first, or row) index q of $\mathbf{M}$, so $\mathcal{S}$ has the same valence as $\mathcal{T}$ but a different range of the k-th index in general (which is now inherited from $\mathbf{M}$, rather than N_k). Let us recall, that although similar to the definition of the contracted product of two tensors, the order of indices in this case is, however, different (compare with Definition 2.3, p. 19).

The *k-mode* product can be equivalently expressed in terms of the flattened matrices $\mathbf{T}_{(k)}$ and $\mathbf{S}_{(k)}$, that is,

$$\text{if } \mathcal{S} = \mathcal{T} \times_k \mathbf{M}, \text{ then } \mathbf{S}_{(k)} = \mathbf{M}\mathbf{T}_{(k)}. \qquad (2.172)$$

The last equation gives us an interface to tensor products expressed in "pure" matrix notation. This property is used in our software framework [88]. Sometimes it is convenient to perform an inverse of tensor flattening (de-flattening, un-folding) which in our framework we will denote with a *negative value* of the k-mode, i.e. given a k-mode flattened version $\mathbf{T}_{(k)}$ of a tensor $\mathcal{T}$, its de-flattened (i.e. original) version is expressed as follows

$$\left(\mathbf{T}_{(k)}\right)_{(-k)} = \mathcal{T}. \qquad (2.173)$$

In other words, $\mathbf{T}_{(0)}$ denotes the original, i.e. un-folded, tensor $\mathcal{T}$. The popular mathematical packages, such as MATLAB®, also offer commands for tensor multiplication, folding and u-nfolding, as well as their basic decompositions. Other packages focused on advanced tensor decompositions are built upon this, such as the ones proposed by Bader and Kolda [122], or Cichocki *et al.* [25], to name a few. These are discussed in Appendix (A.7).

The following useful properties of the k-mode product of a tensor $\mathcal{T}$ and matrices $\mathbf{U}$ and $\mathbf{W}$ can be proved [96]

$$(\mathcal{T} \times_m \mathbf{U}) \times_n \mathbf{W} = (\mathcal{T} \times_n \mathbf{W}) \times_m \mathbf{U} = \mathcal{T} \times_m \mathbf{U} \times_n \mathbf{W}, \tag{2.174}$$

and

$$(\mathcal{T} \times_m \mathbf{U}) \times_m \mathbf{V} = \mathcal{T} \times_m (\mathbf{VU}), \tag{2.175}$$

where $\mathcal{T} \in \mathfrak{R}^{N_1 \times N_2 \times \ldots N_m \times \ldots N_n \times \ldots N_P}$, $\mathbf{U} \in \mathfrak{R}^{U \times N_m}$, $\mathbf{W} \in \mathfrak{R}^{W \times N_n}$, and $\mathbf{V} \in \mathfrak{R}^{V \times U}$ (U, W, V denote free indices).

The k-mode product of a tensor and a matrix can be also denoted without the symbol $\times_k$, i.e. with the help of Einstein's index notation (Section 2.4.1), as follows

$$\mathcal{S}_{n_1 n_2 \ldots n_{k-1} q n_{k+1} \ldots n_P} = (\mathcal{T} \times_k \mathbf{M})_{n_1 n_2 \ldots n_{k-1} q n_{k+1} \ldots n_P} = t_{n_1 n_2 \ldots n_{k-1} n_k n_{k+1} \ldots n_P} m_q^{n_k}. \tag{2.176}$$

It is interesting to observe that the SVD decomposition (2.158) of a matrix $\mathbf{T}$ can be expressed in terms of the k-mode products, as follows

$$\mathbf{T} = \mathbf{SVD}^T = \mathbf{V} \times_1 \mathbf{S} \times_2 \mathbf{D}, \tag{2.177}$$

since from Definition 2.18 we easily obtain that

$$\mathbf{V} \times_1 \mathbf{S} = \mathbf{SV}, \quad \text{and} \quad \mathbf{S} \times_2 \mathbf{D} = \mathbf{SD}^T. \tag{2.178}$$

Thus, the k-mode product allows omission of the transposition in the above formulas, leading to a clearer representation. In Equation (2.177) columns of $\mathbf{S}$ are associated with so called *1-mode* space, while columns of $\mathbf{D}$ are associated with the *2-mode* space.

A special case of the k-mode product of a tensor and a matrix is where the matrix is one-dimensional, i.e. it is a vector. In this case the resulting tensor is of one lower dimension than the original one, since each of its fibers is multiplied by the vector resulting in a single value. In other words, these are inner products of the fibers and the vector.

Finally, let us observe that the k-mode product of a tensor and a matrix allows composition of the summation of the elements in such a way that indices in the two multiplied objects change in the same direction, i.e. alongside the rows. Such an idea of array multiplication was developed in the the 1920s to 1930s by the Polish astronomer and mathematician Tadeusz Banachiewicz. His original *krakowian* calculus[10] was devised to simplify multiplication of the two-dimensional arrays of data [125]. In his definition of a product of krakowians, indices of the arrays of data follow simultaneously alongside their columns. This remains in contrast to the matrix multiplication defined by Hamilton and followed by Cayley in the 19th century. The main advantage of such a scheme is avoiding mixing the directions of indices. Apart from his novel definition of a product, Banachiewicz also introducedthe notion of a division of krakowians. Thanks to this, the algebra of krakowians allows a much more straightforward solution to the set of linear equations and to the problem of least-squares, especially if solved "by hand" or by a mechanical computing machine [125].

---

[10]This mathematical object was named after Kraków where Banachiewicz lived and worked.

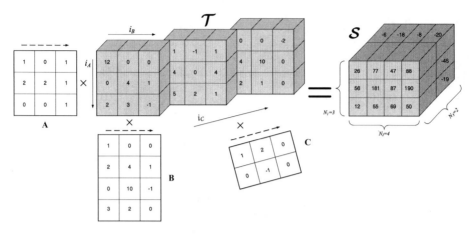

Figure 2.28   Visualization of a 1-mode, 2-mode, and 3-mode products of a $3 \times 3 \times 3$ tensor $\mathcal{T}$ with the three matrices **A**, **B** and **C**. The resulting tensor has dimensions $3 \times 4 \times 2$ imposed by dimensions of the matrices.

To exemplify the *k-mode* products, let us compute the result of the product of the $3 \times 3 \times 3$ tensor $\mathcal{T}$ with the three matrices **A**, **B** and **C**, as shown in Figure 2.28. Their product can be written as

$$\mathcal{S} = \mathcal{T} \times_1 \mathbf{A} \times_2 \mathbf{B} \times_3 \mathbf{C} = [(\mathcal{T} \times_2 \mathbf{B}) \times_1 \mathbf{A}] \times_3 \mathbf{C}.$$

The solid arrows in Figure 2.28 denote the direction of change of a corresponding index in the base tensor. At the same time, the dotted arrows show the direction of change of an index in a matrix used in the multiplication. The obtained innermost partial result of $\times_2$ is

$$\mathcal{S}^* = \left[\begin{array}{cccc|cccc|cccc} 12 & 24 & 0 & 36 & 1 & -1 & -11 & 1 & 0 & -2 & 2 & 0 \\ 0 & 17 & 39 & 8 & 4 & 12 & -4 & 12 & 4 & 48 & 100 & 32 \\ 2 & 17 & 31 & 12 & 5 & 19 & 19 & 19 & 2 & 8 & 10 & 8 \end{array} \right], \quad (2.179)$$

then, after the second product $\times_2$ we have

$$\mathcal{S}^{**} = \left[\begin{array}{cccc|cccc|cccc} 14 & 41 & 31 & 48 & 6 & 18 & 8 & 20 & 2 & 6 & 12 & 8 \\ 26 & 99 & 109 & 100 & 15 & 41 & -11 & 45 & 10 & 100 & 214 & 72 \\ 2 & 17 & 31 & 12 & 5 & 19 & 19 & 19 & 2 & 8 & 10 & 8 \end{array} \right], \quad (2.180)$$

and finally $\times_3$

$$\mathcal{S} = \left[\begin{array}{cccc|cccc} 26 & 77 & 47 & 88 & -6 & -18 & -8 & -20 \\ 56 & 181 & 87 & 190 & -15 & -41 & 11 & -45 \\ 12 & 55 & 69 & 50 & -5 & -19 & -19 & -19 \end{array} \right], \quad (2.181)$$

which part can also be seen in Figure 2.28. However, let us observe that the last product

$$\mathcal{S} = \mathcal{S}^{**} \times_3 \mathbf{C}, \tag{2.182}$$

can be expressed differently as

$$\mathcal{S} = \sum_{i=1}^{3} \mathcal{S}^{**}_{(:,:,i)} \times_3 \mathbf{C}_{(:,i)} = \sum_{i=1}^{3} \mathcal{S}^{**}_{(:,:,i)} \times_3 \mathbf{c}_i, \tag{2.183}$$

where $\mathcal{S}_{(:,:,i)}$ is an i-th slice of the tensor and $\mathbf{C}_{(:,i)} = \mathbf{c}_i$ is an i-th column of a matrix $\mathbf{C}$. The MATLAB® notation was used which sometimes is more legible if one wants to stress the fact of keeping only one index constant while all the others can change in their full extent. For instance, fixing the second index to 4 in (2.179) we obtain

$$\mathcal{S}^*(:,4,:) = \begin{bmatrix} 36 & 1 & 0 \\ 8 & 12 & 32 \\ 12 & 19 & 8 \end{bmatrix},$$

whereas from Equation (2.180) with the third index set to 2 we have

$$\mathcal{S}^{**}(:,:,2) = \begin{bmatrix} 6 & 18 & 8 & 20 \\ 15 & 41 & -11 & 45 \\ 5 & 19 & 19 & 19 \end{bmatrix},$$

or

$$\mathcal{S}^{**}(1,2,:) = \begin{bmatrix} 41 \\ 18 \\ 6 \end{bmatrix}.$$

It is worth noticing that the dimensionality of the objects obtained this way equals the number of variable indices, i.e. those denoted in the expression by the colon. Taking all these into account the results of our example are as follows

$$\mathcal{S} = \mathcal{S}^{**} \times_3 \mathbf{C} = \begin{bmatrix} 14 & 41 & 31 & 48 & 6 & 18 & 8 & 20 & 2 & 6 & 12 & 8 \\ 26 & 99 & 109 & 100 & 15 & 41 & -11 & 45 & 10 & 100 & 214 & 72 \\ 2 & 17 & 31 & 12 & 5 & 19 & 19 & 19 & 2 & 8 & 10 & 8 \end{bmatrix} \times_3 \begin{bmatrix} 1 & 2 & 0 \\ 0 & -1 & 0 \end{bmatrix}$$

$$= \begin{bmatrix} 14 & 41 & 31 & 48 \\ 26 & 99 & 109 & 100 \\ 2 & 17 & 31 & 12 \end{bmatrix} \times_3 \begin{bmatrix} 1 \\ 0 \end{bmatrix} + \begin{bmatrix} 6 & 18 & 8 & 20 \\ 15 & 41 & -11 & 45 \\ 5 & 19 & 19 & 19 \end{bmatrix} \times_3 \begin{bmatrix} 2 \\ -1 \end{bmatrix} + \begin{bmatrix} 2 & 6 & 12 & 8 \\ 10 & 100 & 214 & 72 \\ 2 & 8 & 10 & 8 \end{bmatrix} \times_3 \begin{bmatrix} 0 \\ 0 \end{bmatrix},$$

Let us observe also that in the last equation we encounter an outer product of a two-dimensional matrix with a one-dimensional vector, the result of which is a three-dimensional tensor, so, disregarding the last term, we obtain

$$
\mathbf{S} = \mathbf{S}^{**} \times_3 \mathbf{C} = \begin{bmatrix} 14 & 41 & 31 & 48 \\ 26 & 99 & 109 & 100 \\ 2 & 17 & 31 & 12 \end{bmatrix} \times_3 \begin{bmatrix} 1 \\ 0 \end{bmatrix} + \begin{bmatrix} 6 & 18 & 8 & 20 \\ 15 & 41 & -11 & 45 \\ 5 & 19 & 19 & 19 \end{bmatrix} \times_3 \begin{bmatrix} 2 \\ -1 \end{bmatrix}
$$

$$
= \left[\begin{array}{cccc|cccc} 14 & 41 & 31 & 48 & 0 & 0 & 0 & 0 \\ 26 & 99 & 109 & 100 & 0 & 0 & 0 & 0 \\ 2 & 17 & 31 & 12 & 0 & 0 & 0 & 0 \end{array} \right] + \left[\begin{array}{cccc|cccc} 12 & 36 & 16 & 40 & -6 & -18 & -8 & -20 \\ 30 & 82 & -22 & 90 & -15 & -41 & 11 & -45 \\ 10 & 38 & 38 & 38 & -5 & -19 & -19 & -19 \end{array} \right],
$$

$$
= \left[\begin{array}{cccc|cccc} 26 & 77 & 47 & 88 & -6 & -18 & -8 & -20 \\ 56 & 181 & 87 & 190 & -15 & -41 & 11 & -45 \\ 12 & 55 & 69 & 50 & -5 & -19 & -19 & -19 \end{array} \right],
$$

which is exactly the same as the result (2.181). The property (2.183) will be useful when considering the tensor decompositions in Section 2.12.4.

A special case of a k-mode product given by Definition 2.18 constitutes a k-mode product of a tensor and a vector. In this case we can still apply Definition 2.18, creating from a vector $\mathbf{v}$ of size N_k, a one-row matrix $\mathbf{V}$ of size $1 \times N_k$. A consequence is that the resulting tensor has the same dimensions as the input tensor except for the N_k which now becomes 1. In effect, the output tensor has one dimension less. For example, multiplying a 3D tensor with a vector, if their dimensions match, results in a 2D tensor, i.e. a matrix. It is natural that multiplying this 2D tensor again with a vector gives a vector, and finally if we continue this procedure we obtain a scalar value. Because of this for a k-mode product of a tensor with a vector it is common to use a slightly different operator which shows the dimensionality reduction of the tensor. This is summarized in the following definition.

Definition 2.19   The k-*mode product* $\bar{\times}_k$ of a tensor and a vector

Given a tensor $\mathcal{T} \in \mathfrak{R}^{N_1 \times N_2 \times \dots N_P}$ and a vector $\mathbf{v} \in \mathfrak{R}^{N_k}$, the k-mode product $\mathcal{T} \bar{\times}_k \mathbf{v}$ is a tensor $\mathcal{S} \in \mathfrak{R}^{N_1 \times N_2 \times \dots N_{k-1} \times N_{k+1} \times \dots N_P}$ which elements are given as follows

$$
\mathcal{S}_{n_1 n_2 \dots n_{k-1} n_{k+1} \dots n_P} = (\mathcal{T} \bar{\times}_k \mathbf{v})_{n_1 n_2 \dots n_{k-1} n_{k+1} \dots n_P} = \sum_{n_k=1}^{N_k} t_{n_1 n_2 \dots n_{k-1} n_k n_{k+1} \dots n_P} v_{n_k}. \qquad \square \qquad (2.184)
$$

Let us conclude with an example, which presents multiplication of the tensor from Figure 2.28 with three vectors created from the first rows of each of the matrices presented therein.

In the first stage of computation a 2D tensor, i.e. a matrix, $\mathcal{T}^*$ is obtained. This multiplied with the vector $\mathbf{b}$ gives a 1D tensor, a vector $\mathcal{T}^{**}$. Finally, multiplying two vectors we obtain a zero-order tensor, i.e. a scalar $\mathcal{S}$. It is also not a big surprise that we have just computed one of the values of the output tensor in Figure 2.28.

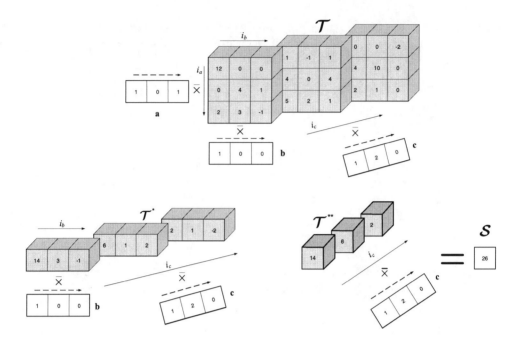

Figure 2.29   An example of a multiplication of a 3D tensor $\mathcal{T}$ with the three vectors **a**, **b** and **c**. Each multiplication lowers tensor dimensionality. The result is a scalar.

2.12.1.4   Ranks of a Tensor

This overview of the basic concepts of multilinear algebra will be concluded with a discussion on ranks of tensors. As alluded to previously, there is no unique way to define the rank of tensors in a general case. As a result, fundamental differences arise when trying to employ matrix analysis to tensors, specifically in contrast to the SVD decomposition of matrices. The definitions of ranks follow.

Definition 2.20   *Rank-1 of a tensor*

A *P-dimensional* tensor $\mathcal{T}$ has *rank* 1 if it is an outer product of *P* vectors $\mathbf{s}_1, \ldots, \mathbf{s}_P$, as follows

$$\mathcal{T} = \mathbf{s}_1 \circ \mathbf{s}_2 \circ \ldots \circ \mathbf{s}_P. \quad \square \tag{2.185}$$

Definition 2.21   *Rank of a tensor*

The rank of a *P-dimensional* tensor $\mathcal{T}$ is the minimal number of *rank-1* tensors that produce $\mathcal{T}$ in their linear combination. $\square$

Definition 2.22   The *k-rank* of a tensor

The *k-rank* R_k of a tensor

$$R_k = rank_k (\mathcal{T}) \tag{2.186}$$

is a *dimension* of the vector space spanned by its *k*-mode vectors (as defined in Definition 2.16).   □

Thus, a *P*-dimensional tensor has *P*, potentially different, *k*-ranks. The *k*-rank R_k of a tensor $\mathcal{T}$ can be expressed from its matrix representation. It is related to the *k*-mode flattening of $\mathcal{T}$, as follows

$$R_k = rank_k (\mathcal{T}) = rank \left(\mathbf{T}_{(k)} \right). \tag{2.187}$$

Unfortunately, different *k*-ranks of tensors of dimensions higher than two are not necessarily the same. It should be also noted that for tensors of an order higher than two, their rank is not necessarily equal to their *nk*-rank. This property holds only for matrices. Thus, tensor $\mathcal{T}$ (2.162) is of *rank-1*, since it is composed as an outer product of three vectors. Its rank is also three.

2.12.1.5   IMPLEMENTATION of Basic Operations on Tensors

The *TensorAlgebraFor* template class relies on the previously discussed *TTensorFor* and *TFlatTensorProxyFor*. The class diagram is shown in Figure 2.30. Template parameters define basic type of tensor elements, as well as a type used to store accumulated values which arise e.g. during multiplication and addition of the elements, etc. The *TensorAlgebraFor* class contains the basic algebraic tensor operations, as well as some of the input/output functions. The *Orphan_Kronecker_Of_Series* function member carries out Kronecker multiplication of a series of matrices, frequently used in tensor calculations, such as in formula (2.201). The Kronecker product can be carried out in the forward or backward modes, as discussed in Section 2.12.1.1. The *Orphan_PROJECTED_Tensor_From* function allows computation of the core tensor in accordance with Equation (2.199). On the other hand, the *Orphan_RECONSTRUCTED_Tensor_From* member allows computation of the tensor $\mathcal{T}$ from its core tensor and series $\mathcal{S}_k$, as shown in Equation (2.189).

The next auxiliary function *Orphan_TensorProjection_Onto_k_Minus_SubSpace* computes a multiple-product of a tensor with a series of matrices, such as in Equation (2.189), except a chosen one *k*-th index, as follows

$$\mathcal{S} = \mathcal{T} \times_1 \mathbf{S}_1 \times_2 \mathbf{S}_2 \ldots \times_{k-1} \mathbf{S}_{k-1} \times_{k+1} \mathbf{S}_{k+1} \ldots \times_P \mathbf{S}_P, \tag{2.188}$$

where $\mathcal{T} \in \mathfrak{R}^{N_1 \times \ldots N_k \times \ldots \times N_P}$ and $\mathcal{S} \in \mathfrak{R}^{R_1 \times \ldots R_{k-1} \times N_k \times R_{k+1} \times \ldots \times R_P}$. Although valences of the two tensor $\mathcal{T}$ and $\mathcal{S}$ in Equation (2.188) are the same as those denoted by *P*, the values of each dimension of the result tensor $\mathcal{S}$ follow consecutive numbers of rows $R_1, \ldots, R_P$ of each of the factor matrices $\mathbf{S}_i$, except for the one at the *k*-th index which is a copy of the *k*-th dimension of $\mathcal{T}$, i.e. N_k. The computation mode is chosen in accordance with the flattening mode of the

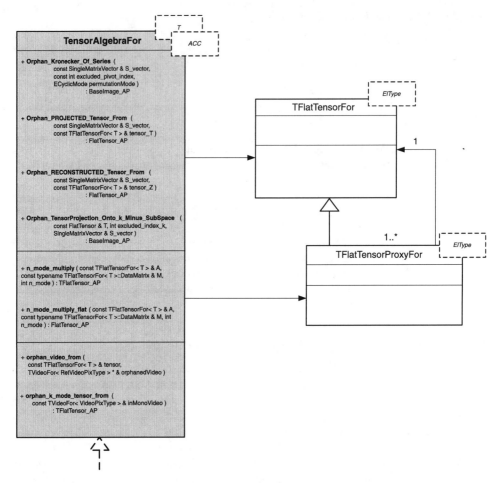

Figure 2.30 Class diagram of the *TensorAlgebraFor* template class which implements the basic algebraic operations.

tensor passed as a first argument to that function. Such products are frequently used in tensor decompositions. Also provided is a version of this function which accepts transposed matrices S_i in a series (2.188). A detailed description can be looked up in the code listing in Algorithm 2.12, while implementation details are in the supplied software platform [28].

The function *n_mode_multiply* carries out the *n*-mode tensor multiplication with a matrix, which proceeds in accordance with Equation (2.172). Its version *n_mode_multiply_flat* returns a tensor which is in the same mode *n* as the mode of this product (i.e. the *n_mode* parameter).

Algorithm 2.12 contains the most important members of the *TensorAlgebraFor* class.

```
////////////////////////////////////////////
// This class implements the basic
// algebraic operations on tensors.
// Template parameter defines type
```

```
// of tensor elements.
//
// T - stands for type of tensor elements
// ACC - is a type used to accumulate
// values in matrix mutliplications, etc.
// Usually it should be capable of storing
// a product of at least two T's.
//
///////////////////////////////////////////
template < typename T = double, typename ACC = double >
class TensorAlgebraFor
{
  public:

        typedef T                                       TensorElem_Type;

        typedef ACC                                     AccumulatorType;


        typedef typename TFlatTensorFor< T >            FlatTensor;

        typedef typename TFlatTensorProxyFor< T >        FlatTensorProxy;

        typedef typename std::auto_ptr< typename
                        TFlatTensorFor< T > >           FlatTensor_AP;

        typedef typename TFlatTensorFor< T >::DataMatrix        SingleMatrix;

        typedef typename safe_pointer_vector< SingleMatrix >
                SingleMatrixVector;


        typedef typename
                        std::auto_ptr< typename TFlatTensorFor< T >::BaseClass >
                        BaseImage_AP;

        typedef typename TFlatTensorFor< T >::ECyclicMode        ECyclicMode;

  public:


        //////////////////////////////////////////////////////////////////////
        // This function creates a tensor from a video.
        // The returned tensor has mode 0 and is in forward cyclic.
        //////////////////////////////////////////////////////////////////////
        //
        // INPUT:
        //                      inMonoVideo - in mono video
        //
        // OUTPUT:
        //                      tensor auto ptr
        //                      0 if failure
        //
        // REMARKS:
```

```
//
//
template < typename VideoPixType >
FlatTensor_AP orphan_k_mode_tensor_from( const TVideoFor< VideoPixType >
                                         & inMonoVideo );


/////////////////////////////////////////////////////////////////
// This function creates a tensor from a video.
// The returned tensor has mode 0 and is in forward cyclic.
/////////////////////////////////////////////////////////////////
//
// INPUT:
//                       inMonoVideo - in mono video
//
// OUTPUT:
//                       tensor auto ptr
//                       0 if failure
//
// REMARKS:
//
//
typename TensorAlgebraFor< T, ACC >::FlatTensor_AP
        orphan_k_mode_tensor_from( const ColorVideo & inColorVideo );


/////////////////////////////////////////////////////////////////
// This function creates and orphans a video object
// from the 3D tensor
/////////////////////////////////////////////////////////////////
//
// INPUT:
//                 tensor - a 3D tensor in the 0 flattening mode
//                 orphanedVideo - a pointer to the orphaned
//                         video object
// OUTPUT:
//                 none
//
// REMARKS:
//                 Number of rows in the output frame is equal
//                     to the 0-th index of the tensor
//                 Number of cols in a frame is from the 1-st index
//                 Number of frames is from the 2-nd index
//
template < typename RetVideoPixType >
bool orphan_video_from( const typename TFlatTensorFor< T > & tensor,
                   TVideoFor< RetVideoPixType > * & orphanedVideo );


/////////////////////////////////////////////////////////////////
// This function creates and orphans a color video object
// from the 4D tensor
/////////////////////////////////////////////////////////////////
//
```

```
            // INPUT:
            //                        tensor - a 4D tensor in the 0 flattening mode
            //                        orphanedVideo - a pointer to the orphaned
            //                            video object
            // OUTPUT:
            //                        true if ok
            //
            // REMARKS:
            //                        Number of rows in the output frame is equal
            //                            to the 0-th index of the tensor
            //                        Number of columns in a frame is from the 1-st index
            //                        Number of colors constitutes the 2-nd index
            //                        Number of frames is from the 3-rd index
            //
            bool orphan_video_from( const typename TFlatTensorFor< T > & tensor,
                        TVideoFor< ColorImage::PixelType > * & orphanedVideo );

    public:

            //////////////////////////////////////////////////////////////
            // Auxiliary function for computation of the common
            // Kronecker product from a series of matrices
            //////////////////////////////////////////////////////////////
            //
            // INPUT:
            //                        S_vector - a vector with the matrices to be
            //                            multiplied
            //                        excluded_pivot_index - an index of the first
            //                            matrix that starts the computations
            //                        permutationMode - forward or backward permutation
            //                            mode which affects the order of
            //                                multiplications
            //
            // OUTPUT:
            //                        a matrix which is an outcome of the series
            //                        of the Kronecker multiplications
            //
            // REMARKS:
            //
            //
            BaseImage_AP Orphan_Kronecker_Of_Series( const SingleMatrixVector
                    & S_vector, const int excluded_pivot_index, ECyclicMode
                    permutationMode );

    public:

            //////////////////////////////////////////////////////////////
            // Computes a projection of a given tensor onto the space
            // spanned by the series of matrices S, i.e.:
            //
            //                        T       T                 T
            //            Z = T   x   S   x   S   x   ...   x   S
            //                        1   1   2   2               P   P
            //
```

```
// Returns a tensor in the corresponding size and mode.
//////////////////////////////////////////////////////////////
//
// INPUT:
//                    S_vector - a vector with the matrices to be
//                               multiplied
//                    tensor_T - the ''original'' tensor T
//
// OUTPUT:
//                    The core tensor of size
//                        col(S) x col(S) x ... x col(S)
//                            1        2           P
//                    and the same mode as the tensor_T
//
// REMARKS:
//
// Forward (preferred) order is as follows:
// Z(k) = S'(k) * T(k) * [ S(k-1) x S(k-2) x ... x S(2) x S(1) x ... x
//                         S(k+1) ]
//
// Backward order is as follows:
// Z(k) = S'(k) * T(k) * [ S(k+1) x S(k+2) x ... x S(1) x S(2) x ... x
//                         S(k-1) ]
//
FlatTensor_AP Orphan_PROJECTED_Tensor_From( const SingleMatrixVector &
                    S_vector, const TFlatTensorFor< T > & tensor_T );


//////////////////////////////////////////////////////////////
// This function computes a tensor T given the tensor Z
// and a series of P matrices {S}, as follows
//
//              T = Z  x  S  x  S  x  ...  x  S
//                  1  1  2  2             P  P
//
//////////////////////////////////////////////////////////////
//
// INPUT:
//                    S_vector - a vector with the matrices to be
//                               multiplied
//                    tensor_Z - the core tensor of the HOSVD
//                               decomposition
//
// OUTPUT:
//                    The product tensor
//
// REMARKS:
//
// Forward (preferred) order is as follows:
// T(k) = S(k) * Z(k) * [ S(k-1) x S(k-2) x ... x S(k) x
//                        S(k-1) x ... x S(k+1) ]'
//
// Backward order is as follows:
// T(k) = S(k) * Z(k) * [ S(k+1) x S(k+2) x ... x S(1) x S(2) x ... x
//                        S(k-1) ]'
```

```
//
FlatTensor_AP Orphan_RECONSTRUCTED_Tensor_From(
        const SingleMatrixVector & S_vector, const TFlatTensorFor< T > &
        tensor_Z );


//////////////////////////////////////////////////////////////////////
//  This function computes the following tensor-matrices product:
//
//         S = T x S(1) ... x S(k-1) x S(k+1) ... x S(P)
/              1          k-1        k+1          P
//
//////////////////////////////////////////////////////////////////////
//
// INPUT:
//                 tensor_T - tensor to be multiplied
//                 excluded_index_k - excluded index (multiplication mode)
//                 S_vector - a complete series of vectors
//
// OUTPUT:
//                 FlatTensor_AP if success,
//                 0 otherwise
// REMARKS:
//
//       In the forward mode this is equivalent to the following
//          matrix product
//       (crosses below denote the Kronecker products):
//
//       S(k) = T(k) * [ S(k-1) x S(k-2) x ... x S(2) x S(1) x ... x
//                      S(k+1) ]'
//
//       in backward
//       S(k) = T(k) * [ S(k+1) x S(k+2) x ... x S(1) x S(2) x ... x
//                      S(k-1) ]'
//
//       The computation mode is chosen in accordance with the mode
//          of the tensor T.
//
//       Here we assume that the vector S is COMPLETE,
//       i.e. it contains also the matrix
//       which actually does not go into multiplication
//          (at index excluded_index_k).
//       The matrix at index excluded_index_k is ignored but
//       ** MUST be present in the series **!
//
FlatTensor_AP          Orphan_TensorProjection_Onto_k_Minus_SubSpace(
        const FlatTensor & T, int excluded_index_k, SingleMatrixVector
        & S_vector );



//////////////////////////////////////////////////////////////////////
// This function does n-mode multiplication of a tensor
// with a matrix. Creates and returns the result n-mode
// tensor.
```

```
//
//                         B = A   x     M
//                                 (n)
//
// in the "flat" matricized form
//
//                         B     =  M    A
//                         (n)            (n)
//
// Remember: mode of the returned tensor is equal to n_mode !!!
//
//////////////////////////////////////////////////////////////////
//
// INPUT:
//                         A - the input tensor
//                         M - a matrix
//                         n_mode - mode of multiplication
//
// OUTPUT:
//                         created and orphaned tensor, or
//                         0 if multiplication cannot be done
//
// REMARKS:
//                         Operation can fail if the provided n_mode
//                         is greater than the mode of a tensor A or
//                         index at n_mode of A is different
//                         than number of columns in the matrix M.
//
FlatTensor_AP n_mode_multiply_flat( const TFlatTensorFor< T > & A,
          const typename TFlatTensorFor< T >::DataMatrix & M,
          int n_mode );


//////////////////////////////////////////////////////////////////
// This function does n-mode multiplication of a tensor
// with a matrix. Creates and returns the result tensor
//
//                         B = A   x     M
//                                 (n)
//
// which is in the SAME mode and permutation cycle as the
// input tensor A. In other words, it does not change
// the mode of the tensors, therefore it is pure tensor
// multiplication times a matrix.
//////////////////////////////////////////////////////////////////
//
// INPUT:
//                         A - the input tensor
//                         M - a matrix
//                         n_mode - mode of multiplication
//
// OUTPUT:
//                         created and orphaned tensor, or
//                         0 if multiplication cannot be done
//
```

```
                // REMARKS:
                //                        Operation can fail if the provided n_mode
                //                        is greater than the mode of a tensor A or
                //                        index at n_mode of A is different
                //                        than number of columns in the matrix M.
                //
                FlatTensor_AP n_mode_multiply( const TFlatTensorFor< T > & A,
                        const typename TFlatTensorFor< T >::DataMatrix & M,
                        int n_mode );


                // =====================================================================



                ///////////////////////////////////////////////////////////////
                // This function returns an orphaned copy of the input
                // n-mode flat tensor, possibly in a different mode.
                // Thus, it can be used for mode conversions of the flat
                // tensors.
                ///////////////////////////////////////////////////////////////
                //
                // INPUT:
                //                        A - the input flat tensor
                //                        k_mode - a mode of the output tensor
                //                        ret_tensor_cyclic_mode - requested cyclic
                //                                mode of the returned tensor
                //
                // OUTPUT:
                //                        New tensor or
                //                        0 if error
                //
                // REMARKS:
                //
                //
                FlatTensor_AP orphan_k_mode_tensor_from( const typename
                                TFlatTensorFor< T > & A,
                                int k_mode, ECyclicMode ret_tensor_cyclic_mode );

                // The same as above but returns a tensor in the same cyclic mode
                // as tensor A
                FlatTensor_AP orphan_k_mode_tensor_from( const typename
                        TFlatTensorFor< T > & A, int k_mode );

};
```

Algorithm 2.12   The *TensorAlgebraFor* template class and its most important members.

There are two sets of format conversion functions, *orphan_video_from* and *orphan_k_mode_tensor_from*, which allow direct interfacing between the 3D tensors (and in the 0-*th* flattening mode) and video objects. These have been shown to be useful in processing video data with tensor methods. These versions with description are shown in Algorithm 2.12.

Finally, to show typical tensor operations in the described framework, let us analyze implementation details of the *Orphan_PROJECTED_Tensor_From* function, listed in Algorithm 2.13.

```
1    ////////////////////////////////////////////////////////////
2    // Computes a projection of a given tensor onto the space
3    // spanned by the series of matrices S, i.e.:
4    //
5    //                     T     T                 T
6    //          Z = T  x  S  x  S  x  ...  x  S
7    //               1     1   2   2             P   P
8    //
9    // Returns a tensor in the corresponding size and mode.
10   ////////////////////////////////////////////////////////////
11   //
12   // INPUT:
13   //         S_vector - a vector with the matrices to be
14   //                 multiplied
15   //         tensor_T - the ''original'' tensor T
16   //
17   // OUTPUT:
18   //         The core tensor of size
19   //                         col(S) x col(S) x ... x col(S)
20   //                               1        2              P
21   //         and the same mode as the tensor_T
22   //
23   // REMARKS:
24   //
25   // Forward (preferred) order is as follows:
26   // Z(k) = S'(k) * T(k) * [ S(k-1) x S(k-2) x ... x S(2) x S(1) x ... x S(k+1) ]
27   //
28   // Backward order is as follows:
29   // Z(k) = S'(k) * T(k) * [ S(k+1) x S(k+2) x ... x S(1) x S(2) x ... x S(k-1) ]
30   //
31   template < typename T, typename ACC >
32   typename TensorAlgebraFor< T, ACC >::FlatTensor_AP
33       TensorAlgebraFor< T, ACC >::Orphan_PROJECTED_Tensor_From(
34           const typename TensorAlgebraFor< T, ACC >::SingleMatrixVector &
35           S_vector,   const TFlatTensorFor< T > & tensor_T )
36   {
37       const int mode = tensor_T.GetTensorMode();
38       REQUIRE( mode < S_vector.size());
39
40       ECyclicMode cyclicMode = tensor_T.GetCyclicMode();
41       REQUIRE( cyclicMode == TFlatTensorFor< T >::kForward
42                       || cyclicMode == TFlatTensorFor< T >::kBackward );
43
44       BaseImage_AP common_Kronecker( Orphan_Kronecker_Of_Series(
45                               S_vector, mode, cyclicMode ) );
46
47       REQUIRE( common_Kronecker.get() != 0 );
48       if( common_Kronecker.get() == 0 )
```

```
49                      return FlatTensor_AP( 0 );

50

51          REQUIRE( S_vector[ mode ] != 0 );

52

53          BaseImage_AP tmp(

54                  Orphan_Mult_Matrix_With_FIRST_Transposed< TFlatTensorFor< T >::

55                  TensorElem_Type, TFlatTensorFor< T >::TensorElem_Type >

                    ( * S_vector[ mode ], tensor_T ) );

56          REQUIRE( tmp.get() != 0 );

57

58

59          ////////////////////////////////////////////////////////////////////

60          // Create an output tensor

61

62          TFlatTensorFor< T >::TensorElem_IndexVector          B_IndexVector;

63

64          Fill_Index_From_Transposed( B_IndexVector, S_vector );

65

66          FlatTensor_AP B( new TFlatTensorFor< T >(

67                  B_IndexVector, mode, tensor_T.GetCyclicMode() ) );

68

69          bool multResult = ::Mult_Matrix< TFlatTensorFor< T >::TensorElem_Type,

70                  TFlatTensorFor< T >::TensorElem_Type >( * B, * tmp,

                    * common_Kronecker );

71          REQUIRE( multResult == true );

72

73          REQUIRE( B->ClassInvariant() );

74

75          return B;

76      }
```

Algorithm 2.13 The *Orphan_PROJECTED_Tensor_From* function.

The function *Orphan_PROJECTED_Tensor_From* (Algorithm 2.13) carries out tensor multiplication with a series of matrices, in accordance with the formula (2.199). Despite a slightly obscure syntax due to a number of long template-like names, the calls to some key functions can be easily distinguished in Algorithm 2.13. Processing starts with a call to the *Orphan_Kronecker_Of_Series* (line <44>) to compute a Kronecker product matrix of a series of supplied matrices, as discussed on p. 115; Then *Orphan_Mult_Matrix_With_FIRST_Transposed* (line <54>) is invoked to compute a product of a matrix $\mathbf{S}_k^T$ with the input tensor. Finally, the output core tensor is created (line <66>) in a form of the auto-pointer *TFlatTensor_AP*, i.e. it is a special object which does not need to be explicitly deleted. It contains an output core tensor after multiplication with the previously computed Kronecker product. This multiplication is done in line <69> as an ordinary matrix product thanks to the internal matrix-like representation of each tensor. Finally, in line <73> the internal consistency of the output tensor is checked for debugging purposes (more discussion on applied debugging techniques can be found in chapter 13 of book [15]).

2.12.2   Higher-Order Singular Value Decomposition (HOSVD)

As alluded to previously, analogously to the SVD decomposition of the matrices, there exists a *P-th* order decomposition of the P dimensional tensor $\mathcal{T}$. This is governed by the following theorem [96].

Theorem 2.3   Higher-Order SVD (HOSVD) decomposition of the P-th order tensors. Each tensor $\mathcal{T} \in \mathfrak{R}^{N_1 \times N_2 \times \ldots N_m \times \ldots N_n \times \ldots N_P}$ can be decomposed as

$$\mathcal{T} = \mathcal{Z} \times_1 \mathbf{S}_1 \times_2 \mathbf{S}_2 \ldots \times_P \mathbf{S}_P, \qquad (2.189)$$

where

$$\mathbf{S}_k = \left[\mathbf{s}_k^{(1)} \mathbf{s}_k^{(2)} \ldots \mathbf{s}_k^{(N_k)} \right] \qquad (2.190)$$

denotes *a unitary mode matrix*[11] of dimensions $N_k \times N_k$, composed of N_k column vectors $\mathbf{s}_k^{(i)}$, spanning the column space of the matrix $\mathbf{T}_{(k)}$ obtained from the *k-mode* flattening of $\mathcal{T}$; $\mathcal{Z} \in \mathfrak{R}^{N_1 \times N_2 \times \ldots N_m \times \ldots N_n \times \ldots N_P}$ is *a core tensor* of the same dimensions as $\mathcal{T}$. It satisfies the following conditions [115]:

1. Two subtensors $\mathcal{Z}_{n_k=a}$ and $\mathcal{Z}_{n_k=b}$, obtained by fixing index n_k to a and b, respectively, are orthogonal in accordance with Equation (2.18), i.e.

$$\mathcal{Z}_{n_k=a} \cdot \mathcal{Z}_{n_k=b} = 0, \qquad (2.191)$$

   for all possible values of k, and $a \neq b$.
2. All subtensors can be ordered according to their Frobenius norms (2.19), as follows:

$$\left\| \mathcal{Z}_{n_k=1} \right\| \geq \left\| \mathcal{Z}_{n_k=2} \right\| \geq \ldots \geq \left\| \mathcal{Z}_{n_k=N_P} \right\| \geq 0, \qquad (2.192)$$

   for all possible values of k.   □

Let us observe that the Frobenius norm

$$\left\| \mathcal{Z}_{n_k=a} \right\| = \sigma_a^k \qquad (2.193)$$

is the *a-mode* singular value of $\mathcal{T}$. Each *i-th* vector of the matrix $\mathbf{S}_k$ is the *i-th k-mode* singular vector.

Visualization of the HOSVD decomposition for 3D tensors is depicted in Figure 2.31. Let us note that if the last column of $\mathbf{S}_1$ is removed (indexed at N_r), then a reconstruction error is introduced which is equal to the Frobenius norm of the 3D subtensor of $\mathcal{Z}$, also at index N_r. This happens since the removed column would multiply with that subtensor.

---

[11] Let us recall that for a unitary $n \times n$ real valued matrix $\mathbf{S}$ it holds that $\mathbf{SS}^T = \mathbf{S}^T \mathbf{S} = \mathbf{I}_n$ (for complex valued matrices transposition should be replaced by conjugation).

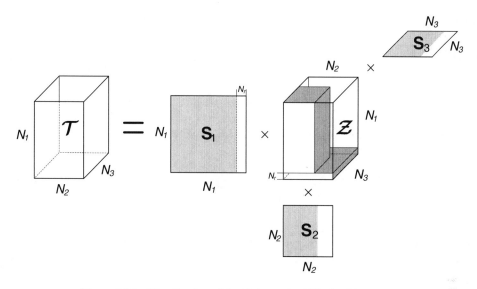

Figure 2.31   Visualization of the higher-order SVD for 3D tensors.

However, contrary to the matrix SVD, the core tensor is usually a full tensor with real[12] entries which can be positive or negative. The role of the singular values from SVD are as Frobenius norms of its $(P\text{-}1)$-*dimensional* subtensors (see Figure 2.31). By this token, they are real and nonnegative. The core tensor in the above orchestrates the influence of the mode matrices $\mathbf{S}_{(k)}$.

It is worth noticing that the highest index r, for which $\|\mathcal{Z}_{n_k=r}\| \geq 0$ in (2.192) is equal to the k-rank of the tensor $\mathcal{T}$, i.e.

$$R_{(k)}(\mathcal{T}) = r. \tag{2.194}$$

Assuming decomposition (2.189) of a tensor $\mathcal{T}$, singular values (2.193) provide a notion of an energy of this tensor in terms of the Frobenius norm, as follows

$$\|\mathcal{T}\|^2 = \sum_{a=1}^{R_1} \left(\sigma_a^1\right)^2 = \cdots = \sum_{a=1}^{R_P} \left(\sigma_a^P\right)^2 = \|\mathcal{Z}\|^2, \tag{2.195}$$

where R_k denotes a k-mode rank of $\mathcal{T}$. Thus, k-mode singular vectors correspond to directions of extreme k-mode energy, whereas squared k-mode singular values correspond to an extreme energy value. In other words, the ordering (2.192) means that the energy (or mass) of the core tensor $\mathcal{Z}$ is concentrated in its "corner" with indices $(1,1,\dots,1)$. This makes possible application of the HOSVD for data compression [126]. The idea is basically the same as for the PCA decomposition, i.e. the core tensor $\mathcal{Z}$ is shrunk to include only the components which

---

[12]These can also be complex if complex tensors are decomposed.

convey the "majority" of the energy in accordance with the ordering (2.192). Equation (2.189) leads to the following approximation

$$\mathcal{T} = \mathcal{Z} \times_1 \mathbf{S}_1 \times_2 \mathbf{S}_2 \ldots \times_P \mathbf{S}_P \approx \hat{\mathcal{Z}} \times_1 \hat{\mathbf{S}}_1 \times_2 \hat{\mathbf{S}}_2 \ldots \times_P \hat{\mathbf{S}}_P = \hat{\mathcal{T}}, \qquad (2.196)$$

where $\hat{\mathcal{Z}} \in \mathfrak{R}^{n_1 \times n_2 \times \ldots n_m \times \ldots n_n \times \ldots n_P}$ for which it holds that $n_k \leq N_k$, and $\mathbf{S}_k$ has dimensions $N_k \times n_k$. The error of the approximation (2.196) can be measured by the norm of the omitted part of the core tensor $\mathcal{Z}$. Thus, from Equation (2.195) it follows that

$$\left\| \mathcal{T} - \hat{\mathcal{T}} \right\|^2 = \sum_{a=k_i}^{R_i} \left(\sigma_a^i \right)^2, \qquad (2.197)$$

where $1 \leq i \leq P$, $n_i < k_i \leq N_i$, and σ_a^i are given by (2.193). This compression method is called HOSVD *truncation*. However, higher compression ratios at the same reconstruction error can be achieved with the best rank-$(R_1, R_2, \ldots, R_P)$ tensor decomposition, as will be discussed in Section 2.12.7.

2.12.3   Computation of the HOSVD

As shown by Lathauwer [96], computation of the HOSVD can be obtained by successive application of the SVD decompositions of the flattened representations of a given tensor. This follows directly from the proof of Theorem 2.3. Based on this observation, the HOSVD of a *P-dimensional* tensor $\mathcal{T}$ is summarized by the following algorithm [11, 96]:

1. **For each $k = 1, \ldots, P$ do:**
   (a) Flatten tensor $\mathcal{T}$ to obtain $\mathbf{T}_{(k)}$;
   (b) Compute $\mathbf{S}_k$ from the SVD decomposition of the flattening matrix $\mathbf{T}_{(k)}$:

$$\mathbf{T}_{(k)} = \mathbf{S}_k \mathbf{V}_k \mathbf{D}_k^T \qquad (2.198)$$

2. Compute the core tensor from the formula:

$$\mathcal{Z} = \mathcal{T} \times_1 \mathbf{S}_1^T \times_2 \mathbf{S}_2^T \ldots \times_P \mathbf{S}_P^T \qquad (2.199)$$

Algorithm 2.14   Computation of the HOSVD for tensors.

The last equation can be easily verified by substitution of above equation into Equation (2.189) and taking into account properties (2.174) and (2.175), as follows:

$$\mathcal{T} = \mathcal{Z} \times_1 \mathbf{S}_1 \times_2 \mathbf{S}_2 \ldots \times_P \mathbf{S}_P = \left(\mathcal{T} \times_1 \mathbf{S}_1^T \times_2 \mathbf{S}_2^T \ldots \times_P \mathbf{S}_P^T \right) \times_1 \mathbf{S}_1 \times_2 \mathbf{S}_2 \ldots \times_P \mathbf{S}_P$$

$$= \mathcal{T} \times_1 \mathbf{S}_1^T \times_1 \mathbf{S}_1 \times_2 \mathbf{S}_2^T \times_2 \mathbf{S}_2 \ldots \times_P \mathbf{S}_P^T \times_P \mathbf{S}_P \qquad (2.200)$$

$$= \mathcal{T} \times_1 \left(\mathbf{S}_1^T \mathbf{S}_1 \right) \times_2 \left(\mathbf{S}_2^T \mathbf{S}_2 \right) \ldots \times_P \left(\mathbf{S}_P^T \mathbf{S}_P \right) = \mathcal{T} \times_1 \mathbf{I} \times_2 \mathbf{I} \ldots \times_P \mathbf{I} = \mathcal{T}.$$

Thus, HOSVD reduces to the P computations of the SVD decompositions of the flattening matrices, each of dimension $N_k \times N_1 N_2 \ldots N_{k-1} N_{k+1} \ldots N_P$. However, in practice direct application of Algorithm 2.14 might be time consuming since each SVD is computed over all elements of the input tensor $\mathcal{T}$. Thus, the computational complexity of HOSVD is $P \cdot O(\text{SVD})$. Nevertheless, we can greatly improve on this basic computation, as will be discussed.

The series of k-mode products, such as in Equation (2.189), can be equivalently represented in matrix notation after tensor flattening (2.172), as follows

$$\mathbf{T}_{(k)} = \mathbf{S}_k \mathbf{Z}_{(k)} \left[\mathbf{S}_{k+1} \otimes \mathbf{S}_{k+2} \otimes \ldots \otimes \mathbf{S}_P \otimes \mathbf{S}_1 \otimes \mathbf{S}_2 \otimes \ldots \otimes \mathbf{S}_{k-1} \right]^T , \qquad (2.201)$$

where $\otimes$ denotes the Kronecker product [17, 31, 127]. The order of indices in Equation (2.201) strictly follows the order of indices assumed for tensor flattening (Section 2.12.1.1). Let us noe that the series index, such as $_k$, is denoted without the parenthesis. On the other hand, the notation with parenthesis, such as $_{(k)}$, refers to the k-mode tensor flattening. This provides us with a convenient link to the matrix representation of tensor equations.

Let us recall that the Kronecker product of two matrices $\mathbf{A} \in \mathfrak{R}^{I \times J}$, and $\mathbf{B} \in \mathfrak{R}^{M \times N}$ is a $\mathbf{C} \in \mathfrak{R}^{IM \times JN}$ matrix defined as follows [127]:

$$\mathbf{A}_{I \times J} \otimes \mathbf{B}_{M \times N} = \mathbf{C}_{(IM) \times (JN)} = \begin{bmatrix} a_{11}\mathbf{B} & \cdots & a_{1J}\mathbf{B} \\ \vdots & \vdots & \vdots \\ a_{I1}\mathbf{B} & \cdots & a_{IJ}\mathbf{B} \end{bmatrix}_{(IM) \times (JN)} \qquad (2.202)$$

$$= \begin{bmatrix} \mathbf{a}_1 \otimes \mathbf{b}_1 & \mathbf{a}_1 \otimes \mathbf{b}_2 & \cdots & \mathbf{a}_J \otimes \mathbf{b}_{N-1} & \mathbf{a}_J \otimes \mathbf{b}_N \end{bmatrix},$$

where $\mathbf{a}_i$, $\mathbf{b}_i$ denotes column vectors of the matrices $\mathbf{A}$ and $\mathbf{B}$, respectively. The Kronecker product can be computed for matrices of any size. For example for the following matrices

$$\mathbf{A}_{2 \times 2} = \begin{bmatrix} 0 & -2 \\ 1 & -1 \end{bmatrix}, \ \mathbf{B}_{1 \times 3} = \begin{bmatrix} 1 & 2 & -7 \end{bmatrix} \qquad (2.203)$$

their Kronecker product is

$$\mathbf{A}_{2 \times 2} \otimes \mathbf{B}_{1 \times 3} = \mathbf{C}_{2 \times 6} = \begin{bmatrix} 0 & 0 & 0 & -2 & -4 & 14 \\ 1 & 2 & -7 & -1 & -2 & 7 \end{bmatrix}. \qquad (2.204)$$

However,

$$\mathbf{B}_{1 \times 3} \otimes \mathbf{A}_{2 \times 2} = \mathbf{C}_{2 \times 6} = \begin{bmatrix} 0 & -2 & 0 & -4 & 0 & 14 \\ 1 & -1 & 2 & -2 & -7 & 7 \end{bmatrix}.$$

This means that in general the Kronecker product is *not* commutative, i.e. $\mathbf{A} \otimes \mathbf{B} \neq \mathbf{B} \otimes \mathbf{A}$. However, it is associative, i.e. $(\mathbf{A} \otimes \mathbf{B}) \otimes \mathbf{C} = \mathbf{A} \otimes (\mathbf{B} \otimes \mathbf{C})$. Let us also observe that the Kronecker product of two vectors is also a vector, for example

$$\mathbf{U} = \begin{bmatrix} 4 \\ 5 \end{bmatrix}, \quad \mathbf{V} = \begin{bmatrix} -11 \\ 2 \\ 0 \end{bmatrix} \tag{2.205}$$

results in

$$\mathbf{U} \otimes \mathbf{V} = \begin{bmatrix} -44 \\ 8 \\ 0 \\ -55 \\ 10 \\ 0 \end{bmatrix}, \quad \text{but} \quad \mathbf{V} \otimes \mathbf{U} = \begin{bmatrix} -44 \\ -55 \\ 8 \\ 10 \\ 0 \\ 0 \end{bmatrix}. \tag{2.206}$$

In this respect the Kronecker product is also different from the tensor product. Finally, let us summarize the following useful properties of the Kronecker product:

$$(\mathbf{A} \otimes \mathbf{B})(\mathbf{C} \otimes \mathbf{D}) = \mathbf{A}\mathbf{C} \otimes \mathbf{B}\mathbf{D}, \tag{2.207}$$

$$(\mathbf{A} \otimes \mathbf{B})^T = \mathbf{A}^T \otimes \mathbf{B}^T, \tag{2.208}$$

$$(\mathbf{A} \otimes \mathbf{B})^{-1} = \mathbf{A}^{-1} \otimes \mathbf{B}^{-1}. \tag{2.209}$$

Comparing the last two equations (2.208) and (2.209) with the transposition and inversion of a "classical" matrix product let us notice a difference in the order of components – the Kronecker product does not reverse the order of the matrices.

From Equation (2.192) we notice that due to the orthogonality of $\mathbf{S}_k$, $\mathbf{T}_{(k)}$ *is spanned by the column vectors* of $\mathbf{S}_k$. The assumed order in the Kronecker product in Equation (2.201) follows the backward permutation of indices (see Figure 2.25(b)). In the forward mode of permutations Equation (2.201) transforms as follows

$$\mathbf{T}_{(k)} = \mathbf{S}_k \mathbf{Z}_{(k)} \left[\mathbf{S}_{k-1} \otimes \mathbf{S}_{k-2} \otimes \ldots \otimes \mathbf{S}_1 \otimes \mathbf{S}_P \otimes \mathbf{S}_{P-1} \otimes \ldots \otimes \mathbf{S}_{k+1} \right]^T. \tag{2.210}$$

Since in image processing we prefer the latter mode of permutations, in our implementations the last form will be predominent, although the two are equivalent, as discussed in Section 2.12.1.1.

As alluded to previously, since S_k are orthogonal, computation of the core tensor $\mathcal{Z}$ in Equation (2.199) can be expressed in backward permutation mode as

$$\mathbf{Z}_{(k)} = \mathbf{S}_k^T \mathbf{T}_{(k)} \, [\mathbf{S}_{k+1} \otimes \mathbf{S}_{k+2} \otimes \ldots \otimes \mathbf{S}_P \otimes \mathbf{S}_1 \otimes \mathbf{S}_2 \otimes \ldots \otimes \mathbf{S}_{k-1}]. \qquad (2.211)$$

Similarly, for the forward mode we obtain

$$\mathbf{Z}_{(k)} = \mathbf{S}_k^T \mathbf{T}_{(k)} \, [\mathbf{S}_{k-1} \otimes \mathbf{S}_{k-2} \otimes \ldots \otimes \mathbf{S}_1 \otimes \mathbf{S}_P \otimes \mathbf{S}_{P-1} \otimes \ldots \otimes \mathbf{S}_{k+1}]. \qquad (2.212)$$

For an example let us compute HOSVD of the tensor $\mathcal{T}$ shown in Figure 2.4. Its three flattening matrices have already been computed in (2.163)–(2.165). Their $\mathbf{S}$ matrices of the SVD decomposition were computed in our software framework, as follows

$$\mathbf{S}_1 = \begin{bmatrix} -0.408716 & 0.8942150 & -0.182567 \\ -0.828351 & -0.447436 & -0.337099 \\ -0.383126 & 0.013452 & 0.923598 \end{bmatrix}, \qquad (2.213)$$

$$\mathbf{S}_2 = \begin{bmatrix} -0.884749 & 0.450728 & -0.118585 \\ -0.458141 & -0.887805 & 0.043689 \\ -0.085589 & 0.092983 & 0.991982 \end{bmatrix}, \qquad (2.214)$$

$$\mathbf{S}_3 = \begin{bmatrix} -0.820518 & 0.552425 & 0.146892 \\ -0.259612 & -0.131196 & -0.956759 \\ -0.509266 & -0.823173 & 0.251066 \end{bmatrix}. \qquad (2.215)$$

Now, from Equation (2.199) the core tensor $\mathcal{Z}$ is computed

$$\mathbf{Z}_{(1)} = \begin{bmatrix} -11.079 & -2.982 & -2.3368 & -3.464 & -3.7869 & 4.1824 & 0.39179 & -0.28565 & 3.5067 \\ 4.893 & -8.0434 & -1.4849 & -6.8268 & 1.0023 & 1.3861 & 2.5922 & 0.55614 & 0.071836 \\ 0.80336 & -0.19291 & 3.5808 & 0.14236 & -2.7883 & 0.83197 & 1.0703 & -0.68003 & 0.7225 \end{bmatrix}.$$

$$(2.216)$$

The same $\mathbf{Z}_{(1)}$ is obtained from (2.211) which for this example has the following form

$$\mathbf{Z}_{(1)} = \mathbf{S}_1^T \mathbf{T}_{(1)} \, (\mathbf{S}_2 \otimes \mathbf{S}_3) .$$

Thus, the 3D tensor $\mathcal{T}$, which components are shown in Figure 2.4, can be decomposed in accordance with Equation (2.189), in which the $3 \times 3 \times 3$ core tensor $\mathcal{Z}$ and the matrices $\mathbf{S}_k$ are given in Equation (2.216) and (2.213)–(2.215), respectively. Actually the core tensor is shown in its 1-mode flattening form, i.e. as $\mathbf{Z}_{(1)}$. Let us notice that in accordance with

Theorem 2.3 all rows of $\mathbf{Z}_{(1)}$ are mutually orthogonal. Moreover, from (2.191) it follows that the triplets of sub-matrices which consist of the rows 1-2-3, 4-5-6 and 7-8-9, then 1-4-7, 2-5-8, and 3-6-9, are orthogonal. Based on Equation (2.192) the core tensor $\mathcal{Z}$ is also ordered, i.e. its submatrices are set in order of decreasing Frobenius norm which provides the k-mode singular values of the tensor $\mathcal{T}$. The three sub-matrices from Equation (2.216), i.e. column 1-2-3, 4-5-6, and 7-8-9, have their Frobenius norms 15.5385, 10.0921, and 4.6579, respectively.

The computational complexity of the SVD decomposition depends on the dimensions of the decomposed matrix and whether all the matrices $\mathbf{S}, \mathbf{V}$, and $\mathbf{D}$ or only some of them are required at output. Since for the HOSVD only the first orthogonal matrix $\mathbf{S}$ needs to be found, computation of the complete SVD is not necessary. The complexity of computation of the matrices $\mathbf{V}$ and $\mathbf{D}$ only, from an $r \times c$ matrix $\mathbf{T}$ is of order $4c^2(r + 2c)$ [31]. Thus, we can use this scheme, decomposing transpose $\mathbf{T}^T$ instead of the original matrix $\mathbf{T}$. In this case more often than not we have $r \gg c$, since r is a product of all P-1 dimensions other than N_c.

Details of implementation of the HOSVD algorithm are provided in the next section. This operates on the tensor objects discussed in Section (2.12.1.2). The HOSVD algorithm presents a key method for analyzing the properties of tensors. However, contrary to the matrix SVD, the HOSVD does not provide strict information on the rank of a tensor. It also does not guarantee nonnegative or sparse components. Finally, its computation can be time and memory demanding. Therefore in some contexts other decompositions can be more useful, for which HOSVD can be used as an initialization step, as will be discussed.

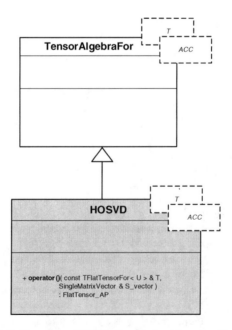

Figure 2.32 Definition of the *HOSVD* class derived from the *TensorAlgebraFor*.

2.12.3.1   Implementation of the HOSVD Decomposition

The *HOSVD* functor class for the HOSVD tensor decomposition is simply derived from the *TensorAlgebraFor* class, discussed in the previous section. This class diagram is depicted in Figure 2.32. Class definition is provided in Algorithm 2.15. As already mentioned, all classes are parameterized by two template types from which the first one, *T*, allows specification of a suitable type for tensor elements, whereas an auxiliary type *ACC* is capable of accumulating some computations on tensor elements.

Apart from the constructor and destructor, the *HOSVD* class implements only one function *operator( )* which carries out the HOSVD decomposition on the supplied tensor. As a result this tensor is decomposed into the core tensor $\mathcal{Z}$ and the series of unitary matrices $\mathbf{S}_k$, as described in Equation (2.189). These are return objects of this function operator.

```
///////////////////////////////////////////
// This class implements the
// Higher-Order Singular Value Decomposition
// of tensors
///////////////////////////////////////////
template < typename T = double >
class HOSVD : public TensorAlgebraFor< T >
{

    public:


            ///////////////////////////////////////////////////////////////
            // This function computes HOSVD
            ///////////////////////////////////////////////////////////////
            //
            // INPUT:
            //                          T - the input flat tensor
            //                          S_vector - the output of the mode matrices
            //
            // OUTPUT:
            //                          The core tensor upon success
            //                          0 otherwise
            //
            // REMARKS:
            //
            //
            template < typename U >
            typename TFlatTensorFor< U >::FlatTensor_AP operator()(
                    const TFlatTensorFor< U > & T, SingleMatrixVector & S_vector );


};
```

Algorithm 2.15   Definition of the *HOSVD* class which implements the HOSVD decomposition of tensors.

Computation of the HOSVD is done with the overloaded function member of the class. The input read-only parameter is a tensor $\mathcal{T}$. If the operation succeeds, the core tensor $\mathcal{Z}$ is returned and the series of mode matrices $\mathbf{S}_k$, as described in Equation (2.189).

```
1  //////////////////////////////////////////////////////////////
2  // This function computes HOSVD
3  //////////////////////////////////////////////////////////////
4  //
5  // INPUT:
6  //       T - the input flat tensor
7  //       S_vector - the output of the mode matrices
8  //
9  // OUTPUT:
10 //       The core tensor upon success
11 //       0 otherwise
12 //
13 // REMARKS:
14 //
15 //
16 template < typename T >
17 template < typename U >
18 typename TFlatTensorFor< U >::FlatTensor_AP HOSVD< T >::operator()(
19             const TFlatTensorFor< U < & T, SingleMatrixVector & S_vector )
20 {
21   REQUIRE( T.GetImageData() != 0 );
22
23   // -----------------------------------------------------
24   // We need to generate all k-modes of the input tensor
25   // and for each compute the SVD, saving S.
26
27   const int T_valence = T.GetIndexVector().size(); // get dimensionality of T
28
29   Compute_SVD svd_calculator;
30
31   S_vector.resize( T_valence );
32
33   for( int k = 0; k < T_valence; ++ k )
34   {
35       // Prepare the k-mode tensor for this run.
36       // In all cases except when k == T_mode we need to generate a new tensor.
37       // For this purpose we use a proxy...
38       TFlatTensorProxyFor< U < proxy_procTensor( (TFlatTensorFor< U > &)T,
39                          k, T.GetCyclicMode() );
40
41       // Now we do SVD on a transposed matrix of the procTensor
               (it should be faster)
42       BaseImage_AP transposedTensorMatrix(
43                            Orphan_Transposed_Matrix( proxy_procTensor ) );
44
45       svd_calculator( * transposedTensorMatrix );
46
47       // We need to save "D" of the SVD decomposition
```

```
48       REQUIRE( svd_calculator.Get_D() != 0 );
49       // matrix D is used over and over,
50       // up to the last iteration in which we can just overtake it
51       S_vector[ k ] = k == T_valence - 1 ? svd_calculator.Release_D()
52           : new SingleMatrix( * svd_calculator.Get_D() );
53
54       REQUIRE( S_vector[ k ] != 0 );
55       REQUIRE( S_vector[ k ]->GetCol() == S_vector[ k ]->GetRow() );
56       // this directly follows the theorem on HOSVD (see de Lathauwer)
57       REQUIRE( S_vector[ k ]->GetCol() == T.GetIndexVector()[ k ] ); // as above
58
59   }
60
61   return Orphan_PROJECTED_Tensor_From( S_vector, T );
62 }
```

Algorithm 2.16   Algorithm for computation of the HOSVD.

The listing of the HOSVD function is shown in Algorithm 2.16. It follows Algorithm 2.14 and relies on the SVD decomposition carried out with the help of the *svd_calculator* object from the *Compute_SVD* class. In each iteration of the loop, which spans the lines <33–59>, matrix SVD decomposition is computed from the input tensor parameter *T*, each time in a different flattening mode *k* thanks to the *proxy_procTensor*, created in line <38>. However, SVD is computed from the transposed version of this matrix, which is created in line <43>. At each iteration of the loop the **D** matrix from the SVD decomposition is stored in the *S_vector* variable (lines <51–52>). *S_vector* is one of the returned parameters. The next returned parameter is the auto pointer to the core tensor which is obtained from the input tensor and the series of matrices $\mathbf{S}_k$ stored in *S_vector*. Computation is done in accordance with Equation (2.211) or (2.212), depending on the index cycling mode of the original tensor.

As we may observe, in many places of the presented code the *REQUIRE* macro is placed with a logical condition as its argument. The purpose is to check that logical condition which must be true in the code all the time and under all circumstances. These are called code invariants and follow the paradigm of programming by contract [128]. In practice this has been shown to be a very useful debugging mechanism which allows early discovery of some logical errors in the process of software construction [15, 129].

2.12.4   HOSVD Induced Bases

The SVD decomposition allows representation of a matrix as a sum of rank one matrices in accordance with Equation (2.158). The summation spans the number of elements, however no more than the rank of the decomposed matrix. A similar sum can be constructed for tensors based on decomposition (2.189), that is

$$\mathcal{T} = \sum_{n=1}^{N_P} \mathcal{T}_n \times_P \mathbf{s}_P^{(n)}, \qquad (2.217)$$

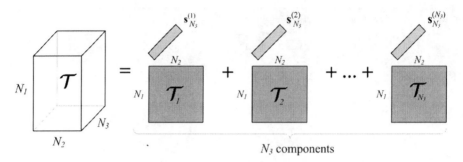

Figure 2.33 Visualization of the decomposition (2.217) for a 3D tensor.

where

$$\mathcal{T}_n = \mathcal{Z}\left(\underbrace{:, :, \ldots, :}_{P-1}, n\right) \times_1 \mathbf{S}_1 \times_2 \mathbf{S}_2 \ldots \times_{P-1} \mathbf{S}_{P-1} \tag{2.218}$$

are the basis tensors and $\mathbf{s}_P^{(n)}$ denotes an n-th column of the unitary matrix $\mathbf{S}_P$. Since $\mathcal{T}_n$ is of dimension P-1 then $\times_P$ in Equation (2.217) is an outer product, i.e. a product of two tensors of dimensions P-1 and 1. The result is a tensor of dimension P, i.e. the same as of $\mathcal{T}$.

Figure 2.33 depicts a visualization of the decomposition (2.217) for a 3D tensor. In this case $\mathcal{T}_n$ becomes two-dimensional, i.e. it is a matrix. Moreover, it is worth noticing that due to orthogonality of the core tensor $\mathcal{Z}$ in (2.189), $\mathcal{T}_n$ are also orthogonal. This can be shown as follows.

$$\langle \mathcal{T}_a, \mathcal{T}_b \rangle = tr\left\{\mathbf{T}_{a(k)}^T \mathbf{T}_{b(k)}\right\} = tr\left\{\left(\mathbf{S}_k \mathbf{Z}_{a(k)}\,[\mathbf{S}_{k-1} \otimes \ldots \otimes \mathbf{S}_1 \otimes \mathbf{S}_{P-1} \otimes \ldots \otimes \mathbf{S}_{k+1}]^T\right)^T\right.$$

$$\left(\mathbf{S}_k \mathbf{Z}_{b(k)}\,[\mathbf{S}_{k-1} \otimes \ldots \otimes \mathbf{S}_1 \otimes \mathbf{S}_{P-1} \otimes \ldots \otimes \mathbf{S}_{k+1}]^T\right)\right\}$$

$$= tr\left\{[\mathbf{S}_{k-1} \otimes \ldots \otimes \mathbf{S}_1 \otimes \mathbf{S}_{P-1} \otimes \ldots \otimes \mathbf{S}_{k+1}]\mathbf{Z}_{a(k)}^T \mathbf{S}_k^T\, \mathbf{S}_k \mathbf{Z}_{b(k)}\,[\mathbf{S}_{k-1} \otimes \ldots \otimes \mathbf{S}_1 \otimes \mathbf{S}_{P-1} \otimes \ldots \otimes \mathbf{S}_{k+1}]^T\right\}$$

$$= tr\left\{[\mathbf{S}_{k-1} \otimes \ldots \otimes \mathbf{S}_1 \otimes \mathbf{S}_{P-1} \otimes \ldots \otimes \mathbf{S}_{k+1}]^T\,[\mathbf{S}_{k-1} \otimes \ldots \otimes \mathbf{S}_1 \otimes \mathbf{S}_{P-1} \otimes \ldots \otimes \mathbf{S}_{k+1}]\mathbf{Z}_{a(k)}^T\, \mathbf{Z}_{b(k)}\right\}$$

$$= tr\left\{[[\mathbf{S}_{k-1}^T \otimes \ldots \otimes \mathbf{S}_1^T \otimes \mathbf{S}_{P-1}^T \otimes \ldots \otimes \mathbf{S}_{k+1}^T]\,[\mathbf{S}_{k-1} \otimes \ldots \otimes \mathbf{S}_1 \otimes \mathbf{S}_{P-1} \otimes \ldots \otimes \mathbf{S}_{k+1}]\mathbf{Z}_{a(k)}^T\, \mathbf{Z}_{b(k)}\right\}$$

$$= tr\left\{[\mathbf{S}_{k-1}^T \mathbf{S}_{k-1} \otimes \ldots \otimes \mathbf{S}_1^T \mathbf{S}_1 \otimes \mathbf{S}_{P-1}^T \mathbf{S}_{P-1} \otimes \ldots \otimes \mathbf{S}_{k+1}^T \mathbf{S}_{k+1}]\mathbf{Z}_{a(k)}^T\, \mathbf{Z}_{b(k)}\right\} = tr\left\{\mathbf{Z}_{a(k)}^T\, \mathbf{Z}_{b(k)}\right\} = 0. \tag{2.219}$$

In the above derivation we used the following matrix properties [17]:

$$\langle \mathbf{A}, \mathbf{B} \rangle = \langle \mathbf{B}, \mathbf{A} \rangle = tr\left(\mathbf{B}^T \mathbf{A}\right) = tr\left(\mathbf{A}^T \mathbf{B}\right), \tag{2.220}$$

$$tr\left(\mathbf{A}_1 \ldots \mathbf{A}_{n-1} \mathbf{A}_n\right) = tr\left(\mathbf{A}_n \mathbf{A}_1 \ldots \mathbf{A}_{n-1}\right), \tag{2.221}$$

as well as rules (2.207) and (2.208) of the Kronecker product, unitary properties of the matrices $\mathbf{S}_k$, and the orthogonality property (2.191) of the core tensors of the HOSVD.

As an example let us check the product rule (2.207) of the Kronecker product for the matrices $\mathbf{A}_{2 \times 2}$ and $\mathbf{B}_{1 \times 3}$, from Equation (2.203), as well as $\mathbf{U}_{2 \times 1}$ and $\mathbf{V}_{3 \times 1}$ from Equation (2.205) (p. 116). In this case the left side of Equation (2.207) becomes

$$(\mathbf{A} \otimes \mathbf{B})(\mathbf{U} \otimes \mathbf{V}) = \begin{bmatrix} 0 & 0 & 0 & -2 & -4 & 14 \\ 1 & 2 & -7 & -1 & -2 & 7 \end{bmatrix} \cdot \begin{bmatrix} -44 & 8 & 0 & -55 & 10 & 0 \end{bmatrix}^T = \begin{bmatrix} 70 \\ 7 \end{bmatrix}, \quad (2.222)$$

whereas the right side evaluates to the following

$$\mathbf{AU} \otimes \mathbf{BV} = \begin{bmatrix} 0 & -2 \\ 1 & -1 \end{bmatrix} \begin{bmatrix} 4 \\ 5 \end{bmatrix} \otimes \begin{bmatrix} 1 & 2 & -7 \end{bmatrix} \begin{bmatrix} -11 \\ 2 \\ 0 \end{bmatrix} = \begin{bmatrix} -10 \\ -1 \end{bmatrix} \otimes [-7] = \begin{bmatrix} 70 \\ 7 \end{bmatrix}. \quad (2.223)$$

As expected, the two results are identical. This property holds for all matrices which dimensions allow their multiplication, as required in Equation (2.207). The proof of this property can be found e.g. in the book by Graham [127].

Because of the above properties, $\mathcal{T}_n$ in decomposition (2.217) *constitutes a basis*. In this light Equation (2.217) means that each pattern contained in the tensor $\mathcal{T}$ can be expressed as *a linear combination* of the basis tensors $\mathcal{T}_n$ weighted by the vectors $\mathbf{s}_P^{(n)}$. This is an important result which allows construction of classifiers based on the HOSVD decomposition. The main idea is first to build a linear space spanned by $\mathcal{T}_n$, and then to project an unknown pattern which we wish to classify. Details of this method, as well as many real examples, are discussed in Section 5.8. As alluded to previously, in this case the series (2.217) is usually truncated to the first $N \le N_P$ components. In other words, a smaller, but dominating, N dimensional subspace is used to approximate $\mathcal{T}$.

2.12.5 Tensor Best Rank-1 Approximation

Definition 2.20, Definition 2.21, and Definition 2.22 provide the basic information on rank for tensors. In this section we deal with the problem of the best *rank-1* and *rank-*$(R_1, R_2, \ldots, R_P)$ approximations. The former can be stated as the following definition [96, 130]:

Definition 2.23 Tensor best rank-1 approximation

Given a tensor $\mathcal{T} \in \Re^{N_1 \times N_2 \times \ldots \times N_P}$ find a scalar ρ and vectors $\mathbf{s}_1, \mathbf{s}_2, \ldots, \mathbf{s}_P$ of unit norms such that the *rank-1* tensor

$$\tilde{\mathcal{T}} = \rho \, \mathbf{s}_1 \circ \mathbf{s}_2 \circ \ldots \circ \mathbf{s}_P \quad (2.224)$$

minimizes the following least-squares functional

$$\Theta(\tilde{\mathcal{T}}) = \|\tilde{\mathcal{T}} - \mathcal{T}\|_F^2, \quad (2.225)$$

in the sense of the Frobenius norm (2.28). □

As shown by Lathauwer *et al.* [131], the above problem can be solved using the method of Lagrange multipliers [132]. The Lagrange equation is built from the minimization problem (2.225) and conditions on unit norms of s_k. Then, after differentiation, this leads to the solution of Equation (2.225) which is summarized by the following theorem.

Theorem 2.4   Best *rank-1* approximation of the tensor

Minimization of the functional Equation (2.225) is equivalent to the maximization of the following functional

$$\Psi\left(s_1, s_2, \ldots, s_P\right) = \left| \mathcal{T} \times_1 s_1^T \times_2 s_2^T \ldots \times_P s_P^T \right|^2 \tag{2.226}$$

over the manifold of unit-norm vectors $s_1, s_2, \ldots, s_P$. □

The above theorem is a constructive one. As a result a higher-order power algorithm was devised which is provided in Algorithm 2.17, based on [100].

---

1. Set initial values $s_k^{(0)}$ to the series of vectors s_k. For instance, initializing $s_k^{(0)}$ to the dominant left singular vector of unfolded matrix $\mathbf{T}_{(k)}$ (for $2 \le k \le P$), as obtained from the HOSVD Algorithm 2.14.
Set the maximal number of iterations t_{max}.

   $t = 0$

2. While not convergent and $t < t_{max}$ **do:**

   3. For each k, such that $1 \le k \le P$, **do:**

$$s_k^{(t+1)} = \mathbf{T}_{(k)} \left[\underbrace{s_1^{(t+1)} \otimes \ldots \otimes s_{k-1}^{(t+1)}}_{F_1} \otimes \underbrace{s_{k+1}^{(t)} \otimes \ldots \otimes s_P^{(t)}}_{F_2} \right] \tag{2.227}$$

$$\rho = \rho_k^{(t+1)} = \left\| s_k^{(t+1)} \right\| \tag{2.228}$$

$$s_k^{(t+1)} = s_k^{(t+1)} \Big/ \rho_k^{(t+1)} \tag{2.229}$$

   $t = t + 1$

3. Compute $\tilde{\mathcal{T}}$ from (2.224) using the set of s_k and ρ from the last iteration.

Algorithm 2.17   Higher-order power method for computation of the best rank-1 approximation of tensors (the Alternating Least-Squares method).

In Equations (2.227)–(2.229) t denotes an iteration (time) step. Let us observe that there are actually two Kronecker factors in (2.227). The first one, F_1 with time step $t + 1$, corresponds to the matrices $\mathbf{s}_k$ already computed in the current iteration. F_1 is not empty for indices $k > 1$. On the other hand, F_2 contains $\mathbf{s}_k$ only from the previous iteration, i.e. at a step t. In each iteration step of (3) only one of the vectors $\mathbf{s}_k$ is optimized while other are kept fixed. Thus the name of this algorithm – the Alternating Least-Squares (ALS). However, the scalar ρ is updated in each step.

Two things in Algorithm 2.17 need further explanation – the initialization process in step (1) and the convergence conditions in step (2). As discussed by Lathauwer et al. [131] Algorithm 2.17 leads to the globally optimal solution only if the starting point is in the region of attraction of this optimum. Usually initial estimates of $\mathbf{s}_k$ obtained with the HOSVD (Algorithm 2.14) belong to the attractors of a global solution. Therefore it is proposed to initialize Algorithm 2.17 with $\mathbf{s}_k$ obtained from HOSVD. Nevertheless, examples can be found in which Algorithm 2.17, initialized in this way, follows other than global attractors. Therefore research on initialization methods of Algorithm 2.17 is still necessary. In this regard, the paper by Regalia and Kofidis can provide more ideas [133]. However, in all our experiments random initialization also led to convergence to the same points as in the case of the HOSVD based initialization. Although this does not prove that this is a better or worse method, in practice it is always a cheaper one, especially if the input tensor is very large. Finally, let us also observe that in the initialization step (1) in Algorithm 2.17 we do not need to initialize the first vector, i.e. $\mathbf{s}_1$, since it is computed in Equation (2.227) exclusively from the remaining vectors $\mathbf{s}_{2 \leq k \leq P}$ in the first iteration of step (3) in Algorithm 2.17.

The termination criterion is usually formed by means of a threshold value on a norm of differences of $\mathbf{s}_k$ and ρ computed in two consecutive iterations. The stop conditions for Algorithm 2.17 are usually specified as a condition of a maximal error of a norm of the difference of any vectors $\mathbf{s}_k^{(t+1)}$ and $\mathbf{s}_k^{(t)}$, computed in the step $t+1$ and t, respectively [131, 134]. Thus, for a given iteration step t the stop condition can be expressed as

$$\max_k \left(\left\| \mathbf{s}_k^{(t+1)} - \mathbf{s}_k^{(t)} \right\| \right) < e_{thresh}, \tag{2.230}$$

where e_{thresh} is a defined accuracy threshold. In our experiments this was in the range of 1e-7 to 1e-3. However, in practice the above stop condition requires computation of Equation (2.230), as well as storage of the current and previous vectors $\mathbf{s}_k$. Therefore in practice we use the following simpler condition

$$\left\| \rho_P^{(t+1)} - \rho_P^{(t)} \right\| < e_{thresh}. \tag{2.231}$$

The above is based solely on the computation of Equation (2.228) and does not require additional storage. Since the value of $\rho_P^{(t+1)}$ depends on all matrices $\mathbf{s}_k^{(t+1)}$ through computation of Equation (2.227), therefore this criteria works well in practice, though this observation does not constitute a formal proof.

Nevertheless, the iteration loop (step (2) in Algorithm 2.17) should be augmented with the counter of iterations, so the loop is guaranteed to stop after a predefined maximal number of iterations regardless of the stop conditions based on the convergence of the vectors. This strategy is a good and practical way to guarantee a fuse-like stop condition for all numerical procedures.

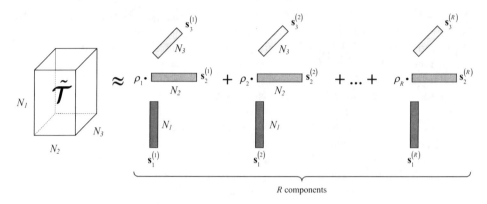

Figure 2.34   Visualization of the rank-1 decomposition of 3D tensors.

2.12.6   Rank-1 Decomposition of Tensors

In many cases of real tensors, application of the rank-1 approximation (2.224) leads to excessive reconstruction error, expressed in Equation (2.225). This error can be reduced following the tensor decomposition outlined in Equation (2.160) for three-dimensional tensors, however. That is, Equation (2.160) can be extended to tensors of arbitrary P-dimensions, which leads to the following decomposition

$$\tilde{\mathcal{T}} = \sum_{r=1}^{R} \rho_r \mathbf{s}_1^{(r)} \circ \mathbf{s}_2^{(r)} \circ \ldots \circ \mathbf{s}_P^{(r)} \tag{2.232}$$

where ρ_r is a scalar and each $\mathbf{s}_k^{(r)}$ denotes an r-th vector of dimension N_k. Let us observe that all vectors $\mathbf{s}_k^{(\cdot)}$, i.e. with the same lower index, have the same dimensions, and thus they can be grouped in a matrix $\mathbf{S}_k \in \mathfrak{R}^{N_k \times R}$ (here we use a notation like in formula (2.190)). Now, in Equation (2.232) there are R times P vectors $\mathbf{s}_k^{(\cdot)}$ of size N_k and scalars ρ_r to be computed. We see that Equation (2.232) is a generalized version of Equation (2.224) which should allow more precise approximation, i.e. the expected error expressed in Equation (2.225) should be lower. Thus, tensor rank-1 decomposition is a generalization of the rank-1 approximation, and the former allows lower reconstruction errors. For completeness, Figure 2.34 visualizes series (2.232) for a three-dimensional case.

For computations it is convenient to represent Equation (2.232) in the following form (assuming the forward cycling mode):

$$\tilde{\mathcal{T}} = \mathcal{K} \times_1 \mathbf{S}_1 \times_2 \mathbf{S}_2 \times \ldots \times_P \mathbf{S}_P \tag{2.233}$$

where $\mathcal{K} \in \mathfrak{R}^{R \times R \times \ldots \times R}$ is a diagonal tensor containing values ρ_r on its diagonal. The matrices $\mathbf{S}_k \in \mathfrak{R}^{N_k \times R}$ for $k = 1, \ldots, P$, are composed of the column vectors $\mathbf{s}_k^{(r)}$ for $r = 1, \ldots, R$, as follows

$$\mathbf{S}_k = \left[\mathbf{s}_k^{(1)} \mathbf{s}_k^{(2)} \ldots \mathbf{s}_k^{(R)} \right] \tag{2.234}$$

It is also possible to use a slightly different notation:

$$\mathbf{T}_{(k)} = \mathbf{S}_k \Upsilon \left[\mathbf{S}_{k-1} \odot \mathbf{S}_{k-2} \odot \ldots \odot \mathbf{S}_1 \odot \mathbf{S}_P \odot \mathbf{S}_{P-1} \odot \ldots \odot \mathbf{S}_{k+1} \right]^T \qquad (2.235)$$

where $\odot$ denotes the Khatri–Rao product of the matrices $\mathbf{S}_k$ defined in Equation (2.234). The diagonal matrix $\Upsilon \in \mathfrak{R}^R$

$$\Upsilon = \begin{bmatrix} \rho_1 & \cdots & 0 \\ \vdots & \ddots & \vdots \\ 0 & \cdots & \rho_R \end{bmatrix} \qquad (2.236)$$

contains the values ρ_r on its diagonal.

The Khatri–Rao product (K–R) is a column-wise matched Kronecker product which for the matrices $\mathbf{P} \in \mathfrak{R}^{I \times M}$, and $\mathbf{Q} \in \mathfrak{R}^{J \times M}$ is defined as follows

$$\mathbf{P}_{I \times M} \odot \mathbf{Q}_{J \times M} = \begin{bmatrix} \mathbf{p}_1 \otimes \mathbf{q}_1 & \mathbf{p}_2 \otimes \mathbf{q}_2 & \cdots & \mathbf{p}_M \otimes \mathbf{q}_M \end{bmatrix}_{(IJ) \times M}, \qquad (2.237)$$

where $\mathbf{p}_i$ and $\mathbf{q}_i$ denote i-th column (a vector) of a corresponding matrix $\mathbf{P}$ or $\mathbf{Q}$, respectively. It is interesting to observe that for vector the Kronecker and Khatri–Rao products are identical.

As an example, for the following matrices

$$\mathbf{P}_{2 \times 2} = \begin{bmatrix} 0 & -2 \\ 1 & -1 \end{bmatrix}, \ \mathbf{Q}_{3 \times 2} = \begin{bmatrix} 2 & -1 \\ 0 & -2 \\ 7 & 0 \end{bmatrix} \qquad (2.238)$$

their K–R product reads as follows

$$\mathbf{P}_{2 \times 2} \odot \mathbf{Q}_{3 \times 2} = \begin{bmatrix} 0 & -2 \\ 1 & -1 \end{bmatrix} \odot \begin{bmatrix} 2 & -1 \\ 0 & -2 \\ 7 & 0 \end{bmatrix} = \begin{bmatrix} \begin{bmatrix} 0 \\ 1 \end{bmatrix} \otimes \begin{bmatrix} 2 \\ 0 \\ 7 \end{bmatrix} & \begin{bmatrix} -2 \\ -1 \end{bmatrix} \otimes \begin{bmatrix} -1 \\ -2 \\ 0 \end{bmatrix} \end{bmatrix} = \begin{bmatrix} 0 & 2 \\ 0 & 4 \\ 0 & 0 \\ 2 & 1 \\ 0 & 2 \\ 7 & 0 \end{bmatrix}_{6 \times 2},$$

$$(2.239)$$

Let us observe that the only condition for the K–R product is that number of columns in the multiplied matrices is the same (there can be any number of rows).

One method of determining the parameters of decomposition (2.232) was proposed by Wang and Ahuja for efficient compression of video sequences [134]. Their main idea is to proceed iteratively, starting from finding a first set of vectors $\mathbf{s}_1^i$ and the scalar ρ_1. This can be done with a method described in the previous Section 2.12.5. Then a difference (residual) tensor is computed subtracting this first approximation from the original tensor, as follows:

$$\tilde{\mathcal{T}}_2 = \mathcal{T} - \tilde{\mathcal{T}}_1 = \mathcal{T} - \left(\rho_1 \mathbf{s}_1^1 \circ \mathbf{s}_2^1 \circ \ldots \circ \mathbf{s}_P^1\right). \qquad (2.240)$$

Now the process is repeated, i.e. a rank-1 decomposition on the residual tensor $\tilde{\mathcal{T}}_2$ is computed, which provides the second set of vectors $\mathbf{s}_i^2$ and the scalar ρ_2, and so on. Algorithm 2.18 shows the stages of computation of the best rank-1 decomposition of tensors. The algorithm follows the higher-order power ALS scheme shown in Algorithm 2.17.

---

1. Choose the number R of elements in Equation (2.232). Choose the accuracy threshold e_{thresh} and the maximum number of iterations t_{max}.

2. **For each** r, such that $1 \leq r \leq R$, **do:**

Set initial values to the matrices $\mathbf{s}_k^{(0)}$ to a dominant left singular vector of $\mathbf{T}_{(k)}$ (for $2 \leq k \leq P$), as obtained from the HOSVD Algorithm 2.14.

$t = 0$

3. **Do:**

$$e = 0$$

4. **For each** k, such that $1 \leq k \leq P$, **do:**

$$\mathbf{s}_k^{(t+1)} = \mathbf{T}_{(k)} \left[\underbrace{\mathbf{s}_1^{(t+1)} \otimes \ldots \otimes \mathbf{s}_{k-1}^{(t+1)}}_{F_1} \otimes \underbrace{\mathbf{s}_{k+1}^{(t)} \otimes \ldots \otimes \mathbf{s}_P^{(t)}}_{F_2} \right] \qquad (2.241)$$

$$\rho = \rho_k^{(t+1)} = \left\| \mathbf{s}_k^{(t+1)} \right\| \qquad (2.242)$$

$$\mathbf{s}_k^{(t+1)} = \mathbf{s}_k^{(t+1)} / \rho_k^{(t+1)} \qquad (2.243)$$

$$e = \max \left(e, \ \left\| \mathbf{s}_k^{(t+1)} - \mathbf{s}_k^{(t)} \right\| \right) \qquad (2.244)$$

$t = t + 1$

while $(e > e_{thresh})$ and $(t < t_{max})$

Store the set of vectors $\{\mathbf{s}_1, \mathbf{s}_2, \ldots, \mathbf{s}_P\}_r$ and the scalar ρ for the parameter r.

$$\mathcal{T} = \mathcal{T} - \left(\rho \mathbf{s}_1^{(t+1)} \circ \mathbf{s}_2^{(t+1)} \circ \ldots \circ \mathbf{s}_P^{(t+1)} \right) \qquad (2.245)$$

Algorithm 2.18 Higher-order power ALS method for computation of the best rank-1 decomposition of tensors.

It is convenient to define the residual tensor $\mathcal{C}_R$ for a chosen rank R in expansion (2.232), as follows

$$\mathcal{C}_R = \mathcal{T} - \tilde{\mathcal{T}}_R = \mathcal{T} - \sum_{r=1}^{R} \rho_r \mathbf{s}_r^{(1)} \circ \mathbf{s}_r^{(2)} \circ \dots \circ \mathbf{s}_r^{(P)}. \qquad (2.246)$$

Its Frobenius norm

$$c_R = \|\mathcal{C}_R\|_F \qquad (2.247)$$

provides information on the accuracy of the approximation. Operation of Algorithm 2.18 for the test tensor shown in Figure 2.4 presents the following example.

In this example we present the results of the rank-1 decomposition of the $3 \times 3 \times 3$ tensor from Figure 2.4. Algorithm 2.18 was run a number of times with different values of the requested rank R, as well as different maximal error Θ. Results are presented in Table 2.2.

Table 2.2 Obtained accuracy and number of elements to store for the rank-1 decomposition of the $3 \times 3 \times 3$ tensor from Figure 2.4 with C++ implementation of Algorithm 2.18 and for different values of the rank R as well as different stopping criteria for each rank-1 approximation.

R	2	3	4	5
c_R, ($\Theta = 1e{-}3$, $t_{max} = 22$)	9.12	5.83	3.39	2.41
c_R, ($\Theta = 1e{-}7$, $t_{max} = 122$)	9.01	5.78	3.33	2.36
E	20	30	40	50

The last row of Table 2.2 contains the number of elements to store E for each setup. We easily see that for $R = 3$ this value exceeds the number of elements of the input tensor (i.e. 27). This means that the rank-1 decomposition does not alway leads to data compression, as one might expect. However, as expected the larger values of R lead to lower reconstruction errors, measured in this case by a norm c_R of the remnant tensor $\mathcal{C}_R$ given in Equation (2.246). For $R = 5$ it is as follows

$$\mathcal{C}_{R=5} = \begin{bmatrix} \begin{array}{ccc|ccc} 0.398935 & -0.131279 & -0.828761 & -0.180412 & 0.362937 & 0.119466 \\ -0.277544 & 0.0575298 & 0.350778 & -0.522589 & -0.0942926 & -0.243519 \\ 0.361135 & -0.107524 & -0.519939 & 0.504322 & 0.296924 & 0.150244 \\[1em] -1.06475 & 0.148109 & -1.03901 \\ -0.69816 & 0.036188 & 0.142997 \\ 0.455812 & 0.13019 & -0.515986 \end{array} \end{bmatrix}.$$

The corresponding approximating tensor in this case is as follows

$$
\tilde{\mathcal{T}} =
\left[
\begin{array}{ccc|ccc}
11.6011 & 0.131279 & 0.828761 & 1.18041 & -1.36294 & 0.880534 \\
0.277544 & 3.94247 & 0.649222 & 4.52259 & 0.0942926 & 4.24352 \\
1.63886 & 3.10752 & -0.480061 & 4.49568 & 1.70308 & 0.849756 \\
\hline
1.06475 & -0.148109 & -0.960993 \\
4.69816 & 9.96381 & -0.142997 \\
1.54419 & 0.86981 & 0.515986
\end{array}
\right].
$$

The above tensor is obtained from Equation (2.232), i.e. by computing the weighted sum of five outer products from the decomposing vectors. The detailed parameters of this decomposition are presented in Table 2.3.

Table 2.3   Vectors and scalars of the rank-1 decomposition of the tensor from Figure 2.4 for $R = 5$, computed with Algorithm 2.18.

$\rho_1 =$ 12.4473		
$\mathbf{s}_1^{(1)} = [$ 0.929136	0.189116	0.317712 $]^T$
$\mathbf{s}_2^{(1)} = [$ 0.988725	0.149298	0.011501 $]^T$
$\mathbf{s}_3^{(1)} = [$ 0.956510	0.258404	0.135339 $]^T$
$\rho_2 =$ 11.3511		
$\mathbf{s}_1^{(2)} = [-0.150061$	0.965055	0.214828 $]^T$
$\mathbf{s}_2^{(2)} = [$ 0.361651	0.924541	0.120139 $]^T$
$\mathbf{s}_3^{(2)} = [$ 0.273127	0.22551	0.935172 $]^T$
$\rho_3 =$ 6.9114		
$\mathbf{s}_1^{(3)} = [$ 0.254734	-0.735678	-0.627606 $]^T$
$\mathbf{s}_2^{(3)} = [-0.893877$	0.266144	-0.360764 $]^T$
$\mathbf{s}_3^{(3)} = [$ -0.60975	0.788795	0.0775115 $]^T$
$\rho_4 =$ 4.72667		
$\mathbf{s}_1^{(4)} = [-0.588996$	-0.674826	0.444628 $]^T$
$\mathbf{s}_2^{(4)} = [$ 0.180321	0.703337	-0.687605 $]^T$
$\mathbf{s}_3^{(4)} = [$ 0.304567	0.831888	-0.463897 $]^T$
$\rho_5 =$ 2.34731		
$\mathbf{s}_1^{(5)} = [-0.195906$	0.657387	0.727643 $]^T$
$\mathbf{s}_2^{(5)} = [-0.084032$	0.696532	0.712588 $]^T$
$\mathbf{s}_3^{(5)} = [$ 0.649671	0.607775	-0.456659 $]^T$

Using the discussed technique for the compression problem for a video sequence of K frames, each of $M \times N$ spatial resoution, we can obtain compression ratios of order (the larger, the better)

$$c_1 = \frac{MNK}{R(1 + M + N + K)}. \tag{2.248}$$

This is a real difference in comparison to the "classical" PCA[13] compression which achieves a compression ratio of at most

$$c_2 = \frac{MNK}{MNQ} = \frac{K}{Q}. \tag{2.249}$$

where Q denotes the number of the used principal components. In the experiments by Wang and Ahuja, a sequence $M \times N \times K = 220 \times 320 \times 20$ was used. In this case, the PCA approach with only one principal component used allows at most $c_2 = 20/1 = 20$. For the rank-1 tensor compression scheme, such a compression ratio c_1 is obtained for $R = 126$, since $c_1 = 2509.8/126 \approx c_2 = 20$. However, at the same compression ratio, the accuracy of the tensor approach is about two times better than for the PCA.

A more general overview on dimensionality reduction with different tensor approximation is discussed in the paper by Wang and Ahuja [135]. They showed superior performance of the rank-R approximation of tensors for the tasks of data representation and also for object classification, as will be discussed in the next section.

2.12.7   Best Rank-(R_1, R_2, . . . , R_P) Approximation

The problem of a best rank-(R_1, R_2, . . . , R_P) approximation can be stated in an analogous way as the best rank-1 approximation, as follows [96]:

Definition 2.24   Best rank-($R_1, R_2, \ldots, R_P$) approximation

Given a tensor $\mathcal{T} \in \Re^{N_1 \times N_2 \times \ldots \times N_P}$ find a tensor $\tilde{\mathcal{T}}$ having $rank_1\left(\tilde{\mathcal{T}}\right) = R_1, rank_2\left(\tilde{\mathcal{T}}\right) = R_2, \ldots, rank_P\left(\tilde{\mathcal{T}}\right) = R_P$, with rank of the tensor defined in Equation (2.186), that minimizies the following least-squares cost function

$$\Theta\left(\tilde{\mathcal{T}}\right) = \left\| \tilde{\mathcal{T}} - \mathcal{T} \right\|_F^2, \tag{2.250}$$

in the sense of the Frobenius norm (2.19).   □

Approximating tensor $\tilde{\mathcal{T}}$ conveys as much of the "energy," in the sense of the squared entries of a tensor, as the original tensor $\mathcal{T}$ given the rank constraints.

It can be easily observed that the assumed rank conditions mean that the approximation tensor $\tilde{\mathcal{T}}$ can be decomposed as

$$\tilde{\mathcal{T}} = \mathcal{Z} \times_1 \mathbf{S}_1 \times_2 \mathbf{S}_2 \ldots \times_P \mathbf{S}_P, \tag{2.251}$$

---

[13]Principal Component Analysis (PCA) is discussed in Section 3.3.1.

where each of the matrices $\mathbf{S}_1 \in \mathfrak{R}^{N_1 \times R_1}$, $\mathbf{S}_2 \in \mathfrak{R}^{N_2 \times R_2}$, ... , $\mathbf{S}_P \in \mathfrak{R}^{N_P \times R_P}$ have *orthonormal* columns (each time, the number of columns for $\mathbf{S}_k$ is given by R_k), whereas the tensor $\mathcal{Z} \in \mathfrak{R}^{R_1 \times R_2 \times \dots \times R_P}$ has dimensions R_1, R_2, ... , R_P. Let us observe that despite an ostensible similarity to Equation (2.189), in Equation (2.251) dimensions of the matrices $\mathbf{S}_k$ are different.

Introducing Equation (2.251) into Equation (2.250) yields

$$\Theta\left(\tilde{\mathcal{T}}\right) = \|\mathcal{T} - \mathcal{Z} \times_1 \mathbf{S}_1 \times_2 \mathbf{S}_2 \dots \times_P \mathbf{S}_P\|_F^2. \tag{2.252}$$

In order to develop a method for the best rank-$(R_1, R_2, \dots, R_P)$ tensor decomposition, first let us observe that for a given input tensor $\mathcal{T}$ and a set of matrices $\mathbf{S}_k$, $\mathcal{Z}$ can be computed from the following equation

$$\mathcal{T} = \mathcal{Z} \times_1 \mathbf{S}_1 \times_2 \mathbf{S}_2 \dots \times_P \mathbf{S}_P, \tag{2.253}$$

solving the "classical" least-squares problem. Now, multiplying P-times both sides of the above equation by transposes of the orthonormal matrices $\mathbf{S}_k$, we come to a conclusion that the tensor $\mathcal{Z}$ can be computed from the original tensor $\mathcal{T}$ as

$$\mathcal{Z} = \mathcal{T} \times_1 \mathbf{S}_1^T \times_2 \mathbf{S}_2^T \dots \times_P \mathbf{S}_P^T. \tag{2.254}$$

This means that to find the best rank-$(R_1, R_2, \dots, R_P)$ approximation of $\mathcal{T}$ it is sufficient to determine only a set of the matrices $\mathbf{S}_k$ in Equation (2.251) since, once known, $\mathcal{Z}$ can be computed from Equation (2.254). This leads to the following theorem [131]:

Theorem 2.5   Best rank-$(R_1, R_2, \dots, R_P)$ approximation of the tensor

Minimization of the functional $\Theta\left(\tilde{\mathcal{T}}\right)$ in (2.250) is equivalent to the maximization of the function

$$\Psi\left(\mathbf{S}_1, \mathbf{S}_2, \dots, \mathbf{S}_P\right) = \|\mathcal{Z}\|^2 = \left\|\mathcal{T} \times_1 \mathbf{S}_1^T \times_2 \mathbf{S}_2^T \dots \times_P \mathbf{S}_P^T\right\|^2 \tag{2.255}$$

over the matrices $\mathbf{S}_1, \mathbf{S}_2, \dots, \mathbf{S}_P$ which have orthonormal columns. The two functions Ψ in Equation (2.255) and Θ in Equation (2.250) are related as follows

$$\Theta\left(\tilde{\mathcal{T}}\right) = \|\mathcal{T}\|^2 - \|\Psi\|^2. \quad \Box \tag{2.256}$$

The above theorem was used to develop a numerical ALS method for computation of the best rank-$(R_1, R_2, \dots, R_P)$ approximation of tensors, as proposed by Lathauwer *et al.* [131]. The ALS approach assumes that in each step only one of the matrices $\mathbf{S}_k$ is optimized, whereas others are kept fixed. In other words, the idea is to represent the function (2.255) as a quadratic expression in the components of the unknown matrix $\mathbf{S}_k$ with orthogonal columns with other matrices kept fixed, i.e.

$$\max_{\mathbf{S}_k} \{\Psi\left(\mathbf{S}_k\right)\} = \max_{\mathbf{S}_k} \left\|\mathcal{T} \times_1 \mathbf{S}_1^T \times_2 \mathbf{S}_2^T \dots \times_P \mathbf{S}_P^T\right\|^2. \tag{2.257}$$

This can be written as

$$\max_{\mathbf{S}_k} \{\Psi(\mathbf{S}_k)\} = \max_{\mathbf{S}_k} \left\| \mathbf{S}_k^T \hat{\mathbf{S}}_k \right\|^2, \tag{2.258}$$

where

$$\hat{\mathbf{S}}_k = \mathcal{T} \times_1 \mathbf{S}_1^T \times_2 \mathbf{S}_2^T \ldots \times_{k-1} \mathbf{S}_{k-1}^T \times_{k+1} \mathbf{S}_{k+1}^T \ldots \times_P \mathbf{S}_P^T. \tag{2.259}$$

Columns of $\mathbf{S}_k$ can be obtained as an orthonormal basis for the dominating subspace of the column space of the matrix in $\hat{\mathbf{S}}_k$. In a pure matrix representation the latter can be expressed as follows (assuming the forward cyclic mode)

$$\hat{\mathbf{S}}_k = \mathbf{T}_{(k)} [\mathbf{S}_{k-1} \otimes \mathbf{S}_{k-2} \otimes \ldots \otimes \mathbf{S}_1 \otimes \mathbf{S}_P \otimes \mathbf{S}_{P-1} \otimes \ldots \otimes \mathbf{S}_{k+1}]. \tag{2.260}$$

Summarizing, for a chosen matrix $\mathbf{S}_k$ and all other matrices fixed, the tensor $\mathcal{T}$ *is projected* onto the $(1, 2, \ldots, k\text{-}1, k+1, \ldots, P)$ dimensional space. As already mentioned, in each step only one matrix $\mathbf{S}_k$ is computed this way, while the others are kept fixed. Such computations are repeated until the stopping condition, expressed by Equation (2.256), is met. As already discussed, in practice we have to be sure that the procedure also stops when the maximal number of allowed iterations is reached. However, in such a case we cannot assume that the solution is in any way optimal. The whole procedure, called the Higher-Order Orthogonal Iteration (HOOI), is outlined in the next section.

The remaining problem is the choice of the best ranks for a given problem. Usually this can be achieved from experimentation with different settings of R_k. However, the HOSVD can give an insight into the distribution of singular values σ_a^i, defined in Equation (2.193), of the original tensor $\mathcal{T}$. Considering the following "energy" measure

$$E_{(k)}(R_k) = \frac{\displaystyle\sum_{i=1}^{R_k} \sigma_i^k}{\displaystyle\sum_{i=1}^{N_k} \sigma_i^k}, \tag{2.261}$$

we can assess transmission of data "energy" for different values of R_k. The closer R_k to the true rank in the k-th direction, the closer $E_{(k)}$ is to 1. Then $E_{(k)}$ can be computed in all dimensions of the data tensor, i.e. for $1 \le k \le P$. Depending on data contents and values of $E_{(k)}$ some dimensions can be truncated more than others.

Another useful measure of a quality of tensor approximation can be obtained from the relative reconstruction error (RRE), as proposed in the paper by Muti and Bourennane in the context of tensor data filtering [113]. RRE is defined as follows

$$RRE = \frac{\left\| \tilde{\mathcal{T}} - \mathcal{T} \right\|^2}{\|\mathcal{T}\|}, \tag{2.262}$$

where $\tilde{\mathcal{T}}$ denotes approximation of the tensor $\mathcal{T}$. In the image compression community other common measures of comparison are the mean-square error (MSE)

$$MSE\left(\tilde{\mathcal{T}}, \mathcal{T}\right) = \sum_{n_1=1}^{N_1} \sum_{n_2=1}^{N_2} \cdots \sum_{n_P=1}^{N_P} \left[\tilde{\mathcal{T}}\left(n_1, n_2, \ldots, n_P\right) - \mathcal{T}\left(n_1, n_2, \ldots, n_P\right)\right]^2, \quad (2.263)$$

and the peak signal to noise ratio (PSNR), which is expressed in dB, as follows

$$PSNR\left(\tilde{\mathcal{T}}, \mathcal{T}\right) = 10 \log_{10} \left(\frac{m^2}{MSE\left(\tilde{\mathcal{T}}, \mathcal{T}\right)}\right), \quad (2.264)$$

where m denotes the maximum value of a pixel (e.g. $m = 255$ for 8 bit images).

2.12.8   Computation of the Best Rank-(R_1, R_2, $\ldots$, R_P) Approximations

A computation method of the best rank-($R_1, R_2, \ldots, R_P$) approximation of tensors is presented in Algorithm 2.19. It follows the aforementioned Higher-Order Orthogonal Iteration method.

---

1. Choose the number of ranks $R_1, R_2, \ldots, R_P$. Choose the accuracy threshold e_{thresh} and the maximum number of iterations t_{max}.

2. Initialize: $\mathbf{S}_k^{(0)} \in \Re^{N_k \times R_k}$ for $1 \leq k \leq P$

   $\qquad\qquad t = 0$

3. **Do:**

   4. **For each** k, such that $1 \leq k \leq P$, **do:**

$$\hat{\mathbf{S}}_k^{(t+1)} = \mathbf{T}_{(k)} \left[\underbrace{\mathbf{S}_{k-1}^{(t)} \otimes \mathbf{S}_{k-2}^{(t)} \otimes \ldots \otimes \mathbf{S}_1^{(t)}}_{F_1} \otimes \underbrace{\mathbf{S}_P^{(t+1)} \otimes \ldots \otimes \mathbf{S}_{k+1}^{(t+1)}}_{F_2}\right] \quad (2.265)$$

$$\mathbf{S}_k^{(t+1)} = svds\left(\hat{\mathbf{S}}_k^{(t+1)}, R_k\right) \quad (2.266)$$

$$\mathcal{Z}_{t+1} = \mathcal{T} \times_1 \mathbf{S}_1^{(t+1)^T} \times_2 \mathbf{S}_2^{(t+1)^T} \ldots \times_P \mathbf{S}_P^{(t+1)^T} \quad (2.267)$$

$$e = \left|\|\mathcal{Z}_{t+1}\|^2 - \|\mathcal{Z}_t\|^2\right| \quad (2.268)$$

$$t = t + 1 \quad (2.269)$$

   while $(e > e_{thresh})$ **and** $(t < t_{max})$

3. Store the last computed $\mathcal{Z}$ and the corresponding set of matrices $\mathbf{S}_k$.

Algorithm 2.19   Best rank-($R_1, R_2, \ldots, R_P$) decomposition of tensors.

In Algorithm 2.19 the function $svds(\mathbf{S}, R)$ returns the R left leading singular vectors of the matrix $\mathbf{S}$, i.e. the ones corresponding to the R largest singular values of $\mathbf{S}$. These vectors are orthogonal. Moreover, it is frequently the case that the matrix $\hat{\mathbf{S}}_k$ in Equation (2.266) has many more columns c than rows r. In such a case we can compute $svds$ from the product $\hat{\mathbf{S}}_k \hat{\mathbf{S}}_k^T$, instead of the $\hat{\mathbf{S}}_k$, taking advantage of the fact that if a matrix $\mathbf{M} = \mathbf{S}\mathbf{V}\mathbf{D}^T$, then $\mathbf{M}\mathbf{M}^T = \mathbf{S}\mathbf{V}^2\mathbf{S}^T$. We also use this property in the software realization of Algorithm 2.19, after checking if fulfilled are the conditions of the prevailing number of columns, i.e. $c \gg r$, as well as of the substantial amount of data, i.e. a large value of the product $c \cdot r$.

Regarding the initialization step (2) in Algorithm 2.19, in some publications it is proposed to obtain these from the HOSVD decomposition [131]. Although such a strategy does not guarantee the optimal solution, in practice it usually leads to good results. However, as already mentioned, HOSVD is computationally demanding, so for larger problems Wang and Ahuja propose initializing $\mathbf{S}_k$, either with constant values, or with uniformly distributed random numbers. These strategies, when applied to image processing tasks, gave almost the same results as initialization with the HOSVD [135]. Such an initialization method is also recommended in the paper by Chen and Saad [136]. In our implementation we also follow this method and initialize $\mathbf{S}_k$ with a uniform random generator, as will be discussed in the next section.

Finally, let us notice that the HOOI approach, used in Algorithm 2.19 as well as in the attached C++ implementation, is not the only possible solution to the best rank-$(R_1, R_2, \ldots, R_P)$ tensor approximations. Other methods can be considered as follows. For large tensors a Tucker-type decomposition method was proposed by Oseledets *et al.* [137]. For large and sparse tensors Savas and Eldén developed a method based on the Krylov subspace projections [138]. Recently, an algorithm based on the trust-region scheme on Riemannian manifolds was proposed by Ishteva *et al.* [139]. It shows quadratic convergence, whereas the convergence speed of HOOI is only linear. As already mentioned, the truncated HOSVD can also be seen as a simpler approach to the the best rank-$(R_1, R_2, \ldots, R_P)$ tensor approximation (Section 2.12.1.5). However, for the majority of problems it provides only suboptimal results and therefore this method can only be recommended as an initialization step to the HOOI and other methods.

As an example using Algorithm 2.19 let us compute rank-$(R_1=2, R_2=2, R_3=1)$ approximation of the 3D tensor in Figure 2.4 which has dimensions $N_1 = 3, N_2 = 3, N_3 = 3$. From Equation (2.251) we obtain the following matrices $\mathbf{S}_k$

$$\mathbf{S}_1 = \begin{bmatrix} 0.76984 & -0.61275 \\ 0.46531 & 0.73033 \\ 0.43684 & 0.30192 \end{bmatrix}, \; \mathbf{S}_2 = \begin{bmatrix} 0.85157 & -0.51297 \\ 0.38702 & 0.47599 \\ 0.35362 & 0.71435 \end{bmatrix}, \; \mathbf{S}_3 = \begin{bmatrix} 0.910751 \\ 0.398345 \\ 0.108879 \end{bmatrix},$$

(2.270)

each of dimensions $N_k \times R_k$ for $k = 1,2,3$, respectively. Let us also observe that the two columns in $\mathbf{S}_1$, as well as in $\mathbf{S}_2$, are orthogonal, as expected. The core tensor

$$\mathcal{Z} = \begin{bmatrix} 12.30200 & 0.00025 \\ 0.00076 & 8.92124 \end{bmatrix},$$

has dimensions $2 \times 2 \times 1$. Finally, the approximating tensor is

$$
\tilde{\mathbf{T}}_{(1)} = \begin{bmatrix} 9.89850 & 4.32941 & 1.18335 & 0.96840 & 0.42356 & 0.11577 & -0.50636 & -0.22147 & -0.06053 \\ 1.39594 & 0.61056 & 0.16688 & 4.84240 & 2.11797 & 0.57890 & 6.08274 & 2.66047 & 0.72718 \\ 2.90962 & 1.27261 & 0.34784 & 3.06201 & 1.33927 & 0.36606 & 3.48330 & 1.52353 & 0.41642 \end{bmatrix}.
$$

which contains quite different values compared to the input tensor (2.166). In this case the approximation error (2.250) is $\Theta = 4.96$.

Wang and Ahuja propose rank-R decomposition, where $R = R_1 = R_2 = \ldots = R_P$, for dimensionality reduction for any kind of multidimensional data, such as video sequences or volume data, which they call Datum-as-Is [135]. As they show, the rank-R decomposition allows the *exploitation of redundancies among data in all dimensions*. Thus, it implicitly encodes the covariances among data. In effect this leads to much higher compression ratio and better data representation for object classification. For the special case of 3D data, such as video, they propose another rank-R tensor approximating algorithm which is based on slice projection of 3D tensors, and which allows faster convergence. Wang and Ahuja also showed the superior results of the low rank tensor approximation approach as compared to the low rank approximation of matrices [140].

To appreciate compression properties of the best rank-$(R_1, R_2, \ldots, R_P)$ decomposition of tensors let us compare the amount of memory required for the original data tensor $\mathcal{T} \in \mathfrak{R}^{N_1 \times N_2 \times \ldots \times N_P}$, as compared to its approximated version (2.251), in which $\mathbf{S}_k \in \mathfrak{R}^{N_k \times R_k}$ and $\mathcal{Z} \in \mathfrak{R}^{R_1 \times R_2 \times \ldots \times R_P}$. Storage of $\mathcal{T}$ requires allocation of memory for the following number of elements

$$
D_0 = N_1 N_2 \ldots N_P. \tag{2.271}
$$

On the other hand $\tilde{\mathcal{T}}$ requires

$$
D_1 = R_1 R_2 \ldots R_P + \sum_{k=1}^{P} N_k R_k. \tag{2.272}
$$

If we assume small values of R_k as compared to N_k, which is frequently the case, then the difference between D_0 and D_1 becomes obvious, i.e. $D_1 \ll D_0$. For instance, for a 128 long sequence of monochrome video of resolution 640×480 $D_0 = 37.5$ MBytes. Assuming that $R_1 = R_2 = \ldots = R_P = 48$ we obtain $D_1 = 0,163$ MBytes. Hence, the compression ratio in this case would be 230:1. However, when considering storage of $\mathbf{S}_k$ and $\mathcal{Z}$ we have to account for the dynamical range of their data which usually requires more than one byte per element.

Finally, let us observe that Algorithm 2.19 is a generalized version of the rank-1 method outlined in Section (2.12.5). That is, it computes the rank-1 approximation if all R_k are set to 1, as in the following example.

Using Algorithm 2.19 let us compute the rank-($R_1 = 1, R_2 = 1, R_3 = 1$) approximation of the tensor in Figure 2.4. We obtain the following results

$$\mathbf{S}_1 = \begin{bmatrix} 0.29523 \\ 0.86353 \\ 0.40885 \end{bmatrix}, \mathbf{S}_2 = \begin{bmatrix} 0.51655 \\ 0.35655 \\ 0.77850 \end{bmatrix}, \mathbf{S}_3 = \begin{bmatrix} 0.61789 \\ 0.77966 \\ 0.10168 \end{bmatrix},$$

each of dimensions $N_k \times R_k$ for $k = 1,2,3$, respectively. This time the core tensor is just a scalar

$$\mathcal{Z} = [12.3706],$$

so it has dimensions $1 \times 1 \times 1$. In this case the approximation error (2.250) is $\Theta = 7.85$.

However, Algorithm 2.17 for rank-1 tensor approximation requires computation of just a norm of a vector in Equation (2.228) while Algorithm 2.19 requires computation of the SVD decomposition in Equation (2.266).

2.12.8.1   IMPLEMENTATION – Rank Tensor Decompositions

Figure 2.35 depicts the hierarchy of template classes for tensor rank related decompositions. The classes *Rank_1_ApproxFor* and its derived *Rank_1_DecompFor* perform rank-1 tensor approximation and decompositions, are presented in Sections 2.12.5 and 2.12.6 respectively. The *Best_Rank_R_DecompFor* class carries out best rank-($R_1, R_2, \ldots, R_P$) decomposition, as discussed in Section 2.12.7.

All classes in Figure 2.35 define functor objects in which their main action, which is related to a tensor decomposition, are implemented with overloaded virtual function *operator()*.The auxiliary hierarchy derived from the base *S_Matrix_Initializer* class is responsible for different types of matrix initialization for decompositions. In this respect we have two classes *OrphanInitializedMatrices_S_DominantLeftSingularVectors* and *OrphanInitializedMatrices_S_UniformRandomGenerator* which do specific initializations, as previously discussed.

From all of the decompositions discussed in this section, special attention will be devoted to the (in some sense) most general type of decomposition, i.e. the best rank decomposition. Algorithm 2.20 contains the code listing with the definition of the *Best_Rank_R_DecompFor* template class.

```
///////////////////////////////////////////////////////
// This class implements the tesor
// best rank-(R1, R2, ..., RP) decomposition of tensors
///////////////////////////////////////////////////////
template < typename T, typename ACC = double >
class Best_Rank_R_DecompFor : public TensorAlgebraFor< T, ACC >
{
        public:
```

```
          typedef TensorAlgebraFor< T, ACC >              BaseClass;

          typedef vector< int >                           RankVector;

          typedef typename TensorAlgebraFor< T, ACC >  TensorAlgebra;

          typedef typename TFlatTensorFor< T >            FlatTensor;

          typedef typename BaseClass::FlatTensor_AP       FlatTensor_AP;

          typedef typename TFlatTensorProxyFor< T >       FlatTensorProxy;

          typedef typename TImageFor< T >                 Matrix;


protected:

          S_Matrix_Initializer< T, ACC >        *        fMatrix_Initializer_Obj;

public:


          ///////////////////////////////////////////////////////////
          // Class constructor
          ///////////////////////////////////////////////////////////
          //
          // INPUT:
          //              S_init_obj - orphaned object used to initialize
          //              the series of decomposing matrices {Si}.
          //              Cannot be 0.
          //
          // REMARKS:
          //              kSquareNessHeuristics and kNumOfElemsHeuristics
          //              set the conditions whether it is beneficial
          //              to compute svd( A ) or svd( AAT )
          //
          Best_Rank_R_DecompFor( S_Matrix_Initializer< T, ACC > * S_init_obj );

public:


          ///////////////////////////////////////////////////////////
          // This function computes the domminant subspace
          ///////////////////////////////////////////////////////////
          //
          // INPUT:
          //          svd_calculator - the object for computation
          //              of the matrix SVD decomposition
          //          S_hat - the input matrix for decomposition
          //              (this contains a projection of a tensor
          //              onto the rest of the decomposing matrices)
          //          S - output matrix with requested_tensor_Rank_k
          //              orthogonal columns, each column has
          //              tensor_index_k elements; However S should
```

```
//                        be supplied in this size.
//                        tensor_index_k - k-th dimension of the tensor (see
//                        above)
//
// OUTPUT:
//            none
//
// REMARKS:
//
//
virtual void FindDominantSubspace( Compute_SVD & svd_calculator,
        const typename FlatTensor::DataMatrix & S_hat,
        typename FlatTensor::DataMatrix & S,
        int requested_tensor_Rank_k, int tensor_index_k );




/////////////////////////////////////////////////////////////
// This function does best rank-(R1, R2, ..., RP)
// decomposition of tensors
/////////////////////////////////////////////////////////////
//
// INPUT:
//            T - tensor to be decomposed
//            requested_ranks - a vector of requested ranks
//                of the approximation. Its size should
//                be the same as number of dimensions of T
//            S_vector - set of output matrices of rank-Rk
//            epsilon - threshold for the convergence error
//            max_iter_counter - a fuse counter for the maximal
//                number of iterations
//            num_of_iterations - is supplied, this conveys
//                the real number of executed iterations
//
// OUTPUT:
//            The core tensor of the decomposition
//
// REMARKS:
//            The S_vector matrices will be intialized with the
//                object supplied to the constructor.
//
virtual FlatTensor_AP operator() ( const FlatTensor & T,
        const RankVector & requested_ranks, typename
        SingleMatrixVector & S_vector,
        const AccumulatorType epsilon = 1e-6, const int
        max_iter_counter = 1000,
        int * num_of_iterations = 0 );

};
```

Algorithm 2.20 Definition of the *Best_Rank_R_DecompFor* class which implements the best rank-(R_1, R_2, ..., R_P) decomposition of tensors.

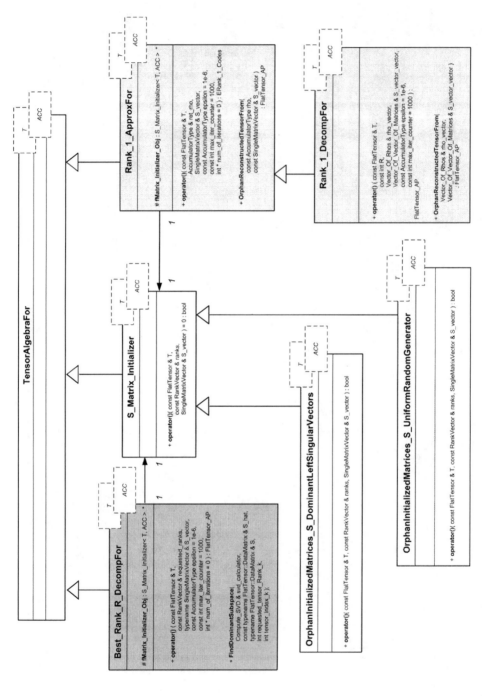

Figure 2.35 Hierarchy of the classes for tensor rank decompositions. The *Best_Rank_R_DecompFor* class carriess out best-rank-R decomposition. The classes *Rank_1_ApproxFor* and its derived *Rank_1_DecompFor* perform rank-one tensor approximation and decompositions, respectively. The auxiliary hierarchy derived from the base *S_Matrix_Initializer* class is responsible for different types of matrix initialization for decompositions.

As alluded to previously, computation of the rank-$(R_1, R_2, \ldots, R_P)$ follows Algorithm 2.19. The main part of the C++ implementation is presented in Algorithm 2.21.

```
1
2     /////////////////////////////////////////////////////////////
3     // This function does best rank-(R1, R2, ..., RP)
4     // decomposition of tensors
5     /////////////////////////////////////////////////////////////
6     //
7     // INPUT:
8     //        T - tensor to be decomposed
9     //        requested_ranks - a vector of requested ranks
10    //           of the approximation. Its size should
11    //           be the same as number of dimensions of T
12    //        S_vector - set of output matrices of rank-Rk
13    //        epsilon - threshold for the convergence error
14    //        max_iter_counter - a fuse counter for the maximal
15    //           number of iterations
16    //        num_of_iterations - is supplied, this conveys
17    //           the real number of executed iterations
18    //
19    // OUTPUT:
20    //        The core tensor of the decomposition
21    //
22    // REMARKS:
23    //        The S_vector matrices will be intialized with the
24    //           object supplied to the constructor.
25    //
26    template < typename T, typename ACC >
27        typename Best_Rank_R_DecompFor< T, ACC >::FlatTensor_AP
28          Best_Rank_R_DecompFor< T, ACC >::operator() ( const FlatTensor & T,
29                      const RankVector & requested_ranks,
30                      typename SingleMatrixVector & S_vector,
31                      const AccumulatorType epsilon = 1e-6,
32                      const int max_iter_counter = 1000,
33                      int * num_of_iterations = 0 )
34    {
35        const int T_valence = T.GetTensorDimension();//get dimensionality of T const
36        FlatTensor::TensorElem_IndexVector & tensorIndices = T.GetIndexVector();
37
38        // Prepare the S vectors
39        (* fMatrix_Initializer_Obj )( T, requested_ranks, S_vector );
40        // initialized the set {Si}
41
42        double prevFrobeniusNorm = 0.0;
43
44        FlatTensor_AP Approx_T( 0 );
45
46        bool stop = false;
47
48        int iter_counter = 0;
49
50        Compute_SVD          svd_calculator;
```

```
51
52       while( iter_counter ++ < max_iter_counter && stop == false )
53       {
54
55         // ----------------------------------
56         for( int k = 0; k < T_valence; ++ k )
57         {
58           // Compute the two Kronecker products
59           BaseImage_AP              common_Kronecker(
60             Orphan_Kronecker_Of_Series( S_vector, k, T.GetCyclicMode() ) );
61
62           // Prepare the k-mode tensor for this run.
63           FlatTensorProxy proxy_procTensor( (RealTensor &)T, k, T.GetCyclicMode() );
64
65           // Multiply with the k-mode flattened version of the input tensor
66           BaseImage_AP          S_hat(   Orphan_TensorProxy_Mult_Matrix(
67                                   proxy_procTensor, * common_Kronecker ) );
68           REQUIRE( S_hat.get() != 0 );
69
70           // +++++++++++++++++
71           // Now compute svds
72
73           if( S_hat->GetElems() >= kNumOfElemsHeuristics &&
74               kSquareNessHeuristics * S_hat->GetRow() <= S_hat->GetCol() )
75           {
76             // only for this conditions svd( S*S' ) makes sense over a simple
77                 svd( S )
             std::auto_ptr< Matrix > S_hat_times_S_hat_transposed(
78               Orphan_Mult_Matrix_With_SECOND_Transposed< TensorElem_Type,
79                   AccumulatorType >( * S_hat, * S_hat ) );
80             S_hat.reset();              // immediately get rid of the S_hat
81             FindDominantSubspace( svd_calculator, *
S_hat_times_S_hat_transposed,
82                     * S_vector[ k ], requested_ranks[ k ], tensorIndices[ k ] );
83           }
84           else
85           {
86             FindDominantSubspace( svd_calculator, * S_hat,
87                 * S_vector[ k ], requested_ranks[ k ], tensorIndices[ k ] );
88           }
89         // +++++++++++++++++
90         }
91         // ----------------------------------
92
93
94         // Compute the approximating tensor from the just computed series
95             of matrices Sk
         Approx_T = Orphan_PROJECTED_Tensor_From( S_vector, T );
96
97         // Compute the Frobenius norm of the approximated
98         Frobenius_MatrixNormFor< TensorElem_Type, AccumulatorType >
                theNormProcessor;
99         ForEachElementDo( * Approx_T, theNormProcessor );
100        AccumulatorType newFrobeniusNorm = theNormProcessor.GetNorm();
```

```
101
102        if( fabs( newFrobeniusNorm - prevFrobeniusNorm ) <= epsilon )
103           stop = true;
104
105        prevFrobeniusNorm = newFrobeniusNorm;        // copy for the next iteration
106      }
107
108      if( num_of_iterations != 0 )
109        * num_of_iterations = iter_counter;
110
111      return Approx_T;
112  }
```

Algorithm 2.21 Definition of the function *operator()* performing the best rank-$(R_1, R_2, \ldots, R_P)$ decomposition of tensors.

Once again, the key steps of the algorithm are shown in bold. At first, in line <39> the decomposition matrices are initialized thanks to the auxiliary object of type *S_Matrix_Initializer* supplied during construction. The main iteration loops span the lines <52–106>. It can be finished under two conditions, reaching the requested accuracy or reaching the maximal number of allowed iterations. In line <60> the product of the matrices is computed with the *Orphan_Kronecker_Of_Series* function belonging to the *TensorAlgebraFor*. Then, in lines <81> or <86>, depending on which branch is the most efficient in the current iteration, the dominating subspace is computed with the *FindDominantSubspace* function, which relies on the SVD matrix decomposition. The current version of the approximating tensor is computed in line <95> invoking the *Orphan_PROJECTED_Tensor_From* function. Its degree of approximation is checked calculating the Frobenius norm in lines <98–103>, and if it is below the requested threshold (which belongs to the one of the parameters of this procedure) the iterations are stopped immediately. Otherwise, the current approximation of the tensor is saved and the next iteration commences.

The best way to see how the above procedure operates is to check it in an application. Below we present a test function which with help of the *Best_Rank_R_DecompFor* class computes a simple rank-1 decomposition.

```
void Test_Best_Rank_R( void )
{
      // Let us create a tensor from a text file
      RealTensor::TensorElem_IndexVector tensor_Indices;
      tensor_Indices.push_back( 3 );
      tensor_Indices.push_back( 3 );
      tensor_Indices.push_back( 3 );

      int input_tensor_mode = 0;

      RealTensor::FlatTensor_AP T(
              gRealTensorAlgebraObj.orphan_k_mode_tensor_from(
              "tensor_2_104_1_forward.txt", tensor_Indices, 0,
```

```
                        RealTensor:: kForward )
                                        );


    // Now compute the best Rank-Ri

    vector< int > ranks;
    ranks.push_back( 1 );      // Setup requested ranks
    ranks.push_back( 1 );      // of the approximation
    ranks.push_back( 1 );


    TensorAlgebraFor< double >::SingleMatrixVector S_vector;



    Best_Rank_R_DecompFor< double, double > obj(
            new OrphanInitializedMatrices_S_UniformRandomGenerator
                    < double, double >( 255.0 ) );


    RealTensor::FlatTensor_AP Approx_T( obj( * T, ranks, S_vector ) );


    /////////////////////
    // Print them out
    save_tensor_data( "tensor_Approx_T.txt", * Approx_T );
    for( int i = 0; i < S_vector.size(); ++ i )
    {
        string outFileName( "S_matrix_" );

        outFileName += '0' + i;
        outFileName += ".txt";

        Save_ASCII_Image( * S_vector[ i ], outFileName.c_str() );
    }
    /////////////////////


    // Test accuracy of the approximation
    RealTensor::FlatTensor_AP
            T_copy( gRealTensorAlgebraObj.Orphan_RECONSTRUCTED_Tensor_From(
                                            S_vector, * Approx_T ) );
    save_tensor_data( "tensor_CHECK_tensor.txt", * T_copy );
    double itBetterBeZero = Get_MSE( * T, * T_copy );
    REQUIRE( itBetterBeZero < 1e-15 );


}
```

Algorithm 2.22 Example of using the *Best_Rank_R_DecompFor* class for tensor decomposition.

In the first part of the function *Test_Best_Rank_R* in Algorithm 2.22 a tensor of dimensions $3 \times 3 \times 3$ is created. Its elements are then read from a text file. The requested ranks for decomposition are prepared and stored in the *ranks* vector. In this case we set all of them to 1, so the best rank-*R* algorithm is used as a substitute for the simpler rank-1 decomposition. Nevertheless, details of implementations are different, so if only rank-1 decomposition is necessary then

Table 2.4   Achieved compression ratios and reconstruction errors for
the sequences in Figure 2.36.

Ranks	Compression ratio	MSE	PSNR [dB]
40-40-15	24:1	40.3	32.1
30-20-10	56:1	83.6	28.9

the *Rank_1_DecompFor* object can perform faster. The next step consists of creating the decomposition object *obj* of type *Best_Rank_R_DecompFor* with the random initialization object to initialize the output matrices, as well as invoking the decomposition operation. Finally, results are saved to the files, and the approximation accuracy is measured with help of the *Get_MSE* function which calculates the mean square error between two tensors.

2.12.8.2   CASE STUDY – Data Dimensionality Reduction

As mentioned in Section 2.12.7, the best rank-$(R_1, R_2, \ldots, R_P)$ decomposition of tensors can be used to approximate any data tensor. This feature can be used to compress video sequences, as will be shown in this short example. The input video sequence will be represented as a 3D tensor $\mathcal{T}$. This after the best rank-(R_1, R_2, R_3) decomposition is represented exclusively by the three matrices $\mathbf{S}_1, \mathbf{S}_2, \mathbf{S}_3$, and the core tensor $\mathcal{Z}$, as expressed in Equation (2.251). The decompositions presented hereafter were computed with the C++ implementation of Algorithm 2.19, as discussed in Section 2.12.8.1. Results for the *Hamburg Taxi* sequence,[14] for different values of the ranks R_1, R_2, R_3, are presented in Figure 2.36.

The original sequence counts 41 monochrome frames, each of resolution 256×191, as shown in the left column of Figure 2.36. This requires at least $D_0 = 2\ 004\ 736$ bytes in an uncompressed representation. The first compressed version, which is shown in the middle column of Figure 2.36, was obtained for the ranks $R_1 = 40, R_2 = 40$, and $R_3 = 15$, respectively. In accordance with Equation (2.272) this requires $D_1 = 42\ 495$ elements to be stored. The second sequence, shown in the rightmost column of Figure 2.36, was compressed for the ranks $R_1 = 30, R_2 = 20$, and $R_3 = 10$. In this case the storage parameter $D_1 = 17\ 910$. Initialization of the matrices in Algorithm 2.19 was done with the random generator with uniform distribution. The stop criterion ε_{ACCEPT} was set to 0.05 to speed up computations. Table 2.4 shows the achieved compression ratios and reconstruction errors for the two sequences in Figure 2.36.

However, we need to remember that in general the computed matrices $\mathbf{S}_1, \mathbf{S}_2, \mathbf{S}_3$, and the core tensor $\mathcal{Z}$, can contain values of high arithmetic dynamics, that is these can be positive or negative, and also can have a large exponent. These can, for instance, be observed in the example on p. 135. Thus, in the general case we need to store floating point values. Usually they require 8 bytes each (in C++ they are denoted with the *double* data type), depending on the computer system. However, in many cases it is possible to use a more efficient representation with the fixed point format. Unfortunately, that is not a built-in type in C++, although it is quite straightforward to create a proper class to achieve this goal, as for instance the

---

[14]These test video sequences were made publicly available by the group of Prof. Dr. H.-H. Nagel from the Fakultät für Informatik of the Karlsruhe Institute of Technology [79].

Frame No	Original sequence 256×191×41	Rank-40-40-15 compressed	Rank-30-20-10 compressed
1			
3			
5			
7			
9			

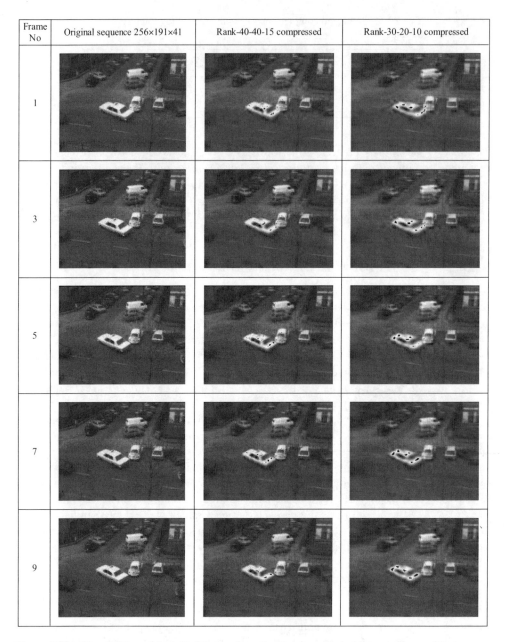

Figure 2.36 Visualization of the initial frames from the Hamburg Taxi video sequence and its two versions compressed with the rank-R decomposition method. Original sequence $256 \times 191 \times 41$ (left column), compressed with rank-40-40-15 decomposition (middle), compressed with rank-30-20-10 decomposition (right).

Table 2.5 Achieved compression ratios and reconstruction errors for
the two sequences in Figure 2.37.

Ranks	Compression ratio	MSE	PSNR [dB]
20-20-3-3	16 : 1	146.87	27.1
10-20-3-3	24 : 1	283.6	26.7

TFixedFor template class from the software framework attached to the book [15]. The tensor
decomposition classes, discussed in Section 1.1, are also template ones which makes them
flexible to use any data type, such as the mentioned *TFixedFor*. Thanks to this, when processing
monochrome images we can limit ourselves to 2 bytes per data, stored as 16 bits integer and 16
bits of fractional. This gives us the precision of $2^{-16} \approx$ 15e-6 and the dynamic range of $\pm 2^{15}$
which more frequently than not is sufficient for the majority of image processing routines.
Thanks to this simple technique we obtain four times the memory reduction as compared to the
most obvious double data representation. This is a substantial gain when processing tensors,
since many algorithms require their representation in memory as one entity.

Figure 2.37 presents the results of the application of the rank-R representation of one of
the color sequences of a person from the Georgia Tech face database which contains RGB
images of fifty people taken in different poses [141]. In this case, the input sequence can be
represented as a 4D tensor. Achieved compression ratios and reconstruction errors for the two
sequences in Figure 2.37 are shown in Table 2.5.

For the discussed data compression task we can go even further with savings on data
representation. For this purpose let us rescale each tensor in Equation (2.251), as follows

$$\tilde{\mathcal{T}} = \lambda \lfloor \mathcal{Z} \rfloor \times_1 \lfloor \mathbf{S}_1 \rfloor \times_2 \lfloor \mathbf{S}_2 \rfloor \times_3 \lfloor \mathbf{S}_3 \rfloor \qquad (2.273)$$

where $\lfloor \mathbf{S}_i \rfloor$ denotes a scaling of values of the matrix $\mathbf{S}_i$ to the range $\pm r$, and λ denotes a scaling
parameter. The scaling of the matrices can be achieved in the following simple steps:

1. Find the maximal absolute value s_{max} of the matrix $\mathbf{S}_i$.
2. Each element multiply by r/s_{max} (assuming $s_{max} \neq 0$).

The same procedure is applied to the core tensor to find out a value of z_{max}. As a result the
additional scalar

$$\lambda = \frac{z_{max}}{r^{P+1}} \prod_{i=1}^{P} s_{max}^{(i)} \qquad (2.274)$$

is introduced which needs to be stored. However, this cost is negligible. Thanks to this, each
element of the tensors in Equation (2.273) can be stored on the reduced number of bytes. For
example, instead of eight bytes used for the floating point we can store the same data using the
mentioned fixed point format, such as 16.16 bits for integer and fractional parts, respectively.
This inevitably adds some quantization errors which, as shown by our experiments, adds only
a small error to the final reconstruction. The results shown in Table 2.4 and Table 2.5 relate to
the aforementioned procedure.

Frame No	Original color[15] sequence 160×120	Rank-20-20-3-3 compressed	Rank-10-20-3-3 compressed

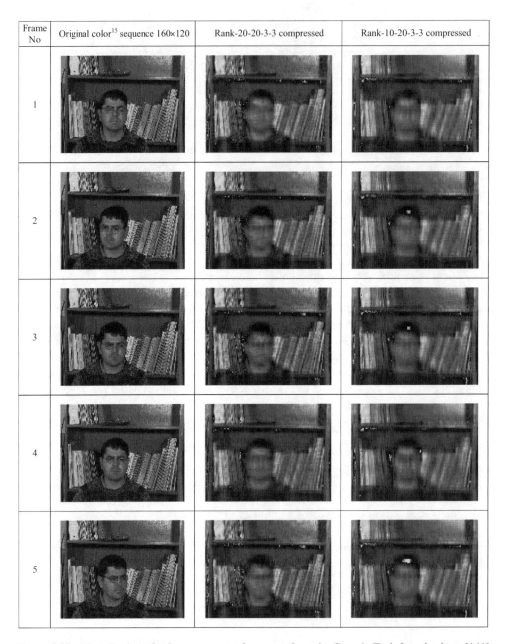

Figure 2.37   Visualization of color sequences of a person from the Georgia Tech face database [141]. Original color sequence (left). Rank-20-20-3-3 compressed version (middle), rank-10-20-3-3 compressed (right). (For a color version of this figure, please see the color plate section.)

---

[15]Color versions of the images are available from the accompanying web page [28].

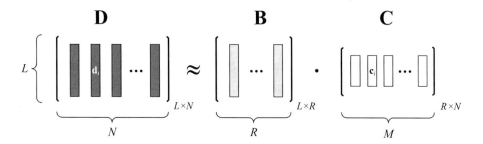

Figure 2.38   Factorization of a data matrix **D** into a product of a base **B** and coefficients **C**.

2.12.9   Subspace Data Representation

In this section we discuss the problem of data representation by means of factorization, i.e. representing collected data points in the form of a matrix product with special properties. We will see that such a transformation allows analysis of the inherent structure of data, in order to gain some insight or for pattern recognition, as well as other useful actions such as data compression, denoising, etc. To this group belongs the already mentioned PCA method, further discussed in Section 3.3.1.

Let us assume that we have collected a series of N measurements, or data, each represented by an L-dimensional column vector $\mathbf{d}_i$. For instance, these can be pixels, each having three values for the red, green, and blue color components, respectively. Then all $\mathbf{d}_i$ are stored in a matrix **D**. Now the question is whether we can find an equivalent representation of **D** which will exhibit some hidden factors, or base information, as well as a number of coefficients. Such a representation can be written in the following factored form

$$\mathbf{D} \approx \mathbf{B}\mathbf{C}, \tag{2.275}$$

where **B** is a hidden matrix, called a *base*, and **C** is a matrix of components (coefficients). The above representation is also called a matrix factorization. Let us now observe that each data point $\mathbf{d}_i$ can be equivalently expressed by the following product

$$\mathbf{d}_i \approx \mathbf{B}\mathbf{c}_i. \tag{2.276}$$

In this product are used all the values from **B** and only one selected column $\mathbf{c}_i$ from the matrix **C**, i.e. the one with the same index i. In other words, each data point can be represented by a linear combination of the base **B**, weighted by the components from **C**. Graphically the factorization is depicted in Figure 2.38. Naturally, we want **B** to convey the internal structure that is latent in data stored in **D**. It would be also good if **B** were as compact as possible, i.e. $R(L + N) < LN$. In such a case the product **BC** in Equation (2.275) can be treated as a compressed form of data in **D**. Because of this the factorizations are frequently used for dimensionality reduction. Now, given a matrix **D** the question is how to compute its factorization (2.275). In this respect the most popular methods are:

1. Vector quantization (VQ) [142].
2. Principal component analysis (PCA) [143].

3. Independent component analysis (ICA) [97, 144].
4. Nonnegative matrix factorization (NMF) [145].

Each of the aforementioned methods comes from different assumptions on data **D** and reveals different structures of data. In the case of VQ each column c_i contains only one value 1 and all the others are 0. This simply means that the product (2.276) will select only one b_j. Because of this each data point d_i is approximated by exactly one base vector b_j. Thus, each b_i becomes a prototype pattern. This type of factorization, strictly related to the k-means methods, is discussed in Section 3.11.1.

On the other hand, PCA restricts each b_i to be orthonormal, as discussed in Section 3.3.1. Vectors b_i are called eigenvectors. Interestingly, this time all rows of **C** are also orthogonal to each other. However, the columns c_i are not unary now, so each pattern d_i is now represented as a linear product of possibly *all* basis vectors b_i. As already mentioned, columns of the input matrix **D** are just data points. For PCA all them must be additionally centered, i.e. from each data point vector the mean of all data points needs to be subtracted. For instance, a well known method for human face recognition that utilizes PCA for appearance-based face recognition with the so called eigenfaces was proposed by Turk and Pentland [146]. In this case, data points are just vectorized images with training faces. These, after being centered, constitute columns of the matrix **D**. PCA decomposition is obtained, e.g. computing the SVD decomposition of **D**, or a product **DD**^{T} [66, 147, 148]. A decade later its multi-linear version in the form of *tensorfaces* was proposed by Vasilescu and Terzopoulos [11], which is discussed in Section 2.12.15.

However, PCA is able to encode only a statistical correlation among pixels from training images. That is, it operates only on the second-order statistics, ignoring all higher-order dependencies. In this respect ICA constitutes a generalization to the PCA since it also captures higher-order statistical dependencies among data in order to learn the statistically independent components of the input. In the case of ICA the main idea is to maximize the statistical independence of the components representing the training data set [97]. ICA can be computed in two different versions [12].

1. In the ICA$_1$ the goal is to find a statistically independent basis **B** that reflect local properties of data in **D**. This is obtained by starting from the PCA and then rotating the principal component vectors such that they become statistically independent. This can be written as follows

$$\mathbf{D}^T = \mathbf{V\Sigma U}^T = \underbrace{(\mathbf{V\Sigma W}^{-1})}_{\mathbf{C}^T}\underbrace{(\mathbf{WU}^T)}_{\mathbf{B}^T} = (\mathbf{BC})^T \qquad (2.277)$$

where $\mathbf{D} = \mathbf{U\Sigma V}^T$ denotes the SVD decomposition[16] of **D**, and **W** is an invertible transformation matrix which should be computed with one of the ICA algorithms [12, 97].

---

[16]In this section matrices of the SVD decomposition are denoted with **U**, Σ, and **V**, to avoid confusion with matrix **D**.

2. In the ICA$_2$ the objective is to find independent coefficients. In this case we also employ rotation of the principal components, this time to make the coefficients **C** statistically independent, however. That is

$$\mathbf{D} = \mathbf{U}\mathbf{\Sigma}\mathbf{V}^T = \underbrace{\left(\mathbf{U}\mathbf{W}^{-1}\right)}_{\mathbf{B}}\underbrace{\left(\mathbf{W}\mathbf{\Sigma}\mathbf{V}^T\right)}_{\mathbf{C}} = \mathbf{B}\mathbf{C} \qquad (2.278)$$

Because **C** now contains statistically independent variables, its coefficients are also known as the factorial code.

Similarly to the multilinear version of the PCA, there is also multilinear version of ICA, in the two aforementioned versions. Such multilinear ICA (MICA) was proposed by Vasilescu and Terzopoulos to be applied to the problem of face recognition [12]. As reported, the experimental results showed to be superior.

2.12.10   Nonnegative Matrix Factorization

In the already presented tensor decompositions we did not assume any constraints on the results of their decomposition. Thus, in general these are real values. However, in many biological, psychological, technical, and other applications to increase a degree of interpretation further constraints can be imposed on the decompositions, such as the following:

1. Nonnegativity.
2. Sparseness.
3. Limited data range.

Nonnegativity is assumed in many disciplines, especially if the input signals are assumed to be always nonnegative, such as in the analysis of signals of brain activity. This is also related to probability distributions or fuzzy membership functions. On the other hand, sparseness relates to feature selection and generalization properties in data classification. Finally, limited data range – which to some extent can be connected to the sparseness property – can help in the processing and storage of the tensors and their decompositions. In many applications it is important to ensure nonnegativity of data before and after decomposition. For instance, since the video scenes are composed of nonnegative values, for proper analysis it is required that a decomposition is also expressed with nonnegative components.

There is strong experimental evidence that humans' abilities of object recognition rely on parts-based representation in the brain [149]. To mimic this process in a computer Lee and Seung proposed application of the nonnegative matrix factorization to learn parts of faces and text semantic features [145]. Parts-based representation remains in contrast to the holistic methods, such as the PCA or vector quantization, discussed in the previous section. Additional constraint on nonnegativity of the signals writes into the framework of additive signals, i.e. all signals in such a system can only be weighted with nonnegative weights and summed up. Subtraction is not allowed at any stage. Interestingly, in this respect this method has some resemblance to the morphological networks which we discuss in Section 3.9.4.

Let us now assume a data model discussed in the previous section. Starting from approximation (2.275) we need to choose a certain measure of similarity which allows us to estimate the accuracy of a particular estimation. As mentioned in the book by Cichocki *et al.* the choice of similarity measure should depend on the probability distribution of the decomposed data or signals [124]. The two most frequently used in computation of NMF are the squared Frobenius norm (i.e. the squared Euclidean distance), as well as the generalized Kullback–Leibler divergence, discussed in Section 3.5. Let us start with the Frobenius norm, which has already been defined in Section 2.5.1. Since the right side of Equation (2.275) denotes only an approximation it can be rewritten as follows

$$\tilde{\mathbf{D}} = \mathbf{BC} \qquad (2.279)$$

We are interested in minimizing the difference in the light of the norm D_F between the original matrix $\mathbf{D}$ and its approximation (2.314), that is

$$\min_{\tilde{\mathbf{D}}} D_F^2\left(\mathbf{D} - \tilde{\mathbf{D}}\right) = \min_{B,C} D_F^2\left(\mathbf{D} - \mathbf{BC}\right) = \min_{B,C} \|\mathbf{D} - \mathbf{BC}\|_F^2$$
$$\text{subject to}: \mathbf{B} \geq 0, \quad \mathbf{C} \geq 0, \qquad (2.280)$$

where condition $\mathbf{B} \geq 0$ means that each element of the matrix $\mathbf{B}$ is not negative. The solution to the above can be found with help of the Karush–Kuhn–Tucker first-order optimality conditions, which lead to the following *multiplicative* update rules, as follows

$$b_{nr} \leftarrow b_{nr}\frac{\left[\mathbf{DC}^T\right]_{nr}}{\left[\mathbf{BCC}^T\right]_{nr} + \mu}, \qquad (2.281)$$

$$c_{rm} \leftarrow c_{rm}\frac{\left[\mathbf{B}^T\mathbf{D}\right]_{rm}}{\left[\mathbf{B}^T\mathbf{BC}\right]_{rm} + \mu}, \qquad (2.282)$$

which follow the work by Lee and Seung [150]. From the above we see that the optimization process is twofold, i.e. in each elements of the matrices $\mathbf{B}$ and $\mathbf{C}$ are obtained alternatively. This is a strategy which does not guarantee a global minimum. However, it converges to a local minimum, similarly to the already presented ALS method (Section 2.12.5). To improve numerical stability the denominators of the above update rules were endowed with a small constant value μ (usually in the range 10^{-12}–10^{-16}). The two updated rules (2.316) and (2.317) can be written in a succinct matrix notation, as follows

$$\mathbf{B} \leftarrow \mathbf{B} \circledast \left[\left(\mathbf{DC}^T\right) \oslash \left(\mathbf{BCC}^T + \mu\right)\right], \qquad (2.283)$$

$$\mathbf{C} \leftarrow \mathbf{C} \circledast \left[\left(\mathbf{B}^T\mathbf{D}\right) \oslash \left(\mathbf{B}^T\mathbf{BC} + \mu\right)\right], \qquad (2.284)$$

where $\circledast$ denotes element-wise multiplication (called also the Hadamard product), and $\oslash$ is an element-wise division of the two matrices which must be of the same dimensions.

The Hadamard product of two matrices is just an element-wise product of their elements. That is, for two matrices $\mathbf{A}, \mathbf{B} \in \mathfrak{R}^{I \times J}$ of the same dimensions, their Hadamard product is a matrix $\mathbf{C} \in \mathfrak{R}^{I \times J}$ defined as follows:

$$\mathbf{A}_{I \times J} \circledast \mathbf{B}_{I \times J} = \mathbf{C}_{I \times J} = \begin{bmatrix} a_{11}b_{11} & \cdots & a_{1J}b_{1J} \\ \vdots & \vdots & \vdots \\ a_{I1}b_{I1} & \cdots & a_{IJ}b_{IJ} \end{bmatrix}. \tag{2.285}$$

Similarly, $\mathbf{A} \oslash \mathbf{B}$ denotes an element-wise division. In this case we assume that $\mathbf{B}$ does not contain elements of value 0. For the following matrices

$$\mathbf{A}_{2 \times 2} = \begin{bmatrix} 0 & -2 \\ 10 & -1 \end{bmatrix}, \quad \mathbf{B}_{2 \times 2} = \begin{bmatrix} 1 & 2 \\ 1 & -1 \end{bmatrix}, \tag{2.286}$$

their Hadamard product and division are

$$\mathbf{A} \circledast \mathbf{B} = \mathbf{B} \circledast \mathbf{A} = \begin{bmatrix} 0 & -4 \\ 10 & 1 \end{bmatrix}, \quad \mathbf{A} \oslash \mathbf{B} = \begin{bmatrix} 0 & -1 \\ 10 & 1 \end{bmatrix},$$

respectively[17]. Let us finally observe that in this fashion we cannot compute $\mathbf{B} \oslash \mathbf{A}$.

Replacing D_F in Equation (2.280) with the Kullback–Leibler divergence D_{KL} the following update rules are obtained [150]

$$b_{nr} \leftarrow b_{nr} \frac{\sum\limits_{m=1}^{M} c_{rm} d_{nm} / (\mathbf{BC})_{nm}}{\sum\limits_{m=1}^{M} c_{rm} + \mu}, \tag{2.287}$$

$$c_{rm} \leftarrow c_{rm} \frac{\sum\limits_{n=1}^{N} b_{nr} d_{nm} / (\mathbf{BC})_{nm}}{\sum\limits_{n=1}^{N} b_{nr} + \mu}, \tag{2.288}$$

where matrix dimensions are as shown in Figure 2.38.

Let us observe useful properties of NMF in the light of object recognition. NMF can be used as a method of discovering the generation mechanism of the observable measurements from a number of some objects' constructive parts and hidden components. The former are contained in the base $\mathbf{B}$, whereas the latter in the $\mathbf{C}$ matrices, respectively. Each hidden component *selects*

---

[17] In MATLAB® the Hadamard multiplication and division is obtained with the .* and ./ operators, respectively [29].

and *additively* combines a subset of objects' parts to generate an observable pattern, such as a pictogram or a face image, etc. The additive nature of this combination gives a useful tool for intuitive analysis of the components and their interaction, especially if data by their nature are nonnegative, such as images, EEG [151], etc.

NMF can be directly used for feature extraction from images. However, the obtained bases are not orthogonal and therefore they are not directly suitable for object recognition with Euclidean distance, in contrast to other methods such as PCA or HOSVD bases discussed in Section 2.12.4. Liu and Zheng propose a method to alleviate this problem with NMF [152]. The first step is to perform vector orthonormalization, e.g. using the Gram–Schmit procedure [66], and then to reproject data onto the new, this time orthogonal, base. The second idea is to employ the Riemannian metric instead of the commonly used Euclidean. As shown in the paper by Liu and Zheng both methods significantly increase the accuracy of object recognition [152]. The other idea for parts-based object representation with NMF, proposed by Guillamet and Vitrià [153], is to use the Earth Mover's Distance (EMD). This type of distance is discussed in Section 3.7.2 of this book. However, the method with EMD has been shown to be time demanding. Nevertheless, no orthonormalizations on NMF guarantee that the new representation will still remain nonnegative. Another possibility is to superimpose an additional orthogonality constraint into the optimization task given in Equation (2.280). This option is discussed in the next section.

Finally, as shown by Ding *et al.* [154], there is equivalence between the NMF and a relaxed form of the k-means algorithm used for data clustering, which we analyze in Section 3.11.1, as well as vector quantization, discussed on p. 149. Considering Equation (2.275), this means that the matrix $\mathbf{B}$ contains cluster centroids whereas $\mathbf{C}$ stores the cluster membership values. Thus, for the crisp k-means each column $\mathbf{c}_i$ contains only one value 1 and all other 0. Kim and Park indicated the importance of the sparsity constraint when considering clustering with help of the NMF [155]. They observed that NMF and k-means are even not equivalent and that NMF outperforms k-means, providing better and more consistent results which can be compared with the recently developed Affinity Propagation method, proposed by Frey and Dueck [156]. Thanks to this property the NMF can be also used to determine the unknown number of clusters in data. A method for band selection in hyperspectral images using clustering by sparse NMF was proposed by Li and Qian [157]. The advantage of sparse NMF is that the distance metric between different spectral bands does not need to be considered, which often constitutes one of the important stages in other cluster-based band selection methods. Finally, cluster assignment of bands is easily indicated by the largest value in each column of the coefficient matrix $\mathbf{C}$. Another example in this area is the method of tumor clustering using NMF with gene selection proposed by Zheng *et al.* [158]. In this case sparseness is reinforced by selecting a subset of genes with the help of independent component analysis. In effect irrelevant or noisy genes are reduced.

As already pointed out, the basic version of NMF belongs to the group of unsupervised methods, that is it does not require prior training. This can be seen as an advantage. However, in some real world problems for which knowledge from domain experts is available this is a limitation since such additional information on data cannot be directly used in the factorization. On the other hand, it was shown that if machine learning from unlabeled data is supported even with a limited number of labeled data, then better results can be obtained [159]. In order to take advantage of this possibility NMF was recently extended by Liu *et al.* into the semi-supervised domain and named *Constrained* NMF (CNMF) [160]. The main idea is that data from the same class should be kept together in the new representation space too. In the result, the obtained parts based representation after the NMF decomposition stay consistent to the original data. The method proposed by Liu *et al.* realizes this task, taking the available labeled data as an additional hard constraint to the NMF. As demonstrated by the authors, the CNMF with the Kullback–Leibler divergence outperforms other versions of NMF in data clustering tested on four popular data sets: ORL, Yale, Corel, and Caltech-101 [161].

2.12.11 Computation of the Nonnegative Matrix Factorization

Both of the multiplicative NMF algorithms presented in the previous section belong to the general group of multiplicative Image Space Reconstruction Algorithms (ISRA), as well as multiplicative Expectation Maximization Maximum Likelihood (EMML) methods (sometimes also called the Richardson–Lucy RLA method). The above multiplicative update rules for computation of the NMF work well in practice. However, we need to remember that in the general case they do not assure reaching the globally optimal solution and can frequently even get bogged down in spurious local minima. This is caused by the fact that the optimization problem (2.280) is strictly convex with respect to either **B** or **C**, but not to both. Also, the convergence process can be slow, i.e. the number of iteration can be substantial. Therefore, the multiplicative update rule algorithm has been modified many times and by many authors. A quite uniform and general approach is presented in the already recommended book by Cichocki *et al.* [124]. They introduce the generalized similarity measures and divergences, called alpha- and beta-divergences, respectively. Also, they comprise many other divergence measures, such as the previously considered Frobenius or Kullback–Leibler ones. Then, the modifications of the NMF algorithms are considered in the light of these alpha- and beta-divergences. The required sparsity and smoothness constraints are enforced by adding proper regularization or penalty terms to the chosen divergences, such as D_{KL}. This and other techniques, as well as suitable algorithms are presented in [124].

Apart from the multiplicative approach to the NMF there is also a group of additive update methods which follow the Projected Gradient optimization algorithm. Also the methods that rely on higher order derivatives can be employed, such as the Projected Quasi-Newton or Conjugate Gradients, which we will not discuss here (more can be found e.g. in [124]). However, what is very promising is yet another group of methods which follow the Alternating Least-Squares (ALS) methodology, already discussed in Section 2.12.5. Since solutions of the NMF problem with the presented methods are not usually unique, it is therefore necessary to impose some additional constraints, such as on sparsity, orthogonality, or smoothness. Using the constrained and regularization methodology Cichocki *et al.* propose the two ALS based algorithms called robust ALS (RALS) and hierarchical ALS (HALS) which show superior convergence properties over simple multiplicative counterparts and are especially suitable for large-scale problems [124].

In general, selection of the proper method for the NMF mostly depends on the size and type of the problem to be solved. If a distribution of noise can be assumed to be known, then for the Gaussian noise the D_F in Equations (2.281)–(2.282), and for the Poisson noise the D_{KL} in Equations (2.287)–(2.288), can be chosen, respectively. In this section we present a modification of the basic version of the multiplicative NMF methods, given by Equations (2.283) and (2.284). This modification was proposed by Cichocki *et al.* [124] to incorporate over-relaxation and sparsity control (ISRA-NMF) to the standard multiplicative method by Lee and Seung. This version modifies the optimization problem (2.280) and requires an additional three scalars, as will be discussed. The optimization problem with additional control parameters is as follows

$$\min D_F \left(\mathbf{D} - \tilde{\mathbf{D}} \right) = \min_{\mathbf{B}, \mathbf{C}} \left\{ \tfrac{1}{2} \|\mathbf{D} - \mathbf{BC}\|_F^2 + s_\mathbf{B} J_\mathbf{B} (\mathbf{B}) + s_\mathbf{C} J_\mathbf{C} (\mathbf{C}) \right\}$$

$$\text{subject to}: \mathbf{B} \geq 0, \quad \mathbf{C} \geq 0 \tag{2.289}$$

where $J_\mathbf{B}(\mathbf{B})$ and $J_\mathbf{C}(\mathbf{C})$ are penalty functions that enforce some additional requirements on the matrices $\mathbf{B}$ and $\mathbf{C}$, while $s_\mathbf{B}$ and $s_\mathbf{C}$ are constant values that control their influence on the solution. The higher these constants, the higher the degree of the corresponding constraints also. Using the standard gradient descent method we obtain the following multiplicative update rules for the NMF problem

$$b_{nr} \leftarrow b_{nr} \left[\frac{\left[\mathbf{DC}^T \right]_{nr} - s_\mathbf{B} J'_\mathbf{B} (\mathbf{B})}{\left[\mathbf{BCC}^T \right]_{nr} + \mu} \right]_+^q, \tag{2.290}$$

$$c_{rm} \leftarrow c_{rm} \left[\frac{\left[\mathbf{B}^T \mathbf{D} \right]_{rm} - s_\mathbf{C} J'_\mathbf{C} (\mathbf{C})}{\left[\mathbf{B}^T \mathbf{BC} \right]_{rm} + \mu} \right]_+^q, \tag{2.291}$$

where q is a positive parameter (usually in the range 0.5–2) whose role is to speed up the convergence, and $[.]_+$ denotes an operation of enforced nonnegativity, which can be defined as follows

$$[x]_+ = \max (x, \mu), \tag{2.292}$$

for sufficiently low positive threshold μ (usually in the range 10^{-9}–10^{-12}). This is necessary due to the subtraction in the numerator. In the above we have used the element wise derivatives of the control functions

$$J'_\mathbf{B} (\mathbf{B}) = \frac{\partial J_\mathbf{B} (\mathbf{B})}{\partial b_{nr}} \quad \text{and} \quad J'_\mathbf{C} (\mathbf{C}) = \frac{\partial J_\mathbf{C} (\mathbf{C})}{\partial c_{rm}}. \tag{2.293}$$

As alluded to previously, the functions $J_\mathbf{B}(\mathbf{B})$ and $J_\mathbf{C}(\mathbf{C})$ are chosen to superimpose additional properties on the output matrices. Exemplary types and their derivatives are presented in Table 2.6.

In Table 2.6 we assume a matrix $\mathbf{M}$ of size $R \times S$, $\mathbf{1}_{S \times S}$ is a matrix of dimensions $S \times S$ with all ones, and $\mathbf{I}$ is an identity matrix. In the case of the smoothness constraint $\mathbf{L}$ can be

Table 2.6 Control functions and their derivatives for different constraints of the multiplicative update NMF.

Condition	Function $J\,(\mathbf{M}_{R\times S})$	Derivative $J'\,(\mathbf{M}) = \partial J_{\mathbf{M}}\,(\mathbf{M})\big/\partial m_{rs}$
Sparsity	$J\,(\mathbf{M}) = \|\mathbf{M}\|_1 = \sum_{r,s} |m_{rs}|$	$J'\,(\mathbf{M}) = \mathbf{1}_{R\times S}$
Orthogonality	$J\,(\mathbf{M}) = \sum_{i=1}^{S-1}\sum_{j=i+1}^{S} \mathbf{m}_i^T\mathbf{m}_j$	$J'\,(\mathbf{M}) = \mathbf{M}\,(\mathbf{1}_{S\times S} - \mathbf{I})$
Smoothness	$J\,(\mathbf{M}) = \tfrac{1}{2}\,\|\mathbf{LM}\|_F^2$	$J'\,(\mathbf{M}) = \mathbf{L}^T\mathbf{LM}$

a Laplace matrix $\mathbf{L}_2$ of dimensions $(R\text{-}2) \times R$ which defines second-order differences among values of $\mathbf{M}$. It can be defined as follows [124]

$$\mathbf{L}_2 = \begin{bmatrix} 1 & -2 & 1 & & & & 0 \\ & 1 & -2 & 1 & & & \\ & & \ddots & \ddots & \ddots & & \\ & & & 1 & -2 & 1 & \\ 0 & & & & 1 & -2 & 1 \end{bmatrix}_{(R-2)\times R}. \tag{2.294}$$

We easily notice that the basic update rules in Equations (2.281)–(2.282), originally proposed by Lee and Seung [150], are special versions of Equations (2.290)–(2.291) for $s_B = s_C = 0$ and $q = 1$. The algorithm for computation of the nonnegative matrix factorization with the multiplicative update method is presented in Algorithm 2.23.

---

1. Choose the number of bases R.
   Choose the accuracy threshold e_{thresh} and the maximum number of iterations t_{max}.
   Choose the control parameters
   - J (see Table 2.6),
   - s_B, s_C (proposed range: 0.01–0.5, higher values denote higher sparsity),
   - q (proposed range: 0.05–2),
   - μ (typically 10^{-9}–10^{-12}).

2. Initialize: $\mathbf{B}^{(0)}, \mathbf{C}^{(0)}$

   $t = 0$

3. Do:

   (3.1) Compute $J'_B\,(\mathbf{B})$ and $J'_C\,(\mathbf{C})$ based on (2.294) and Table 2.6.

   (3.2) Compute $\mathbf{B}$ from (2.290).

   (3.3) Compute $\mathbf{C}$ from (2.291).

   (3.4) Normalize each column of $\mathbf{C}$.

(3.5) Compute stop condition e from (2.296).

$t = t + 1$

while $(e > e_{thresh})$ **and** $(t < t_{max})$.

3. Store the last computed **B, C**.

Algorithm 2.23 Multiplicative update method for computation of the nonnegative matrix factorization.

The NMF algorithms are very sensitive to the initialization of the matrices **B** and **C**. This can be done with the uniform random generator. However, when starting from a random initialization a more practical approach is to run a given NMF algorithm a number of times (for example 10), each time with different initialization, and each time saving the obtained factoring matrices. Then what is returned is the estimate which corresponds to the lowest achieved cost function, i.e. the best fit to **D**. An initialization method based on the SVD was proposed by Boutsidis and Gallopoulos [162]. Instead of a random initialization they proposed employing two SVD decompositions: The first is conducted for approximation of the data matrix, and the other for approximation of the positive sections of the resulting partial SVD factors, taking advantage of the properties of the unit rank matrices. As a result better approximation accuracy was reported based on some numerical examples.

Regarding the stopping criteria there are also many possibilities. The most obvious one is computation of the fitness accuracy value in Equation (2.280), i.e.

$$D_F^2 \left(\mathbf{D} - \tilde{\mathbf{D}} \right) < e_{thresh} \tag{2.295}$$

where e_{thresh} defines an acceptable accuracy level. However, sometimes guessing such a threshold is not easy and if it is chosen to be too low, then the convergence cannot be reached. Therefore it is much better to check the difference of successive computations of the matrices **B** and **C**, e.g.

$$e = \max \left[D_F^2 \left(\mathbf{B}^{(t+1)} - \mathbf{B}^{(t)} \right), \quad D_F^2 \left(\mathbf{C}^{(t+1)} - \mathbf{C}^{(t)} \right) \right] < e_{thresh} \tag{2.296}$$

at iteration steps $t + 1$ and t. However, fulfillment of Equation (2.296) does not imply Equation (2.295).

The final remark concerns computation of NMF for matrices of a very large size. It can be shown that in such a case we can split the computations into a number of NMF processes on a reduced data matrix **D** which can be obtained by a few strategies [124].

Finally, the problem of interpretation of the components obtained with NMF does not need to be a simple one, as indicated by many authors, e.g. Chu *et al.* [163]. This is also due to the global approach undertaken with NMF, in which all patterns are arbitrarily vectorized into the input matrix, and internal interrelations among data are discarded. The nonnegative tensor extension allows us to facilitate this problem, as will be discussed in the next section.

Using Algorithm 2.23 let us now compute NMF for the following positive valued matrix

$$\mathbf{D} = \begin{bmatrix} 12 & 1 & 0 & 0 & 1 & 0 & 0 & 1 & 2 \\ 0 & 4 & 4 & 4 & 0 & 10 & 1 & 4 & 0 \\ 2 & 5 & 2 & 3 & 2 & 1 & 1 & 1 & 0 \end{bmatrix}_{3\times9} \tag{2.297}$$

and for a few different optimization constraints and iteration parameters. Exemplary results are presented in Table 2.7, in which each value was averaged from ten runs of the Algorithm 2.23 with the given settings.

Table 2.7 Results of the NMF computations for the matrix D in Equation (2.297) for different constraint settings.

B	Averaged reconstruction accuracy (2.295)	Averaged iterations t	Conditions
$\begin{bmatrix} 0.0002 & 0.9999 & 0 \\ 0 & 0 & 0.9091 \\ 0.9998 & 0 & 0.0909 \end{bmatrix}$	1.17e-08	9 170	No constraints $s_{\mathbf{B}} = s_{\mathbf{C}} = 0$ $q = 1.2$ $e_{thresh} = 10^{-7}$ $\mu = 10^{-12}$
$\begin{bmatrix} 0 & 0.8597 & 0.1125 \\ 0.8891 & 0 & 0 \\ 0.1109 & 0.1404 & 0.8875 \end{bmatrix}$	0.0545	521	Sparsity constraint on **B** and **C** $s_{\mathbf{B}} = s_{\mathbf{C}} = 0.25$ $q = 1.2$ $e_{thresh} = 10^{-7}$ $\mu = 10^{-12}$
$\begin{bmatrix} 0.2025 & 0.0171 & 0.8612 \\ 0 & 0.7978 & 0 \\ 0.7975 & 0.1848 & 0.1388 \end{bmatrix}$	0.28	3 359	Orthogonality constraint on **B** and **C** $s_{\mathbf{B}} = s_{\mathbf{C}} = 0.2$ $q = 1.2$ $e_{thresh} = 10^{-7}$ $\mu = 10^{-12}$
$\begin{bmatrix} 0 & 1 & 0 \\ 0 & 0 & 1 \\ 1 & 0 & 0 \end{bmatrix}$	2.6e-16	3 630	Orthogonality constraint exclusively on **B** $s_{\mathbf{B}} = 0.2, s_{\mathbf{C}} = 0$ $q = 1.2$ $e_{thresh} = 10^{-7}$ $\mu = 10^{-12}$

These experiments indicate that the method is indeed slowly convergent. More severely, even in this simple case a number of iterations can be quite different from run to run and this depends heavily on initialization (random). Also interesting is comparing the average accuracy of the obtained approximations. This is best if there are no additional constraints or only a constraint on one of the matrices. Also, some constraint settings cannot be fulfilled with a reasonable accuracy and number of iterations.

$\mathbf{D} =$

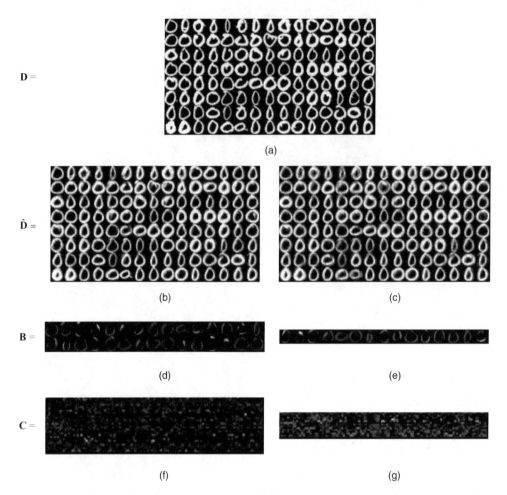

(a)

$\tilde{\mathbf{D}} =$

(b)   (c)

$\mathbf{B} =$

(d)   (e)

$\mathbf{C} =$

(f)   (g)

Figure 2.39 Application of the NMF to the MNIST handwritten digits set. The first hundred of the MNIST training images are used, each of dimensions 16×16, of a handwritten digit "0". Each column of the data matrix $\mathbf{D}$ contains 256 values row-by-row scanned of a training image. Visualization of $\mathbf{D}$ (a). Visualizations of the reconstructed matrices $\tilde{\mathbf{D}}$, where $\mathbf{D} \approx \tilde{\mathbf{D}} = \mathbf{BC}$, for two different numbers of bases; $R = 30$ (b) and for $R = 15$ (c). Visualization of the base matrices $\mathbf{B}$ for $R = 30$ (d) and for $R = 15$ (e). Visualization of the coefficient matrices $\mathbf{C}$ for $R = 30$ (f) and for $R = 15$ (g).

2.12.12   Image Representation with NMF

In this section we present experiments showing properties of the NMF decompositions applied to two different data sets. Figure 2.39 depicts the results of NMF applied to first data set of training images which are handwritten digits from the MNIST database [164]. Each training pattern in the MNIST is a 16×16 monochrome image. Such an image is then row scanned to form a successive 256 element column vector of the corresponding data matrix $\mathbf{D}$. In this experiment we constrain ourselves to the first hundred handwritten "0s", shown in

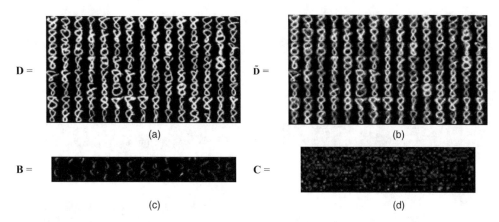

$D =$

$\tilde{D} =$

(a)

(b)

$B =$

$C =$

(c)

(d)

Figure 2.40 Visualization of the input data D containing the first hundred handwritten "8"s (a). Visualization of the approximated data matrix $\tilde{D}$ (b). Visualization of the base matrix B (c). Visualization of the component matrix C (d).

Figure 2.39a, as well as "8"s, shown in Figure 2.40a, respectively. Thus, each D is of dimensions 256×100. Then each D was decomposed with the NMF algorithm, listed in Algorithm 2.23, using different parameters. One of them is the number of assumed bases R, which corresponds to the number of columns of the matrix B and the number of rows in the components matrix C. Multiplication of B and C results in the reconstruction matrix $\tilde{D}$, as shown in Equation (2.279). Visualizations of $\tilde{D}$ for two different numbers of bases $R = 30$ and $R = 15$ are shown in Figure 2.39b and Figure 2.39c, respectively. Their corresponding matrices B and C are shown in Figure 2.39d–g. Under closer observation we notice that $\tilde{D}$ differs from D, especially with respect to some details. The differences are even more pronounced for $R = 15$, as depicted in Figure 2.39c. The accuracies when measured with *RRE* in (2.262) are *RRE* = 3.45% and *RRE* = 5.99%, respectively. In these experiments, however, no constraints were used. This can be interpreted as setting the parameters s_B and s_C in Equations (2.290)–(2.291) to 0. The exponent parameter was set as $q = 1.2$.

As alluded to previously, Figure 2.40 shows the experiments with the handwritten patterns of "8"s. In this case however, the orthogonality constraint was superimposed on B and sparsity on C. The two multiplicative parameters s_B and s_C were set to 0.25. In this case the obtained accuracy is *RRE* = 5.71%. However, the average number of iterations to achieve even a moderate convergence requirement set to $e_{thresh} = 0.01$ is high and in the range of 10–20k iterations. In this respect other algorithms offer better performance [19, 124].

The last experiments were conducted on a data set containing 12 pictograms of prohibition road signs, taken from the Polish regulations [165]. These are shown in Figure 2.41a. The factoring matrices B and C for $R = 8$ are visualized in Figure 2.41c and Figure 2.41d, respectively. However, the reconstruction matrix shown in Figure 2.41b shows a high degree of distortion, called ghosting. This reveals a serious problem with the NMF formulation when applied to image representation. Namely, the assumption of only additive influence of the decomposing factors (bases) does not allow sufficiently high accuracy due to a common bias.

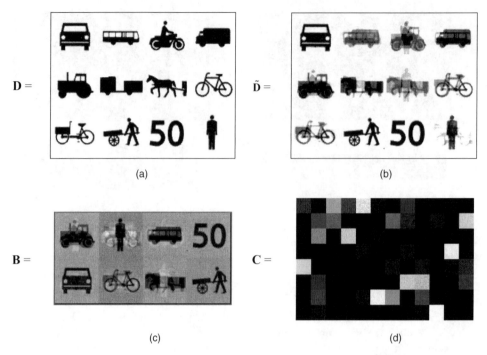

Figure 2.41   Application of the NMF to coding of the road sign pictograms: $\mathbf{D} \approx \tilde{\mathbf{D}} = \mathbf{BC}$. Visualization of the input data $\mathbf{D}$ (a). Approximated data matrix (b). Visualization of the base matrix $\mathbf{B}$ (c). Visualization of the component matrix $\mathbf{C}$ (d).

However, this can be amended with a slightly modified formulation to Equation (2.279) called the affine NMF, stated as follows [124]:

$$\tilde{\mathbf{D}} = \mathbf{BC} + \mathbf{r1}^T \tag{2.298}$$

where $\mathbf{r}$ contains the constant bias and $\mathbf{1}$ is a matrix with all ones. Such formulation allows absorption of the constant bias out of the data matrix and as a result improves factorization.

The other way to overcome the problem of bias is to employ the tensor approach to nonnegative factorization, as will be discussed in the next section.

2.12.13   Implementation of the Nonnegative Matrix Factorization

The nonnegative matrix decomposition can be computed in accordance with the discussed multiplicative update method which is outlined in Algorithm 2.23. Our implementation is realized by the class hierarchy shown in Figure 2.42.

The basic version of the multiplicative NMF can be computed with the help of the functional *operator*() from the *Multiplicative_Frobenius_NMF* base class. Its main input argument is a matrix $\mathbf{D}$ matrix, as well as a number of control parameters which role was explained in the previous section. The main output parameters are pointers to the decomposing matrices

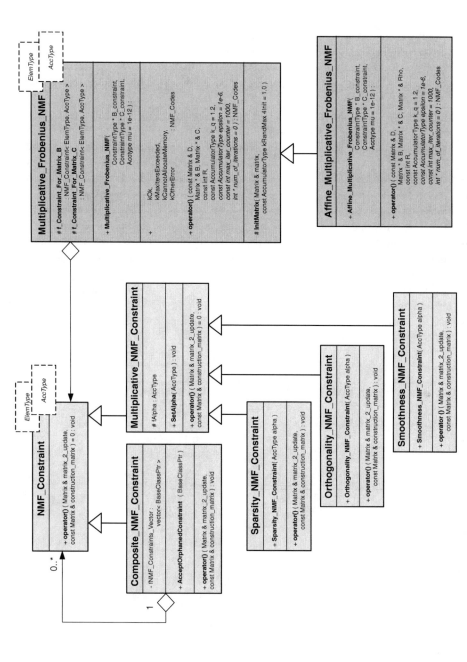

Figure 2.42 A class hierarchy for nonnegative matrix decompositions. The left hierarchy based on the *NMF_Constraint* base class implements optional sparsity, orthogonality, or smoothness constraints. NMF can be computed with the *Multiplicative_Frobenius_NMF* class. The affine version is implemented by its derived class the *Affine_Multiplicative_Frobenius_NMF*.

created by this function, **B** and **C**, respectively. An important aspect of this mechanism is that the returned objects, i.e. the matrices **B** and **C**, need to be deleted by the calling function. In C++ these are called orphaned objects. If not deleted, they will cause system memory leaks. This can be handled by adding the auto pointer objects, as will be shown in the following example. The *operator*() returns also one of the *NMF_Codes*, shown in Figure 2.42, which indicates the operation status. Another important aspect of NMF computations is addition of one of the constraints, such as sparsity or orthogonality, discussed in the previous section. These can be added in the form of special objects supplied during the construction of the *Multiplicative_Frobenius_NMF* object. From this point on they are handled and deleted by this object. The mentioned constraints can be created from the *Sparsity_NMF_Constraint*, *Orthogonality_NMF_Constraint*, or *Smoothness_NMF_Constraint* classes. They simply compute the formulas presented in Table 2.6. Each is accompanied by a multiplicative variable corresponding to s_B and s_C in the formulas (2.290)–(2.291), respectively. For this purpose the base class *Multiplicative_NMF_Constraint* was introduced into the hierarchy. The final class that needs explanation is the *Composite_NMF_Constraint* which realizes the so called composite design pattern. Thanks to this, many constraints can be joined and behave as one. In other words, one can join e.g. sparsity and orthogonality constraints, and such an object can be supplied in the construction of the *Multiplicative_Frobenius_NMF* class. Let us also observe that all the classes from the hierarchy in Figure 2.42 are templates, parameterized by two meta-types: The first, denoted as *ElemType*, directly corresponds to the type of element stored in the matrices. Most often than not this is a floating point type represented in C++ by *double*. However, it can be also a fixed point type realized, for example, with the help of the *TFixedFor* class from the HIL framework [15]. This consumes less memory while still allowing the necessary

```
template < typename T, typename ACC = double >
class Sparsity_NMF_Constraint : public Multiplicative_NMF_Constraint< T, ACC >
{
        public:

                Sparsity_NMF_Constraint( AccType alpha )
                        : Multiplicative_NMF_Constraint< T, ACC >( alpha ) {}
                virtual ~Sparsity_NMF_Constraint() {}

        public:

                virtual void operator() ( Matrix & matrix_2_update,
                                          const Matrix & construction_matrix )
                {
                        // for sparsness constraint we just subtract the constant
                        matrix_2_update -= fAlpha;
                }

};
```

Algorithm 2.24 Implementation of the *Sparsity_NMF_Constraint* class responsible for the sparsity constraint in NMF computation.

computations, however in the reduced dynamical range. The second type *AccType* denotes the type used for auxiliary computations, such as sums of elements, etc. Again, the most obvious choice for this one is *double*. The *Affine_Multiplicative_Frobenius_NMF* class, derived from the *Multiplicative_Frobenius_NMF* base class, implements the affine version of the NMF, as denoted in Equation (2.298).

Let us now take a closer look at the implementation of the short smoothness constraint which presents the following code listing.

The role of an object of the *Sparsity_NMF_Constraint* class, shown in Algorithm 2.24, is to update the *matrix_2_update* in the functional operator to obtain some sparsity control in accordance with the formula from the first row of Table 2.6. In this case computations are very simple and do not even require utilization of the input matrix *construction_matrix*. Application of a given constraint consists of its computation, multiplication by a constant – named here *fAlpha* – and finally subtraction from the *matrix_2_update* matrix. This exactly follows formulas (2.290)–(2.291).

Algorithm 2.25 presents the most important parts of the *Multiplicative_Frobenius_NMF* class. NMF computations start from creating an object of this class with optional constraint objects. Then the functional operator needs to be called.

```
/////////////////////////////////////////////////////////////////
// This class implements nonnegative matrix factorization (NMF)
/////////////////////////////////////////////////////////////////
template < typename T, typename ACC = double >
class Multiplicative_Frobenius_NMF
{
    public:

            typedef T                       MatrixElemType;
            typedef ACC                     AccumulatorType;
            typedef TImageFor< T >          Matrix;

    protected:

            const ACC       k_Mu;           // all values below are threated as 0.0

            typedef         NMF_Constraint< T, ACC >                 ConstraintType;

            // an object for constraints for the matrix B
            ConstraintType *            f_Constraint_For_Matrix_B;
            // an object for constraints for the matrix C
            ConstraintType *            f_Constraint_For_Matrix_C;

    public:

            // Message codes for the members of this class
            enum NMF_Codes { kOk, kMaxItersExceeded, kCannotAllocateMemory,
                        kOtherError };

    public:
```

```
//////////////////////////////////////////////////////////////
// Class constructor
//////////////////////////////////////////////////////////////
//
// INPUT:
//            B_constraint, C_constraint - orphaned object used
//               to compute optimization constraint for the matrices
//               B and C, respectively. If supplied 0 then
//               no constraint is assumed.
//            mu - a computational threshold on a minimal allowed
//               value to be conceived different than 0; All below
//               that are treated as 0.
//
// REMARKS:
//            k_Mu - defines a threshold on "very small values"
//            to avoid division by 0, etc.
//
Multiplicative_Frobenius_NMF( ConstraintType * B_constraint,
                ConstraintType * C_constraint, ACC mu = 1e-12 );
public:
//////////////////////////////////////////////////////////////
// This function does nonnegative matrix factorization:
//                      D = B * C
// where
// D, B, C >= 0
//////////////////////////////////////////////////////////////
//
// INPUT:
//            D - a matrix with input data; Each data object
//                is stored in one column of D.
//            B - output orphaned base matrix
//            C - output orphanded component matrix
//            R - number of requested bases
//            k_q - an exponent used to rise expressions
//                in each iteration
//            epsilon - a threshold on change of B and C
//                If changes are below epsilon then iterations
//                are stopped
//            max_iter_counter - a maximal allowed number of iterations
//            num_of_iterations - a pointed to the counter which
//                gets real number of executed iterations (can be 0)
//
// OUTPUT:
//            NMF_Codes - operation status
//
// REMARKS:
//            Given a matrix D, matrices B and C will be created
//            by this function; This is a caller responsibility
//            to delete them.
//
//            If size of D is NxM, then B is NxR and C is RxM
//
virtual NMF_Codes operator()( const Matrix & D, Matrix * & B,
                          Matrix * & C, const int R,
```

```
                                        const AccumulatorType k_q = 1.2,
                                        const AccumulatorType epsilon = 1e-6,
                                        const int max_iter_counter = 1000,
                                        int * num_of_iterations = 0 );
        protected:
                //////////////////////////////////////////////////////////////
                // This function fills a matrix with random values
                // of the uniform distribution. Used to initialize NMF
                //////////////////////////////////////////////////////////////
                //
                // INPUT:
                //                 matrix - the matrix to be filled
                //                 kRandMax4Init - max range of random values
                // OUTPUT:
                //                 none
                //
                // REMARKS:
                //
                //
                virtual void InitMatrix( Matrix & matrix,
                                const AccumulatorType kRandMax4Init = 1.0 );
};
```

Algorithm 2.25   The *Multiplicative_Frobenius_NMF* class and its members.

The best comprehension of the above discussed classes can be gained from a simple example, such as the one shown in Algorithm 2.26.

```
typedef Multiplicative_Frobenius_NMF< double, double > ISRA_NMF;
// ...

void Multiplicative_Frobenius_NMF_TEST( void )
{
1       // Create a composite to two constraints for matrix B
2       Composite_NMF_Constraint< double, double > * theConstraintComposite_B
3         = new Composite_NMF_Constraint< double, double >();
4
5       double alpha_B_sparsity = 0.2, alpha_B_orthogonality = 0.2;
6
7       theConstraintComposite_B->AcceptOrphanedConstraint(
8        new Sparsity_NMF_Constraint< double, double >( alpha_B_sparsity ) );
9       theConstraintComposite_B->AcceptOrphanedConstraint(
10       new Orthogonality_NMF_Constraint< double, double >( alpha_B_orthogonality ) );
11
12      // Create a composite to two constraints for matrix C
13      Composite_NMF_Constraint< double, double > * theConstraintComposite_C
14        = new Composite_NMF_Constraint< double, double >();
15
16      double alpha_C_sparsity = 0.2, alpha_C_orthogonality = 0.2;
17
18      theConstraintComposite_C->AcceptOrphanedConstraint(
```

```
19      new Sparsity_NMF_Constraint< double, double >( alpha_C_sparsity ) );
20    theConstraintComposite_C->AcceptOrphanedConstraint(
21      new Orthogonality_NMF_Constraint< double, double >( alpha_C_orthogonality ) );
22
23    // Now create the NMF object which will do the decomposition
24    ISRA_NMF NMF_Obj( theConstraintComposite_B, theConstraintComposite_C );
25
26    // Read matrix D to be decomposed
27    int D_cols = 9, D_rows = 3;
28    ISRA_NMF::Matrix D( D_cols, D_rows );
29
30    if( Load_ASCII_Image( D, "D.txt" ) == false )
31     return;
32
33    // Create some auxiliary variables
34    ISRA_NMF::Matrix * B = 0;
35    ISRA_NMF::Matrix * C = 0;
36
37    int R = 3;
38
39    ISRA_NMF::AccumulatorType k_q = 1.2;
40    ISRA_NMF::AccumulatorType epsilon = 1e-7;
41
42    const int max_iter_counter = 10000;
43    int num_of_iterations = 0;
44
45    // Do the NMF
46    ISRA_NMF::NMF_Codes theCode =
47        NMF_Obj( D, B, C, R, k_q, epsilon, max_iter_counter, & num_of_iterations );
48
49    REQUIRE( B != 0 );
50    REQUIRE( C != 0 );
51
52    // This will automatically delete the returned matrices
53    std::auto_ptr< ISRA_NMF::Matrix > B_AP( B ), C_AP( C );
54
55    // Store the decomposed matrices
56    Save_ASCII_Image( * B, "B.txt" );
57    Save_ASCII_Image( * C, "C.txt" );
58
59    // Test accuracy
60    std::auto_ptr< ISRA_NMF::Matrix > D_Approx(
61            Orphan_Mult_Matrix< double, double >( * B, * C ) );
62    Save_ASCII_Image( * D_Approx, "D_Approx.txt" );
63
64    double accuracy = Get_MSE( D, * D_Approx );
}
```

Algorithm 2.26   Implementation of the *Multiplicative_Frobenius_NMF_TEST* function showing basic functionality of the *ISRA_NMF* class.

In lines <2–21> of Algorithm 2.26 two groups of constraints are created, one for **B**, the second for the **C** matrix. Thanks to the objects *theConstraintComposite_B* and *theConstraintComposite_C*, which follow the composite pattern, the two sparsity and orthogonality constraints can be grouped together and treated by the rest of the algorithms as a single one. Also, each of the elementary constraints is supplied with its multiplicative constant. Then, in line <24>, the *NMF_Obj* of the class *Multiplicative_Frobenius_NMF* instantiated with *double* type is created with the two compound constraints supplied to it. Then the parameters are defined that control the NMF computations, as already described in previous sections.

The main decomposition of the matrix **D** of fixed dimensions 3×9, previously loaded from a file, takes place in lines <46–47>. To automatically delete the orphaned objects of representing matrices **B** and **C**, corresponding auto pointer objects are created in line <53>. These, being automatic variables, will be destroyed automatically by the system when control goes out of this function. Since they will be killed, their destructors will be invoked, which in turn will destroy the contained objects. This technique has been shown to be more safe than directly entering *delete*, since the process is automatic and all paths of execution will result in eventual object deletion. Then, after saving matrices **B** and **C** to the files *B.txt* and *C.txt*, respectively, in lines <60–64> we test the accuracy of the decomposition. Algorithm 2.26 can serve as a startup for other projects involving NMF computations by simply changing the way the matrix **D** is entered. For example, in the case of pictogram factorization already presented in Section 2.12.12, each column of **D** was composed of the vectorized version of consecutive pictogram, and so on.

2.12.14 Nonnegative Tensor Factorization

The requirement of nonnegativity can be superimposed on any decomposition, such as the already discussed HOSVD (Section 2.12.2) and rank-1 (Section 2.12.5). The resulting tensor factorization, either for HOSVD or rank-1, is called the *nonnegative tensor factorization* (NTF). This is an extension to the nonnegative matrix factorization problem discussed in the previous section.

For detection of the local part features of images stacked into a tensor Hazan *et al.* propose a tensor decomposition method with the nonnegative rank-1 tensors [166]. In this section we briefly outline this method, called rank-1 nonnegative tensor factorization (R1NTF), for the case of 3D tensors. The decomposition is similar to the one already discussed in Section 2.12.6 and visualized in Figure 2.21 (p. 77). However, this time an additional nonnegativity constraint is superimposed upon the solution. That is, for a given 3D tensor $\mathcal{T}$ of dimensions $N_1 \times N_2 \times N_3$, we start from Equation (2.160) and form the following approximation

$$\tilde{\mathcal{T}} = \sum_{r=1}^{R} \mathbf{a}^{(r)} \circ \mathbf{b}^{(r)} \circ \mathbf{c}^{(r)}, \tag{2.299}$$

with the elementwise nonnegativity constraint

$$\forall_{r} : \quad \mathbf{a}^{(r)}, \mathbf{b}^{(r)}, \mathbf{c}^{(r)} \geq 0. \tag{2.300}$$

To find $\tilde{\mathcal{T}}$ in Equation (2.299) with constraint (2.300) we consider minimization of the following function based on Equation (2.225)

$$\Theta\left(\tilde{\mathcal{T}}\right) = \left\|\tilde{\mathcal{T}} - \mathcal{T}\right\|_F^2.$$

Thus, the optimization problem reads

$$\underset{\mathbf{a}^{(r)},\mathbf{b}^{(r)},\mathbf{c}^{(r)}\geq 0}{\arg\min}\ \Theta\left(\tilde{\mathcal{T}}\right) = \underset{\mathbf{a}^{(r)},\mathbf{b}^{(r)},\mathbf{c}^{(r)}\geq 0}{\arg\min}\ \left\|\mathcal{T} - \sum_{r=1}^{R}\mathbf{a}^{(r)}\circ\mathbf{b}^{(r)}\circ\mathbf{c}^{(r)}\right\|_F^2. \quad (2.301)$$

This can be approached with a mixture of the Jacobi and Gauss–Seidel update scheme with a positive preserving update rule. In consequence the solution to the above comes as the following update rules [166]

$$a_i^{(j)} \leftarrow a_i^{(j)} \frac{\sum_{m=1}^{N_2}\sum_{n=1}^{N_3}t_{imn}b_m^{(j)}c_n^{(j)}}{\sum_{r=1}^{R}a_i^{(r)}\left[\mathbf{b}^{(r)^T}\mathbf{b}^{(j)}\right]\left[\mathbf{c}^{(r)^T}\mathbf{c}^{(j)}\right]+\mu}, \quad (2.302)$$

$$b_i^{(j)} \leftarrow b_i^{(j)} \frac{\sum_{m=1}^{N_1}\sum_{n=1}^{N_3}t_{min}a_m^{(j)}c_n^{(j)}}{\sum_{r=1}^{R}b_i^{(r)}\left[\mathbf{a}^{(r)^T}\mathbf{a}^{(j)}\right]\left[\mathbf{c}^{(r)^T}\mathbf{c}^{(j)}\right]+\mu}, \quad (2.303)$$

$$c_i^{(j)} \leftarrow c_i^{(j)} \frac{\sum_{m=1}^{N_1}\sum_{n=1}^{N_2}t_{mni}a_m^{(j)}b_n^{(j)}}{\sum_{r=1}^{R}c_i^{(r)}\left[\mathbf{a}^{(r)^T}\mathbf{a}^{(j)}\right]\left[\mathbf{b}^{(r)^T}\mathbf{b}^{(j)}\right]+\mu}, \quad (2.304)$$

where $a_i^{(j)}$ denotes an i-th component in the j-th vector $\mathbf{a}^{(j)}$, t_{imn} is an element of the tensor $\mathcal{T}$ at position (i,m,n), and μ denotes a small constant value ensuring a nonzero denominator in each iteration. Number R of components in Equation (2.299) is also one of the parameters of R1NTF which need to be *a priori* chosen. The updates (2.302)–(2.304) lead to a usually local minimum of Equation (2.301) and preserve nonnegativity of $\mathbf{a}^{(j)}$, $\mathbf{b}^{(j)}$, $\mathbf{c}^{(j)}$ under the condition that their initial values were also nonnegative.

Application of R1NTF to image representation starts from stacking a series counting N_3 images of resolution $N_1 \times N_2$ into a tensor $\mathcal{T}$, as depicted in Figure 2.43.

It is interesting to observe the relationship between 2D images $\mathbf{I}_i$ and the right side of the decomposition in Figure 2.43. For each k an outer product of a pair of vectors $\mathbf{a}^{(k)}$, $\mathbf{b}^{(k)}$ forms a nonnegative matrix

$$\mathbf{J}_k = \mathbf{a}^{(k)}\mathbf{b}^{(k)^T} \quad (2.305)$$

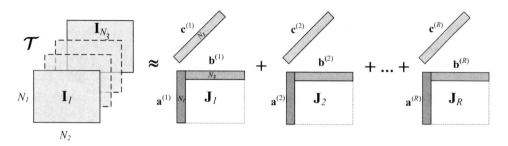

Figure 2.43 Visualization of the R1NTF decomposition of 3D tensors composed of the stacked image $\mathbf{I}_i$.

of rank 1. All these matrices, multiplied by a chosen set of corresponding components from all $\mathbf{c}^{(j)}$ and summed up, give respectively an approximation to each of the images $\mathbf{I}_i$ that form the input tensor $\mathcal{T}$. Hence, each image $\mathbf{I}_i$ can be expressed as follows

$$\mathbf{I}_i = \mathbf{A}\mathbf{X}_i\mathbf{B}^T \tag{2.306}$$

where

$$\mathbf{A}_{N_1 \times R} = \left[\mathbf{a}^{(1)}\mathbf{a}^{(2)}\ldots\mathbf{a}^{(R)}\right], \ \mathbf{B}_{N_2 \times R} = \left[\mathbf{b}^{(1)}\mathbf{b}^{(2)}\ldots\mathbf{b}^{(R)}\right], \ \mathbf{C}_{N_3 \times R} = \left[\mathbf{c}^{(1)}\mathbf{c}^{(2)}\ldots\mathbf{c}^{(R)}\right],$$

$$\tag{2.307}$$

and $\mathbf{X}_i$ is an $R \times R$ diagonal matrix created from an i-th row of $\mathbf{C}$ as follows

$$\mathbf{X}_i = \begin{bmatrix} c_i^{(1)} & 0 & \cdots & 0 \\ 0 & c_i^{(2)} & \cdots & 0 \\ \vdots & \vdots & \ddots & \vdots \\ 0 & 0 & \cdots & c_i^{(R)} \end{bmatrix}_{R \times R} \tag{2.308}$$

In other words, each vectorized version of an input image $\mathbf{I}_i$ is a linear combination of all $\mathbf{J}_k$ and coefficients taken from the i-th row of the matrix $\mathbf{C}$.

Thus, from R1NTF we obtain R nonnegative rank 1 matrices $\mathbf{J}_k$, which constitute *a basis* (features) for representation of object(s) conveyed in the images $\mathbf{I}_i$. Coefficients $\mathbf{C}$ can be discarded. However, the bases need not be orthogonal. Nevertheless, as shown by Hazan *et al.* in the case of face detection, the obtained features are highly discriminatory which is important to achieve a high accuracy of pattern classification. The difference between matrix- and tensor- nonnegative factorization is a qualitative one. Once again, the tensor approach inherently takes into account spatial relations among pixels, which is not the case for matrices in which each column is a vectorized image in some arbitrary order. We have already observed a similar phenomenon in the case of video compression discusses in Section 2.12.8.2. In this case tensor based approach allows more efficient representation and higher compression rates as a result as compared to the matrix based solutions such as NMF or "classical" PCA.

Other tensor decompositions can also be constrained to the positive values, especially if dealing with positive valued signals, and also to facilitate interpretation of the decomposed factors. For this purpose methods were developed that are called nonnegative or semi-nonnegative Tucker decompositions.

Yet a simple nonnegative approximation can be achieved starting from the HOSVD algorithm presented in Section 2.12.1.5. Thus, a tensor $\mathcal{T} \in \Re^{N_1 \times N_2 \times \dots N_m \times \dots N_n \times \dots N_P}$ can be decomposed as

$$\mathcal{T} = \mathcal{Z} \times_1 \mathbf{S}_1 \times_2 \mathbf{S}_2 \dots \times_P \mathbf{S}_P,$$

with the nonnegativity constraint on the mode matrices $\mathbf{S}_k$ and the core tensor $\mathcal{Z}$

$$\mathcal{Z}, \mathbf{S}_1, \mathbf{S}_2, \dots, \mathbf{S}_P \geq 0, \tag{2.309}$$

using Algorithm 2.14 with an additional positive update step. That is, step (2.198) reads now as follows

$$\mathbf{T}_{(k)} = \mathbf{S}_k \mathbf{V}_k \mathbf{D}_k^T \tag{2.310}$$

$$\mathbf{T}_{(k)} = \left[\mathbf{T}_{(k)}\right]_+$$

Algorithm 2.27   Updated step of the HOSVD computation in Algorithm 2.14 for the nonnegativity constraint.

Let us recall that $[.]_+$ denotes an operation of enforced nonnegativity, as defined in Equation (2.292).

A similar update can be applied to the HOOI method, presented in Algorithm 2.19. The step (2.266) is now updated to take account of the nonnegative constraint as follows.

$$\mathbf{S}_k^{(t+1)} = svds\left(\hat{\mathbf{S}}_k^{(t+1)}, R_k\right) \tag{2.311}$$

$$\mathbf{S}_k^{(t+1)} = \left[\mathbf{S}_k^{(t+1)}\right]_+$$

Algorithm 2.28   Updated step of the HOOI computation in Algorithm 2.19 for the nonnegativity constraint.

However, the above usually do not lead to optimal tensor approximation in the sense of Equation (2.225), and do not produce an optimal lower rank approximation since the algorithm optimizes each mode separately without accounting for interaction among them. For this purpose another method need to be employed. For instance Mørup *et al.* proposed an algorithm for sparse nonnegative Tucker decompositions, called SN-TUCKER [168]. Their method follows the multiplicative update rule, already presented in the case of NMF (Section 2.12.10). The authors also discuss how sparse coding can help to identify the best decomposition (rank-1 vs. Tucker) and how to choose a sufficient number of components in the decomposition. Their MATLAB® implementation of SN-TUCKER is provided in [168]. A different optimization strategy based on the damped Gauss–Newton method was recently proposed by Phan *et al.* [169]. The method shows outstanding convergence speed thanks to the connection of the Levenberg–Marquardt method with fast construction of the approximate Hessian, as well as computation of gradients without the need for the large-scale Jacobian. Thanks to this computation of the Kronecker products is avoided too. Their method was tested for clustering on the ORL face database [170]. This database contains 400 faces of 40 persons, from which Phan *et al.* used 100 faces of 10 persons for their tests. Then, for each face a tensor of Gabor features, computed for 8 orientations and 4 scales, was created. All these were then down sampled to form $16 \times 16 \times 8 \times 4$ tensors for each face, and finally packed into one 5D tensor. This was then decomposed along the first 4 modes which resulted in the core tensor of dimensions $(3 \times 3 \times 2 \times 2) \times 100$. Thus, each face is now represented by only 36 features which were then directly fed to the k-means clustering algorithm, which we discuss in Section 3.11.1. For this setup Phan *et al.* report 92% accuracy which increases to 99% if the number of features is doubled.

Last but not least are the memory requirements when processing and decomposing tensors. There are many works which aim towards the development of methods which do not require manipulation of the whole structure at once. For instance Phan and Cichocki propose a block decomposition for very large-scale NTF [171]. This problem has also been addressed in the classification framework based on tensor decomposition. Instead of a single large classifier, an ensemble of classifiers, each operating on a smaller portion of data from the original data sets, is proposed in the paper by Cyganek [172]. Training partitions are obtained by data bagging. Additionally, the method achieves higher accuracy than a single classifier due to diversity, as will be discussed in Section 5.8.3. Practical methods of computing nonnegative tensor factorization are also discussed in the technical report by Friedlander and Hatz [173].

2.12.15 *Multilinear Methods of Object Recognition*

As alluded to previously, multilinear tensor algebra constitutes a very powerful tool for analysis of variability of multidimensional signals, such as the digital images which we mostly focus on. A tensor can comprise such factors as spatial variability of an image, variability of pixels, illumination, pose of objects, etc. Nevertheless, components of such tensors are usually highly correlated. This, in turn, allows their analysis in a lower dimensional subspace, i.e. they can be analyzed in manifolds of intrinsically lower dimensions. However, concurrently to the reduction of tensor space the most important information conveyed by tensors is retained.

As discussed in Section 3.3.1, a very popular unsupervised method for dimensionality reduction is the principal component analysis (PCA). The method assumes correlation of data which are then transformed to a new system spanned by uncorrelated and ordered principal components. Due to this ordering it is possible to retain and analyze only the most important (principal) components of decomposed data. As already mentioned, a PCA based method for face recognition was proposed by Turk and Pentland [146]. Their idea of *eigenfaces* was extended to *tensorfaces* by Vasilescu and Terzopoulos.

In their original work Vasilescu and Terzopoulos proposed using multilinear tensor analysis – called tensorfaces – for recognition of human faces from a series of different images, taken under different illuminations and view positions [11, 12]. Their tensor $\mathcal{D}$ which collects all face data with its internal aspects has five dimensions. In their experiments there were 28 humans, visualized in images of $I_{pixel} = 7943$ pixels, each expressed in 5 different poses, times 3 different illuminations, and finally 3 different expressions. Thus the tensor $\mathcal{D}$ is five-dimensional of dimensions $28 \times 5 \times 3 \times 3 \times 7943$. Taking Equation (2.189) into an account, $\mathcal{D}$ can be decomposed as follows[18]

$$\mathcal{D} = \mathcal{Z} \times_1 \mathbf{U}_{faces} \times_2 \mathbf{U}_{views} \times_3 \mathbf{U}_{illum} \times_4 \mathbf{U}_{expressions} \times_5 \mathbf{U}_{pixels}. \tag{2.312}$$

The core matrix $\mathcal{Z}$ orchestrates interaction among the factors represented in the five mode matrices. The first mode matrix $\mathbf{U}_{faces}$, of dimensions 28×28, spans the space of human face parameters, the 5×5 mode $\mathbf{U}_{views}$ spans the space of viewpoints, the two 3×3 matrices $\mathbf{U}_{illuminations}$ and $\mathbf{U}_{expressions}$ – the spaces of illumination and face expression parameters, respectively. The last matrix $\mathbf{U}_{pixels}$ has dimensions 7943×7943 and orthonormally spans the space of images. Let us observe that *each column of this matrix* is an *eigenimage*, or more specifically an *eigenface* in this example.

Flattening of $\mathcal{D}$ in respect to the "pixels" dimension can be obtained directly from Equations (2.312) and (2.201), as follows

$$\underbrace{\mathbf{D}_{(pixels)}}_{\substack{image \\ data}} = \underbrace{\mathbf{U}_{pixels}}_{\substack{base \\ vectors}} \underbrace{\mathbf{Z}_{(pixels)} \left[\mathbf{U}_{expressions} \otimes \mathbf{U}_{illum} \otimes \mathbf{U}_{views} \otimes \mathbf{U}_{faces} \right]^T}_{coefficients}. \tag{2.313}$$

In other words, the above equation denotes a "classical" linear combination of the ensemble of images with the $\mathbf{U}_{pixels}$ matrix containing *basis vectors*, as well as the matrix of coefficients, composed of a *product* of the pixel-mode flattened core tensor $\mathbf{Z}_{pixel}$ and the Kronecker series of all other mode matrices $\mathbf{U}$. As pointed out in the paper by Vasilescu and Terzopoulos, the real advantage of the tensor approach over the classical PCA is that the core tensor $\mathcal{Z}$ transforms the eigenimages from the $\mathbf{U}_{pixels}$ matrix into the *eigenmodes*. The latter represent the principal axes of variations across other modes, such as people, illumination, etc. Furthermore, by projecting the core tensor onto the $\mathbf{U}_{pixels}$, i.e. creating the product $\mathcal{Z} \times_5 \mathbf{U}_{pixels}$, they also give an insight into how these different factors interact with each other to create an image. On the other hand, the "classical" PCA, which operates on only one matrix containing all available images, provides basis vectors (eigenimages) which show only principal axes of variation across *all* images with no control over the particular factors, such as illumination, view, etc.

---

[18]In this section for better tracking we follow the same notation and tensor modes as in the paper by Vasilescu and Terzopoulos [11].

Now let us observe that based on Equation (2.201) and Equation (2.312) a training image for the f-th face, v-th view, i-th illumination, and e-th expression can be represented as the following $1 \times 1 \times 1 \times 1 \times I_{pixel}$ tensor

$$\mathbf{d}^{(f,v,i,e)} = \left(\boldsymbol{\mathcal{Z}} \times_5 \mathbf{U}_{pixels}\right) \times_1 \mathbf{u}_{faces}^{(f)\mathrm{T}} \times_2 \mathbf{u}_{views}^{(v)\mathrm{T}} \times_3 \mathbf{u}_{illum}^{(i)\mathrm{T}} \times_4 \mathbf{u}_{express}^{(e)\mathrm{T}}. \qquad (2.314)$$

In the above $\mathbf{u}_{faces}^{(f)\mathrm{T}}$ is a row vector of dimensions $1 \times I_{faces}$, which controls inclusion of a concrete face, i.e. a person identity, to the image $\mathbf{d}^{(f,v,i,e)}$. It is the f-th row of the orthogonal $\mathbf{U}_{faces}$ matrix, since $\mathbf{d}^{(f,v,i,e)}$ concerns only the f-th face. Similarly, $\mathbf{u}_{views}^{(v)\mathrm{T}}$ is the v-th row of the $\mathbf{U}_{views}$, and so on. Hence, each image is a fiber in $\mathcal{D}$, which is a five-dimensional tensor (2.312) or, equivalently, is a column vector (with I_{pixel} elements) in $\mathbf{D}_{(pixel)}$, given in Equation (2.313). The last property follows Equation (2.167).

Let us also observe that the coefficients in Equation (2.314) can be seen as *independent sources* contributing to the data point $\mathbf{d}^{(f,v,i,e)}$. On the other hand, interaction of these sources is controlled by the tensor $\boldsymbol{\mathcal{Z}} \times_5 \mathbf{U}_{pixels}$. This factorization property is therefore used to decompose any multidimensional data set (a tensor) into perceptually independent contributing sources. In this respect is has some resemblance to the subspace decompositions discussed in Section 2.12.9.

However, let us now reverse a little the above flow of reasoning and consider that we have a new test image $\mathbf{d}_x$ which parameters, such as person, view, etc. are unknown. We can assume that it can also be expressed in the same way as all the train images in Equation (2.314) were. If so, then

$$\mathbf{d}_x = \boldsymbol{\mathcal{Z}} \times_5 \mathbf{U}_{pixels} \times_1 \mathbf{u}_{faces}^{\mathrm{T}} \times_2 \mathbf{u}_{views}^{\mathrm{T}} \times_3 \mathbf{u}_{illum}^{\mathrm{T}} \times_4 \mathbf{u}_{express}^{\mathrm{T}}. \qquad (2.315)$$

However, this time the parameters of the above decomposition are unknown. Based on Equation (2.313), Equation (2.315) can be expressed in the flattened form as

$$\begin{aligned} \mathbf{d}_x &= \underbrace{\mathbf{U}_{pixels}\mathbf{Z}_{pixels}}_{\mathbf{A}_{pixels}} \left(\mathbf{u}_{faces}^{\mathrm{T}} \otimes \mathbf{u}_{views}^{\mathrm{T}} \otimes \mathbf{u}_{illum}^{\mathrm{T}} \otimes \mathbf{u}_{express}^{\mathrm{T}}\right)^T \\ &= \mathbf{A}_{pixels} \left(\mathbf{u}_{faces} \otimes \mathbf{u}_{views} \otimes \mathbf{u}_{illum} \otimes \mathbf{u}_{express}\right). \end{aligned} \qquad (2.316)$$

In the above known are $\mathbf{d}_x$ and $\mathbf{A}_{pixels}$, whereas $\mathbf{u}_{faces}, \ldots, \mathbf{u}_{express}$ are unknown. The latter need to be determined through the process of *tensor factorization*. This means a decomposition of a rank-1 tensor into a number of vectors. The most prominent is $\mathbf{u}_{faces}$ since it allows person identification. However, this is not an easy task since all $\mathbf{u}_{faces}, \ldots, \mathbf{u}_{express}$ need to be computed together. This problem has been addressed in many publications. For example Tenenbaum and Freeman propose a bilinear model which allows face analysis in the context of two unknown factors, such as people and views, peoples and expressions, or peoples and illumination conditions [174]. In this light Lin *et al.* present a solution if only two factors $\mathbf{u}_{faces}$ and $\mathbf{u}_{illum}$ are unknown [175]. Their method resembles an alternating update and consists of changing one factor while the other is kept fixed. Then the process is reversed and the second factor is updated while the first is kept unchanged. The method starts from some initial values and proceeds iteratively until convergence. A similar approach is undertaken by Peng and Qian in their method for online gesture spotting from visual hull data in which pose features are extracted using HOSVD decomposition [176]. Namely, given a visual hull, its pose and orientation features are computed with the alternating least-squares algorithm.

A different approach was undertaken by Park and Savvides [177] who proposed to reduce a number of factors by individual subspace modeling of each of the faces, each belonging to a different person. In their work the tensor factorization problem was first represented as the least-squares problem for the Kronecker product of all of the unknown parameters with the quadratic equality constraint. The number of these parameters was restricted to three, i.e. to the face, expression, and illumination. The individual subspace method writes into the framework of object classification in the individual PCA spaces. Such a strategy had already been pointed out by Wold [178], and was then used by many other authors, such as Savas and Eldén [126].

A reconstruction of the mode parameters of a test image can be represented as follows

$$\left(\hat{\mathbf{u}}_{faces}, \hat{\mathbf{u}}_{views}, \hat{\mathbf{u}}_{illum}, \hat{\mathbf{u}}_{express}\right) = \underset{\left(\mathbf{u}_{faces}, \mathbf{u}_{views}, \mathbf{u}_{illum}, \mathbf{u}_{express}\right)}{\arg\min} \|\mathbf{d}_t - \mathbf{d}_x\|^2, \qquad (2.317)$$

where $\hat{\mathbf{u}}_{faces}, \hat{\mathbf{u}}_{views}, \hat{\mathbf{u}}_{illum}, \hat{\mathbf{u}}_{express}$ are estimators of the true parameters $\mathbf{u}_{faces}, \mathbf{u}_{views}, \mathbf{u}_{illum}, \mathbf{u}_{express}$, $\mathbf{d}_t$ is one of the training images, $\mathbf{d}_x$ is an unknown image which parameters we wish to determine (e.g. is this a face, anyway?).

Introducing (2.316) into Equation (2.317) yields

$$\left(\hat{\mathbf{u}}_{faces}, \hat{\mathbf{u}}_{views}, \hat{\mathbf{u}}_{illum}, \hat{\mathbf{u}}_{express}\right) = \underset{\left(\mathbf{u}_{faces}, \mathbf{u}_{views}, \mathbf{u}_{illum}, \mathbf{u}_{express}\right)}{\arg\min} \left\|\mathbf{d}_t - \mathbf{A}_{pixels}\mathbf{u}_{faces} \otimes \mathbf{u}_{views} \otimes \mathbf{u}_{illum} \otimes \mathbf{u}_{express}\right\|^2,$$

$$(2.318)$$

with an additional constraint on the unit norm

$$\left\|\mathbf{u}_{faces}\right\|^2 = \left\|\mathbf{u}_{views}\right\|^2 = \left\|\mathbf{u}_{illum}\right\|^2 = \left\|\mathbf{u}_{express}\right\|^2 = 1, \qquad (2.319)$$

which automatically follows from Equation (2.312), since $\mathbf{u}_{(\dots)}$ are rows of the unitary mode matrices $\mathbf{U}_{(\dots)}$.

If, now, taking advantage of the properties of the Kronecker product, we join all $\mathbf{u}_{faces} \otimes \mathbf{u}_{views} \otimes \mathbf{u}_{illum} \otimes \mathbf{u}_{express}$ in Equation (2.318) into one long column vector[19]

$$\mathbf{u} = \mathbf{u}_{faces} \otimes \mathbf{u}_{views} \otimes \mathbf{u}_{illum} \otimes \mathbf{u}_{express} \qquad (2.320)$$

of size $I_{faces} \cdot I_{views} \cdot I_{illum} \cdot I_{express} \times 1$ then Equation (2.318) can be expressed in a simpler form as follows

$$\hat{\mathbf{u}} = \underset{(\mathbf{u})}{\arg\min} \left\|\mathbf{d}_t - \mathbf{A}_{pixels}\mathbf{u}\right\|^2, \text{ subject to } \|\mathbf{u}\|^2 = 1. \qquad (2.321)$$

Now it is easy to observe that Equation (2.321) is a linear-regression problem with a least-squares estimator $\hat{\mathbf{u}}$. In other words, the generalized tensor-factorization can be stated as the

---

[19]Since in (2.314) $\mathbf{u}_{(\dots)}^{(f)\mathrm{T}}$ are row vectors.

least-squares problem with a quadratic equality constraint (LSQE). In the abovementioned work by Park and Savvides [177], a solution to such a defined optimization problem was obtained by first considering separately each set of images for a single person, i.e. the individual multilinear approach, and then for each individual model applying the iterative Projection method. However, the authors showed that prior transformation of the input patterns into the feature space with the kernel technique, which we discuss in Section 3.10, leads to much better results. This method, called Kernel-Tensor-Faces, does not require solving the tensor factorization problem for face recognition. However, in the method by Park and Savvides, only three factors were used (no *views* factor).

The concept of individual tensor spaces is very similar to the individual PCA spaces that will be discussed in Section 3.3.1. Namely, each pattern, such as face or a road sign, is treated separately, and for each of them a separate space is built which is spanned by its specific bases. When considering an unknown test pattern, its projections into each of the previously built spaces is checked and the one which fits the best in terms of the reconstruction error is reported, i.e. an object is classified into one of the classes. Thanks to this approach, the dimension of the input tensor space is lowered by one dimension. The limitation of this approach lies in the assumption that there are enough training patterns to construct representative and well discriminative spaces for each of the training patterns. Although the two approaches are similar with respect to the individual pattern spaces, the method by Park and Savvides [177] is more practical since it allows optimization with respect to all of the unknown factors, such as $\mathbf{u}_{faces}, \ldots, \mathbf{u}_{express}$ in Equation (2.316). The method also carries out classification in the kernel spaces, in which better separation between classes can be obtained due to nonlinear kernel mapping. This usually leads to superior results when compared to the scalar product in the linear spaces, as discussed in Section 3.10. Nevertheless, we are usually interested in the answer of a *single* parameter, such as whose is that face or what is that road sign regardless of its other modes, such as lighting conditions. In such cases, the two discussed methods lead to similar results considering only the linear spaces, although they do this using different algorithms. In this respect the method by Savas and Eldén is the more preferred since the algorithm is given in a close algebraic form (5.37), as discussed in Section 5.8.

Until recently color information was not widely considered important when designing methods for face recognition. This follows humans' abilities easily recognize faces using only monochrome images, such as black-and-white TV, etc. However, recent work by Wang *et al.* shows that color image discriminant (CID) information can play an important role in this process [179]. The goal of developing new CID models is to find a meaningful color space for accurate object recognition in color images. Usually the CID models assume evaluation of two transformations – the color space transformation and the projection transformation which will discriminate objects of interest [180]. In the paper by Wang *et al.* the tensor based discriminant color space is discussed in the context of face recognition. A color image of spatial pixel resolution $I_1 \times I_2$ is naturally interpreted as a 3D tensor of dimensions $I_1 \times I_2 \times I_3$, where obviously $I_3 = 3$. Then the discriminant projection matrices are computed: two related to the spatial variables I_1 and I_2, and the other related to the color dimension I_3. The main concept of this technique is that the projected tensors of the images of the same person are as close as possible, whereas the projected tensors of the images of different persons are spread out. As shown in the paper by Wang *et al.* the tensor discriminant color space (TDCS) allows the uniform treatment of the underlying structure of an image, which leads to higher accuracy of face recognition as compared to other state-of-the-art methods of CID.

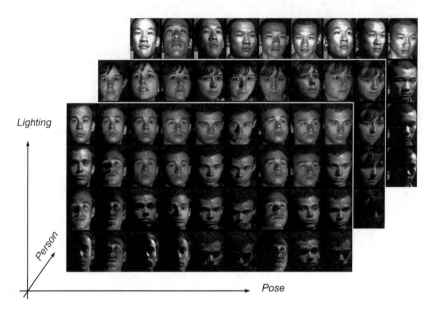

Figure 2.44 Images of different people in different lighting and pose conditions built up a tensor (adapted from Georghiades, Belhumeur and Kriegman [167]).

Recently Duchenne *et al.* presented a tensor based algorithm for high-order graph matching which can be used to derive point correspondences between two sets of points representing some visual features in images [181]. The method is characteristic of building higher order constraints of point settings instead of the previously proposed unary or up to pairwise relations. In the work by Duchenne *et al.* hypergraph point matching is formulated as the maximization of a multilinear function over all permutations of the image features. This function is formulated with the help of a tensor which conveys affinity between tuples of image features. This way the stated optimization problem is solved by a multidimensional power method (Section 2.12.5). The solution is then proposed to be projected onto the closest assignment matrix. Going into detail, let us assume that we have two images, say no.1 and no. 2, in which number N_1 and N_2 (feature) points were respectively selected. In consequence, all point indices can be indexed from 1 to N_1, for image no. 1, and from 1 to N_2 for image no. 2, respectively. The two numbers N_1 and N_2 can even be different. The mentioned objective function is defined as follows [181]

$$S(\mathbf{X}) = \sum_{i_1,i_2,j_1,j_2,k_1,k_2} H_{i_1,i_2,j_1,j_2,k_1,k_2} X_{i_1,i_2} X_{j_1,j_2} X_{k_1,k_2} \qquad (2.322)$$

where $X_{...}$ in the above are elements of the respective assignment matrices. That is, the aforementioned problem of matching points from image no. 1 to points in image no. 2 can be represented by an assignment matrix $\mathbf{X}$, for which if an element $X_{i_1,i_2} \in \{0, 1\}^{N_1 \times N_2}$ is 1 then a point P_{i_1} from image no. 1 is matched to a point P_{i_2} from no. 2. All other elements X_{i_1,i_2} are 0. Additionally it is assumed that all points from the first image are matched to exactly one point in the second, but the relation is not a symmetric one. That is, a point from the second image can be matched to an arbitrary number of points in the first image. This means that all rows of **X** sum up to 1, i.e.

$$\sum_{i_2} X_{i_1,i_2} = 1.$$

Let us observe that in Equation (2.322) we deal with three such assignments as X_{i_1,i_2} at a time, since we deal with triplets of points. Such an approach allows extraction of more pronounced point configurations which can deal with the matching of scale-invariant, rotation-invariant, and even projectively distorted images of objects.

Tensor **H** conveys information on the sets of features. That is, each of its elements $H_{i_1,i_2,j_1,j_2,k_1,k_2}$ conveys a triple potential corresponding to the triplets of feature nodes (points) $(P_{i_1}, P_{j_1}, P_{k_1})$ of image no. 1, and $(P_{i_2}, P_{j_2}, P_{k_2})$ of image no. 2. Thus, $H_{i_1,i_2,j_1,j_2,k_1,k_2}$ corresponds to a similarity measure for each triplet of nodes in the hyper-graph, and in the method by Duchenne *et al.* it is defined as follows

$$H_{i_1,i_2,j_1,j_2,k_1,k_2} = \exp\left(-\gamma \left\| f_{i_1,j_1,k_1} - f_{i_2,j_2,k_2} \right\|^2\right), \tag{2.323}$$

for all possible permutations of the indices i_1,i_2,j_1,j_2,k_1, and k_2, whereas γ denotes a parameter that controls the area of point influence, as discussed in Section 3.10.1. Elements of the feature vectors are denoted by f_{i_1,j_1,k_1} and f_{i_2,j_2,k_2}. These can be any sufficiently discriminative features, such as e.g. the SIFT operator [182]. In practice, not all elements of **H** need to be computed. Instead, for each tuple of points in the first image only features in the neighborhood controlled by parameter γ need to be considered. Further details of this multilinear point matching method, as well as examples of successful matches of projectively distorted views, can be found in the cited paper by Duchenne *et al.* [181].

In this context it is worth noticing the work by Liu *et al.* who address the problem of understanding images based on the annotated databases [183]. Their method is called non-parametric scene parsing via label transfer. At first, for an input image its nearest neighbors from the database are searched. Then the dense correspondences are computed with help of the SIFT flow method. This aligns the images on their local structures. Finally, the test image is segmented with the Markov random field method using annotations of the found nearest images from the database.

2.13   Closure

2.13.1   Chapter Summary

As we can easily see, even a simple color image is a three-dimensional data structure, whereas face recognition entails the space of patterns influenced by different views, poses, illuminations, or persons. Thus, computer vision is about multidimensional data processing. The purpose of this chapter was to present a variety of tensor methods in application to computer vision and pattern recognition tasks. As expected, we started with definitions of tensors, first viewed as mathematical objects, especially stressing their transformation laws. However, as discussed, in the context of multidimensional data processing, tensors can be seen as multi-dimensional arrays with well defined algebraic operations and most of all with the primary concept in pattern recognition – definitions of distances between tensors.

As shown, tensor fields can be filtered. For this purpose we discussed filters based on ordered statistics, as well as those relying on anisotropic diffusion. Then the structural tensor was discussed which allows analysis of a type of local neighborhood in multidimensional images. This will prove to be very useful in object detection and recognition tasks, as will be discussed in subsequent chapters. Yet another representation of objects can be achieved with the help of statistical moments, derived from the inertia tensor. Then tensor eigendecompositon was analyzed with an important conclusion on tensor representation as a weighted sum of stick, plate, and ball components. These, as well the other tensor invariants discussed, have found many applications in pattern recognition in tensor data, as discussed in the next chapters.

As already mentioned, tensors have proved to be very useful in definitions of multiview relations, as discussed in the section devoted to the geometry of multiple views, ending with definitions of the multifocal tensors. Here the second part of the chapter commenced – an analysis of multilinear tensor methods with the three most important types of tensor decomposition: higher-order singular value (HOSVD), best rank-1, as well as rank-$(R_1, R_2, \ldots , R_P)$ decompositions. As shown, these have found broad applications in computer vision, starting from signal compression, spanning orthogonal tensor bases for subspace data representation, up to the framework of multiview–pose–illumination based face recognition.

Another important group of discussed methods falls into the category of nonnegative matrix and tensor factorizations. In many cases these allow intuitive interpretation and processing of the factored out components.

For all of the aforementioned methods, not only were their properties derived and analyzed but also detailed algorithms were provided with working implementations in C++. These will hopefully be useful for further exploration of this rapidly developing domain.

2.13.2   Further Reading

An ample source of information on tensors in computer vision and image processing is the book edited by Aja-Fernández et al. [184]. It is organized into five main parts, starting from an introduction to tensors, tensor field processing, tensors in computer vision, following then to diffusion tensor imaging and medical applications, as well as tensor storage, visualization, and interfaces. Another two books, edited by Weickert and Hagen [64], as well as that edited by Laidlaw and Weickert [185], contain many useful and up-to-date articles on different aspects of tensor processing. All three can be recommended as references for almost all aspects of modern tensor methods in computer vision, image processing, and pattern recognition.

An in-depth treatment on structural tensors and on other aspects of vision with direction can be accessed in the book by Bigun [10].

A highly recommended reading on moments and moment invariants in CV is the latest book by Flusser et al. [90]. It contains theory, but also numerous examples of object recognition based on moment invariants in 2D and 3D images.

Regarding basic mathematical properties of tensors and their decompositions, as well as their applications in pattern recognition in digital signals, the recommended reading is the thesis by de Lathauver [96], as well as the papers by de Lathauwer et al. [115, 131]. Very readable and recommended papers on basic tensor definitions, as well as their decompositions, are the papers by the team of Bader and Kolda [20, 122]. Another recommended source of different methods of NMF and NTF with numerous MATLAB® implementations and examples is the the book by Cichocki *et al.* [124]. These publications also provide the necessary mathematical background which helps in understanding the main thesis. They are also an ample resource of further references.

An interesting application of multilinear methods to an analysis of relations among web links is proposed in the paper by Kolda *et al.* [186]. In this area, Hongli *et al.* propose a method for discovering potential web communities from tensor decomposition [187].

Much interesting information on multiway analysis with a special stress on applications in chemistry can be found in the book by Smilde *et al.* [121]. It contains a very intuitive introduction to many aspects of tensor processing and analysis, as well as being endowed with many examples. A slightly different approach and information about multiway data analysis can be found in the book by Kroonenberg [21]. Here one can find useful information on data preprocessing, the problem of missing data in multiway analysis, but also on interpreting and visualizing components in different tensor decomposition models.

Good reading on independent component analysis can be found in two books, by Hyvärinen *et al.* [97], as well as that by Stone [144]. The first offers a very in-depth introduction into estimation and information theory, as well as optimization methods, PCA, tensors, and applications. Conversely, the book by Stone presents ICA in a much more condensed, but useful, fashion.

An important issue in data mining and machine learning is dimensionality reduction, which we only touched upon when discussing PCA in this section. These issues are further explored in Section 3.3.1 of this book. The more advanced approach relies on nonlinear methods, called manifold learning. In this respect graphs are frequently used to represent the topology of a manifold with geodesic distances. More on this important domain can be accessed in numerous publications. A very intuitive introduction in this respect is provided by the book by Lee and Verleysen [188]. Last but not least are numerous papers published in leading journals and conferences on computer vision and pattern recognition. These are cited in particular sections of this chapter.

2.13.3   Problems and Exercises

1. Check all the values of the tensor $\mathcal{U}_{3\times 3}$ on p. 20. Compute a contraction of tensors $\mathcal{S}$ and $\mathcal{T}$ in respect to only their first common index K_1.
2. Show that for a positive definite and symmetrical matrix its eigenvalues are nonnegative and its eigenvectors are orthogonal.
3. Show that the triangular inequality condition (2.22) leads to the nonnegativity and symmetry conditions of the distance function D in Definition 2.7.
4. Prove the Kronecker properties (2.207) and (2.208).
5. Derive the formula (2.122) for statistical moments. Using software implementation find numerical differences between moments computed with (2.122) as compared to (2.120).
6. Derive Equation (2.199) on core tensor $\mathcal{Z}$ from (2.189).
7. Based on the implementation of the *Orphan_Log_Of_Matrix* function, listed in Algorithm 2.2, create a more general function object for computation of other matrix functions, such

as *exp*, etc. The function object should accept a function as its parameter (e.g. in the form of another function object) to be computed from the matrix.

8. From Equation (2.160) show that each element of a tensor $\mathcal{T}$ at a position (i_1, i_2, i_3) is a product of three components of the vectors **a**, **b**, and **c** at corresponding indices.

9. With the help of Algorithm 2.14 and its implementation a computed HOSVD of a color image is seen as a 3D tensor. What we can tell about its singular values and mode matrices?

10. Perform rank-1 and best-rank decompositions on a video sequence. For both test a number of components and reconstruction errors.

11. Use the NMF procedure to segment a color image, as discussed in Section 2.12.10. The base matrix should contain cluster centroids and membership values will be encoded into the component matrix. How do we choose the number of centroids?

References

[1] Mitra S.K.: Digital Signal Processing. McGraw-Hill, 2000.

[2] Oppenheim A.V., Schafer R.W.: Discrete-Time Signal Processing. Prentice-Hall, 1989.

[3] Einstein A., Davis F.A.: The Principle of Relativity. Dover Publications, 1952.

[4] Bishop R.L., Goldberg S.I.: Tensor Analysis on Manifolds. Dover Publications, Inc. 1980.

[5] Penrose R.: The Road to Reality. A Complete Guide to the Laws of the Universe. Alfred A. Knopf, 2005.

[6] Dimitrienko, I.: Tensor Analysis and Nonlinear Tensor Functions, Springer, 2003.

[7] Hartley R.I., Zisserman A.: Multiple View Geometry in Computer Vision. 2nd edition, Cambridge University Press, 2003.

[8] Faugeras O.D., Luong Q.-T.: The Geometry of Multiple Images. MIT Press, 2001.

[9] Bigün, J., Granlund, G.H., Wiklund, J., Multidimensional Orientation Estimation with Applications to Texture Analysis and Optical Flow. IEEE PAMI 13(8), pp. 775–790, 1991.

[10] Bigun J.: Vision with Direction. A Systematic Introduction to Image Processing and Computer Vision. Springer, 2006.

[11] Vasilescu M.A., Terzopoulos D.: Multilinear analysis of image ensembles: TensorFaces. Proceedings of Eurpoean Conference on Computer Vision, pp. 447–460, 2002.

[12] Vasilescu M.A., Terzopoulos D.: Multilinear Independent Component Analysis. IEEE Conference on Computer Vision and Pattern Recognition, CVPR 2005, Vol. 1, pp. 547–553, 2005.

[13] Triggs B.: The Geometry of Projective Reconstruction I: Matching Constraints and the Joint Image. International Journal of Computer Vision, July, 1995.

[14] Triggs B.: Plane + Parallax, Tensors and Factorization. INRIA Technical Report, 2000.

[15] Cyganek B., Siebert J.P.: An Introduction to 3D Computer Vision Techniques and Algorithms, Wiley, 2009.

[16] Korn G.A., Korn T.M.: Mathematical Handbook for Scientists and Engineers. Dover Publications, 2000.

[17] Moon T.K., Stirling W.C.: Mathematical Methods and Algorithms for Signal Processing. Prentice-Hall 2000.

[18] Kay D.C.: Theory and Problems of Tensor Calculus. Schaum's Outlines Series. McGraw-Hill 1988.

[19] Cichocki A., Zdunek R., Amari S.: Nonnegative Matrix and Tensor Factorization. IEEE Signal Processing Magazine, Vol. 25, No. 1, 2008, pp. 142–145.

[20] Kolda T.G., Bader B.W.: Tensor Decompositions and Applications. SIAM Review, 2008.

[21] Kroonenberg P.M.: Applied Multiway Data Analysis, Wiley, 2008.

[22] Montoto L.J.V.: Maximum Likelihood Methods for Three-Way Analysis in Chemistry. PhD Thesis. Dalhousie University, Halifax, Nova Scotia, 2005.

[23] Simard D., LeCun Y, Denker J.: Efficient pattern recognition using a new transformation distance. Advances in Neural Information Processing Systems, Morgan Kaufman, pp. 50–58, 1993.

[24] Gilchrist J.R., Karthick S.: Hyperspectral imaging: seeing the invisible. EETimes Europe, pp. 36–37, 2011.

[25] Camps-Valls G., Bruzzone L.: Kernel Methods for Remote Sensing and Data Analysis, Wiley, 2009.

[26] http://aviris.jpl.nasa.gov/

[27] Plaza A., Plaza J., Paz A., Sánchez S.: Parrallel Hyperspectral Image and Signal Processing. IEEE Signal Processing Magazine, Vol. 28, No. 3, pp. 119–126, 2011.

[28] http://www.wiley.com/go/cyganekobject

[29] http://www.mathworks.com

[30] Borisenko A.I., Tarapov I.E.:Vector and Tensor Analysis with Applications. Dover Publications, Inc. 1979.

[31] Golub G.H., Van Loan C.F.: Matrix Computations. The Johns Hopkins University Press, 3rd edition 1996.

[32] Meyer C.D.: Matrix Analysis and Applied Linear Algebra. SIAM, 2000.

[33] Arsigny V.: Processing Data in Lie Groups: An Algebraic Approach. Application to Non-Linear Registration and Diffusion Tensor MRI. PhD Thesis, INRIA, 2006.

[34] Peeters, T., Rodrigues, P., Vilanova, A., ter Haar Romeny, B.: Analysis of distance/similarity measures for diffusion tensor imaging. In: Visualization and Processing of Tensor Fields: Advances and Perspectives. Springer, Berlin, pp. 113–136, 2008.

[35] Pennec, X., Fillard, P., Ayache, N.: A Riemannian framework for tensor computing. International Journal of Computer Vision, Vol. 66, No. 1, pp. 41–66, 2006.

[36] Wang, Z., Vemuri, B.C.: DTI segmentation using an information theoretic tensor dissimilarity measure. IEEE Transactions on Medical Imaging, Vol. 24, No. 10, pp. 1267–1277, 2005.

[37] Lenglet C., Campbell J.S.W., Descoteaux M., Haro G., Savadjiev P., Wassermann D., Anwander A., Deriche R., Pike G.B., Sapiro G., Siddiqi K., Thompson M: Mathematical methods for diffusion MRI processing. Neuroimage, Vol. 45, No 1, pp. 111–122, 2009.

[38] Batchelor P.G., Moakher M., Atkinson D., Clamante F., Connelly A.: A rigorous framework for diffusion tensor calculus. Magnetic Resonance in Medicine, Vol. 53, pp. 221–225, 2005.

[39] Arsigny V., Fillard P., Pennec X., Ayache N.: Log-Euclidean Metrics for Fast and Simple Calculus on Diffusion Tensors. Magnetic Resonance in Medicine, Vol. 56, No. 2, pp. 411–421, 2006.

[40] Rittner L., Flores F.C., Lotufo R.A.: A tensorial framework for color images. Pattern Recognition Letters, Vol. 31, No. 4, pp. 277–296, 2010.

[41] Tuzel O., Porikli F., Meer P.: Pedestrian Detection via Classification on Riemannian Manifolds. IEEE Transactions on Pattern Analysis and Machine Intelligence, Vol. 30, No. 10, pp. 1713–1727, 2008.

[42] Moler C., Van Loan C.: Nineteen Dubious Ways to Compute the Exponential of a Matrix, Twenty-Five Years Later. SIAM REVIEW, Vol. 45, No. 1, pp. 3–49, 2003.

[43] Higham N.J.: The Scaling and Squaring Method for the Matrix Exponential Revisited. SIAM REVIEW, Vol. 51, No. 4, pp. 747–764, 2009.

[44] Cheng S.H., Higham N.J., Kenney C.S., Laub A.J.: Approximating the logarithm of a matrix to specified accuracy. SIAM Journal of Matrix Analysis and Applications, Vol. 22, No. 4, pp. 1112–1125, 2001.

[45] Higham N.J.: Functions of matrices. Theory and computation. SIAM, 2008.

[46] Wang L., Zhang Y., Feng J.: On the Euclidean Distance of Images. IEEE Transactions on Pattern Analysis and Machine Intelligence, Vol. 27, No. 8, pp. 1334–1339, 2005.

[47] Sun B., Feng J.: A Fast Algorithm for Image Euclidean Distance. Chinese Conference on Pattern Recognition CCPR '08, pp. 1–5, 2008.

[48] Liu Y., Liu Y., Chan K.C.C.: Tensor Distance Based Multilinear Locality-Preserved Maximum Information Embedding. IEEE Transactions on Neural Networks, Vol. 21, No. 11, pp. 1848–1854, 2010.

[49] Cyganek B.: Embedding of the Extended Euclidean Distance into Pattern Recognition with Higher-Order Singular Value Decomposition of Prototype Tensors. Lecture Notes in Computer Science Vol. 7564, Springer, pp. 180–190, 2012.

[50] Pitas I., Venetsanopoulos A.N.: Nonlinear Digital Filters. Principles and Applications. Kluwer Academic Publishers 1990.

[51] Cyganek B.: Computational Framework for Family of Order Statistic Filters for Tensor Valued Data, Lecture Notes in Computer Science, Vol. 4141, Springer, pp. 156–162, 2006.

[52] Welk, M., Feddern, C., Burgeth, B., Weickert, J.: Median Filtering of Tensor-Valued Images. Lecture Notes in Computer Science, Vol. 2781, Springer, pp. 17–24, 2003.

[53] Mitra S.K., Sicuranza G.L.: Nonlinear Image Processing. Academic Press, 2000.

[54] Zhang F., Hancock E.R.: New Riemannian technique for directional and tensorial image data. Pattern Recognition, Vol. 43, No. 4, pp. 1590–1606, 2010.

[55] Halliday D., Resnick R., Walker J.: Principles of Physics. Wiley 2010.

[56] Shapira Y.: Solving PDEs in C++. Numerical Methods in a Unified Object-Oriented Approach. SIAM, 2006.

[57] Weickert J.: Anisotropic Diffusion in Image Processing. B.G. Teubner Stuttgart, 1998.

[58] Jähne B.: Digital Image Processing. 6th edition, Springer-Verlag, 2005.

[59] Perona P., Malik J.: Scale-Space and Edge Detection Using Anisotropic Diffusion. IEEE Transactions On Pattern Analysis and Machine Intelligence. Vol. 12, No. 7, pp. 629–639. July 1990.

[60] Weickert J., Brox T.: Diffusion and Regularization of Vector- and Matrix- Valued Images. Technical Report, Universität des Saarlandes, 2002.

[61] Sapiro G.: Geometric Partial Differential Equations and Image Analysis. Cambridge University Press, 2001.

[62] Gerig G. Kübler O., Kikinis R., Jolesz F.A.: Nonlinear anisotropic filtering of MRI data. IEEE Transactions on Medical Imaging, Vol. 11, No. 2, pp. 221–232, 1992.

[63] Brox T., Weickert J., Burgeth B., Mrázek, P.: Nonlinear Structure Tensors. Universität des Saarlandes, Nr. 113, 2004.

[64] Weickert J., Hagen H. (Editors): Visualization and Processing of Tensor Fields. Springer, 2006.

[65] Cyganek B.: Traffic Scene Segmentation and Robust Filtering for Road Signs Recognition. The International Conference on Computer Vision & Graphics L. Bolc et al. (Eds.): ICCVG 2010, Part I, Lecture Notes in Computer Science, Vol. 6374, 2010. Springer-Verlag Berlin Heidelberg, pp. 292–299, 2010.

[66] Press W.H., Teukolsky S.A., Vetterling W.T., Flannery B.P.: Numerical Recipes in C. The Art of Scientific Computing. Third Edition. Cambridge University Press, 2007.

[67] Farid H., Simoncelli E.P.: Differentiation of Discrete Multidimensional Signals. IEEE Transactions on Image Processing, Vol. 13, No. 4, April 2004, pp. 496–508.

[68] http://www.wiley.com/CyganekSiebertBook.html

[69] http://www.wiley.com/legacy/wileychi/cyganek3dcomputer/supp/HIL_Manual_01.pdf

[70] Brox T., Boomgaard van den R., Lauze F., Weijer van de J., Weickert J., Mrázek P., Kornprobst P.: Adaptive Structure Tensors and their Applications, in Visualization and Processing of Tensor Fields edited by Weickert, J., Hagen, H., Springer, pp. 17–47, 2006.

[71] Jähne B.: Spatio-Temporal Image Processing, Springer-Verlag, 1993.

[72] Gonzalez R.C, Woods R.E.: Digital Image Processing, 2nd Edition, Prentice Hall, 2002.

[73] Jähne B.: Practical Handbook on Image Processing for Scientific Applications, CRC Press 1997.

[74] Pratt W.K.: Digital Image Processing. Third Edition. Wiley, 2001.

[75] http://en.wikipedia.org/wiki/Atan2

[76] Cyganek B.: Depth Recovery with an Area Based Version of the Stereo Matching Method with Scale-Space Tensor Representation, Lecture Notes in Computer Science, Vol. 3037, pp. 548–551, 2004.

[77] Fillard P., Arsigny V., Ayache N., Pennec X.: A Riemannian Framework for the Processing of Tensor-Valued Images. Lecture Notes in Computer Science, Vol. 3753, pp. 112–123, 2005.

[78] Forsyth D.A., Ponce J.: Computer Vision: A Modern Approach, Prentice Hall, 2003.

[79] http://i21www.ira.uka.de/image_sequences/

[80] Horn B, Schunck B.: Determining optical flow. Artificial Intelligence, Vol. 17, pp. 185–203, 1981.

[81] Di Zenzo S.: A note on the gradient of a multi-image. Computer Vision, Graphics and Image Processing, 33, pp. 116–125, 1986.

[82] Sochen N., Kimmel R., Malladi R.: A General Framework for Low Level Vision. IEEE Transactions on Image Processing, Vol. 7, No. 3, pp. 310–318, 1998.

[83] Brox T., Rousson M., Derich, R., Weickert J.: Unsupervised Segmentation Incorporating Colour, Texture, and Motion. INRIA Technical Report No 4760, 2003.

[84] Cyganek B.: Object Detection in Multi-Channel and Multi-Scale Images Based on the Structural Tensor. Lecture Notes in Computer Science, Vol. 3691, Springer, pp. 570–578, 2005.

[85] Luis-García R., Deriche R., Rousson M., Alberola-López C.: Tensor Processing for Texture and Colour Segmentation. Lecture Notes in Computer Science, Vol. 3540, pp. 1117–1127, 2005.

[86] Heinbockel J.H.: Introduction to Tensor Calculus and Continuum Mechanics. Trafford Publishing, 2001.

[87] Grimmet G., Stirzaker D.: Probability and Random Processes. Third Edition, Oxford University Press, 2001.

[88] Papoulis A.: Probability, Random Variables, and Stochastic Processes. Third Edition, McGraw-Hill, 1991.

[89] Orkisz M., Valencia L.F., Hoyos M.H.: Models, Algorithms and Applications in Vascular Image Segmentation. Machine Graphics & Vision, Vol. 17, No. 1/2, pp. 5–33, 2008.

[90] Flusser J., Suk T., Zitová B.: Moments and Moment Invariants in Pattern Recognition. Wiley, 2009.

[91] Lin, W. G. and Wang, S.: A note on the calculation of moments. Pattern Recognition Letters, Vol. 15, No. 11, pp. 1065–1070, 1994.

[92] Hu, M.K.: Visual pattern recognition by moment invariants. IRE Transactions on Information Theory, No. 8, pp. 179–187, 1962.

[93]   Flusser, J., Suk, T.: Pattern recognition by affine moments invariants. Pattern Recognition, Vol. 26, No. 1, pp. 167–174, 1993.

[94]   Flusser, J., Suk, T.: Affine moment invariants: A new tool for character recognition. Pattern Recognition Letters, Vol. 15, No. 4, pp. 433–436, 1994.

[95]   Sühling M., Arigovindan M., Hunziker P. Unser M.: Multiresolution Moment Filters: Theory and Applications. IEEE Transactions on Image Processing, Vol. 13, No. 4, pp. 484–495, 2004.

[96]   Lathauwerde L.: Signal Processing Based on Multilinear Algebra. PhD dissertation, Katholieke Universiteit Leuven, 1997.

[97]   Hyvärinen A., Karhunen J., Oja E.: Independent Component Analysis. Wiley, 2001.

[98]   Trefethen L.N., Bau D.: Numerical Linear Algebra. SIAM 1997.

[99]   Sierra R.: Nonrigid registration of diffusion tensor images. Master Thesis, Swiss Federal Institute of Technology (ETHZ), Computer Vision Laboratory, 2001.

[100] Mordohai P., Medioni G.: Tensor Voting. A Perceptual Organization Approach to Computer Vision and Machine Learning, Morgan & Claypool Publishers, 2007.

[101] Desolneux A., Moisan L. Morel J-M.: From Gestalt Theory to Image Analysis. Springer, 2008.

[102] Kindlmann G: Tensor Invariants and their Gradients in Visualization and Processing of Tensor Fields, ed. Weickert J., Hagen H., pp. 215–224, 2006.

[103] Yoo T.S.: Insight into Images. Principles and Practice for Segmentation, Registration, and Image Analysis. A.K. Peters, 2004.

[104] Bahn M.M.: Comparison of Scalar Measures Used in Magnetic Resonance Diffusion Tensor Imaging. Journal of Magnetic Resonance, 139, pp. 1–7, 1999.

[105] Pajevic S., Aldroubi A., Basserz P.J.: A Continuous Tensor Field Approximation of Discrete DT-MRI Data for Extracting Microstructural and Architectural Features of Tissue. Journal of Magnetic Resonance, Vol. 154, No. 1, pp. 85–100, 2002.

[106] Westin C.-F., Maier S.E., Mamata H., Nabavi A., Jolesz F.A., Kikinis R.: Processing and visualization for diffusion tensor MRI. Medical Image Analysis, Vol. 6, No. 2, pp. 93–108, 2002.

[107] Ma Y., Soatto S., Košecká J., Sastry S.S.: An Invitation to 3-D Vision. From Images to Geometrical Models. Springer, 2004.

[108] Heyden A.: A Common Framework for Multiple View Tensors. ECCV' 98, 5th European Conference on Computer Vision, Proceedings, Vol. 1, pp. 3–19, 1998.

[109] Torr P.H.S, Zisserman A.: Robust Parametrization and Computation of the Trifocal Tensor. Department of Engineering Science, University of Oxford, Hilary Term, 1998.

[110] Tucker L.R.: Some mathematical notes of three-mode factor analysis. Psychometrika, 31, 1966, pp. 279–311.

[111] Bro R.: Multi-Way Analysis in The Food Industry. Models, Algorithms & Applications. PhD Thesis. Department of Dairy and Food Science, Royal Veterinary and Agricultural University, Denmark. 1998.

[112] le Bihan N.: Traitement Algebrique des Signaux Vectoriels. Application en Separation d'Ondes Sismiques. Docteur These. Institut National Polytechnique de Grenoble, 2001.

[113] Muti D., Bourennane S.: Survey on tensor signal algebraic filtering. Signal Processing 87, pp. 237–249, 2007.

[114] Vasilescu M.A., Terzopoulos D.: Multilinear (Tensor) Image Synthesis, Analysis, and Recognition. IEEE Signal Processing Magazine, Vol. 24, No. 6, pp. 118–123, 2007.

[115] Lathauwer de L., Moor de, B., Vandewalle J.: A Multilinear Singular Value Decomposition. SIAM Journal Matrix Analysis and Applications, Vol. 21, No. 4, pp. 1253–1278, 2000.

[116] Kolda T.G.: Orthogonal tensor decompositions. SIAM Journal on Matrix Analysis and Applications, Vol. 23, No. 1, 2001, pp. 243–255.

[117] Franc A.: Etude Algebrique des Multitableaux: Apports de l'Algebre Tensorielle. These de Dimplome de Doctorat. Universite de Montpellier II, 1992.

[118] Kilmer M.E., Martin C.D.M.: Decomposing a Tensor. SIAM News, Vol. 37, No. 9, pp. 19–20, 2004.

[119] Harshman, R.A.: Foundations of the PARAFAC procedure: Models and conditions for an "explanatory" multimodal factor analysis. UCLA Working Papers in Phonetics, 16, pp. 1–84, 1970.

[120] Kiers H.A.L.: Towards a standardized notation and terminology in multiway analysis. Journal of Chemometrics, Vol. 14, No. 3, pp. 105–122, 2000.

[121] Smilde A., Bro R., Geladi P.: Multi-way Analysis: Applications in the Chemical Sciences. Wiley, 2004.

[122] Bader W.B., Kolda T.G.: MATLAB® Tensor Classes for Fast Algorithm Prototyping. ACM Transactions on Mathematical Software. Vol. 32, No. 4, pp. 635–653, 2006.

[123] Gamma E., Helm R., Johnson R., Vlissides J.: Design Patterns. Elements of Reusable Object-Oriented Software. Addison-Weseley, 1995.

[124] Cichocki A., Zdunek R., Phan A.H., Amari S-I.: Nonnegative Matrix and Tensor Factorizations. Applications to Exploratory Multi-way Data Analysis and Blind Source Separation. Wiley, 2009.

[125] Banachiewicz T.: Krakowian Calculus with Applications (in Polish: Rachunek krakowianowy z zastosowaniami), PWN, 1959.

[126] Savas B., Eldén L.: Handwritten digit classification using higher order singular value decomposition. Pattern Recognition, Vol. 40, No. 3, pp. 993–1003, 2007.

[127] Graham A.: Kronecker Products and Matrix Calculus: with Applications. Ellis Horwood Ltd., 1981.

[128] Meyer B.: Applying "Design by Contract". IEEE Computer, Vol. 25, No. 10, pp. 40–51, 1992.

[129] McConnell S.: Code Complete. 2nd edition. Microsoft Press, 2004.

[130] Kofidis E., Regalia P.A.: On the best rank-1 approximation of higher-order supersymmetric tensors. SIAM Journal on Matrix Analysis and Applications, Vol. 23, No. 3, pp. 863–884, 2002.

[131] Lathauwer de L., Moor de, B., Vandewalle J.: On the Best Rank-1 and Rank-$(R_1, R_2, \ldots, R_N)$ Approximation of Higher-Order Tensors, Vol. 21, No. 4, pp. 1324–1342, 2000.

[132] Bertsekas D.P.: Constraint Optimization and Lagrange Multiplier Methods. Athena Scientific, 1996.

[133] Regalia P.A., Kofidis E.: The higher-order power method revisited: convergence proofs and effective initialization. IEEE International Conference on Acoustics, Speech, and Signal Processing ICASSP '00, Vol.5, pp. 2709–2712, 2000.

[134] Wang H., Ahuja N.: Compact Representation of Multidimensional Data Using Tensor Rank-One Decomposition. Proceedings of the 17th International Conference on Pattern Recognition, Vol. 1, pp. 44–47, 2004.

[135] Wang H., Ahuja N.: A Tensor Approximation Approach to Dimensionality Reduction. International Journal of Computer Vision, Vol. 76, pp. 217–229, 2008.

[136] Chen J. Saad Y.: On the Tensor SVD and the Optimal Low Rank Orthogonal Approximation of Tensors. SIAM Journal on Matrix Analysis and Applications, Vol. 30, No. 4, pp. 1709–1734, 2009.

[137] Oseledets I. V., Savostianov D. V., Tyrtyshnikov E. E.: Tucker dimensionality reduction for three-dimensional arrays in linear time. SIAM Journal of Matrix Analysis and Applications, Vol. 30, No. 3, pp. 939–956. 2008.

[138] Savas B., Eldén L.: Krylov subspace methods for tensor computations. Technical Report LITH-MAT-R-2009-02-SE, Department of Mathematics, Linköpings Universitet, 2009.

[139] Ishteva M, Lathauwer L. de, Absi P.-A., Huffel S. van: Best low multilinear rank approximation of higher-order tensors, based on the Riemannian trust-region scheme. SIAM Journal on Matrix Analysis and Applications, Vol. 32, No. 1, pp. 115–135, 2011.

[140] Ye J.: Generalized low rank approximation of matrices. Machine Learning, Vol. 61, pp. 167–191, 2005.

[141] http://www.anefian.com/research/face_reco.htm

[142] Gersho A., Gray R.M.: Vector Quantization and Signal Compression. The Springer International Series in Engineering and Computer Science, Springer, 1991.

[143] Jolliffe I.T.: Principal component analysis. Springer, 2002.

[144] Stone J.V.: Independent Component Analysis. A Tutorial Introduction. MIT Press, 2004.

[145] Lee D.D, Seung H.S.: Learning the parts of objects by non-negative matrix factorization. Nature, Vol. 401, No. 6755, pp. 788–791, 1999.

[146] Turk M.A., Pentland A.P.: Face recognition using eigenfaces. IEEE Conference on Computer Vision and Pattern Recognition, pp. 586–590, 1991.

[147] Shakhnarovich G., Moghaddam B.: Face Recognition in Subspaces. Mitsubishi Electric Research Laboratories (http://www.merl.com), Technical Report TR2004-041, 2004.

[148] Tuncer Y, Tanik M.M., Allison D.B.: An overview of statistical decomposition techniques applied to complex systems. Computational Statistics & Data Analysis, Vol. 52, No. 5, pp. 2292–2310, 2008.

[149] Földiák P.: Sparse Coding in the Primate Cortex in the Handbook of Brain Theory and Neural Networks, edited by M.A.Arbib, 2nd edition, MIT Press, pp. 1064–1068, 2003.

[150] Lee D.D, Seung H.S.: Algorithms for Non-negative Matrix Factorization. Advances in Neural Information Processing Systems 13: Proceedings of the 2000 Conference. MIT Press. pp. 556–562, 2001.

[151] Li J., Zhang L., Tao D., Sun H., Zhao Q.: A prior neurophysiologic knowledge free tensor-based scheme for single trial EEG classification, IEEE Transactions on Neural Systems and Rehabilitation Engineering, Vol. 17, No. 2, pp. 107–115, 2009.

[152] Liu W, Zheng N.: Non-negative matrix factorization based methods for object recognition. Pattern Recognition Letters, Vol. 25, No. 8, pp. 893–897, 2004.

[153] Guillamet D., Vitrià J.: Evaluation of distance metrics for recognition based on non-negative matrix factorization. Pattern Recognition Letters, Vol. 24, No. 9–10, pp. 1599–1605, 2003.

[154] Ding C., He X., Simon H.D.: On the Equivalence of Nonnegative Matrix Factorization and Spectral Clustering. SIAM International Conference on Data Mining, pp. 606–610, 2005.

[155] Kim J., Park H.: Sparse Nonnegative Matrix Factorization for Clustering. Georgia Institute of Technology Technical Report GT-CSE-08–01, 2008.

[156] Frey B. J., Dueck D.: Clustering by passing messages between data points. Science 315, No. 5814, pp. 972–976, 2007.

[157] Li J-M., Qian Y-T: Clustering-based hyperspectral band selection using sparse nonnegativematrix factorization. Journal of Zhejiang University-SCIENCE C (Computers & Electronics). Vol. 12, No. 7, pp. 542–549, 2011.

[158] Zheng C-H., Huang D-S., Zhang L., Kong X-Z.: Tumor Clustering Using Nonnegative Matrix Factorization With Gene Selection. IEEE Transactions on Information Technology in Biomedicine, Vol. 13, No. 4, pp. 599–607, 2009.

[159] Chapelle O., Schölkopf B., Zien A.: Semi-Supervised Learning. MIT Press, 2006.

[160] Liu H., Wu Z., Li X., Cai D., Huang T.S.: Constrained Nonnegative Matrix Factorization for Image Representation. IEEE Transactions on Pattern Analysis and Machine Intelligence, Vol. 34, No. 7, pp. 1299–1311, 2012.

[161] Fei-Fei L., Fergus R., Perona P.: One-Shot Learning of Object Categories. IEEE Transactions on Pattern Analysis and Machine Intelligence, Vol. 28, No. 4, pp. 594–611, 2006.

[162] Boutsidis C., Gallopoulos E.: SVD based initialization: A head start for nonnegative matrix factorization. Pattern Recognition, Vol. 41, No. 4, pp. 1350–1362, 2008.

[163] Chu M., Diele F., Piemmons R., Ragni S.: Optimality, computation and interpretation of nonnegative matrix factorizations. Available at http://citeseerx.ist.psu.edu/viewdoc/summary?doi=10.1.1.61.5758, 2004.

[164] http://yann.lecun.com/exdb/mnist/

[165] Polish Road Signs and Signalization. Directive of the Polish Ministry of Infrastructure (in Polish), Dz. U. Nr 170, poz. 1393, 2002.

[166] Hazan T., Polak S., Shashua A.: Sparse Image Coding using a 3D Non-negative Tensor Factorization. ICCV 2005 10th IEEE International Conference on Computer Vision, Vol. 1, pp. 50–57, 2005.

[167] Georghiades A.S., Belhumeur P.N., Kriegman D.J.: From Few to Many: Illumination Cone Models for Face Recognition under Variable Lighting and Pose. IEEE Trans. IEEE Transactions on Pattern Analysis and Machine Intelligence, Vol. 23, No. 6, pp. 643–660, 2001.

[168] Mørup M., Hansen L.K., Arnfred S.M.: Algorithms for Sparse Nonnegtive Tucker Decompositons. Neural Computations, Vol. 20, No. 8, pp. 2112–2131, 2008.

[169] Phan A.H., Tichavskỳ P., Cichocki A.: Damped Gauss-Newton Algorithm for Nonnegative Tucker Decomposition. IEEE Statistical Signal Processing Workshop, pp. 669–672, 2011.

[170] Samaria F., Harter A.C.: Parameterisation of a stochastic model for human face identification. Proceedings of the Second IEEE Workshop on Applications of Computer Vision, 1994.

[171] Phan A.H., Cichocki A.: Block Decomposition for Very Large-Scale Nonnegative Tensor Factorization. IEEE International Workshop on Computational Advances in Multi-Sensor Adaptive Processing, pp. 316–319, 2009.

[172] Cyganek B.: Ensemble of Tensor Classifiers Based on the Higher-Order Singular Value Decomposition. Springer, Lecturer Notes in Computer Science, Vol. 7209, pp. 578–589, 2012.

[173] Friedlander M.P., Hatz K.: Computing nonnegative tensor factorization. Department of Computer Science, University of British Columbia, Technical Report TR-2006–21, 2007.

[174] Tenenbaum J. B., Freeman W. T.: Separating style and content with bilinear models. Neural Computations, Vol. 12, No. 6, pp. 1246–1283, 2000.

[175] Lin D., Xu Y., Tang X., Yan S.: Tensor-based factor decomposition for relighting. Proceedings of the IEEE International Conference on Image Process, Vol. 2, pp. 386–389, 2005.

[176] Peng B., Qian G.: Online Gesture Spotting from Visual Hull Data. IEEE Transactions on Pattern Analysis and Machine Intelligence, Vol. 33, No. 6, pp. 1175–1188, 2011.

[177] Park S.W., Savvides M.: Individual Kernel Tensor-Subspaces for Robust Face Recognition: A Computationally Efficient Tensor Framework Without Requiring Mode Factorization. IEEE Transactions on Systems, Man, and Cybernetics – Part B: Cybernetics, Vol. 37, No. 5, pp. 1156–1166, 2007.

[178] Wold, S.: Pattern recognition by means of disjoint principal components models. Pattern Recognition, Vol. 8, No. 3, pp. 127–139, 1976.

[179] Wang S-J., Yang J., Zhang N., Zhou C-G.: Tensor Discriminant Color Space for Face Recognition. IEEE Transactions on Image Processing, Vol. 20, No. 9, pp. 2490–2501, 2011.

[180] Yang J., Liu C.: Color image discrimination models and algorithms for face recognition. IEEE Transactions on Neural Networks, Vol. 19, No. 12, pp. 2088–2098, 2008.

[181] Duchenne O., Bach F., Kweon I-S, Ponce J.: A Tensor-Based Algorithm for High-Order Graph Matching. IEEE Transactions on Pattern Analysis and Machine Intelligence, Vol. 33, No. 12, pp. 2383–2395, 2011.

[182] Lowe D.: Distinctive Image Features from Scale-Invariant Keypoints. International Journal of Computer Vision, Vol. 60, No. 2, pp. 91–110, 2004.

[183] Liu C., Yuen J., Torralba A.: Nonparametric Scene Parsing via Label Transfer. IEEE Transactions on Pattern Analysis and Machine Intelligence, Vol. 33, No. 12, pp. 2368–2382, 2011.

[184] Aja-Fernández, S, de Luis García, R., Tao, D., Li, X. (eds): Tensors in Image Processing and Computer Vision. Springer, 2009.

[185] Laidlaw D.H., Weickert J. (Editors): Visualization and Processing of Tensor Fields. Advances and Perspectives. Springer, 2009.

[186] Kolda T.G., Bader B.W., Kenny J.P.: Higher-Order Web Link Analysis Using Multilinear Algebra. Sandia Report, 2005.

[187] Hongli Y., Guoping H., Yongshan W.: A Novel Model and Algorithm for Discovering the Potential Web Communities Based on Tensor Decomposition, pp. 5189–5194, 2008.

[188] Lee J.A., Verleysen M.: Nonlinear Dimensionality Reduction. Springer, 2007.

3

Classification Methods and Algorithms

3.1 Abstract

Classification is the process of naming an object based on observation of it. This is a very "human" like definition, however it reflects our daily experience. For instance, the way we read a text or how we pick a ripe plum, or how we sometimes have the impression that we know the person in a car that has just passed – all of these are examples of our daily judgments, i.e. classifications. What we are mostly interested in this book are the methods that can allow a computer to automatically classify objects seen in images. Thus, classifiers are the key concept of computer vision. In this section we present a very brief, but hopefully useful, overview of some of them. The main assumption is to present the main ideas behind each method, first starting from the theory, then presenting its implementation. Thus, this section presents the tools for computer vision and pattern recognition, with majority of them presented to practical problems of object detection and recognition in the next two chapters of this book. An attempt was also made to indicate the connections between different approaches. Nevertheless, the subject is very large and for more information the interested reader is referred to the cited publications.

We start with a presentation of the classification framework followed by a series of subsections devoted to one of the most important concepts of data investigation, called principal component analysis, which has its roots in a statistical framework. Then the two fundamental concepts of pattern recognition are analyzed – the Bayes decision rule and the maximum *a posteriori* classification method. Next, parametric and nonparametric methods are discussed, followed by a discussion of the two methods with primary applications to object tracking, that is, the Kalman filter and the mean shift method. Subsequent subsections are devoted to three types of neural network, as well as to their implementations. What follows is a presentation on the realm of kernel based methods moving slightly towards the domain of classification and clustering. Finally, cooperating groups of classifiers, called ensembles, are discussed. The presented concepts are underpinned by working code examples, case studies of some applications, as well as numerous literature references.

Object Detection and Recognition in Digital Images: Theory and Practice, First Edition. Bogusław Cyganek.
© 2013 John Wiley & Sons, Ltd. Published 2013 by John Wiley & Sons, Ltd.

3.2   Classification Framework

In the classification process it is common to assume that objects can be assigned to one of the C classes. As we already mentioned, classification consists of "naming" an unknown object based on a measurement, so it becomes obvious to which class it belongs. It is also common to name each i-th class of objects as ω_i which then form a set of labels

$$\Omega = \{\omega_1, \omega_2, \ldots, \omega_C\}. \tag{3.1}$$

On the other hand, the measurement is described by a *feature vector* $\mathbf{x}$. With these prerequisites a classifier can be defined as a function [1, 2, 3, 4, 5]:

Definition 3.1   A classifier is a function

$$f: \quad \mathbf{X} \rightarrow \Omega, \tag{3.2}$$

where $\mathbf{X} \in \Re^L$ denotes an L-dimensional feature space.   ☐

In other words, a classifier function returns a class name ω_i providing a feature vector $\mathbf{x}$. In practice a classification process is defined indirectly, providing a discriminant function for each pattern class which returns a score for that class. This can be written as follows

Definition 3.2   A discriminant function

$$g_c: \quad \mathbf{X} \rightarrow \Re, \text{ for } 1 \leq c \leq C \text{ and } \mathbf{X} \in \Re^L. \quad ☐ \tag{3.3}$$

Classification with the help of discriminant functions is defined as choosing a class ω_c for which the returned score is maximal, as follows:

Definition 3.3   A maximum membership rule

For a given feature vector $\mathbf{x}$, a classifier f returns a class[1] $\omega_{c'}$ for which its discriminant function $g_{c'}(\mathbf{x})$ is maximized, as follows

$$f(\mathbf{x}) = \omega_{c'}: \quad g_{c'}(\mathbf{x}) = \max_{1 \leq c \leq C} \{g_c(\mathbf{x})\}. \quad ☐ \tag{3.4}$$

The design of classifier f thus consists of choosing the functions g_c in Equation (3.4). These, naturally also depend on the type of features $\mathbf{x}$ selected to characterize the classes Ω. Classification with the help of discriminant functions is depicted in Figure 3.1. If we reverse the way of thinking slightly and start with a number N of measurements (features) $\mathbf{x}_n$, each

---

[1]In some places we use a simplified notation for the pattern classes and instead of writing ω_c we simply write its index c. However, if a concrete class is mentioned, such as ω_2, its full name is used (and not just a digit "2").

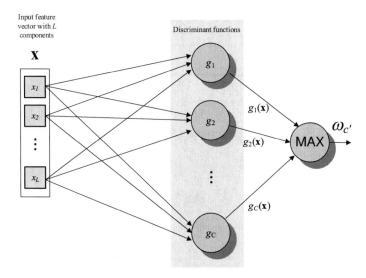

Figure 3.1 Explanation of the maximum membership rule. A feature vector **x** is scored by C discriminant functions g_c. A class $\omega_{c'}$ with maximal score is returned.

belonging to a different class from the set of all classes Ω, application of each of discriminant function to the whole set of points $\mathbf{x}_n \in \mathbf{X}$ and choice of the best scores results in the partition of the feature space into usually separate *classification regions*. Their boundaries are called *classification boundaries*. The points on the boundaries, i.e. those which belong to more than one class simultaneously, are called *tie points*. These can be resolved using some other rule, e.g. picking by random, etc.

An interesting and frequently used type of discriminant function is a binary version of the function (3.3), which returns 1/0 or a positive/negative output to differentiate between two classes. Let us observe that such a binary function can be easily created from two discriminant functions obeying the maximum membership rule

$$g(\mathbf{x}) = g_1(\mathbf{x}) - g_2(\mathbf{x}). \tag{3.5}$$

In the above, if **x** belongs to the class ω_1 then g_1 is maximized and g is positive; otherwise, the output of g_2 is the greatest which results in **g** being negative which indicates class ω_2.

The other important characteristic of a discriminant function is whether it is linear or nonlinear. A data set is linearly separable if the boundary regions can be represented in the form of a *hyperplane*, which is a single point for 1D, a line for 2D, and a plane for 3D cases, respectively.

3.2.1 *IMPLEMENTATION Computer Representation of Features*

What are features in terms of data structures? Any ordered series of values of any type. Simple, but such a broad definition can cause problems when trying to express it in computer languages. Using the generic paradigms a universal definition of the feature structure was created, as shown in Algorithm 3.1. This will unify further manipulations of features in various programming constructs.

```
// ---------------------------------------------------------------
// This class defines a concept of a feature with an arbitrary number of components
template < typename F >
class FeatureSpaceFor
{
    public:


        typedef typename F                        FeatureType;

        typedef typename vector< FeatureType >    DataPoint;

        typedef typename vector< DataPoint >      DataRepository;
};
// ---------------------------------------------------------------


typedef FeatureSpaceFor< int >      FeatureSpaceForInt;
typedef FeatureSpaceFor< double >   FeatureSpaceForDouble;


// ---------------------------------------------------------------
```

Algorithm 3.1 Generic definition of a data structure to represent the features.

The *FeatureSpaceFor* template class defines three types: The *FeatureType* for a single component of a feature, the *DataPoint* for a single data point, and finally the *DataRepository* for a collection of data points. The last two are based on STL vectors thanks to which their length can change dynamically in the run-time. Thus, it is easy to add new features to a repository or use features with different lengths from task to task. In the last two lines of code in Algorithm 3.1 two useful *typedefs* are introduced, namely *FeatureSpaceForInt* and *FeatureSpaceForDouble*. These define features with integer and floating-point components, respectively.

Apart from the bare definitions to represent the features we add some useful operations on them which treat features as mathematical vectors. Thus the methods allow feature addition, subtraction, the inner and outer products, as well as their normalization.

```
/////////////////////////////////////////////////////////////////
// This function adds two features (vectors)
/////////////////////////////////////////////////////////////////
//
// INPUT:
//            a, b - data points to be added; Can be of different
//                   size - in this case the minimal length is chosen.
```

```
//
// OUTPUT:
//                  a is updated by values from b
//
// REMARKS:
//
//
template < typename F >
inline typename FeatureSpaceFor< F >::DataPoint & operator += (
                          typename FeatureSpaceFor< F >::DataPoint & a,
                          typename const FeatureSpaceFor< F >::DataPoint & b )
{
      REQUIRE( a.size() == b.size() );// anyway, it is weird to add vectors of
      different size
      const int commonSize = min( a.size(), b.size() );
      for( register int i = 0; i < commonSize; ++ i )
            a[ i ] += b[ i ];
      return a;
}
/////////////////////////////////////////////////////////////////
// This function computes an inner product of two features
/////////////////////////////////////////////////////////////////
//
// INPUT:
//                  a, b - data points to be multiplied; Must
//                     be of the same size!
//
// OUTPUT:
//                  an inner product (a scalar)
//
// REMARKS:
//
//
template < typename F >
inline typename FeatureSpaceFor< F >::FeatureType InnerProduct(
            typename const FeatureSpaceFor< F >::DataPoint & a,
            typename const FeatureSpaceFor< F >::DataPoint & b )
{
    REQUIRE( a.size() == b.size() );    // must be of the same size
    FeatureSpaceFor< F >::FeatureType result = (FeatureSpaceFor< F > ::FeatureType)0;
    for( register int i = 0; i < a.size(); ++ i )
            result += a[ i ] * b[ i ];
    return result;
}
/////////////////////////////////////////////////////////////////
// This function computes an outer product of two features
/////////////////////////////////////////////////////////////////
//
// INPUT:
//            a, b - data points to be multiplied
//            result - a result matrix which must be already
//            of the size (at least) a.size x b.size (rows x cols)
```

```
//
// OUTPUT:
//              none
//
// REMARKS:
//
//
template < typename F >
inline void OuterProduct(
            typename const FeatureSpaceFor< F >::DataPoint & a,
            typename const FeatureSpaceFor< F >::DataPoint & b,
            typename TImageFor< F > & result )
{
        register int i, j;
        for( i = 0; i < a.size(); ++ i )
            for( j = 0; j < b.size(); ++ j )
                result.SetPixel( j, i, a[ i ] * b[ j ] );
}
```

Algorithm 3.2 Definitions of the functions operating on feature data represented by vector objects from STL.

Algorithm 3.2 shows code examples of the functions operating on feature objects. The first is an increment operator which simply increments elements of one feature object with elements of the second in an element by element fashion. The other two functions *InnerProduct* and *OuterProduct* compute the inner and outer products out of the two feature objects, respectively. However, in the first one the computation result is a simple scalar, so it is returned as a return value of the function. Conversely, in the case of *OuterProduct* function, its result is a matrix which for efficiency reasons is returned through the reference to the third argument of this function. Last but not least are the streaming operations which allow data transfer to and from any C++ stream, such as a screen or a file. These can be looked up in the attached code [6].

Presented feature definitions with the STL vector prove to be very useful in practice. Let us note however, that in some situations such a representation cannot be optimal since a vector object stores more data than a sum of feature components. Thus, in the case where a huge number of features need be stored at once, a simple C++ array with a fixed number of elements is chosen. In such cases we can also consider a more suitable memory allocator.

3.3 Subspace Methods for Object Recognition

Measurements or features which describe objects form an L-dimensional vector space. In this section we discuss its basic properties and methods for its reduction, as well as subspace based methods for object recognition.

3.3.1   *Principal Component Analysis*

The Principal Component Analysis (PCA) is one of the most important methods in data analysis, having its roots in statistics [2, 5, 7]. Having a set of N data points $\{x_i\}$, each represented as a vector of dimension L, the main idea of PCA is to extract important information conveyed, but usually hidden, in the data set $\{x_i\}$. This is achieved by finding a suitable *transformation* of possibly correlated values from the set $\{x_i\}$ into a new orthogonal space of uncorrelated variables, which are called *principal components*. Their number does not exceed the number of original data points, i.e. the cardinality of $\{x_i\}$. From the statistical point of view the principal components have a fundamental property – they are *mutually uncorrelated random variables*. However, they are also independent only if $\{x_i\}$ is jointly normally distributed.

Intuitively, finding the mutually uncorrelated components out of the unknown $\{x_i\}$ leads to retrieval of the intrinsic information conveyed by $\{x_i\}$. This happens since some data points can convey some information on others. An example is "creating" new variables from sums of the variables already included in the data set, and so forth.

To derive the main formulas explaining how PCA operates, at the start let us first transform each vector x_i from the set $\{x_i\}$ into the zero mean samples, as follows

$$\bar{x}_i = x_i - m_x, \tag{3.6}$$

where

$$m_x = E\left(x\right) = \frac{1}{N}\sum_{i=1}^{N}x_i \tag{3.7}$$

is a mean of data $\{x_i\}$ and E denotes the statistical expectation.

Now, in accordance with the aforementioned description we are looking for the following transformation

$$y = T\bar{x}, \tag{3.8}$$

which transforms x into *mutually uncorrelated variables* y. However, for y to be uncorrelated the following condition has to be met

$$E\left(y_i y_j\right) = 0, \text{ for all } i \neq j, \tag{3.9}$$

where y_i and y_j denote i-th and j-th $(i,j \leq L)$ components of a data point y. This means that

$$E\left(yy^T\right) = \Lambda, \tag{3.10}$$

where Λ is a diagonal matrix of dimensions $L \times L$. Because $E(\bar{x}) = 0$ it holds also that $E(y) = 0$.

Now substituting Equation (3.8) into Equation (3.10) we obtain

$$\underbrace{E\left(yy^T\right)}_{C_y} = E\left(T\bar{x}\bar{x}^T T^T\right) = T\underbrace{E\left(\bar{x}\bar{x}^T\right)}_{C_x} T^T = \Lambda, \tag{3.11}$$

where $\mathbf{C}_x$ and $\mathbf{C}_y$ are correlation matrices of the variables $\mathbf{x}$ and $\mathbf{y}$, each of dimensions $L \times L$, respectively. Now, substituting Equation (3.6) into Equation (3.11) yields

$$\mathbf{T} \underbrace{E\left((\mathbf{x} - \mathbf{m}_x)(\mathbf{x} - \mathbf{m}_x)^T\right)}_{\Sigma_x} \mathbf{T}^T = \Lambda, \tag{3.12}$$

$$\mathbf{T}\Sigma_x\mathbf{T}^T = \Lambda, \tag{3.13}$$

where Σ_x denotes the covariance matrix of data $\{\mathbf{x}_i\}$. From the definition of Σ_x it follows that it is a positive definite and symmetrical matrix, so its eigenvalues are nonnegative and its eigenvectors are orthogonal, as already discussed in Section 2.9. Thus, based on the properties of the matrix algebra, if $\mathbf{T}$ is chosen so that its *rows are formed from the orthonormal eigenvectors* of Σ_x, then (3.13) is fulfilled, i.e. $\mathbf{T}\Sigma_x\mathbf{T}^T$ is a diagonal matrix. Thus, $\mathbf{T}$ is a square matrix of dimensions $L \times L$. Additionally, Equation (3.13) can be rearranged, so the eigenvalues in Λ are set in order, from the highest to the lowest.

Now, once we know how to construct the matrix $\mathbf{T}$ (i.e. as orthonormal eigenvectors of Σ_x), from Equations (3.8) and (3.6) we get

$$\mathbf{y} = \mathbf{T}(\mathbf{x} - \mathbf{m}_x). \tag{3.14}$$

Since the rows of $\mathbf{T}$ are orthonormal, $\mathbf{T}^T = \mathbf{T}^{-1}$ and any vector $\mathbf{x}$ can be recovered from Equation (3.14) as follows

$$\mathbf{x} = \mathbf{T}^T\mathbf{y} + \mathbf{m}_x. \tag{3.15}$$

Vector $\mathbf{y}$ in Equation (3.14) can be seen as a new and *uncorrelated* feature of the vector $\mathbf{x}$. This property of PCA is frequently used in pattern recognition for "feature generation," as will be discussed.

Figure 3.2 depicts two exemplary point distributions with eigenvectors of their covariance matrices. The first set in Figure 3.2(a) contains a highly directional distribution. The first eigenvector y_1 points in the direction of the maximal variance of data. The second one, y_2, orthogonal to y_1, points in the direction of the minimal variance of data. Therefore if points are expressed in the new coordinate system $y_1 - y_2$, then only first coordinate, being a projection onto the y_1 axis, shows high variability, as shown for points $\mathbf{A}$ and $\mathbf{B}$. This makes possible space reduction from 2D to 1D.

Conversely, the second set shown in Figure 3.2(b) spreads almost uniformly in all directions. Thus, there is no dominating direction of data variance, so the two eigenvectors are of the same magnitude. Hence, any set of orthogonal vectors can be used as a new coordinate system. Also, in this case, it is not possible to represent data with only one coordinate, since the variance of the two coordinates is the same.

This illustrates a key feature of PCA – namely, when constructing the matrix $\mathbf{T}$ in Equation (3.14) from the data covariance matrix Σ_x, we do not need to take all of the eigenvectors. Instead, only the first $R \leq L$ of them, corresponding to the R *largest* eigenvalues, can be used. In this way, a new matrix $\mathbf{T}_R$ of dimensions $R \times L$ is obtained, which is simply a minor of

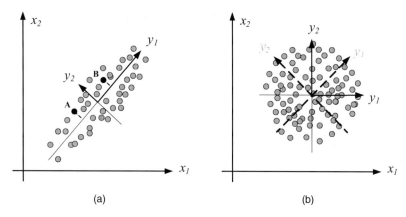

(a)   (b)

Figure 3.2 Two point distributions with eigenvectors of their covariance matrices. The first set exhibits a highly linear structure (a). The first vector y_1 (which is a doubled eigenvector) points in the direction of the maximal variance of data. The second eigenvector, y_2, orthogonal to y_1, points in the direction of the minimal variance of data. Therefore if points are expressed in the new coordinate system $y_1 - y_2$, then only the first coordinate, being a projection onto the y_1 axis, shows high variability, as shown for points **A** and **B**. This means a space reduction from 2D to 1D. The second set does not have a dominating direction (b). Thus, there is no dominating direction of data variance, so the two eigenvectors are of the same magnitude. Additionally, any set of orthogonal vectors can be used as a new coordinate system. In this case, it is not possible to represent data with only one coordinate.

T. However, this also means that **y** in Equation (3.14) is less dimensional (i.e. shorter) and actually holds only R first elements of **y**, which is denoted as $\hat{\mathbf{y}}$, that is

$$\hat{\mathbf{y}} \equiv \mathbf{y}\,(1 : R),\qquad\qquad(3.16)$$

and the following holds

$$\hat{\mathbf{y}} = \mathbf{T}_R\,(\mathbf{x} - \mathbf{m}_x).\qquad\qquad(3.17)$$

Under these conditions, the reconstruction (3.15) becomes now an approximation $\hat{\mathbf{x}}$ of **x**, i.e.

$$\hat{\mathbf{x}} = \mathbf{T}_R^T\hat{\mathbf{y}} + \mathbf{m}_x.\qquad\qquad(3.18)$$

In other words, $\hat{\mathbf{x}}$ is *a subspace projection of* **x** *onto the subspace spanned* by the chosen R eigenvectors (both $\hat{\mathbf{x}}$ and **x** are vectors of length L). This is where the method took its name from – the principal components correspond to the chosen eigenvectors.

This fundamental property of PCA makes it one of the most popular methods of *dimensionality reduction* and to search for the intrinsic dimensionality of the original space. The latter denotes a number of free parameters which are sufficient to describe the data set $\{\mathbf{x}_i\}$. If we assume that all remaining L-R eigenvalues are 0, then the intrinsic space of $\{\mathbf{x}_i\}$ is R. However, PCA is a linear method. This means that it works best if the data set is distributed throughout a hyperplane. For other data distributions other nonlinear dimensionality reduction methods should be used, such as the kernel PCA (KPCA) [8, 9, 10], or a tensor-based dimensionality reduction [11].

It is also worth noting that the eigenvalues of the matrix $\mathbf{\Sigma}_x$ are equal to the variances of the variables $\mathbf{y}$. Therefore chosing R dominating components, maximizes the total variance associated with the original data set $\{\mathbf{x}_i\}$. This proves to be a desirable property e.g. in pattern recognition based on PCA since the chosen features are the most discriminating ones. It can be also demonstrated that the choice of R dominating eigenvectors maximizes the entropy of the random process (3.14) [5]. Thus, PCA can be also used to filter data, assuming that the cut $L\text{-}R$ smallest components are due to noise.

To assess how good the approximation (3.18) is, the mean square error can be computed, as follows

$$\varepsilon = E\left(\|\mathbf{x} - \hat{\mathbf{x}}\|^2\right). \tag{3.19}$$

For this purpose let us rewrite Equations (3.15) and (3.18), as follows

$$\mathbf{x} = \mathbf{T}^T \mathbf{y} + \mathbf{m}_x = \sum_{i=1}^{L} y_i \mathbf{t}_i^T + \mathbf{m}_x. \tag{3.20}$$

$$\hat{\mathbf{x}} = \mathbf{T}_R^T \hat{\mathbf{y}} + \mathbf{m}_x = \sum_{i=1}^{R} y_i \mathbf{t}_i^T + \mathbf{m}_x. \tag{3.21}$$

where $\mathbf{t}_i$ denotes an i-th row of the matrix $\mathbf{T}$, which as we remember is the i-th eigenvector of the covariance matrix $\mathbf{\Sigma}_x$. Now subtracting Equation (3.21) from Equation (3.20) and substituting the result into Equation (3.19) yields

$$\varepsilon = E\left(\|\mathbf{x} - \hat{\mathbf{x}}\|^2\right) = E\left[(\mathbf{x} - \hat{\mathbf{x}})^T (\mathbf{x} - \hat{\mathbf{x}})\right] = E\left[\left(\sum_{i=R+1}^{L} y_i \mathbf{t}_i^T\right)^T \left(\sum_{i=R+1}^{L} y_i \mathbf{t}_i^T\right)\right]$$

$$= E\left[\left(\sum_{i=R+1}^{L} y_i \mathbf{t}_i\right)\left(\sum_{i=R+1}^{L} y_i \mathbf{t}_i^T\right)\right] = E\left[\sum_{i=R+1}^{L} \sum_{j=R+1}^{L} y_i \mathbf{t}_i \mathbf{t}_j^T y_j\right].$$

Considering now that $\mathbf{t}_i \mathbf{t}_j^T$, being a product of the orthonormal eigenvectors, is 1 only for $i = j$ and 0 otherwise, we finally obtain

$$\varepsilon = E\left[\sum_{i=R+1}^{L} y_i^2\right] = \sum_{i=R+1}^{L} E\left(y_i^2\right) = \sum_{i=R+1}^{L} \lambda_i \tag{3.22}$$

In the above we used Equation (3.10) which directly relates the expected value of y_i^2 with eigenvalue λ_i. Thus, the mean square error of PCA is equal to the sum of the left $L\text{-}R$ eigenvalues λ_i.

A practical difficulty is the choice of the number R of dominating components. A simple and sometimes useful hint is to find a certain index i of eigenvalues from which point on they start to become significantly smaller, say an order of magnitude. As suggested in the book by Lee and Verleysen, in some cases a better method is to check the behavior of the minus logarithm of the normalized eigenvalues [12], as follows

$$d\left(i\right) = -\log\frac{\lambda_i}{\lambda_1} \tag{3.23}$$

A sudden ascent of $d(i)$ in Equation (3.23) for a certain index i, if existing, can indicate the intrinsic dimension of the data.

It is worth noting that PCA is also known as the Hotelling transformation, the discrete Karhunen–Loève transform (KLT), or the proper orthogonal decomposition (POD). These names reflect independent derivations of the aforementioned transformation by different authors and at different times. Literature on PCA is very large. The recommended reading on PCA is, for example the paper, by Abdi [13], as well as the books by Duda *et al.* [2], Theodoridis and Koutroumbas [5], and Gonzalez and Woods [14], while an in-depth treatment can be found in the book by Jolliffe [7]. On the other hand, covariance matrices, computed from selected vectors of features, were shown to be very effective object descriptors, as proposed in the paper by Tuzel *et al.* on pedestrian detection [15]. The problem of segmenting an unknown number of subspaces of unknown and changing dimensions from a supplied data set is addressed in the paper by Vidal *et al.* [16]. Their approach is called *generalized* PCA.

3.3.1.1 Computation of the PCA

Since the covariance matrix is of dimensions $L \times L$, its direct computation can be intractable for large dimensional data. Thus, basically there are two implementations of the PCA decomposition depending on a number of data points N, as well as their dimensionality L. Both are discussed in this section.

Low dimensional features ($L \ll N$)
Algorithm 3.3 presents a simple algorithm for computation of the PCA which follows from the discussion in the previous section.

1. Compute the covariance matrix Σ_x (3.12) from the data set $\{\mathbf{x}_i\}$:

$$\Sigma_x = \frac{1}{N}\sum_{i=1}^{N}\mathbf{x}_i\mathbf{x}_i^T - \mathbf{m}_x\mathbf{m}_x^T, \tag{3.24}$$

 where N is number of points $\mathbf{x}_i$ and $\mathbf{m}_x$ is a mean from (3.7);
2. Compute the eigendecomposition of Σ_x;
3. Choose a number of dominating components R;
4. Arrange the eigenvectors $\mathbf{e}_j$ of Σ_x in decreasing order in accordance with their eigenvalues $\lambda_1 \le \lambda_j \le \lambda_R$ and form the matrix $\mathbf{T}_R$ which rows are first R eigenvectors;

Algorithm 3.3 Computation of the PCA for a data set $\{\mathbf{x}_i\}$ suitable for problems for which the number of data points is much larger than their dimensionality ($L \ll N$).

In the presented method, the main computational task is eigendecomposition of the covariance matrix, in step 2. In the presented software framework this is obtained computing the SVD decomposition of $\mathbf{\Sigma}_x$ [17, 18]. Since the matrix is symmetrical, if $\mathbf{\Sigma}_x = \mathbf{SVD}^T$ then it holds also that $\mathbf{S} = \mathbf{D}$. Thus, the eigenvectors are columns of the matrix $\mathbf{S}$ (or $\mathbf{D}$). This property is used in the presented implementation.

```
/////////////////////////////////////////////////////////
// This class implements
// the Principal Component Decomposition (PCA)
/////////////////////////////////////////////////////////
template < typename IN_DATA_TYPE, typename NUMERIC_TYPE >
class PCA_For
{
   public:

      // Class inherent variables

      typedef NUMERIC_TYPE                               NumericType;

      typedef IN_DATA_TYPE                               InDataType;

      typedef typename
         FeatureSpaceFor< NumericType >::FeatureType     FeatureType;

      typedef typename
         FeatureSpaceFor< NumericType >::DataPoint       DataPoint;

      typedef typename
         FeatureSpaceFor< NumericType >::DataRepository  DataRepository;

      typedef TVideoFor< InDataType >                    Video;

      typedef TImageFor< NumericType >                   Matrix;

   protected:

      DataPoint    fMean;

      Matrix *     fPrincipalCompMatrix;

   public:

      /////////////////////////////////////////////////////////////
      // These functions compute the covariance matrix sigma
      /////////////////////////////////////////////////////////////
      //
      // INPUT:
      //          inVideo - reference to the series of frames (a video)
      //                 in which each frame contains one feature
      //                 of data (i.e. it is only one component
      //                 of the feature vector)
      //          theMean - on success contains the mean data point
```

```
//            orphanCovarMatrix - on success this is computed
//                 covariance matrix (an orphaned object)
//
//
// OUTPUT:
//            true on success
//            false otherwise
//
// REMARKS:
//            orphanCovarMatrix should be deleted by the calling
//            function
//
virtual bool Compute_CovarianceMatrix( const Video & inVideo,
                                       DataPoint & theMean,
                                       Matrix * & orphanCovarMatrix );


/////////////////////////////////////////////////////////
// These functions compute the principal component
// decomposition (called: PCA)
/////////////////////////////////////////////////////////
//
// INPUT:
//      inVideo - reference to the series of frames (a video)
//            in which each frame contains one feature
//            of data (i.e. it is only one component
//            of the feature vector)
//      orphanCovarMatrix (optional) - on successful exit this points at an
//      orphaned covariance matrix object of input data
//      orphanEigenvalues (optional) - on success points at orphaned matrix
//      with eigenvalues set in descending order (actually this
//      is a one row image)
//
// OUTPUT:
//            true on success
//            false otherwise
//
// REMARKS:
//      It computes and saves the fMean member which contains the mean point.
//
//      orphanEigenvalues, orphanCovarMatrix should be deleted by the caller
//
virtual bool Compute_PCA( const Video & inVideo );
virtual bool Compute_PCA( const Video & inVideo, Matrix * & orphanCovarMatrix );
virtual bool Compute_PCA( const Video & inVideo, Matrix * & orphanCovarMatrix,
                          Matrix * & orphanEigenvalues );


/////////////////////////////////////////////////////////
// This function projects a given point into the already
// computed PCA space. Due to properties of the PCA space
// it can also compress data if the number of requested
// components (in numOfFirstComponents) is less than
// the space dimensionality.
/////////////////////////////////////////////////////////
//
```

```
// INPUT:
//      in - the input point
//      out - the output point which size is numOfFirstComponents which is
//            also less or equal to the size of the "in" point
//      numOfFirstComponents - defines number of principal components used in
//            computation of the "out" point; If less than size of "in" then
//            we have data compression.
//
// OUTPUT:
//        none
//
// REMARKS:
//        Strictly requires prior call of the Compute_PCA() !!!
//
//        Just subtract the mean and multiply with fPrincipalCompMatrix
//
//             out = D( in - mx )
//        where
//        D is a matrix of the principal components, mx denotes a mean
//        of the input population {in}
//
virtual void Project_FeatureVector( const DataPoint & in,
                                    DataPoint & outPt,
                                    int numOfFirstComponents );


/////////////////////////////////////////////////////////////
// This function takes a point from the PCA space to the
// input space. In some sense this is an inverse of
// the Project_FeatureVector.
/////////////////////////////////////////////////////////////
//
// INPUT:
//        in - the input point which size can be equal
//             or less than the PCA space
//        outPt - the output point which size
//             is equal to the size of the PCA space
//
// OUTPUT:
//        none
//
// REMARKS:
//        Strictly requires prior call of the Compute_PCA() !!!
//
//        It adds the mean and multiplies with fPrincipalCompMatrix
//
//             <outPt> = D<k> * in + mx
//        where
//          D<k> denotes k principal components (k == size of in),
//
//          mx - a mean of the population {in}
//
virtual void Reconstruct_FeatureVector( const DataPoint & in,
                                        DataPoint & outPt );
```

```
/////////////////////////////////////////////////////////////
// This function converts data space into its PCA space.
/////////////////////////////////////////////////////////////
//
// INPUT:
//       inVideo - the input data space (each frame conveys
//             only one feature component)
//       out_PCA_Space - an orphaned output space of
//             size numOfImportantComponents which can
//             be equal or less (for compression) than
//             size of the input space (inVideo)
//
// OUTPUT:
//       power spectrum of the PCA transformation
//       (i.e. a ratio of the sum of
//       numOfImportantComponents
//       first components to the sum of all components)
//
//
// REMARKS:
//
//
virtual double Space_Project_To_PCA( const Video & inVideo,
                               Video * & out_PCA_Space,
                               int numOfImportantComponents );


/////////////////////////////////////////////////////////////
// This function reconstructs a PCA transformed video
// into its original space
/////////////////////////////////////////////////////////////
//
// INPUT:
//       inVideo - the input data space (each frame conveys
//             only one feature component)
//       out_PCA_Space - an orphaned output space
//
// OUTPUT:
//       none
//
// REMARKS:
//
//
virtual void Space_Reconstruct_From_PCA( const Video & in_PCA_Space,
                                   Video * & out_Space );


/////////////////////////////////////////////////////////////
// This function computes the power spectrum of the PCA
// transformation which is a ratio of the sum of
// numOfImportantComponents first components to the sum
// of all components.
/////////////////////////////////////////////////////////////
//
// INPUT:
//         eigenvalues - the row matrix with eigenvalues
```

```
//                    sorted in the decending order
//               numOfImportantComponents - number of components
//                    used in the denominator of the computed ratio
//
// OUTPUT:
//               PCA power spectrum
//
// REMARKS:
//
//
    double ComputePowerSpectrum(  const Matrix & eigenvalues,
                                  int numOfImportantComponents );
};
```

Algorithm 3.4 Implementation of the Principal Component Decomposition with the *PCA_For* template class (showing most important members).

Algorithm 3.4 presents a definition of the *PCA_For* template class implementing PCA decomposition which follows the previous discussion and Algorithm 3.3. The class has two template parameters, *IN_DATA_TYPE,* and *NUMERIC_TYPE*, defining the type of processed features (pixels) and data type for computation of the covariance matrix (usually *double*), respectively. There are at least two important issues regarding this class. The first is the generic definition of the feature data structure *FeatureSpaceFor* used in a number of the initial *typedefs*. Thanks to this we can define any feature space, i.e. of any dimension and with data of any type. Details are discussed in Section 3.2.1. The second important issue concerns data organization for PCA. In the presented implementation we decided to use the data structure for storing video streams, as depicted in Figure 3.3.

Each feature component occupies a separate frame from the video. The reason being that our framework was designed mostly for image processing. However, a data structure for video

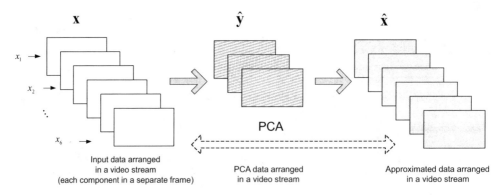

Figure 3.3 Organization of data structures for PCA implementation. Data stored in a data structure for video. Each feature component is placed in a separate frame (matrix).

streams is also generic. Thus, it allows data storing of any type which is accessed by three indices. Another advantage is the possibility of adding or removing new components without affecting placement of the previous ones.

However, if such organization of data is not convenient then it is easy to apply a different data structure in a class derived from PCA which implements its own (specialized) version of the virtual functions *Compute_PCA* and others if necessary. For instance, in some applications (such as eigenbackground subtraction, eigenfaces, etc.) it is convenient to treat the whole image as a data point. Then, when constructing the input matrix, each image becomes its single column. This approach will be discussed in Section 3.3.1.3.

```
/////////////////////////////////////////////////////////////
// These functions compute the covariance matrix sigma
/////////////////////////////////////////////////////////////
//
// INPUT:
//          inVideo - reference to the series of frames (a video)
//                in which each frame contains one feature
//                of data (i.e. it is only one component
//                of the feature vector)
//          theMean - on success contains the mean data point
//          orphanCovarMatrix - on success this is computed
//                covariance matrix (an orphaned object)
//
//
// OUTPUT:
//          true on success
//          false otherwise
//
// REMARKS:
//          orphanCovarMatrix should be deleted by the calling
//          function
//
template < typename IN_DATA_TYPE, typename NUMERIC_TYPE >
bool PCA_For< IN_DATA_TYPE, NUMERIC_TYPE >::Compute_CovarianceMatrix(
          const Video & inVideo, DataPoint & theMean, Matrix * & orphanCovarMatrix )
{
    // denotes number of components of each data point
    const int kFeatureSize = inVideo.GetNumOfFrames();

    if( kFeatureSize < 2 )
      return false;    // for features with only one component there is nothing to do

    const int kFrameCols = inVideo.GetFrameAt( 0 )->GetCol();  // get dimensions
    const int kFrameRows = inVideo.GetFrameAt( 0 )->GetRow();  // of the frame

    const int kTotalElems = kFrameCols * kFrameRows;    // total number of data points

    REQUIRE( kTotalElems != 0 );
    if( kTotalElems < 1 )
```

```
                    return false;                    // nothing to do


      theMean.resize( kFeatureSize );    // prepare the mean point
      fill( theMean.begin(), theMean.end(), 0.0 );


      DataPoint x( theMean );            // temporary vector for current feature


      orphanCovarMatrix = new Matrix( kFeatureSize, kFeatureSize, 0.0 );
      Matrix & covarMatrix = * orphanCovarMatrix;


      Matrix tmpMatrix( kFeatureSize, kFeatureSize, 0.0 );


      register int i = 0, f = 0;


      vector< Video::PixelType * > framesDataPtrVec;
      for( f = 0; f < kFeatureSize; ++ f )
       framesDataPtrVec.push_back( inVideo.GetFrameAt( f )->GetImageData() );


      // First run is to compute the mean point
      for( i = 0; i < kTotalElems; ++ i )
      {
            // Collect next feature across the frames
            for( f = 0; f < kFeatureSize; ++ f )
                  x[ f ] = * framesDataPtrVec[ f ] ++;
            theMean += x;


            // Compute an outer product of a feature vector with itself
            OuterProduct( x, x, tmpMatrix );


            covarMatrix += tmpMatrix;


      }
      theMean /= (NumericType)kTotalElems;


      covarMatrix /= (NumericType)kTotalElems;


      OuterProduct( theMean, theMean, tmpMatrix )    // computes: mean * mean'


      // Finally let's assembly all together
      covarMatrix -= tmpMatrix;
}
```

Algorithm 3.5 Implementation of the *Compute_CovarianceMatrix* function for computation of the mean data point and covariance matrix.

Algorithm 3.5 shows implementation of the *Compute_CovarianceMatrix* member which computes the mean data point and the covariance matrix in accordance with the step 1 of Algorithm 3.3. This is a prerequisite for all further computations of the PCA decomposition. Additionally, this function can be called independently by other components which need only the mean and/or the covariance matrix of the supplied data points.

The set of overloaded members *Compute_PCA* is devoted to the PCA computation from the initial data set, organized in a video data structure. As shown, it calls the *Compute_CovarianceMatrix* from Algorithm 3.5 member to compute the mean vector (data point) and the covariance matrix. Then the eigendecomposition is performed. Results are stored in the internal data members *fMean* and *fPrincipalCompMatrix*, respectively. These are used in subsequent operations over the PCA decomposition. For this reason *Compute_PCA* should be called first to construct the PCA space. Thus, Algorithm 3.6 actually realizes steps 1 and 2 of Algorithm 3.3. As already mentioned, the main data structures follow from the image processing library, which details can be found in [19].

```
/////////////////////////////////////////////////////////////
// These functions compute the principal component
// decomposition (called: PCA)
/////////////////////////////////////////////////////////////
//
// INPUT:
//          inVideo - reference to the series of frames (a video)
//                 in which each frame contains one feature
//                 of data (i.e. it is only one component
//                 of the feature vector)
//          orphanCovarMatrix (optional) - on successful exit this
//                 points at an orphaned covariance matrix
//                 object of input data
//          orphanEigenvalues (optional) - on success points at orphaned
//                 matrix with eigenvalues set in descending order
//                 (actually this is a one row image)
//
// OUTPUT:
//          true on success
//          false otherwise
//
// REMARKS:
//          It computes and saves the fMean member
//          which contains the mean point.
//
//          orphanEigenvalues, orphanCovarMatrix should be deleted
//                 by the caller
//
template < typename IN_DATA_TYPE, typename NUMERIC_TYPE >
bool PCA_For< IN_DATA_TYPE, NUMERIC_TYPE >::Compute_PCA(
    const Video & inVideo, Matrix * & orphanCovarMatrix, Matrix *
    & orphanEigenvalues )
{
    orphanCovarMatrix = 0;   // precondition
    if( Compute_CovarianceMatrix( inVideo, fMean, orphanCovarMatrix ) == false )
      return false;        // return if error computing covariance matrix

    REQUIRE( orphanCovarMatrix != 0 );   // postcondition
```

```
    // Decompose the covariance matrix
    Compute_SVD       svd_calculator;

    svd_calculator ( * orphanCovarMatrix );

    fPrincipalCompMatrix = svd_calculator.Release_D();
    // this is ok, both should be (and are) COLUMN ORTHONORMAL (!)
    REQUIRE( fPrincipalCompMatrix != 0 );

    orphanEigenvalues = svd_calculator.Release_V();

    REQUIRE( orphanEigenvalues!= 0 );

    return true;
}
```

Algorithm 3.6 Implementation of the *Compute_PCA* function which computes PCA decomposition.

Once the main data structures of PCA are created, these can be used to transform the data points. The *Project_FeatureVector* function carries out projection of a data point **x** into the feature space in accordance with Equation (3.17). Not surprisingly, this function requires passing the number of important components R. However, this means that once PCA data structures are computed, the parameter R can be chosen in run time. An opposite action can be invoked by calling the *Reconstruct_FeatureVector* member whose role is to synthesize a data point given its PCA projection. This follows Equation (3.18). Interestingly, the number of components is indirectly deduced from the size of the supplied data point.

The pair of members *Space_Project_To_PCA* and *Space_Reconstruct_From_PCA* is responsible for converting the whole data space into the PCA representation. Once again, the former takes as an argument the number of important components R. This, in turn, can be assessed thanks to the last member of the *PCA* class, the *ComputePowerSpectrum* function which computes the power spectrum of the PCA transformation which is a ratio of the sum of R first eigenvalues to the sum of all eigenvalues. The power spectrum is computed from the following formula

$$\theta = \frac{\sum_{h=1}^{R} \lambda_h}{\sum_{h=1}^{L} \lambda_h}. \tag{3.25}$$

The power spectrum can be used to assess the number R of components required for an optimal decomposition. Our goal is to find the smallest possible value of R, for which θ is close to 1 to the desired degree. Implementation details can be easily analyzed in the attached source code [6]. An example of application of the presented PCA class is presented in Section 3.3.1.2.

High dimensional features (L≫N)

For L dimensional data, their covariance matrix is of dimensions $L \times L$. If L is high, e.g. it is a total number of all pixels in an image, then memory requirements become prohibitive. For instance, if data points are images of resolution 256×256, then their covariance matrix would be 65536×65536, which, assuming 8 bytes per element, results in 32 GB. Even more problematic is then eigendecomposition of such a matrix. However, as presented by Turk and Pentland, in the case of a much lower number of data patterns N, computations can be greatly reduced if constrained to the problem of finding only N eigenvectors of a much smaller matrix [20]. To show this, from the points $\bar{x}_i$ let us create an $L \times N$ matrix $\bar{X}$, which columns are set in accordance with Equation (3.6). From this the low dimensional $N \times N$ product matrix $\bar{X}^T \bar{X}$ can be computed, which can be eigendecomposed next, as follows

$$\bar{X}^T \bar{X} \, e_k = \mu_k \, e_k, \tag{3.26}$$

where e_k are respectively eigenvectors ($N \times 1$) and μ_k eigenvalues of the matrix $\bar{X}^T \bar{X}$. Left multiplying both sides of (3.26) by $\bar{X}$ yields

$$\underbrace{\bar{X}\bar{X}^T}_{\Sigma_x} \underbrace{\left(\bar{X} \, e_k\right)}_{q_k} = \mu_k \underbrace{\left(\bar{X} \, e_k\right)}_{q_k}. \tag{3.27}$$

From the above it is easy to see that

$$q_k = \bar{X} \, e_k, \tag{3.28}$$

are eigenvectors of dimensions $L \times 1$ of the covariance matrix, expressed as

$$\Sigma_x = \bar{X}\bar{X}^T. \tag{3.29}$$

So, if e_k are known, then based on Equation (3.28) the matrix of eigenvectors of Σ_x can be easily computed as follows

$$Q = \bar{X} E, \tag{3.30}$$

where E is a matrix with columns of e_k. However, eigenvectors contained in Q need to be normalized to form an orthonormal basis, so for instance the reconstruction equation (3.15) still holds. Thus, T in Equation (3.13) is obtained as follows

$$T = \hat{Q}^T, \tag{3.31}$$

where $\hat{Q}$ denotes a column-wise normalization of Q. These steps can be arranged as shown in Algorithm 3.7.

1. Compute the mean vector $\mathbf{m}_x$ out of the data set$\{\mathbf{x}_i\}$;
2. Compute the matrix $\bar{\mathbf{X}}$ with columns composed of points $\mathbf{x}_i - \mathbf{m}_i$, for all $1 \leq i \leq N$;
3. Form the matrix $\bar{\mathbf{X}}^T\bar{\mathbf{X}}$ of dimensions $N \times N$ and compute its eigenvectors $\mathbf{e}_k$ (e.g. from its SVD);
4. From (3.30) compute matrix $\mathbf{Q}$;
5. Compute $\mathbf{T}$ from column normalized $\mathbf{Q}$, in accordance with (3.31);

Algorithm 3.7 Computation of the PCA for a data set $\{\mathbf{x}_i\}$ suitable for problems for which the dimensionality of data points is much larger than number of points $(L \gg N)$.

It is interesting to notice that from Algorithm 3.7 at most N eigenvectors is obtained, whereas Algorithm 3.3 outputs L eigenvectors. Implementation of this version of the PCA decomposition for eigenimages is realized by the *EigenImageFor* class, which code details can be looked up in the attached software framework [6]. *EigenImageFor* is derived from the base *PCA_For*, which definition was presented in Algorithm 3.4. It overrides two member functions. The first is *Compute_CovarianceMatrix* whose role is to compute the covariance matrix. However, in this case this is the covariance matrix of the product $\bar{\mathbf{X}}^T\bar{\mathbf{X}}$. *Compute_PCA* is the second of the overriden members of the *EigenImageFor* class. It simply computes the R principal components, which in this case can be called eigenimages in accordance with the methodology outlined in Algorithm 3.7.

Section 2.12.3 contains a discussion on different aspects of the SVD decomposition with respect to the tensor decomposition. However, for larger data sets other and more effective methods for eigendecomposition can be considered, such as the iterative method or MRRR developed by Dhillon *et al.* [21]. An overview of direct solvers for symmetric eigenvalue problems can be found in the paper by Lang [22].

3.3.1.2   PCA for Multi-Channel Image Processing

Let us not employ PCA to decompose a color image into three components. In the case of a color image, such as the one presented in Figure 3.4(a), each pixel is a vector $\mathbf{x}$ of three components R, G, and B. The three components for the original image in Figure 3.4(a), ordered in accordance with the lowering eigenvalues, are depicted in Figure 3.4(b)–(d), respectively. We see that the first component conveys the monochromatic information of the color image.

The computed eigenvectors for Figure 3.4a are as follows

$$\mathbf{t}_1 = \begin{bmatrix} 0.591421 & -0.697373 & 0.4048360 \end{bmatrix}^T,$$

$$\mathbf{t}_2 = \begin{bmatrix} 0.577595 & 0.0160436 & -0.816166 \end{bmatrix}^T,$$

$$\mathbf{t}_3 = \begin{bmatrix} 0.562677 & 0.7165291 & 0.4122881 \end{bmatrix}^T.$$

(a)   (b)

(c)   (d)

Figure 3.4   A color image (a) and its three principal components arranged in order of lowering eigenvalues (b,c,d).

The corresponding eigenvalues are respectively

$$\lambda_1 = 12135.7, \quad \lambda_2 = 269.902, \quad \lambda_3 = 38.7778.$$

Now, in accordance with Equation (3.21) the three possible reconstructions are computed. These are shown in Figure 3.5.[2] We see that using all components the reconstruction is almost perfect, as shown in Figure 3.5(a). In this case PSNR between the original image in Figure 3.4(a) and Figure 3.5(a) is 53.13dB. Using only two components 32.36 dB is obtained, as shown in Figure 3.5(b). Finally, for only one component we obtain a monochrome-like version (containing three, but the same, channels), shown in Figure 3.5(c), for which PSNR = 27.08 dB.

Another interesting observation is that the ordering of the eigenvalues obtained during the PCA decomposition *induces an ordering* on the components of data vectors. This can be used to *impose an ordering* on nonscalar data, such as the RGB images discussed in this section.

---

[2]Color versions of the images are available on the accompanying web page [6].

(a)   (b)   (c)

Figure 3.5 Reconstructed images from the three (a), two (b), and one (c) principal components of the original image from Figure 3.4a. Reconstruction PSNR values: 53.13 dB, 32.36 dB, and 27.08 dB, respectively. (For a color version of this figure, please see the color plate section.)

Such ordering is required in some of the image processing methods, such as morphological operations or morphological neural networks, discussed in Section 3.9.4. We will illustrate this technique with an example of the morphological gradient on the RGB color images. To compute such a gradient an order on the three RGB component needs to be established. One way, is to assume a *lexicographic order*. For 8-bit components this is equivalent to composition of a long value R·256^2 + G·256 + B, i.e. to form a bit concatenation of the R, B, and G values, in a presumed order. However, without further information we cannot be sure what component should be chosen as the most important one, and which as the least important one. In other words, all 3! permutations, such as B-G-R, R-B-G, and so on, are equally possible. In this case, instead of assuming an arbitrary order in the RGB space, PCA transformation can be first perfomed, which provides a more natural ordering in accordance with the eigenvalues of each of the components. The above idea can be outlined as follows:

1. Perform PCA decomposition of data;
2. Perform an operation requiring an order relation assuming a lexicographic order of the components imposed by the ordered eigenvalues;
3. Do reconstruction from the PCA components;

Algorithm 3.8 Computations on multidimensional data in lexicographic order implied by PCA.

The above idea is applied to computation of the morphological gradient of a color RGB image. It is defined as follows [23, 24]

$$g = d - e \tag{3.32}$$

where

$$d(\mathbf{x}) = \max_{y \in S} \left[f(\mathbf{x} + \mathbf{y}) + s(\mathbf{y}) \right], e(\mathbf{x}) = \min_{y \in S^*} \left[f(\mathbf{x} + \mathbf{y}) - s^*(\mathbf{y}) \right] \tag{3.33}$$

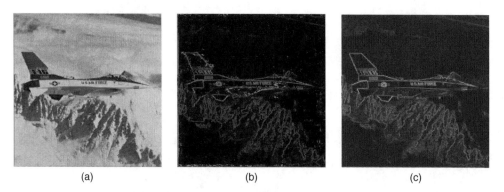

(a)   (b)   (c)

Figure 3.6 A test color image (a) and its morphological gradients computed in each RGB color channel with arbitrary order (b). The same operation but in the PCA decomposed space in order of eigenvalues and after reconstruction (c). (For a color version of this figure, please see the color plate section.)

denotes morphological dilation and erosion operations on a signal f with the structural element S denoted by $s(\mathbf{y})$ [23]. The morphological gradient was computed for a color image as shown in Figure 3.6(a). In the first case a simple RGB lexicographic order was used, which results are visible in Figure 3.6(b). In the second case, the described PCA modified ordering was additionally used. Results are shown in Figure 3.6(c).

It is clear that direct application of the morphological gradient does not lead to good results, as depicted in Figure 3.6(b). In this case one, but an arbitrarily chosen, color component dominates. Also choosing different dominating colors leads to different results. However, the same computation but applied to the PCA components, ordered in accordance with the eigenvalues, and then reconstructing the output image, gives more consistent results, as visible in Figure 3.6(c).

```
CIAP LexicographicMorphology_PCA( const ColorImage & inImage )
{
    // ========= 1 =========
    RealVideo * inVideo = OrphanRealVideo( inImage );
    RealVideo * out_PCA_Space = 0;

    PCA obj;
    int numOfImportantComponents = 3;
    double spectrumRatio = obj.Space_Project_To_PCA( * inVideo, out_PCA_Space,
                                                     numOfImportantComponents );
    delete inVideo;

    // ========= 2 =========
    RealColorImage * realColorImage = OrphanRealColorImageFrom( * out_PCA_Space );
    delete out_PCA_Space;

    MorphologyFor< RealTriple > morpher;

    const int kStructElem_Cols = 3;
```

```
const int kStructElem_Rows = 3;
const RealTriple kStructElem_InitVal( 0, 0, 0 );

TMaskedImageFor< RealTriple > theStructuralElement( kStructElem_Cols,
kStructElem_Rows,
       kStructElem_InitVal, true );    // Create the structural element

std::auto_ptr< TImageFor< RealTriple > > transformedImage(
         morpher.Gradient( * realColorImage, theStructuralElement ) );
delete realColorImage;

// ========= 3 =========
RealVideo * processedButPCAVideo = OrphanRealVideoFrom( * transformedImage );

RealVideo * reconstructed_space = 0;

obj.Space_Reconstruct_From_PCA( * processedButPCAVideo, reconstructed_space );

delete processedButPCAVideo;
CIAP retImage( 0 );
if( reconstructed_space != 0 )
{
      ChangeImageRange( * reconstructed_space, 0.0, 255.0 );
      retImage = CIAP( OrphanColorImage( * reconstructed_space ) );
}
delete reconstructed_space;
return retImage;
}
```

Algorithm 3.9 Code listing for computation of the morphological gradient in multivalued images and ordering induced by the PCA decomposition.

The C++ code for the *LexicographicMorphology_PCA* function which exemplifies application of the PCA class from Algorithm 3.4 to the problem of computation of the morphological gradient in color images is shown in Algorithm 3.9. The code is organized in the three steps 1-2-3, as described in Algorithm 3.8.

3.3.1.3   PCA for Background Subtraction

Background subtraction is the process of detecting moving parts in a video sequence. The simplest method is to subtract the two frames and threshold all values which are different from 0. However, this simple algorithm does not work well in practice due to local pixel variations and noise. Therefore many other methods have been developed for this purpose, all called background subtraction, although their methodology is usually more sophisticated. Many of them rely on statistical modeling of the static background with the MoG or mean-shift approaches. An overview of the most popular background subtraction methods can be found in the paper by Piccardi [25]. A recent tensor based approach to foreground segmentation can be found in the paper by Hu *et al.* which deals mostly with incremental tensor subspace learning [26], as well as in the one by Caseiro *et al.* in which the nonparametric Riemannian framework on the tensor field is used, to name a few [27].

Nevertheless, relatively good results with reasonable computation requirements can be obtained with the eigenbackground detection, a method originally proposed by Oliver *et al.* [28]. The method starts building the PCA model of a static background from a chosen sequence of frames. The basic idea is that the obtained eigenspace well describes large and rather static areas, contrary to moving objects, since they do not appear in the same locations and usually are not too large. Now, to discover moving objects in a new frame F it is first projected into the PCA eigenspace, and then reconstructed to the image $\hat{F}$. The last step is to compute and threshold the Euclidean distance between F and F^*, which corresponds to the DFFS distance, as will be discussed in Section 3.5. Details of this procedure are outlined in Algorithm 3.10.

1. Select the static frames and compute their PCA decomposition (e.g. with Algorithm 3.7); Select and store R dominant eigenimages (the eigenspace);

2. From (3.17) compute the projection F_R of a new frame F onto the eigenspace;

3. Based on (3.18) reconstruct the projected image F_R to $\hat{F}$;

4. Create the background map B of points:

$$B_i = \begin{cases} 0, & for\ |F_i - \hat{F}_i| < \tau \\ 1, & otherwise \end{cases} \tag{3.34}$$

where i denotes pixel index and τ is a threshold;

Algorithm 3.10   Eigenbackground subtraction routine.

Figure 3.7 shows eight frames from the *Hamburg taxi* sequence [29] used to build the eigenbackground model in accordance with Algorithm 3.10. The eigenimages of the eigenbackground model of the frames in Figure 3.7 are depicted in Figure 3.8.

Finally, Figure 3.9 shows results of the eigenbackground subtraction for three other frames of the *Hamburg taxi* sequence. The upper row of Figure 3.9 depicts visualization of the image differences, i.e. $|F_i - \hat{F}_i|$. Then, the lower row contains a binary background mask obtained after thresholding in accordance with Equation (3.34) and for $\tau = 44$.

For the eight images in Figure 3.7 of resolution 256×191 pixels, the average time of the eigenbackground computation is 25 ms in a serial software implementation.

3.3.2   Subspace Pattern Classification

As alluded to previously, PCA can be used for dimensionality reduction and filtering of the input data. It can be also used for direct classification of multiple patterns. In this case, C different classes are assumed, each represented by a data set of its exemplar patterns $\{\mathbf{x}_i^{(c)}\}$,

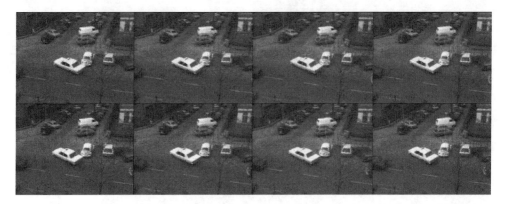

Figure 3.7   Frames from the *Hamburg taxi* sequence [29] used to build the eigenbackground model.

where $1 \leq c \leq C$. Each data set is then used to independently built its PCA: First the covariance matrix Σ_{x_c} is computed from the corresponding $\{\mathbf{x}_i^{(m)}\}$, and then the matrix $\mathbf{T}_m$ is constructed from the eigenvectors of Σ_{x_m}, as described in Section 3.14. Now, during the classification an unknown pattern is projected onto each of the PCA spanned spaces, and is classified to the class c to which it fits the best. The method was proposed almost forty years ago by Wold and named Soft Independent Modeling of Class Analogies (SIMCA) [30].

To understand this idea let us observe an exemplary data set and its PCA induced subspace, as shown in Figure 3.10, which can be obtained with the procedure described in Section 3.3.1.

Let us analyze the representation of point **B** which belongs to the data set used to compute PCA. From basic vector algebra this point in the original (measurement) coordinate system can be expressed as follows

$$\mathbf{B} = B_1 \mathbf{e}_1 + B_2 \mathbf{e}_2, \tag{3.35}$$

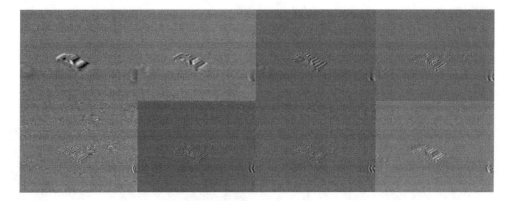

Figure 3.8   Eigenbackground images from the frames in Figure 3.7.

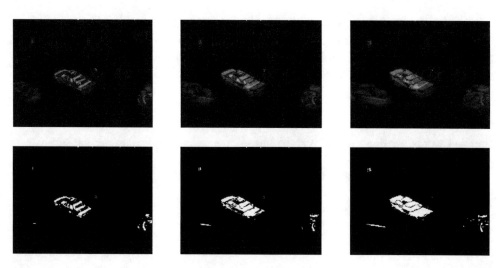

Figure 3.9 Results of the eigenbackground subtraction for three different frames of the *Hamburg taxi* sequence. Visualization of the image difference $|F_i - \hat{F}_i|$ (upper row). Binary background mask obtained after thresholding (3.34) (lower row).

where B_1 and B_2 are coordinates of **B**, i.e. these are projections (inner, dot products) of **B** onto base vectors $\mathbf{e}_1$ and $\mathbf{e}_2$, that is

$$B_i = \langle \mathbf{B}, \mathbf{e}_i \rangle = \mathbf{B} \cdot \mathbf{e}_i. \tag{3.36}$$

The same also holds for the subspace representation $\mathbf{B}'$ of the point **B**, that is

$$\mathbf{B}' = \mathbf{B} - \mathbf{m}_x = B_1'\mathbf{v}_1 + B_2'\mathbf{v}_2, \tag{3.37}$$

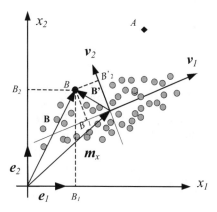

Figure 3.10 Visualization of the PCA subspace of a data set. Point *B* belongs to the data set. Point *A* belongs to other data set and was not used in computation of the PCA subspace.

where $\mathbf{v}_1$ and $\mathbf{v}_2$ are base vectors of the subspace which were obtained as dominating eigenvectors of the covariance matrix of the data set. However, we cannot say the same about the point $\mathbf{A}$, since it was not taken into the PCA computation of this subspace, and therefore $\mathbf{A}$ does not belong to it. Thus, in general the following will hold

$$\mathbf{A} \neq A_1 \mathbf{v}_1 + A_2 \mathbf{v}_2. \tag{3.38}$$

In other words, we don't expect $\mathbf{A}$ to be linearly dependent on the basis of the subspace to which $\mathbf{B}$ belongs.

The above strategy can be used to check the points if they belong or not to a given linear subspace, this time not necessarily obtained just with the PCA method. Namely, for any vector $\mathbf{x}$ its degree of membership to a subspace can be measured as *an orthogonal residual vector* $\tilde{\mathbf{x}}$ of that vector $\mathbf{x}$ and its projection to that subspace $\mathbf{x}'$, that is

$$\tilde{\mathbf{x}} = \mathbf{x} - \mathbf{x}'. \tag{3.39}$$

Returning to our previous example, in the subspace shown in Figure 3.10, we can assume that for a vector $\mathbf{B}'$ a module of its residual vector $\|\tilde{\mathbf{B}}\| = 0$, whereas for $\mathbf{A}$ we expect $\|\tilde{\mathbf{A}}\| > 0$.

The above holds in any linear subspace, which can be obtained in a number of ways, one of which is the already discussed PCA decomposition. In general, a linear subspace S is spanned by a set of R linearly independent base vectors $\mathbf{v}_i \in \Re^L$, for $1 \leq i \leq R$. This means that any vector $\mathbf{x}$ belonging to S can be expressed as a linear combination of the basis

$$S = S(\mathbf{v}_1, \mathbf{v}_2, \ldots, \mathbf{v}_R) = \left\{ \mathbf{x} \mid \mathbf{x} = \sum_{r=1}^{R} w_r \mathbf{v}_r \right\}, \tag{3.40}$$

where w_r are scalar values. Based on the above we can say that a linear subspace is defined by its set of base vectors $\mathbf{v}_i$. For a finite dimensional space, these can be arranged in the following matrix of dimensions $L \times R$

$$\mathbf{V} = [\mathbf{v}_1, \mathbf{v}_2, \ldots, \mathbf{v}_R]. \tag{3.41}$$

For these reasons $\mathbf{V}$ can be seen as a linear operator which transforms a point $\mathbf{x}$ as follows

$$\mathbf{Vx} = x_1 \mathbf{v}_1 + x_2 \mathbf{v}_2 + \cdots + x_R \mathbf{v}_R. \tag{3.42}$$

which is a linear combination of the columns of $\mathbf{V}$. The range, i.e. the set of vectors that can be reached by linear combinations of columns of $\mathbf{V}$, is called the column space of $\mathbf{V}$. Finally, from the matrix $\mathbf{V}$ the corresponding projection matrix $\mathbf{P}_\mathbf{V}$ that projects the points orthogonally onto the column space of $\mathbf{V}$ can be defined as follows [31]

$$\mathbf{P}_\mathbf{V} = \mathbf{V} \left(\mathbf{V}^T \mathbf{V} \right)^{-1} \mathbf{V}^T. \tag{3.43}$$

Analogously to Equation (3.18) a projection of a vector $\mathbf{x}$ in a space spanned by $\mathbf{V}$ is defined as follows

$$\mathbf{x}' = \mathbf{P}_\mathbf{V} \mathbf{x}. \tag{3.44}$$

In a special, but very practical, case when all base vectors of $\mathbf{V}$ are orthonormal, Equation (3.44) can be written as follows

$$\mathbf{x}' = \sum_{r=1}^{R} \underbrace{\langle \mathbf{x}, \mathbf{v}_r \rangle}_{x_r} \mathbf{v}_r = \sum_{r=1}^{R} \underbrace{(\mathbf{x}^T \mathbf{v}_r)}_{x_r} \mathbf{v}_r. \tag{3.45}$$

Based on the above discussion a suitable measure of the distance of a vector $\mathbf{x}$ to the space S can be defined as follows [32, 33]

$$d_S(\mathbf{x}, S) \equiv \frac{\|\tilde{\mathbf{x}}\|^2}{\|\mathbf{x}\|^2}, \tag{3.46}$$

where $\tilde{\mathbf{x}}$ denotes a residual vector given in Equation (3.39). In the above definition, division by the vector module makes the measure independent of the magnitude of the vector $\mathbf{x}$. Thus, our multiclass classification problem can be resolved as follows: given a number C of subspaces S_c, where $1 \leq c \leq C$, a test pattern $\mathbf{x}$ is classified to the class $\omega_{c'}$ for which the following holds

$$\omega_{c'} = \arg \min_{1 \leq c \leq C} [d_S(\mathbf{x}, S_c)]. \tag{3.47}$$

In other words, a class $\omega_{c'}$ is chosen for which its corresponding $d_S(\mathbf{x}, S_{c'})$ is the lowest among all subspaces.

Let us now insert Equation (3.39) into (3.46) so the following is obtained

$$d_S(\mathbf{x}, S) \equiv \frac{\|\mathbf{x} - \mathbf{x}'\|^2}{\|\mathbf{x}\|^2}. \tag{3.48}$$

Inserting Equation (3.44) into (3.48) yields

$$d_S(\mathbf{x}, S) \equiv \frac{\|\mathbf{x} - \mathbf{x}'\|^2}{\|\mathbf{x}\|^2} = \frac{(\mathbf{x} - \mathbf{x}')^T (\mathbf{x} - \mathbf{x}')}{\mathbf{x}^T \mathbf{x}} = \frac{\mathbf{x}^T \mathbf{x} - \mathbf{x}^T \mathbf{x}' - \mathbf{x}'^T \mathbf{x} + \mathbf{x}'^T \mathbf{x}'}{\mathbf{x}^T \mathbf{x}}$$

$$= \frac{\mathbf{x}^T \mathbf{x} - \mathbf{x}^T (\mathbf{P_V x}) - (\mathbf{P_V x})^T \mathbf{x} + (\mathbf{P_V x})^T (\mathbf{P_V x})}{\mathbf{x}^T \mathbf{x}} = \frac{\mathbf{x}^T \mathbf{x} - \mathbf{x}^T \mathbf{P_V x} - \mathbf{x}^T \mathbf{P_V}^T \mathbf{x} + \mathbf{x}^T (\mathbf{P_V}^T \mathbf{P_V}) \mathbf{x}}{\mathbf{x}^T \mathbf{x}}$$

$$= \frac{\mathbf{x}^T \mathbf{x} - \mathbf{x}^T \mathbf{P_V x} - \mathbf{x}^T \mathbf{P_V}^T \mathbf{x} + \mathbf{x}^T \mathbf{P_V x}}{\mathbf{x}^T \mathbf{x}} = \frac{\mathbf{x}^T \mathbf{x} - \mathbf{x}^T \mathbf{P_V x}}{\mathbf{x}^T \mathbf{x}} = 1 - \frac{\mathbf{x}^T \mathbf{P_V x}}{\mathbf{x}^T \mathbf{x}} = 1 - \hat{\mathbf{x}}^T \mathbf{P_V} \hat{\mathbf{x}}$$

where $\hat{\mathbf{x}} = \mathbf{x}/\|\mathbf{x}\|$ denotes a normalized vector $\mathbf{x}$. In the above, the idempotent properties of the projective matrix were used, i.e. $\mathbf{P_V}^2 = \mathbf{P_V P_V} = \mathbf{P_V}$, as well as the property $\mathbf{P_V}^T = \mathbf{P_V}$ which directly follows from (3.43). Hence, it holds that $\mathbf{P_V}^T \mathbf{P_V} = \mathbf{P_V}$. Substituting the above into Equation (3.47) a new, but equivalent, classification rule is obtained

$$\omega_{c'} = \arg \max_{1 \leq c \leq C} [\delta_S(\mathbf{x}, S_c)]. \tag{3.49}$$

where

$$\delta_S\left(\mathbf{x}, S_i\right) = \hat{\mathbf{x}}^T \mathbf{P}_{\mathbf{V}_i} \hat{\mathbf{x}}. \tag{3.50}$$

In the case of orthonormal base vectors, Equation (3.50) can be further simplified. Namely, inserting Equation (3.45) into (3.50) yields

$$\delta_S\left(\mathbf{x}, S_i\right) = \hat{\mathbf{x}}^T \left(\sum_{r=1}^R \left(\hat{\mathbf{x}}^T \mathbf{v}_r^{(i)}\right) \mathbf{v}_r^{(i)}\right) = \sum_{r=1}^R \left(\hat{\mathbf{x}}^T \mathbf{v}_r^{(i)}\right) \hat{\mathbf{x}}^T \mathbf{v}_r^{(i)} = \sum_{r=1}^R \left(\hat{\mathbf{x}}^T \mathbf{v}_r^{(i)}\right)^2 = \sum_{r=1}^R \left\langle \hat{\mathbf{x}}, \mathbf{v}_r^{(i)} \right\rangle^2, \tag{3.51}$$

where $\mathbf{V}_i = [\mathbf{v}_1^{(i)}, \mathbf{v}_2^{(i)}, \dots, \mathbf{v}_R^{(i)}]$ are base vectors for the class i.

It is interesting to observe that the decision surfaces of the classifier defined by Equations (3.49)–(3.51) are quadratic. However, in some pattern recognition tasks a more flexible approach would be desirable. In this respect Prakash and Murty proposed a modification which consists of additional clustering of the data sets for each of the classes and then keeping the same number of principal components in all clusters [33]. Thanks to these, their model does a piecewise linear approximation which helps overcome limitations of the quadratic surfaces.

As already mentioned, in the case of subspaces created with the PCA method the R most important components are used for classification, in accordance with Equations (3.49) and (3.51). This would suggest that the remaining L-R components can be freely thrown out as useless. However, interestingly enough, these remaining components can become even more efficient for the same task if the so called *dual space* is used [32]. Going into detail, for any linear subspace and any data set representing a class we have two choices for representing that class: The first is to use the R basis vectors, as already discussed. The other, however, is to use the remaining L-R basis vectors

$$\bar{\mathbf{V}} = [\mathbf{v}_{R+1}, \mathbf{v}_{R+2}, \dots, \mathbf{v}_L] \tag{3.52}$$

of the subspace $\bar{S}$ which is *complementary to the subspace S*, defined in Equation (3.40). For the two subspaces the following hold

$$\bar{S} \perp S, \text{ and } \bar{S} \cup S = \Re^L. \tag{3.53}$$

Figure 3.11 depicts an orthogonal decomposition into the principal subspace S and its orthogonal complement $\bar{S}$. Prototypical patterns are embedded entirely in S. The component embedded in $\bar{S}$ can be due to observation noise [34].

With this new set of base vectors $\bar{\mathbf{V}}$ and analogously to Equation (3.43) a projection matrix of the complementary space can be defined as follows

$$\bar{\mathbf{P}}_{\mathbf{V}} = \bar{\mathbf{V}} \left(\bar{\mathbf{V}}^T \bar{\mathbf{V}}\right)^{-1} \bar{\mathbf{V}}^T, \tag{3.54}$$

for which we have

$$\mathbf{x}' = \mathbf{P}_{\mathbf{V}}\mathbf{x} = \left(\mathbf{I} - \bar{\mathbf{P}}_{\mathbf{V}}\right)\mathbf{x}. \tag{3.55}$$

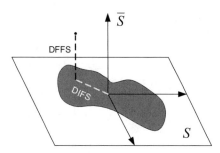

Figure 3.11 Orthogonal decomposition into the principal subspace S and its orthogonal complement $\bar{S}$. Prototypical patterns are embedded entirely in S. The component embedded in $\bar{S}$ can be due to observation noise (after [34]).

From the above we see that the spaces $\mathbf{P_V}$ and $(\mathbf{I} - \bar{\mathbf{P}}_\mathbf{V})$ are equivalent. Therefore, based on Equation (3.46) it also holds that

$$d_S\left(\mathbf{x}, \bar{S}\right) = 1 - d_S\left(\mathbf{x}, S\right). \tag{3.56}$$

Based on the above motivations the classification rule (3.49) for the complementary space transforms to the following

$$\omega_{c'} = \arg\min_{1 \le c \le C}\left[\delta_S\left(\mathbf{x}, \bar{S}_c\right)\right], \tag{3.57}$$

where

$$\delta_S\left(\mathbf{x}, \bar{S}\right) = \begin{cases} \hat{\mathbf{x}}^T \bar{\mathbf{P}}_\mathbf{V} \hat{\mathbf{x}} & \text{for nonorthonormal base} \\ \sum_{r=R+1}^{L}\left(\hat{\mathbf{x}}^T \mathbf{v}_r\right)^2 & \text{for orthonormal base.} \end{cases} \tag{3.58}$$

Summarizing, the own subspace S is a better choice for representing a class if the number of dominating components is less than half of all possible components, i.e. $R \le L/2$. In the opposite case the complementary subspace $\bar{S}$ would be more useful. Taking advantage of this property Xu *et al.* proposed splitting all data sets of multiple classes into two disjoint groups: The first one containing only those spaces in which the number of important components is $R \le L/2$, i.e. these will be classified in the subspaces S_i. In the second group are those classifiers for which $R > L/2$. These will be described using their complementary subspaces $\bar{S}_i$. This hybrid method allows more efficient multiclass pattern recognition and is called a *dual subspace* representation method [32].

Recently robust methods have gained great attention. Their main goal is to improve, usually the well known methods, so they can deal better in cases of outliers in data sets, e.g. erroneous measurements, as well as missing data. Robust classification in high dimensions based on the SIMCA method is discussed in the paper by Branden and Hubert [35]. Yet another approach is presented in the work by Bischof *et al.* [36]. Ritter and Gallegos discuss outliers in statistical pattern recognition in the context of automatic chromosome classification [37].

Linear subspaces obtained with PCA decomposition were used for face recognition by Turk and Pentland [20]. Their approach is based on "eigenfaces" which are eigenvectors of training face patterns. This concept was then extended with the help of the multilinear analysis, and called "tensorfaces," as discussed in Section 2.12.15. In this context, details of face detection in dual subspaces are further discussed in Section 3.5.

A survey of face recognition methods in subspaces contains the report by Shakhnarovich and Moghaddam [38], as well as the paper Raoa and Noushath [39]. Both are highly recommended reading. The former discusses subspace methods in the chronological order, starting from PCA (eigenfaces), probabilistic PCA, Fisher linear discriminant (Fisherfaces), Bayesian approach, ICA, up to the kernel and tensor methods. The important conclusion of this paper is that such an ostensibly complex task of face recognition, when represented in a high-dimensional space, reveals its intrinsically low-dimensional nature which allows accurate and computationally efficient solution. On the other hand, the recent paper by Raoa and Noushath contains a comparison of 25 different subspace algorithms and 4 databases: ORL, Yale, FERET, as well as COIL-20.

Finally, let us indicate that the probabilistic PCA is discussed in Section 3.5. Additionally, let us note that concepts of pattern recognition in the subspaces can be extended into tensor bases, as discussed in Section 2.12.4. Examples of these follow in Section 5.8.

3.4   Statistical Formulation of the Object Recognition

3.4.1   Parametric and Nonparametric Methods

The problem of pattern classification can be viewed in terms of information available for the classification process. If a classified phenomenon is known in terms of its distribution then such *a priori* knowledge directly leads to the parameters of the model. Thus, in this case we consider a parametric approach. On the other hand, if no *a priori* knowledge is assumed then a nonparametric approach can be considered. Both have their advantages and disadvantages depending on the specific case [2, 3, 40, 41].

The Gaussian and mixture of Gaussians (MoG) are the most common examples belonging to the group of parametric methods. This requires an indirect assumption that the classified phenomenon can be modeled with the Gaussian distribution which parameters have to be computed.

In practice however, such *a priori* knowledge on a type of distribution is rarely known with high confidence and therefore nonparametric methods are more common. These are, for instance, neural networks [42], SVMs [43, 135], or fuzzy methods [44], to name a few. Based on the available examples of known patterns, called prototypes or training sets, these classifiers are able to learn an unknown decision function which allows them to determine a class of a new and unknown patterns, also called also test patterns.

3.4.2   Probabilistic Framework

Measurements and judgments are inevitably burdened with errors and uncertainty. Therefore the probabilistic framework proved to be a convenient tool in classifier design. In this approach a class label ω is a random variable taking its values in the set of class names Ω, as shown in Section 3.1. In effect, with each class ω_c a probability mass function $P(\omega_c)$ can be associated which fulfills

$$\sum_{c=1}^{C} P(\omega_c) = 1. \tag{3.59}$$

However, $P(\omega_c)$ only conveys information on a distribution of the probabilities of the classes within Ω. In order to join information on a class with the characteristics of its objects, expressed with the feature (measurement) vectors, the class-conditional probability density function $p(\mathbf{x}|\omega_c)$ can be used. It fulfills the following condition

$$\int_{\mathbf{X}} p(\mathbf{x}|\omega_c)\,d\mathbf{x} = 1, \text{ for each } 1 \leq c \leq C. \tag{3.60}$$

In consequence, the class probability $P(\omega_c)$ and the class-conditional probability $p(\mathbf{x}|\omega_c)$ lead to the unconditional probability of observing a feature $\mathbf{x}$, as follows

$$p(\mathbf{x}) = \sum_{c=1}^{C} p(\mathbf{x}|\omega_c)\,P(\omega_c). \tag{3.61}$$

However, what we are really interested in pattern classification is the probability of a class ω_c given a measurement $\mathbf{x}$, that is a conditional probability $P(\omega_c|\mathbf{x})$. This can be estimated with help of the Bayes rule, which is discussed in the next section.

3.4.3 Bayes Decision Rule

The Bayes theorem provides one of the most important inferring methods in pattern recognition which is based on the probability theory. No surprise then that the method has gained broad applications in all areas of science and technology involved in making decisions on the type of an object, having observations of its features. Literature on this subject is ample [2, 3, 40, 41].

The Bayes pattern recognition framework can be used for patterns for which it is known that each belongs exclusively to one class ω_c, for $1 \leq c \leq C$. Let us assume that we know *a priori* probabilities $P(\omega_c)$ of each class ω_c, as well as conditional densities $p(\mathbf{x}|\omega_c)$, i.e. the probability of observation $\mathbf{x}$ if the class of an object is known to be ω_c, as discussed in the previous sections. Having this information the inverse problem of deciding *a posteriori* on a class ω_c based on the observation $\mathbf{x}$ can be approached with the Bayes rule

$$P(\omega_c|\mathbf{x}) = \frac{p(\mathbf{x}|\omega_c)\,P(\omega_c)}{p(\mathbf{x})}. \tag{3.62}$$

Inserting Equation (3.61) into the denominator of Equation (3.62) yields

$$P(\omega_c|\mathbf{x}) = \frac{p(\mathbf{x}|\omega_c)P(\omega_c)}{\sum_{c=1}^{C} p(\mathbf{x}|\omega_c)P(\omega_c)}. \tag{3.63}$$

In the case of discrete feature variables $\mathbf{x}$ in Equation (3.63) a conditional probability mass function has to substitute for $p(\mathbf{x}|\omega_c)$ which leads to the following

$$P(\omega_c|\mathbf{x}) = \frac{P(\mathbf{x}|\omega_c)P(\omega_c)}{\sum\limits_{c=1}^{C} P(\mathbf{x}|\omega_c)P(\omega_c)}. \tag{3.64}$$

In the above formulas $P(\omega_c|\mathbf{x})$ is called the *posterior class probability*.

3.4.4 *Maximum* a posteriori *Classification Scheme*

After the introduction in the previous section it is no surprise that the posterior class probability, given in formulas (3.63) or (3.64), can be used as a discriminant function defined in Equation (3.3), as follows

$$g_c(\mathbf{x}) = P(\omega_c|\mathbf{x}). \tag{3.65}$$

Under this assumption the maximum membership rule in Definition 3.3 leads to the *maximum likelihood* (ML) classification method, as follows

$$\omega_{c'} = \arg\max_{1 \le c \le C} \{g_c(\mathbf{x})\} = \arg\max_{1 \le c \le C} \{P(\omega_c|\mathbf{x})\}. \tag{3.66}$$

However, because the denominator in Equation (3.63) and (3.64) stays the same for all $1 \le c \le C$, it has no effect on the maximization process (3.66). Therefore inserting just a nominator of a Bayes formula (3.64) into Equation (3.66) the following is obtained

$$\omega_{c'} = \arg\max_{1 \le c \le C} \{P(\mathbf{x}|\omega_c)P(\omega_c)\}. \tag{3.67}$$

The above is known as *the maximum a posteriori classification rule* (MAP) [2, 40]. The same holds also for the continuous case (3.63), changing probability mass to the density function, as discussed.

The classification rule (3.67) can be further simplified if it is known that each class is equally probable, i.e. $P(\omega_c)$ is the same for all c. In such a case the factor $P(\omega_c)$ can be eliminated from Equation (3.67), which yields the simpler classification rule

$$\omega_{c'} = \arg\max_{1 \le c \le C} \{P(\mathbf{x}|\omega_c)\}. \tag{3.68}$$

The above can sometimes be simplified even further assuming that the features x_i in the feature vector $\mathbf{x}$ are conditionally independent.[3] Hence

$$P(\mathbf{x}|\omega_c) = \prod_{\ell=1}^{L} P(x_\ell|\omega_c), \tag{3.69}$$

where L denotes the dimension of the feature space (i.e. the size of $\mathbf{x}$). Considering now Equations (3.68) and (3.69), as well as taking advantage of the monotonicity of the logarithmic function, the following is obtained

$$\omega_{c'} = \arg\max_{1 \le c \le C} \{\log P(\mathbf{x}|\omega_c)\} = \arg\max_{1 \le c \le C} \left\{ \sum_{\ell=1}^{L} \log P(x_\ell|\omega_c) \right\} \tag{3.70}$$

which is known as a *naïve Bayesian classifier* [2].

As presented, for a given feature vector $\mathbf{x}$ the maximum *a posteriori* rule consists of computation of a class for which its conditional *a posteriori* probability is maximized, as outlined by Equation (3.67). However, for this to work the $P(\mathbf{x}|\omega_c)$ needs to be determined. Depending on the problem this can be done in a number of ways from which the two groups emerge: The parametric and nonparametric methods. We will discuss both in the subsequent sections.

On the other hand, the Bayes classification process can also be viewed in terms of the costs of misclassification. For this purpose let as assume that k_{bc} denotes a cost of misclassification of an object c into the class b. Then the expected conditional cost becomes

$$K(\omega_b|\mathbf{x}) = \sum_{c=1}^{C} k_{bc} P(\omega_c|\mathbf{x}). \tag{3.71}$$

Now the classification rule can be stated in terms of minimization of the total conditional cost, that is MAP becomes

$$\arg\min_{1 \le b \le C} \{K(\omega_b|\mathbf{x}|)\} = \arg\min_{1 \le b \le C} \left\{ \sum_{c=1}^{C} k_{bc} P(\omega_c|\mathbf{x}) \right\}. \tag{3.72}$$

The last expression is convenient if one wants to control potential misclassification errors. For instance if there are two groups of mushrooms, the edible and the poisonous ones, then the cost of misclassification of a poisonous mushroom to the edible group should be set to a higher value than classifying vice versa. Let us observe that if

$$k_{bc} = \begin{cases} 1 & \text{for} \quad\quad b \ne c, \\ 0 & \text{otherwise} \end{cases} \tag{3.73}$$

---

[3]This assumption of feature independence is sometimes overused. In practice, however, it can be justified by good experimental results and a much shorter computation time (for example in the feature tracking task).

then

$$K(\omega_b|\mathbf{x}) = \sum_{c=1, b \neq c}^{C} k_{bc} P(\omega_c|\mathbf{x}) = 1 - P(\omega_b|\mathbf{x}) \qquad (3.74)$$

and the two classification schemes (3.66) and (3.72) become identical.

3.4.5 Binary Classification Problem

A frequent classification task in computer vision is binary classification, for instance in CV to distinguish an object from the background. The binary classification problem assumes the existence of two classes, ω_0 for objects of the first class and ω_1 for objects of the second class, respectively. In this case the classification rule for an object of class ω_0 can be stated simply as (also known as a minimum error rate [2])

$$P(\omega_0|\mathbf{x}) > P(\omega_1|\mathbf{x}). \qquad (3.75)$$

Considering now the Bayes formula (3.64), the above equation can be written as follows

$$P(\mathbf{x}|\omega_0)P(\omega_0) > P(\mathbf{x}|\omega_1)P(\omega_1). \qquad (3.76)$$

After simple rearrangement, this leads to the following binary classification rule

$$\frac{P(\mathbf{x}|\omega_0)}{P(\mathbf{x}|\omega_1)} > \frac{P(\omega_1)}{P(\omega_0)} = \kappa. \qquad (3.77)$$

where κ is a decision threshold, assuming that the denominator is greater than zero (which condition should be always checked during computations).

Further, assuming independence of all features x_i, for $1 \leq i \leq L$, and taking the logarithm of the above yields

$$\sum_{k=1}^{L} \log \frac{P(x_k|\omega_0)}{P(x_k|\omega_1)} > \kappa'. \qquad (3.78)$$

It is proposed in many publications to consider $P(\omega_1) \approx P(\omega_0)$. Under this assumption the constant κ' in Equation (3.78) is set to 0. Thus, if the sum of logarithms of pairs of probabilities object/background in Equation (3.78) is above 0 then possibly a pixel with feature $\mathbf{x}$ belongs to an object; otherwise it is considered background. Nevertheless more often than not background pixels in images are more numerous, so in the above $P(\omega_1) \gg P(\omega_0)$ and therefore κ' should be greater than 0. This also remains in agreement with our experiments performed on the system for road sign detection and tracking [45], which is also discussed in further sections of this book. However, an exact setting of κ' depends on the contents of an image. Therefore much better results with this simple classification rule (3.78) are obtained for tasks in which only a peak in the log likelihood needs to be estimated, such as in the case of the mean shift method [46, 47, 48, 49], discussed in Section 3.8.

3.5 Parametric Methods – Mixture of Gaussians

The probability density function (PDF) of the L-dimensional Gaussian distribution is given as follows [2, 40, 50]

$$G(\mathbf{x}, \mathbf{m}_x, \boldsymbol{\Sigma}_x) = \frac{1}{(2\pi)^{L/2} \sqrt{\det(\boldsymbol{\Sigma}_x)}} e^{-\frac{1}{2} D_M(\mathbf{x}, \mathbf{m}_x)}. \qquad (3.79)$$

D_M in the above is known as the Mahalanobis distance from $\mathbf{x}$ to $\boldsymbol{\mu}$ [51]:

$$D_M(\mathbf{x}, \mathbf{m}_x) = (\mathbf{x} - \mathbf{m}_x)^T \boldsymbol{\Sigma}_x^{-1} (\mathbf{x} - \mathbf{m}_x), \qquad (3.80)$$

where $\mathbf{x}$ is an L-dimensional data sample, $\mathbf{m}_x$ denotes a mean value of the population, and $\boldsymbol{\Sigma}_x$ is the $L \times L$ covariance matrix computed from the training set of samples $\{\mathbf{x}_i\}$ which empirically denote a class ω of our objects (such as for instance faces or cars in images). A plot of one-dimensional Gaussian functions is presented in Figure 3.24 on p. 297. Thus Equation (3.79) can represent a conditional probability of a feature $\mathbf{x}$ when observing an object of class ω, that is

$$P(\mathbf{x}|\omega) = G(\mathbf{x}, \mathbf{m}_x, \boldsymbol{\Sigma}_x). \qquad (3.81)$$

This constitutes a base for inferring a class given an observation $\mathbf{x}$ by means of one of the variations of the Bayes scheme (discussed in Section 3.4.3).

A common practice is to represent Equation (3.79) in the logarithmic form

$$\ln[G(\mathbf{x}, \mathbf{m}_x, \boldsymbol{\Sigma})] = -\frac{L}{2} \ln(2\pi) - \frac{1}{2} \ln[\det(\boldsymbol{\Sigma}_x)] - \frac{1}{2} D_M(\mathbf{x}, \mathbf{m}_x). \qquad (3.82)$$

The logarithmic representation (3.82) takes advantage again of the monotonicity of the logarithmic function. Therefore it can be used instead of a direct representation (3.79) when comparing the Gaussian distribution or looking for a maximum likelihood, requiring less computations at the same time.

Mixtures of Gaussians (MoG), which are weighted series of Equation (3.79), can be used to model broader groups of complex multidimensional distributions which arise in many physical phenomena. These can be used to represent even arbitrarily complex (and usually unknown) distributions. Because of this, MoG finds broad applications in scientific modeling. A MoG composed of J components can be expressed as follows

$$M(\mathbf{x}) = \sum_{j=1}^{J} w_j G(\mathbf{x}, \mathbf{m}_j, \boldsymbol{\Sigma}_j), \qquad (3.83)$$

where $\mathbf{\Sigma}_j$ denotes a covariance matrix $\mathbf{\Sigma}_x$ for the j-th data set $\{\mathbf{x}_i\}_j$, and w_j are nonnegative weights that sum up to 1, that is

$$\underset{1 \leq j \leq J}{\forall} : w_j \geq 0, \text{ and } \sum_{j=1}^{J} w_j = 1. \qquad (3.84)$$

Thus, a model described with help of Equation (3.83) is governed by the number J of Gaussians, as well as by all their parameters $\mathbf{m}_j$ and $\mathbf{\Sigma}_j$. To find these from the test data the iterative *expectation maximization* (EM) algorithm can be used [2, 50]. However, the number of Gaussians J has to be set *a priori*. An alternative method to EM is the *elliptical k-means* algorithm used for instance by Sung and Poggio [52].

There are dozens of examples of using the normal distribution (3.79) and mixture of Gaussians (3.83) in computer vision to model statistics of sought phenomena. For instance lines can be efficiently detected after analyzing the Laplacian of the Gaussian of the intensity signal [14, 53]. Segmentation of RGB color images can be easily accomplished with the method of thresholded Mahalanobis distance, as presented by Gonzalez and Woods [14]. Yet more sophisticated segmentation method for human skin detection in color images was developed by Jones and Rehg [54]. Their model consists of $J = 16$ Gaussians which were computed from almost one billion labeled pixels from images gathered mostly from the Internet. Their method operates in the RGB color space, achieving a detection rate of 80% with about 9% false positives. For years their human skin model has served as a referential method for comparison with other approaches, such as the Bayes or neural based solutions. Similar methodology based on application of a mixture of Gaussians was undertaken by Yang and Ahuja in [55]. Surveys on pixel based skin detection, such as the one by Vezhnevets *et al.* [56] or by Phung *et al.* [57], provide further details on this group of methods.

In the work of D'Orazio *et al.* [58] MoG was used to model the normal behavior of a driver. The method plays an important role in the system for visual inspection of driver conditions during long driving. It can help in detecting dangerous situations related to the driver's inattention, drowsiness, etc. Thus, the goal is to prevent road accidents by controlling driver vigilance. D'Orazio *et al.* proposed using a neural network operating with Haar wavelets to recognize eyes of a driver. Then the analysis of the eye occurrence in video is carried out with the help of the probabilistic model (3.79), containing $J = 3$ Gaussian components. Their parameters were found with the help of the EM algorithm. Their system reached a true-positive (TP) factor of 95% at a rate of 15 frames of 320×240 pixels per second (see also Section A.5). However, what is really important in systems like this is the factor of false positives (FP), that is the number of eyes recognized by the algorithm in wrong areas (i.e. eyes not existing in reality). This is important since the algorithm can be fooled by the existence of virtually open eyes which would indicate that a driver has awoken, though the real situation might be different. The FP is in the order of 1% in the mentioned system [58].

Application of the Gaussian distribution (3.79) to modeling of upright frontal faces is proposed in the system for face detection by Shih and Liu [59]. However, in this case the dimensionality L of the samples is very high. It is because for each input image (a face candidate) of $p \times q$ pixels its two Haar representations are computed. These are horizontal and vertical difference images, respectively. Then all three images are concatenated row by row to yield a feature vector of dimensions $L = 3pq$. Finally, the feature vectors are normalized to zero mean and unit variance and in this form are used to build a statistical face model. However, even for very small test patterns, such as 16×16 used in experiments by Shih and Liu [59], L is very large. In this example $L = 768$ which results in a 768×768 covariance matrix $\boldsymbol{\Sigma}$. Therefore, to avoid direct manipulations of such large data structures they propose taking advantage of the symmetric nature of the covariance matrix and decompose it in the way we already discussed in Section 3.3.1 [34, 60]. Thus, based on Equation (3.13) the following can be written

$$\boldsymbol{\Sigma}_x = \mathbf{T}^T \boldsymbol{\Lambda} \mathbf{T}, \text{ and } \boldsymbol{\Sigma}_x^{-1} = \mathbf{T}^T \boldsymbol{\Lambda}^{-1} \mathbf{T} \tag{3.85}$$

where

$$\mathbf{T}\mathbf{T}^T = \mathbf{T}^T \mathbf{T} = \mathbf{I}_{L \times L}, \text{ and } \boldsymbol{\Lambda} = diag\,[\lambda_1, \lambda_2, \ldots, \lambda_L]. \tag{3.86}$$

Let us recall, that in the above $\mathbf{T} \in \Re^{L \times L}$ is an orthonormal matrix of eigenvectors of $\boldsymbol{\Sigma}_x$ and $\boldsymbol{\Lambda}$ is a diagonal matrix of its eigenvalues, which elements are set in decreasing order, i.e. $\lambda_1 \geq \lambda_2 \geq \ldots \geq \lambda_L$. Introducing (3.85) into (3.80) yields

$$D_M\,(\mathbf{x}, \mathbf{m}_x) = (\mathbf{x} - \mathbf{m}_x)^T \left(\mathbf{T}^T \boldsymbol{\Lambda}^{-1} \mathbf{T} \right) (\mathbf{x} - \mathbf{m}_x) = \underbrace{[\mathbf{T}\,(\mathbf{x} - \mathbf{m}_x)]}_{\mathbf{y}}^T \boldsymbol{\Lambda}^{-1} \underbrace{[\mathbf{T}\,(\mathbf{x} - \mathbf{m}_x)]}_{\mathbf{y}} = \mathbf{y}^T \boldsymbol{\Lambda}^{-1} \mathbf{y}$$

$$\tag{3.87}$$

where the vector $\mathbf{y} = \mathbf{T}(\mathbf{x} - \mathbf{m}_x)$, $\mathbf{y} \in \Re^{L \times 1}$, denotes a new variable in the space obtained by the PCA transformation, as already derived in Equation (3.14). Since $\boldsymbol{\Lambda}$ is diagonal D_M in (3.80) it can be written as follows

$$D_M\,(\mathbf{x}, \mathbf{m}_x) = \mathbf{y}^T \boldsymbol{\Lambda}^{-1} \mathbf{y} = \sum_{j=1}^{L} \frac{y_j^2}{\lambda_j}. \tag{3.88}$$

where y_j are elements of $\mathbf{y}$. This result indicates that the Mahalanobis distance can be represented as *a weighted sum of uncorrelated components* (signal energies). The sum also means that the Gaussian can be now *partitioned into a series of independent Gaussian* densities.

However, Equation (3.88) still does not relieve the computational burden due to the high dimensionality of the input features. Taking advantage of the decreasing order of λ_i usually only a much smaller subset $R \ll L$ of them is retained, whereas all remaining, i.e. for $i > R$, are set to 0. In this case it is well known that PCA allows an optimal reconstruction of the decomposed signal in terms of the minimum mean-square error (Section 3.3.1). Thus Equation (3.88) can be decomposed into two components

$$D_M\,(\mathbf{x}, \mathbf{m}_x) = \sum_{j=1}^{R} \frac{y_j^2}{\lambda_j} + \sum_{j=R+1}^{L} \frac{y_j^2}{\lambda_j}. \tag{3.89}$$

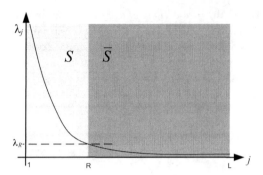

Figure 3.12   Illustration of an eigenvalue spectrum and its division at *R-th* value into the two orthogonal subspaces.

The first term in the above equation corresponds to the principal R-dimensional subspace S (a feature space), whereas the second one denotes its orthogonal $(L-R)$-dimensional complement $\bar{S}$. The latter is usually related to the minor statistical components or noise in data, as already discussed in Section 3.3.2. A typical eigenvalue spectrum is illustrated in Figure 3.12. However, it is easy to show that the last term in Equation (3.89) can be expressed in terms of so called residual PCA reconstruction error (Section 3.3.1)

$$\varepsilon^2(\mathbf{x}) = \sum_{j=R+1}^{L} y_j^2 = \|(\mathbf{x} - \mathbf{m}_x)\|^2 - \sum_{j=1}^{R} y_j^2. \tag{3.90}$$

The above is known as the *distance from feature space* (DFFS), as a complement to the *distance in feature space* (DIFS), as already shown in Figure 3.11. DFFS can be computed from the first R principal components. Thus Equation (3.89) can be estimated as follows

$$\tilde{D}_M(\mathbf{x}, \mathbf{m}_x) = \sum_{j=1}^{R} \frac{y_j^2}{\lambda_j} + \underbrace{\sum_{j=R+1}^{L} \frac{y_j^2}{\rho}}_{} = \sum_{j=1}^{R} \frac{y_j^2}{\lambda_j} + \underbrace{\frac{1}{\rho} \sum_{j=R+1}^{L} y_j^2}_{} = \sum_{j=1}^{R} \frac{y_j^2}{\lambda_j} + \underbrace{\frac{\varepsilon^2(\mathbf{x})}{\rho}}_{}, \tag{3.91}$$

where ρ is a parameter which can be seen as an estimate of *an average* λ_i for $R < i \leq L$. This property will be used in further derivations. Inserting the estimate $\tilde{D}_M$ (3.91) into the Gaussian (3.79) the following density is obtained

$$\tilde{G}(\mathbf{x}, \mathbf{m}_x, \mathbf{\Sigma}_x) = \left[\frac{e^{-\frac{1}{2} \sum_{j=1}^{R} \frac{y_j^2}{\lambda_j}}}{(2\pi)^{R/2} \prod_{j=1}^{R} \sqrt{\lambda_j}} \right] \cdot \left[\frac{e^{-\frac{1}{2} \frac{\varepsilon^2(\mathbf{x})}{\rho}}}{(2\pi\rho)^{(L-R)/2}} \right] = \tilde{G}_S(\mathbf{x}) \tilde{G}_{\bar{S}}(\mathbf{x}). \tag{3.92}$$

Considering now Equation (3.81) the following estimate of the conditional probability for a class ω is obtained

$$\tilde{P}(\mathbf{x}|\omega) = \tilde{G}(\mathbf{x}, \mathbf{m}_x, \boldsymbol{\Sigma}_x) = \tilde{G}_S(\mathbf{x})\tilde{G}_{\bar{s}}(\mathbf{x}). \tag{3.93}$$

Interestingly, the same scheme can be used for the multimodal densities modeled with the MoG [34]. In this case, however, the in-feature-space factor $\tilde{G}_S(\mathbf{x})$ of Equation (3.93) is replaced with the estimate of (3.83). Thus

$$\tilde{P}(\mathbf{x}|\omega) = \tilde{M}(\mathbf{x})\,\tilde{G}_{\bar{s}}(\mathbf{x})\,. \tag{3.94}$$

That is, the nonfeature space is still modeled with the Gaussian $\tilde{G}_{\bar{s}}(\mathbf{x})$.

What is left is to determine an estimation $\tilde{\rho}$ of the parameter ρ in Equation (3.92). For this purpose Moghaddan and Pentland propose finding $\tilde{\rho}$ as a parameter that minimizes the cost function resulting from the Kullback–Leibler relative entropy between the true density (3.79) and its estimate (3.92). It was found to be the arithmetic average of the eigenvalues in the orthogonal subspace, that is [34]

$$\tilde{\rho} = \frac{1}{L - R} \sum_{j=R+1}^{L} \lambda_j. \tag{3.95}$$

In practice, however, $\tilde{\rho}$ cannot be computed directly since only the first R eigenvalues were computed for the principal subspace. Instead it can be approximated by fitting a function (usually it is $f(x) = 1/x$ or other fractals [38]) to the R known eigenvalues. In this procedure we take advantage of the decaying energy of the signal for the higher eigenvalues which usually correspond to the intrinsic noise in an image. Moghaddan and Pentland stress the "critical" importance of the second factor $\tilde{G}_{\bar{s}}(\mathbf{x})$ in Equation (3.92) which is sometimes (incorrectly) omitted. This is directly related to the problem of selecting the number R of principal components for a given problem, as discussed in Section 3.3.1.

On the other hand, if we insert Equation (3.90) into (3.91), and then their result into Equation (3.82), the following is obtained

$$\ln\left[\tilde{P}(\mathbf{x}|\omega)\right] = \ln\left[\tilde{G}(\mathbf{x}, \mathbf{m}_x, \boldsymbol{\Sigma}_x)\right] =$$

$$-\frac{L}{2}\ln(2\pi) - \frac{1}{2}\ln\left(\prod_{j=1}^{R}\lambda_j\right) - \frac{1}{2}(L-R)\ln(\tilde{\rho}) - \frac{1}{2}\left(\sum_{j=1}^{R}\frac{k_j^2}{\lambda_j} + \frac{\|(\mathbf{x}-\mathbf{m}_x)\|^2 - \sum_{j=1}^{R}y_j^2}{\tilde{\rho}}\right). \tag{3.96}$$

In the above we used the fact that $\mathbf{\Sigma}_x$ is a diagonal matrix, so

$$\ln\left[\det\left(\mathbf{\Sigma}\right)\right] = \ln\left(\prod_{j=1}^{R}\lambda_j \cdot \prod_{j=R+1}^{L}\lambda_j\right) = \ln\left(\prod_{j=1}^{R}\lambda_j\right) + \ln\left(\prod_{j=R+1}^{L}\lambda_j\right).$$

The last term in the above can be estimated knowing that $\tilde{\rho}$ estimates $\bar{\lambda}$

$$\ln\left(\prod_{j=R+1}^{L}\lambda_j\right) = \ln\left(\underbrace{\lambda_{R+1}\lambda_{R+2}\ldots\lambda_L}_{L-R}\right) \approx \ln\left(\tilde{\rho}^{L-R}\right) = (L-R)\ln\left(\tilde{\rho}\right).$$

The distribution (3.96) was used by Shih and Liu in their face detector to the initial face/non-face classification task [59]. During training the R principal components with corresponding eigenvalues, as well as the mean face $\mathbf{m}_x$, have to be computed. Now we see, however, that the three first components in Equation (3.96) are constant for a chosen set of training patterns and therefore *can be omitted* in the classification stage. Finally, $\tilde{\rho}$ can be set as a fraction ($\sim$0.1) of the smallest eigenvalue.

We see that the ML type of classification can be carried out solely based on the third term of Equation (3.96). However, it is very important to note that even for the binary classification problem, i.e. deciding whether an object belongs or not to a class, there can be objects which are not easy to reliably classify to either one. Such a situation was solved by Shih and Liu in their face detection system by assigning an uncertainty range to the conditional probability values. Thus, the classifier of type (3.96) can be used to partition objects into three classes: objects, nonobjects, and undecided, as follows

$$\omega = \begin{cases} \omega_f, & \ln\left[P(\mathbf{x}|\omega)\right] \geq \kappa_f \\ \omega_u, & \kappa_n < \ln\left[P\left(\mathbf{x}|\omega\right)\right] < \kappa_f \\ \omega_n, & otherwise \end{cases} . \tag{3.97}$$

Then the other classifier is used to disambiguate the class of an object (SVM in the work by Shih and Liu). However, the scheme (3.97) requires determination of the two threshold values κ_f and κ_n, which can be done by means of the minimization of the classification error function with respect to these thresholds.

Literature on Gaussian processes in pattern recognition is ample. For instance the book by Rasmussen and Williams [61] is a very good source of information on Gaussian processes (GP) in the context of machine learning. It deals with regression and classification carried out with the help of GP. Model selection, relationships with other methods (such as kernel based, SVM, etc.), and approximation methods for large databases are also discussed in that book.

3.6   The Kalman Filter

The Kalman filter[4] is a mathematical model for recursive estimation of the state of a process based on the series of its previous observations, as well as measurements [62]. It belongs to the broader group of Gaussian filters in which the primary assumption is representation of beliefs exclusively with multivariate normal distributions, as described by Equation (3.79). Estimation is done in such a way as to minimize the mean of the square error. Since its inception by Kalman over 50 years ago, the Kalman filter has become an important mechanism for *fault-tolerant computing* and *data fusion*. Not surprisingly, the Kalman filter has found many applications, especially in positioning and tracking systems, such as the famous Apollo vehicle, phase locked loops, data smoothing, and many others. The Kalman filter is also an important method in computer vision for image filtering and object tracking.

The Kalman filter model describes a process of estimating the state of a discrete-time process based on observations of its state at time step k, as described by the following linear stochastic difference equation [63]

$$\mathbf{x}_k = \mathbf{F}_k \mathbf{x}_{k-1} + \mathbf{B}_k \mathbf{u}_k + \mathbf{w}_{k-1}, \tag{3.98}$$

and the measurement equation

$$\mathbf{z}_k = \mathbf{H}_k \mathbf{x}_k + \mathbf{v}_k, \tag{3.99}$$

where the meaning of the symbols used is explained in a common Table 3.1. Equation (3.98) is called linear Gaussian since it is linear in its arguments with the addition of Gaussian noise. Although the matrices $\mathbf{F}_k$, $\mathbf{B}_k$, and $\mathbf{H}_k$ can change at each time step, frequently they are assumed to be constant. In such a case their subscript k can be omitted.

Since in practice state variables are not known, in derivation of the basic equations governing the Kalman filter it is convenient to define the so called *a priori* and *a posteriori* state estimates, denoted with symbols $\hat{\mathbf{x}}_k^-$ and $\hat{\mathbf{x}}_k$ respectively (see also their definitions in Table 3.2). Then the two estimation errors are defined as follows

$$a\ priori: \quad \mathbf{e}_k^- = \mathbf{x}_k - \hat{\mathbf{x}}_k^-, \tag{3.100}$$

$$a\ posteriori: \quad \mathbf{e}_k = \mathbf{x}_k - \hat{\mathbf{x}}_k. \tag{3.101}$$

From these the *a priori* and *a posteriori* error covariance matrices are derived, as follows

$$a\ priori: \quad \mathbf{P}_k^- = E\left[\mathbf{e}_k^- \mathbf{e}_k^{-T}\right], \tag{3.102}$$

$$a\ posteriori: \quad \mathbf{P}_k = E\left[\mathbf{e}_k \mathbf{e}_k^{T}\right]. \tag{3.103}$$

Based on Equation (3.99) let us now define the difference between an actual measurement $\mathbf{z}_k$ and a measurement prediction, formed from the *a priori* state estimate $\hat{\mathbf{x}}_k^-$ and the state transformation matrix $\mathbf{H}_k$, as follows

$$\mathbf{r}_k = \mathbf{z}_k - \mathbf{H}_k \hat{\mathbf{x}}_k^-. \tag{3.104}$$

---

[4] Also known as a linear quadratic estimator (LQE).

Table 3.1 Explanation of the symbols used in the Kalman filter.

Parameter	Description	Properties
$\mathbf{x}_k$	State vector containing important parameters of the system (such as position, acceleration, etc.)	$\mathbf{x}_k \in \Re^L$
$\mathbf{F}_k$	State transition matrix that joins the state at the previous time step k-1 with the current state at step k	$\mathbf{F}_k \in \Re^{L \times L}$
$\mathbf{B}_k$	Control input matrix which transforms the control vector to affect the state vector (e.g. transforms force value into acceleration, etc.)	$\mathbf{B}_k \in \Re^{L \times C}$
$\mathbf{u}_k$	Control input vector (such as applied force, throttle settings, etc.)	$\mathbf{u}_k \in \Re^C$
$\mathbf{w}_k$	Process noise for each state component. It is assumed to be from the zero mean multivariate normal distribution with covariance matrix $\mathbf{Q}_k$.	$p(\mathbf{w}_k) \sim N(0, \mathbf{Q}_k)$
$\mathbf{v}_k$	Measurement noise for each measurement component. It is assumed to be from the zero mean multivariate normal distribution with covariance matrix $\mathbf{R}_k$.	$p(\mathbf{v}_k) \sim N(0, \mathbf{R}_k)$
$\mathbf{H}_k$	State transformation matrix which maps state into measurement domain	$\mathbf{H}_k \in \Re^{M \times L}$
$\mathbf{z}_k$	Measurement vector (can be a direct or indirect measurement of the state vector $\mathbf{x}_k$)	$\mathbf{z}_k \in \Re^M$

The basic idea behind the Kalman filter is to compute an *a posteriori* state estimate $\hat{\mathbf{x}}_k$ expressed as a linear combination of an *a priori* estimate $\hat{\mathbf{x}}_k^-$ and a weighted measurement residual $\mathbf{r}_k$, that is

$$\hat{\mathbf{x}}_k = \hat{\mathbf{x}}_k^- + \mathbf{K}_k\, \mathbf{r}_k, \tag{3.105}$$

Table 3.2 Explanation of the symbols used in the Kalman filter equations.

Parameter	Description	Properties
$\hat{\mathbf{x}}_k^-$	*A priori* estimate of the state vector $\mathbf{x}_k$ at step k given knowledge of the process prior to step k (notice custom superscript minus)	$\hat{\mathbf{x}}_k^- \in \Re^L$
$\hat{\mathbf{x}}_k$	*A posteriori* estimate of the state vector $\mathbf{x}_k$ at step k given measurement $\mathbf{z}_k$ at step k	$\hat{\mathbf{x}}_k = E[\mathbf{x}_k]$ $\hat{\mathbf{x}}_k \in \Re^L$
$\mathbf{P}_k^-$	*A priori* estimate error covariance matrix	Equation (3.102) $\mathbf{P}_k^- \in \Re^{L \times L}$
$\mathbf{P}_k$	*A posteriori* estimate error covariance matrix	Equation (3.103) $\mathbf{P}_k \in \Re^{L \times L}$
$\mathbf{K}_k$	Kalman gain (blending factor)	Equation (3.106) $\mathbf{K}_k \in \Re^{L \times C}$
$\mathbf{r}_k$	Residual (measurement innovation, measurement update) conveys a discrepancy between the actual $\mathbf{z}_k$ and predicted measurement $\mathbf{H}_k \hat{\mathbf{x}}_k^-$	Equation (3.104) $\mathbf{r}_k \in \Re^M$

where the matrix $\mathbf{K}_k$ describes a filter gain. The second basic idea of the Kalman filter is to perform minimization of the *a posteriori* error covariance (3.103). For this purpose, and to derive an expression on $\mathbf{K}_k$, Equations (3.104)–(3.105) are substituted into (3.100)–(3.101) to obtain errors, and then these are introduced into expectations (3.102)–(3.103). Computing the derivative of the trace of the result with respect to $\mathbf{K}_k$ and equating that to zero, the following formula is obtained [63]

$$\mathbf{K}_k = \mathbf{P}_k^- \mathbf{H}_k^T \left(\mathbf{H}_k \mathbf{P}_k^- \mathbf{H}_k^T + \mathbf{R} \right)^{-1}. \qquad (3.106)$$

Analyzing $\mathbf{K}_k$ in the context of the filter equation (3.106) we easily see that if the measurement error $\mathbf{R}$ tends to 0, then the actual measurement $\mathbf{z}_k$ is more "trusted" than the measurement prediction $\mathbf{H}_k \hat{\mathbf{x}}_k^-$. On the other hand, if the *a priori* error covariance matrix $\mathbf{P}_k^-$ tends to 0, then the situation is reversed.

Operation of the Kalman filter can be split into two stages:

1. Time update stage (predictor).
2. Measurement update stage (corrector).

These are shown as a state diagram in Figure 3.13. The time update projects forward the current state and error covariance estimates to get the *a priori* estimates for the next step of operation. On the other hand, the measurement stage acts as a feedback, i.e. a new measurement is incorporated into the *a priori* estimate in order to compute the new value of the *a posteriori* estimate.

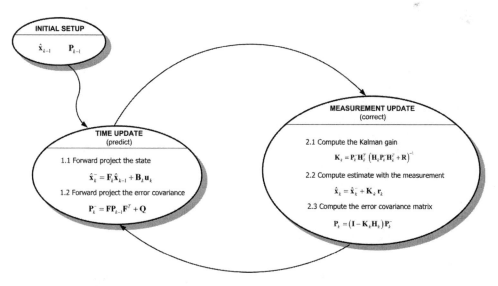

Figure 3.13   State diagram of the Kalman filter. Two stages are repeated: prediction and correction.

The recursive nature of the Kalman filter is very useful in its practical realization. However, in practice determination of its parameters, such as noise covariances $\mathbf{R}$ and $\mathbf{Q}$, can be difficult.

A nice and concise introduction to the Kalman filter theory is provided in the paper by Welch and Bishop [63]. Another very intuitive introduction is provided in the paper by Faragher [64], as well as in the book on OpenCV by Bradski and Kaehler [65]. Software package OpenCV also contains implementation of the Kalman filter. On the other hand Piovoso and Laplante discuss application of the Kalman filter to a variety of filtering problems in real-time image processing [66]. Their paper contains a description of the class hierarchy implementing the Kalman filter in an object-oriented fashion. More in-depth mathematical treatment on Kalman, and in general Gaussian, filters can be found in the book by Thrun *et al.* [67].

As alluded to previously, although the Kalman filter is reliable for many tracking applications, it is constrained to a relatively restricted group of linear Gaussian processes. In the case of nonlinear processes, the Kalman filter can be extended to cope with nonlinearities (EKF). It is done locally, linearizing approximations about the current mean and covariance. However, such a procedure is slightly ad hoc, since random variables are no longer normal Gaussian distributions after the nonlinear transformations. Therefore in the case of nonlinear processes, other methods such as particle filters should be considered. Recommended reading on these groups of filters is, for example, the book by Ristic *et al.* [68]. Particle filters have found broad applications in object tracking in digital images. Examples of such systems with further references can be found in publications [69, 70].

3.7 Nonparametric Methods

In Section 3.4.4 the maximum *a posteriori* classification rule was presented (3.67) which for a given feature vector $\mathbf{x}$ finds a class for which its conditional *a posteriori* probability is maximized. This requires computation of the conditional probability $P(\mathbf{x}|\omega_c)$, which can be done if a model is assumed. This usually leads to the parametric methods, as discussed in Section 3.5. However, if a model is not applicable then the nonparametric approach can be undertaken.

In consecutive subsections the pattern recognition methods that rely on estimation of the probabilistic density function $P(\mathbf{x}|\omega_c)$ in the nonparametric way are discussed. The main advantage of such an approach is more automatic and direct expression of the feature densities through lack of an *a priori* insight into the nature of an observed phenomenon. As a consequence parameters of a model need not be estimated either. However, the price to pay is usually much higher memory and computational demands. The histogram based methods, Parzen density estimation, as well as the k-nearest-neighbors method are discussed. All of these are frequently used in computer vision tasks.

3.7.1 Histogram Based Techniques

Histograms are data structures that store a number of occurrences of certain values (features) such as pixel intensity or phase of a local structure in an image (discussed in Section 5.2). Due to straightforward computations they find broad application in different areas of image processing and computer vision [14, 53, 71, 72]. The values which are registered in a histogram can be one (1D) or multidimensional, which leads to one or multidimensional histograms, respectively. A single entry in a histogram is called a bin. Usually the input values are further quantized to save the space required by a histogram. However, quantization is much more than this since it introduces a kind of nonlinear filtering of data (clipping). Regarding the implementation issues, in the case of 1D histograms usually a simple vector is an option. However, in the case of multidimensional histograms rather sparse representation and hashing tables should be considered.

As already mentioned, histograms are data structures which collect counters of numbers of occurrences of the specific values of features. Thus, they can be directly used to estimate a probability associated with the features, as will be discussed. This comes from the simple reasoning that features that are more frequent are more probable. However, details of the organization of these counters are also important. Let us consider a 1D case of counting occurrences of 8-bit intensity I in a monochrome image. Apparently, there are $N = 256$ possible values (data points) of I, so we can go ahead and count each of them in a separate counter, i.e. a bin. Such a scenario means that each value of I is assigned to exactly one bin. However, we might be interested in a rather coarser treatment of intensities, either due to the allowed number of bins, noise or computation time. Therefore, we can assign say only $\rho = 32$ bins, now each responsible for $\Delta = 256/32 = 8$ consecutive values of intensity I. Thus, the parameter Δ denotes the width of the bin. Figure 3.14 depicts three different 1D histograms of a variable x in the range from 0 to x_{max}. The first histogram, shown in Figure 3.14(a), contains $\rho = 8$ bins. Figure 3.14(b) shows the histogram with the same number of bins but with different boundaries of the bins which can be defined specifying their offset from the origin t. On the other hand, a histogram with $\rho = 16$ bins is shown in Figure 3.14(c) which exhibits a higher variability of bins. Thus, the histogram is characterized by its bin size Δ and offset t.

In a more general case Δ can be different for each bin. However, let us observe also that if $\Delta > 1$ then we have different possibilities for splitting the input values of I. In other words, the bins can be placed on different boundaries. For instance, a bin of 8 values can span intensities $0 \ldots 7$ or from $3 \ldots 10$. Hence, each time a different histogram is obtained with a different offset. Choosing all these parameters is cumbersome in practice, so usually the offset $t = 0$ is set and the same width for all bins is assumed which is then chosen experimentally. It is also obvious that a choice of Δ leads to smoothing properties of the histogram. In our example we have two extremes in this respect. The first one is to set $\Delta = 1$, so each intensity values has its own bin. At the other end one can consider setting $\Delta = 256$, so there will be only one bin for all intensity values. Apparently, the former is preferred since the latter loses all information on intensity distribution among the pixels. This example gives us insight into the way probability distribution of the samples can be estimated from the histograms.

For a given variable x which reaches values in the range $0 \ldots x_{max}$, the number of bins can be computed as follows

$$\rho = \left\lfloor \frac{x_{max}}{\Delta} \right\rfloor, \tag{3.107}$$

where Δ is the chosen common width of a bin and $\lfloor x \rfloor$ denotes a ceiling operation which for a real argument x returns the smallest possible integer value which is equal or greater than x.

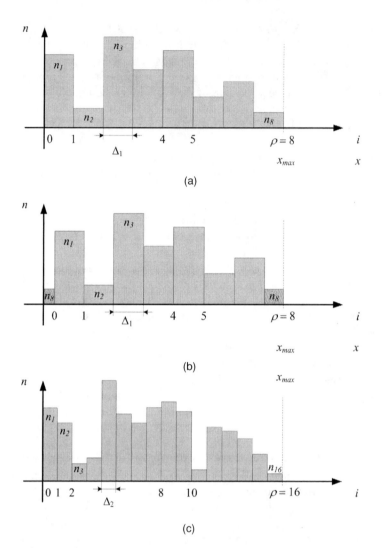

Figure 3.14   One-dimensional histograms of a variable x. Bins (in gray) contain occurrences a of values of x falling in the range defined by a bin position and its width Δ. An example of $\rho = 8$ bins (a). The same number of bins but placed at different boundaries (b). A histogram with $\rho = 16$ bins (c).

Thus, in the histogram we have ρ bins, so instead of values of x we can operate with the consecutive bin indices i. For a given value of x its corresponding bin index is simply

$$i = \left\lceil \frac{x}{\Delta} \right\rceil \text{ for } 0 \leq i < r, \tag{3.108}$$

where $\lceil x \rceil$ denotes the floor operation which returns the largest possible integer which is equal to or smaller than the real argument x.

Each bin contains an integer counter $n_i = n(i)$ of occurrences of x values falling inside that bin, as described by Equation (3.108). As already mentioned, histograms can be used to estimate the probability P_i of a value of x in a range

$$P_i \left(\Delta \cdot i \leq x < \Delta \cdot (i+1) \right) \approx \frac{n_i}{N}, \tag{3.109}$$

where

$$N = \sum_{i=1}^{\rho} n_i, \tag{3.110}$$

denotes the total number of occurrences.

Interesting further reading is the paper by Scott on the averaged shifted histogram which is a nonparametric method for density estimation from a collection of histograms [73].

3.7.2   Comparing Histograms

As already mentioned, histograms can be used to estimate probabilities in a nonparametric way. However, in our context these are estimated for the purpose of classification in the probabilistic framework, as discussed in Sections 3.4.2–3.4.5. Thus, we are interested in the ways of comparing histograms, which we shall discuss. As also mentioned, a histogram can be treated as a vector of ρ integers, each representing the frequency of occurrences of a given discrete random event (such as x falling in a defined range, etc.). A number ρ is determined from Equation (3.107). Let us assume that two histograms are available, each represented by vectors $\mathbf{a}$ and $\mathbf{b}$ consisting of ρ data, i.e. $\mathbf{a} = \{a_i\}_{1 \leq i \leq \rho}$ and $\mathbf{b} = \{b_i\}_{1 \leq i \leq \rho}$. Analogously to Equation (3.109), the two probabilities associated with $\mathbf{a}$ and $\mathbf{b}$ can be approximated as follows

$$P(a_i) \approx \frac{a_i}{\sum_{i=1}^{\rho} a_i} = \frac{a_i}{A} \text{ and } P(b_i) \approx \frac{b_i}{\sum_{i=1}^{\rho} b_i} = \frac{b_i}{B}, \tag{3.111}$$

assuming that the total numbers of points used to build each histogram, denoted in (3.111) by A and B, respectively, are different from 0.

To measure how close together the two distributions are, represented in this nonparametric way, two histograms represented by the vectors $\mathbf{a}$ and $\mathbf{b}$, need to be compared. This can be done bin-by-bin or in a cross-bin fashion. In the former only the corresponding bins are compared at each step, i.e. bin a_i will be compared only with b_i and never with b_j if only $j \neq i$. On the other hand, in the cross-bin method, comparison of noncorresponding bins is also allowed. Thus this approach is more versatile since in practice the histograms of similar objects can be slightly shifted, although if aligned the two would produce a strong match. Such a situation can happen when comparing histograms of gradient orientations of objects which are slightly rotated, for instance. Table 3.3 and Table 3.4 present the most common histogram matching measures for the bin-by-bin and cross-bin versions, respectively.

Table 3.3   Common bin-by-bin histogram matching measures.

Simple bin matching D_{nm}

$$D_{nm}(\mathbf{a}, \mathbf{b}) = \sum_{i=1}^{\rho} \frac{|P(a_i) - P(b_i)|^n}{[P(a_i) + P(b_i)]^m}. \tag{3.112}$$

For $n = 2$ and $m = 0$ the simple versions of the Minkowski's measure are obtained. Setting $n = 2$ and $m = 1$ leads to the χ^2 statistics distance. However, in some applications better results are obtained for $n = m = 2$ which promotes matches of bins with higher probability (for this effect one can set $m \geq n$) [45]. In the above $P(a_i) + P(b_i) \neq 0$ is assumed for $m \neq 0$. For m = 0 the denominator is assumed to be 1.

Bhattacharyya distance D_B

$$D_B = \sum_{i=1}^{\rho} \sqrt{P(a_i)P(b_i)}. \tag{3.113}$$

D_B follows from the definition of the scalar product for the two vectors a and b representing *square roots* of probabilities of the ρ-dimensional multinomial populations [74], i.e.

$$D_B = \cos \angle (\mathbf{a}, \mathbf{b}) = \left(\sum_{i=1}^{\rho} \sqrt{P(a_i)} \sqrt{P(b_i)} \right) \Big/ \left(\sqrt{\sum_{i=1}^{\rho} P(a_i)} \sqrt{\sum_{i=1}^{\rho} P(b_i)} \right) = \sum_{i=1}^{\rho} \sqrt{P(a_i)P(b_i)}$$

Thus D_B can be interpreted as a cosine of an angle between two vectors in a ρ dimensional space of an orthogonal coordinate system. Since probabilities are positive, the Bhattacharyya measure (3.113) takes values from 0 to 1. The higher the value, the closer the two probabilities. Thus, if $D_B = 1$, then the two populations are the same. In the continuous case the summation in (3.113) has to be changed into integration.

It is easy to observe that the value of $p_i = P(a_i)$ can be seen as a squared directional cosine (thus taking values 0 to 1) of a population vector to the *i-th* axis x_i of the coordinate system, since

$$\cos \angle (\mathbf{a}, \mathbf{x}_i) = \sqrt{P(a_i)} \Big/ \sqrt{\sum_{i=1}^{\rho} P(a_i)} = \sqrt{P(a_i)}, \text{ thus } P(a_i) = \cos^2 \angle (\mathbf{a}, \mathbf{x}_i). \tag{3.114}$$

Histogram intersection D_i

$$D_i(\mathbf{a}, \mathbf{b}) = 1 - \frac{\sum_{i=1}^{\rho} \min(a_i, b_i)}{B}. \tag{3.115}$$

where $B \neq 0$ is given in Equation (3.111). D_i can be used even for partial matching of the histograms, i.e. to handle situations in which $A \neq B$. In general, however, D_i is not symmetrical.

The Kullback–Leibler divergence D_{KL}

$$D_{KL}(\mathbf{a}, \mathbf{b}) = \sum_{i=1}^{\rho} P(a_i) \log \frac{P(a_i)}{P(b_i)} = \log \frac{B}{A} + \frac{1}{A} \sum_{i=1}^{\rho} a_i \log \frac{a_i}{b_i}, \text{ where } a_i, b_i \neq 0. \tag{3.116}$$

D_{KL} is also a nonsymmetrical distance in the general case, thus it is not a metric [75].

Jeffrey divergence D_J (Symmetrical Kullback–Leibler)

$$D_J(\mathbf{a}, \mathbf{b}) = D_{KL}(\mathbf{a}, \mathbf{b}) + D_{KL}(\mathbf{b}, \mathbf{a}) = \frac{1}{A} \sum_{i=1}^{\rho} a_i \log \frac{a_i}{b_i} + \frac{1}{B} \sum_{i=1}^{\rho} b_i \log \frac{b_i}{a_i} \tag{3.117}$$

assuming $a_i, b_i \neq 0$. D_J this defines a true metric (in contrast to D_{KL}).

Table 3.4 Common cross-bin histogram matching measures.

Quadratic form distance D_Q

$$D_Q(\mathbf{a}, \mathbf{b}) = \sqrt{(\mathbf{a} - \mathbf{b})^T \mathbf{Q}(\mathbf{a} - \mathbf{b})}, \tag{3.118}$$

where the matrix Q defines the similarity degree between pairs of bins, i.e. q_{ij} is a similarity between bin number i and j, respectively. A useful approach is to set $q_{ij} = 1 - d_{ij}/d_{max}$ where d_{ij} denotes the so called ground distance between the bins i and j, i.e. the further apart the bins the larger d_{ij} and smaller q_{ij} in consequence. d_{max} is the maximum distance between the bins. However, d_{ij} (and q_{ij} in general) does not need to be a geometrical distance between the bins (i.e. $|i\text{-}j|$); Instead it can reflect some *a priori* knowledge of a distance between the features represented by the bins i and j, respectively.

Cumulative (match) distance D_C

$$D_C(\mathbf{a}, \mathbf{b}) = \sum_{i=1}^{\rho} |\bar{A}_i - \bar{B}_i| \tag{3.119}$$

where $\bar{A}_i = \sum_{k \leq i} a_k$ is *a cumulative histogram* up to the index i (the same holds for B).

Kolmogorov–Smirnov D_{KS} distance

$$D_{KS}(\mathbf{a}, \mathbf{b}) = \max_i |\bar{A}_i - \bar{B}_i| \tag{3.120}$$

where $\bar{A}_i$ and $\bar{B}_i$ are cumulative histograms as in the case of D_C.

When using the measures in Table 3.3 it should be noticed that some measures attain 0 for the perfect match (for instance D_{nm}) and some complementary 1 in the same case (e.g. D_B).

Special attention should be devoted to the Bhattacharyya measure D_B in (3.113) since it exhibits many advantages over other histogram matching measures. These can be summarized as follows [76]:

1. D_B can be used irrespective of the type of distributions from which data is sampled;
2. D_B is superior when compared to the χ^2 measure because it can capture differences in the shapes of two distributions rather than their means (such as in a case of the Gaussian distribution);
3. D_B is independent of the widths of the bins of the histograms;
4. D_B is affinely related to the Matusita measure $\sum_i (\sqrt{P(a_i)} - \sqrt{P(b_i)})^2$.

In Section 5.2.2 yet another method of cross-bin matching is presented to match histograms of the local phases of objects obtained from the structural tensor. In this method we take advantage of the fact that each rotation of a rigid object causes a modulo shift of its phase histogram. This means that a histogram simply warps, i.e. when shifted by one bin to the right, then the last bin becomes the first one, and so on. However, each of the shifted versions is matched with one of the bin-by-bin measures contained in Table 3.3.

Details on signal processing using information theory with a special stress on computation of the enhanced mutual information is presented in the paper by Hudson [77]. An in-depth treatment of information theory applied to computer vision and pattern recognition can be accessed in the recent book by Escolano *et al.* [78].

An improvement in image content description and object recognition can be achieved with *signatures* instead of simple histogram matching. A signature S_I conveys information on an image I in a form of a series of clusters $\mathbf{s}_i$

$$S_I = \{\mathbf{s}_i\} = \{(\mathbf{m}_i, w_i)\}. \tag{3.121}$$

Each cluster $(\mathbf{m}_i, w_i)$, in turn, is represented by a single mean (or mode) vector $\mathbf{m}_i$ of a distribution of the characteristic features (e.g. color, texture, etc.) and the weight w_i which denotes the fraction of pixels that belong to that cluster (i.e. this is a relative size of a cluster). The pairs $(\mathbf{m}_i, w_i)$ are obtained from the original distribution of an image as a result of the operation of a clustering algorithm (for instance k-means, discussed in Section 3.11.1). The index i in Equation (3.118) spans up to the required number of clusters which, in general, is also unknown *a priori* and reflects the complexity of an image.

Let us observe that the already discussed histograms are special cases of the signatures in which $\mathbf{m}_i$ corresponds to an index i of a feature (or a vector of indices for the multidimensional histograms), whereas w_i is simply the frequency of occurrences of that feature in a histogram (denoted as a_i in Equation (3.111), for instance).

A notion of the distance of such a sparse representation as signatures requires the definition of a distance of basic features (e.g. between colors) which is called a ground distance [79]. The distance of basic features is a prerequisite to the notion of a distance between distributions of such basic features, which are usually computed over an entire image. Rubner *et al.* [79] state this task as a transportation problem which they coined an *Earth Mover's Distance* (EMD), denoted here as D_{EM}. Informally this can be explained as the problem of the minimal work to fill specifically distributed holes in a garden with a specifically distributed mass of earth. Thus it is the minimal cost that has to be paid to transform one distribution into the other.

More formally, however, this can be stated as the transportation problem which can be solved with the methods of linear optimization [17, 80, 81]. The task of computing D_{EM} is to find the least expensive way to transform one signature (a supplier) into the other (a consumer). The ground distance d_{jk} between simple features, represented by the mean vectors $\mathbf{m}_j$ and $\mathbf{n}_k$ of chosen image features (such as color), is proposed as

$$d_{jk} = 1 - e^{-\sigma \|\mathbf{m}_j - \mathbf{n}_k\|}, \tag{3.122}$$

where σ is a module of a vector composed of all standard deviations σ_i in each i-*th* dimension of the features. With the above definition D_{EM} between two signatures S and T, each containing J and K clusters, respectively, can be defined as follows

$$D_{EM}(S, T) = \frac{\sum\limits_{j=1}^{J} \sum\limits_{k=1}^{K} d_{jk} f_{jk}}{\sum\limits_{j=1}^{J} \sum\limits_{k=1}^{K} f_{jk}}, \tag{3.123}$$

where f_{jk} are the components of the optimal flow between the two signatures, which are known after solving the transportation problem [79].

Such a statement allows also partial matching, since the number of clusters can be different in the matched signatures, and usually leads to a better perceptual similarity, e.g. as perceived by humans. The other advantage is that distributions obtained in high dimensional feature space can be matched more efficiently since computational complexity depends mostly on the number of clusters and less on the dimensionality of $\mathbf{m}_i$ in Equation (3.118). Also, it can be shown that when the signatures hold equal weights w_i and the ground distance d_{ij} is a metric, then D_{EM} also is a true metric. An algorithmic statement of the transportation problem pertaining to the signature matching, as well as experimental results, can be found in [79].

Practical aspects of application of the D_{EM} into the problems of image matching and retrieval mostly involve selection of the features and then the clustering method to obtain their signatures. For instance, in the cited paper by Rubner *et al.* [79] *k-d tree* clustering was employed for object retrieval based on color and texture features. More specifically in the former case the joint distribution of color *and position* gave the best results. Thus the feature space is five-dimensional in which the CIE-Lab color space was used. The ground distance in this case is defined as follows

$$d_{jk} = \sqrt{\left(\Delta L_{jk}\right)^2 + \left(\Delta a_{jk}\right)^2 + \left(\Delta b_{jk}\right)^2 + \gamma \left[\left(\Delta x_{jk}\right)^2 + \left(\Delta y_{jk}\right)^2\right]}, \qquad (3.124)$$

where each $(\Delta._{jk})^2$ denotes a scalar difference between corresponding color components (L,a,b) or spatial (x, y) features of the clusters j and k, respectively, and γ is a parameter that controls the influence of the positional term. For matching with texture, Gabor filters were proposed resulting in a 24-dimensional feature vector for each pixel.

The Earth Mover's Distance D_{EM} was also proposed by Grauman and Darrell to match contours of human figures and handwritten digits [82]. The descriptive local contour features are used to build signatures around each "significant" point of a silhouette of a person. However, an additional embedding of the signatures into the normed space L_1 is applied to further reduce the complexity of matching. Through this method the distance in the L_1 subspace corresponds to the D_{EM} distance in the higher dimensional signature space. Then the approximated nearest-neighbor search is employed to retrieve the most similar objects from the database of prototypes. This is achieved via the Locality-Sensitive Hashing (LSH) method which allows sublinear time (in the number of the prototypes) of response of the system. Further details are provided in [82].

More information on the nearest-neighbor search in high dimensional spaces, *kd*-trees, and Locality-Sensitive Hashing can be found in literature on the subject, for instance in the book by Shakhnarovich *et al.* [83].

3.7.3 IMPLEMENTATION – Multidimensional Histograms

Implementation of the 1D histogram is straightforward. Therefore in Algorithm 3.11 the _3D_Histogram_For template class is presented which implements dense 3D histograms. Since there are three dimensions we have three, possibly different, values of the parameter Δ in Equation (3.107). These are called *kFirstQuant*, *kSecondQuant*, and *kThirdQuant*, as shown in Algorithm 3.11. Not surprisingly these are also the startup parameters when creating a new histogram object. However, to determine dimensions of a suitable data structure for the histogram bins in Equation (3.107) we need the value of x_{max}. In _3D_Histogram_For this is set as a template parameter _kEntryValueExtend. The other template parameter indicates the data type capable of storing the number of occurrences $n(i,j,k)$ of each data, analogously to n_i in Equation (3.109) for the 1D case. The other computations in the initializer of the _3D_Histogram_For constructor are to compute the ceiling function, as in Equation (3.107). After determining dimensions, a 3D array is allocated (in this case the TVideoFor plays its role well).

```
//////////////////////////////////////////////////////////////////////
template < typename _CounterType = unsigned long,
    int _kEntryValueExtend = 1 << 8 * sizeof( MonochromeImage::PixelType ) >
class _3D_Histogram_For
{
   public:

      enum { kEntryValueExtend = _kEntryValueExtend };

      typedef _CounterType CounterType;

      typedef TVideoFor< CounterType > HistDataStruct;


   protected:

      HistDataStruct  * fHistDataStruct;   // here we will store histogram values


      // Quantization values, separate for each dimension of the cube
      // The maximum dimension is divided into quant values
      const int     kFirstQuant;
      const int     kSecondQuant;
      const int     kThirdQuant;

      // These are actual dimensions of the histogram cube
      const int     kFirstDimension;
      const int     kSecondDimension;
      const int     kThirdDimension;

   public:

      //////////////////////////////////////////////////////////////
      // Class constructor
      //////////////////////////////////////////////////////////////
      //
      // INPUT:
      //           firstQuant
      //           secondQuant
      //           thirdQuant - quantization factors for the
      //                  three dimensions, respectively
      //
```

```
    // OUTPUT:
    //
    //
    // REMARKS:
    //
    //
    _3D_Histogram_For( int firstQuant = 1, int secondQuant = 1, int thirdQuant = 1 )
    :    kFirstQuant(firstQuant),kSecondQuant(secondQuant),kThirdQuant(thirdQuant),
            kFirstDimension    ( _kEntryValueExtend / firstQuant +
                        ( _kEntryValueExtend % firstQuant != 0 ? 1 : 0 ) ),
            kSecondDimension( _kEntryValueExtend / secondQuant +
                        ( _kEntryValueExtend % secondQuant  != 0 ? 1 : 0 ) ),
            kThirdDimension    ( _kEntryValueExtend / thirdQuant +
                        ( _kEntryValueExtend % thirdQuant   != 0 ? 1 : 0 ) )
    {
        REQUIRE( kFirstDimension > 1 );
        REQUIRE( kSecondDimension > 1 );
        REQUIRE( kThirdDimension > 1 );

        fHistDataStruct = new HistDataStruct( kFirstDimension, kSecondDimension,
                                    kThirdDimension, (_CounterType) 0 );
        REQUIRE( fHistDataStruct != 0 );
    }

public:

    //////////////////////////////////////////////////////////////
    // This function increments an entry at the position (f,s,t)
    //////////////////////////////////////////////////////////////
    //
    // INPUT:
    //                f, s, t - three data values
    //
    // OUTPUT:
    //                true if ok, false otherwise
    //
    // REMARKS:
    //                The entries must not exceed the _kEntryValueExtend
    //
    bool    IncrementHistogramEntry( int f, int s, int t )
    {
        REQUIRE( fHistDataStruct != 0 );

        int dim_1 = f / kFirstQuant;
        REQUIRE( dim_1 < kFirstDimension );
        int dim_2 = s / kSecondQuant;
        REQUIRE( dim_2 < kSecondDimension );
        int dim_3 = t / kThirdQuant;
        REQUIRE( dim_3 < kThirdDimension );

        ++ fHistDataStruct->GetRefPixel( dim_1, dim_2, dim_3 );
        return true;
    }
    //////////////////////////////////////////////////////////////
```

```
      // This function returns a value of a histogram at a position.
      //////////////////////////////////////////////////////////////////
      //
      // INPUT:
      //                    f, s, t - three values of a point in the
      //                        histogram
      //
      // OUTPUT:
      //                    histogram counter at position (f,s,t)
      //
      // REMARKS:
      //                    The entries must not exceed the _kEntryValueExtend
      //
      CounterType GetHistogramEntry( int f, int s, int t ) const
      {
            REQUIRE( fHistDataStruct != 0 );
            int dim_1 = f / kFirstQuant;
            REQUIRE( dim_1 < kFirstDimension );
            int dim_2 = s / kSecondQuant;
            REQUIRE( dim_2 < kSecondDimension );
            int dim_3 = t / kThirdQuant;
            REQUIRE( dim_3 < kThirdDimension );
            return fHistDataStruct->GetPixel( dim_1, dim_2, dim_3 );
      }
};
//////////////////////////////////////////////////////////////////////////////
```

Algorithm 3.11   Definition of the _3D_Histogram_For class implementing 3D histograms in dense representation (the most important members are shown).

The main two members reflecting semantics of the _3D_Histogram_For class are *Increment HistogramEntry* and *GetHistogramEntry*, respectively. The former is to increment, the latter to read histogram entries. Both accept three components of a feature which are converted (quantized) into corresponding bin indices in accordance with the 3D version of Equation (3.108), as shown in the code of Algorithm 3.11. It is left to a caller to ensure that the provided components are in the range of allowable values. For this purpose the *REQUIRE* fuses help to ensure this condition in the debug version of software.

A final remark concerns data representation. In the presented implementation we choose dense representation, which is practical especially if large widths of bins (*kFirstQuant*, ... , *kThirdQuant*) are chosen. However, this strategy becomes less useful for dimensions higher than three. In such cases sparse representations should be considered [84].

3.7.4   Parzen Method

Analyzing the histograms in Section 3.7.1 we noted that the probability that a random variable is in a certain range can be estimated in a nonparametric way by counting the number of occurrences of its values, as in Equation (3.109). However, the question was how to choose such range for a given random variable. In the case of histograms this boils down to choosing a certain width Δ_i of each of its bins. Thus, the concept of counting the number of occurrences needs to be analyzed in the context of a certain neighborhood of points. Starting from Equation (3.109) of a probability $P(n_i)$ corresponding to an i-th bin of a 1D continuous random variable, its probability density function (PDF) is obtained after dividing $p(n_i)$ by the range of the argument, which is a width Δ_i of the bin, i.e.

$$p(n_i) \approx \frac{P(n_i)}{\Delta_i} = \frac{n_i}{N\Delta_i},\qquad(3.125)$$

where i spans all bins of the histogram. Obviously the following holds

$$\sum_{i=1}^{\rho} p(n_i)\Delta_i = 1.\qquad(3.126)$$

More often than not Δ_i is the same for all bins. So, to estimate the PDF from a histogram a certain number of locally neighboring samples need to be gathered in a bin. As alluded to previously, all of them count for that single bin, so the width of the bin is also a smoothing parameter. The locality of the samples is indirectly measured by their Euclidean distance. Certainly, for other domains, such as a histogram of similar words, other distance measures should be considered (linguistic in this case). This simple idea of measuring PDF from the histograms is extended by other two methods: the *nearest neighbors* and the *Parzen window* (known also as the kernel estimator) [3, 40, 41].

Let us extend the idea expressed in Equation (3.125) to an L-dimensional space. In this case observations are drawn from the unknown probability density $p(\mathbf{x})$ which can be estimated as follows

$$p(\mathbf{x}) \approx \frac{n}{NV},\qquad(3.127)$$

where n denotes the number of samples that lie inside a sufficiently chosen volume V, whereas N is the total number of observations. From the above we have two ways to proceed. We can keep the required number of observations n constant and try to find a volume V, or we can do the opposite, i.e. fix V and look for the number of observations n that fall inside the fixed V. The former method is known as the k-nearest-neighbors (actually n-nearest-neighbors in our discussion). The latter is specific to the domain of kernel methods, as will be discussed in the next section.

Figure 3.15 visualizes the nonparametric classification problem for the two classes, denoted as "○" and "□". Each object is characteristic of a feature vector (such as color, texture, tensor components, etc.). Test object "**x**" becomes erroneously classified to the class "○" if only one nearest neighbor is considered (since it can be an outlier of "○"). However, if more nearest neighbors vote for "**x**" then it can be correctly classified to the class "□". Also, to assess the probability density around objects T_0 and T_1 in Figure 3.15, objects of the same type can be counted in the same size neighborhood V_0. It is evident that the "density" of points around T_0 is greater than for T_1. Alternatively to obtain the same number of objects around T_0 and T_1 the neighborhood V_0 has to be extended to V_1. As already mentioned, such an approach is characteristic to the k-nearest-neighbors method.

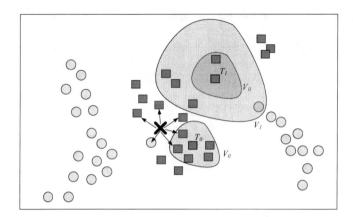

Figure 3.15 The nonparametric classification problem for two classes. Test object "**x**" erroneously classified to the class "O" if only one nearest neighbor is considered. If more nearest neighbors vote for "**x**" then it is correctly classified to the class "□". The density of points around objects T_0 and T_1 are assessed in the same size neighborhood V_0. To obtain the same density for T_0 and T_1 the neighborhood V_0 has to be extended to V_1.

3.7.4.1 Kernel Based Methods

Methods discussed in this section rely on the notion of *a kernel* which, in this context,[5] is a function with a highly localized response. In the simplest scalar case the kernel $K(x)$ can be seen as a function returning 1 if $|x| \leq 1$, or 0 otherwise. Thus, for vectors a value $K((\mathbf{x}-\mathbf{x}_i)/h)$ will be 1 if the data point $\mathbf{x}_i$ lies inside a hypercube of radius h_i, centered at $\mathbf{x}$. A distance between points has to be measured by the kernel function K. Thus, the total number of samples that lie in the hypercube around a point $\mathbf{x}$ is given as follows

$$n = \sum_{i=1}^{N} K\left(\frac{\mathbf{x} - \mathbf{x}_i}{h_i}\right), \qquad (3.128)$$

where K is a kernel function and N denotes the total number of points. It is interesting to observe that K takes as its argument a distance vector, in the form of a difference of two vectors, normalized by the parameter h_i. Thus, h_i controls an allowable range of distances among the points. Inserting now (3.128) into (3.127) yields

$$p(\mathbf{x}) = \frac{1}{N} \sum_{i=1}^{N} \frac{1}{h_i^L} K\left(\frac{\mathbf{x} - \mathbf{x}_i}{h_i}\right), \qquad (3.129)$$

in which $V = h_i^L$ is a volume of the hypercube in an L-dimensional space. In the general case h_i can be different for each point i and therefore in Equation (3.129) it was left under the summation sign. The above is called the Parzen method for nonparametric density estimation.

---

[5] For kernel based classification methods (such as SVM) a kernel function is defined differently, as discussed in Section 3.10.

More formally, kernel should be a function which satisfies the following conditions

$$K(\mathbf{x}) \geq 0, \text{ and } \int_{-\infty}^{+\infty} K(\mathbf{x})\,d\mathbf{x} = 1. \tag{3.130}$$

In the simplest approach the kernel K can be a square window. However, a common practice for K is to choose the Gaussian (3.79) which allows proper conditions for the boundary values, i.e. points which lie near the border of the hypercube have less influence. Notice, however, that such a choice of a kernel does not indicate that $p(\mathbf{x})$ follows a Gaussian nor another type of distribution. Assuming that h is constant Equation (3.129) reduces in this case to

$$p(\mathbf{x}) = \frac{1}{Nh\sqrt{2\pi}} \sum_{i=1}^{N} e^{-\frac{\|\mathbf{x}-\mathbf{x}_i\|^2}{2h^2}}, \tag{3.131}$$

where in $\|.\|$ the Euclidean distance between points is assumed. Comparing Equation (3.131) with (3.83) we easily see that the former is a mixture of Gaussians with weights $w_i = 1/N$.

The other popular kernels are the already presented Tukey biweight (2.70), as well as the Epanechnikov kernel

$$K_E(\mathbf{x}) = \begin{cases} \frac{L+2}{2V}\left(1 - \mathbf{x}^T\mathbf{x}\right), & if \quad \mathbf{x}^T\mathbf{x} < 1 \\ 0, & otherwise \end{cases}, \tag{3.132}$$

which is characteristic of a minimum mean integrated square error (MISE) [2].

However, the main problem of the kernel method is proper selection of the kernel size h in Equation (3.128). In Equation (3.131) the parameter h is equivalent to the variance of the Gaussian kernel or width Δ of a bin of a histogram (3.7.1). Generally h depends on data and therefore its choice demands some insight into data whose probability one wishes to determine. If h is too large, then a density estimated based on Equation (3.129) or (3.131) will be oversmoothed, in effect losing important information on the shape of data. On the other hand, if h is too small, then the estimations will be noisy. However, there are many methods of its approximation, such as experimental or based on some heuristics of the processed data [85]. For instance h is chosen to give optimal results on the training partition of data samples. If the training set can be split into the training and testing partitions, then the former set can be used to build a classifier with an *a priori* set parameter h, then the latter partition is used to test the classifier, and so on. The best value of h is that which gives the best classification results. However, in practice very rarely is there enough samples for such partitioning which would not negatively affect quality of classification. Therefore the other approach, called a *jackknife method*, can be used instead [2]. It also relies on data partitioning but this time the testing partition consists of only *one sample*. Then the process is repeated with another sample selected, and so on [2]. We will return to the analogous problem of choosing the control parameter in the context of the width of kernel functions in Section 3.10.1.

A neural realization of the C class classifier based on the density estimator (3.131) was proposed by Specht [86, 87] as the *Probabilistic Neural Network* (PNN), discussed in Section 3.9.1. A version of PNN was used for classification of traffic signs by means of their affine moment invariants – discussed in Section 5.3.1.

3.7.4.2   Nearest-Neighbor Method

Explicit choice of the kernel width h in Equation (3.129) can be somewhat alleviated in the *k-nearest-neighbor* method. However, the free parameter is k, i.e. the number of neighbors considered for classification of each new point. In this method each new test point can be classified as belonging to one of the C known classes. For a test point, such as "$\mathbf{x}$" in Figure 3.15, the neighborhood V is extended so as to comprise exactly k neighbors regardless of their class. Thus Equation (3.127) can be written to describe the conditional probability

$$p\left(\mathbf{x}|\omega_i\right) = \frac{n_i}{N_i V},\qquad(3.133)$$

where $n_i \leq k$ is the number of neighbors of "$\mathbf{x}$" which are of the class ω_i, whereas N_i is the total number of patterns of the class ω_i in the whole population. Thus, $\sum_i^C A_i = A$, i.e. the total number of patterns of all classes, and the probability of encountering the class ω_i among all others is simply

$$P\left(\omega_i\right) = \frac{N_i}{N}.\qquad(3.134)$$

Finally, the unconditional density $p(\mathbf{x})$ can be estimated from Equation (3.127) as follows

$$p\left(\mathbf{x}\right) = \frac{k}{NV}\qquad(3.135)$$

in which k counts all points in the neighborhood of $\mathbf{x}$ irrespective of their class. Inserting Equations (3.133)–(3.135) into the Bayes scheme (3.62) yields the posterior probability

$$P\left(\omega_i|\mathbf{x}\right) = \frac{p\left(\mathbf{x}|\omega_i\right) P\left(\omega_i\right)}{p\left(\mathbf{x}\right)} = \frac{n_i}{k}.\qquad(3.136)$$

Equation (3.136) denotes what is known as the *k-nearest-neighbor* (*k*-NN) method. Thus, a test object "$\mathbf{x}$" is assigned to such a class ω_i for which Equation (3.136) is maximized, i.e. for which n_i is *the largest*.

Nevertheless the simplest approach is to set $k = 1$, in which case the so called *one-nearest-neighbor* method is obtained. In this case a test object is assigned to the first found nearest prototype pattern [3]. Such a classification scheme is frequently used in computer vision especially if gathering a significant number of prototypes is not feasible. An example here is the classification of pictograms of the road signs done by the Hamming neural network, discussed in Section 3.9.3. In this experiment it was assumed that each road sign was represented only by one prototype coming from the formal specification of the road signs, i.e. the same way driving students learn the road signs from the driving handbook.

The k-NN is one of the cornerstone methods of pattern classification. However, regarding the implementation issues, a main burden is computation of the distances among the points. Thus, when considering k-NN we have to ensure a fast method of distance computation among the points. Also problematic can be the choice of a suitable number k of neighbors for a given classification task. The next problem is that all points from the population need to be stored since they are used over and over in computations of the distances. Nevertheless, there are many advanced computational techniques that facilitate computations of the k-NN. The book by Shakhnarovich *et al.* is a good source of further information on these techniques [83].

3.8   The Mean Shift Method

The mean shift is a nonparametric method of *tracing the mode of a distribution*, i.e. a location of the maximal probability density, as proposed by Fukunaga and Hostetler [47]. For this purpose the mean shift algorithm climbs the gradient of the probability density function [88]. The method has been of interest in computer vision mostly for the tasks of clustering and object tracking. In this respect Shimshoni *et al.* proposed the adaptive mean shift for data clustering in high dimensions [89]. In their approach the locality-sensitive hashing technique is used, which allows significant reduction of computations (as already discussed in Section 3.5). On the other hand, Comaniciu *et al.* propose the mean shift for real time tracking of color videos [90]. Kim *et al.* propose the mean shift method for tracking text in images, first detected with the SVM[6] classifier trained with text examples [91]. Then the mean-shift is used to find text chips. Processing is done in the scale space (three level image pyramids) to facilitate detection of fonts of different size. Another method that exploits the mean shift is the road signs tracking system proposed by Cyganek [92]. In this approach the method was reformulated to operate in the domain of fuzzy membership fields, rather than probability, which usually allows more convenient definition and operations on tracked features.

3.8.1   Introduction to the Mean Shift

The estimation of a PDF from a sample $\{\mathbf{x}_i\}$ of N points, in an L dimensional space $\mathcal{R}^L$, can be obtained with a nonparametric kernel based method, discussed in Section 3.7.4.1. Based on Equation (3.129), a PDF can be estimated by means of the kernel K with the L_2 distance between vectors, as follows

$$P_K(\mathbf{x}) = \frac{1}{N} \sum_{l=1}^{N} \frac{1}{h_l^L} k \left(\left\| \frac{\mathbf{x} - \mathbf{x}_l}{h_l} \right\|^2 \right), \qquad (3.137)$$

where k is *a profile of the kernel* K. It is a *scalar* function which fulfills the following condition

$$K(\mathbf{x}) = ck \left(\|\mathbf{x}\|^2 \right), \qquad (3.138)$$

---

[6] A similar method is discussed in Section 3.8.4.

for a constant c, which assures that K complies with the second assumption in Equation (3.130), i.e. that it integrates to 1. In the above h_l denotes *a variable* size of a kernel window (i.e. its bandwidth). In other words, h_l can be locally adapted to data.

In the mean shift method the goal is to *trace the gradient* of the PDF, that is

$$\nabla P(\mathbf{x}) = \frac{1}{N} \sum_{l=1}^{N} \frac{1}{h_l^L} \nabla k \left(\left\| \frac{\mathbf{x} - \mathbf{x}_l}{h_l} \right\|^2 \right). \tag{3.139}$$

With the assumed Euclidean distance L_2 between vectors, the above expands to

$$\nabla P(\mathbf{x}) = \frac{1}{N} \sum_{l=1}^{N} \frac{1}{h_l^L} \nabla k \left(\frac{(\mathbf{x} - \mathbf{x}_l)^T (\mathbf{x} - \mathbf{x}_l)}{h_l^2} \right) = \frac{1}{N} \sum_{l=1}^{N} \frac{2(\mathbf{x}_l - \mathbf{x})}{h_l^{L+2}} g \left(\frac{\|\mathbf{x} - \mathbf{x}_l\|^2}{h_l^2} \right), \tag{3.140}$$

where

$$g(x) = -k'(x), \tag{3.141}$$

denotes a derivative of the profile of the kernel K. For example, for the Epanechnikov kernel defined in Equation (3.132) its profile is given as follows

$$k_E(x) = \begin{cases} 1 - x, & if \quad x < 1 \\ 0, & otherwise \end{cases}, \tag{3.142}$$

whereas its negative derivative

$$g_E(x) = \begin{cases} 1, & if \quad x < 1 \\ 0, & otherwise \end{cases}. \tag{3.143}$$

The steepest ascent follows the direction toward a stationary point, i.e. the one for which $\nabla P(\mathbf{x})$ in Equation (3.140) tends toward 0, that is

$$\nabla P(\mathbf{x}) = \frac{1}{N} \sum_{l=1}^{N} \frac{2(\mathbf{x}_l - \mathbf{x})}{h_l^{L+2}} g \left(\frac{\|\mathbf{x} - \mathbf{x}_l\|^2}{h_l^2} \right) = 0,$$

$$\sum_{l=1}^{N} \frac{\mathbf{x}_l}{h_l^{L+2}} g \left(\frac{\|\mathbf{x} - \mathbf{x}_l\|^2}{h_l^2} \right) - \sum_{l=1}^{N} \frac{\mathbf{x}}{h_l^{L+2}} g \left(\frac{\|\mathbf{x} - \mathbf{x}_l\|^2}{h_l^2} \right) = 0,$$

which yields the vector $\mathbf{x}_m$ of an extreme value of $P(\mathbf{x})$, as follows

$$\mathbf{x}_m = \frac{\displaystyle\sum_{l=1}^{N} \frac{\mathbf{x}_l}{h_l^{L+2}} g\left(\frac{\|\mathbf{x} - \mathbf{x}_l\|^2}{h_l^2}\right)}{\displaystyle\sum_{l=1}^{N} \frac{1}{h_l^{L+2}} g\left(\frac{\|\mathbf{x} - \mathbf{x}_l\|^2}{h_l^2}\right)}. \tag{3.144}$$

From the above we can easily see that $\mathbf{x}_m$ is actually *a weighted mean vector* of the data samples with the weights being computed *from the kernel g*. Iteratively computed Equation (3.144) constitutes a basis for the mean shift algorithm, as will be shown later in this section.

Let us observe that Equation (3.140) can be rewritten in the following form

$$
\begin{aligned}
\nabla P(\mathbf{x}) &= \frac{2}{N} \left\{ \sum_{l=1}^{N} \frac{\mathbf{x}_l}{h_l^{L+2}} g\left(\frac{\|\mathbf{x} - \mathbf{x}_l\|^2}{h_l^2}\right) - \sum_{l=1}^{N} \frac{\mathbf{x}}{h_l^{L+2}} g\left(\frac{\|\mathbf{x} - \mathbf{x}_l\|^2}{h_l^2}\right) \right\} = \\
&= \frac{2}{N} \left\{ \sum_{l=1}^{N} \frac{\mathbf{x}_l}{h_l^{L+2}} g\left(\frac{\|\mathbf{x} - \mathbf{x}_l\|^2}{h_l^2}\right) - \mathbf{x} \sum_{l=1}^{N} \frac{1}{h_l^{L+2}} g\left(\frac{\|\mathbf{x} - \mathbf{x}_l\|^2}{h_l^2}\right) \right\} = \\
&= \frac{2}{N} \underbrace{\left[\sum_{l=1}^{N} \frac{1}{h_l^{L+2}} g\left(\frac{\|\mathbf{x} - \mathbf{x}_l\|^2}{h_l^2}\right) \right]}_{\hat{P}_G(\mathbf{x})} \underbrace{\left\{ \underbrace{\frac{\displaystyle\sum_{l=1}^{N} \frac{\mathbf{x}_l}{h_l^{L+2}} g\left(\frac{\|\mathbf{x} - \mathbf{x}_l\|^2}{h_l^2}\right)}{\displaystyle\sum_{l=1}^{N} \frac{1}{h_l^{L+2}} g\left(\frac{\|\mathbf{x} - \mathbf{x}_l\|^2}{h_l^2}\right)}}_{\mathbf{x}_m} - \mathbf{x} \right\}}_{\mathbf{m}}
\end{aligned}
\tag{3.145}
$$

Thus, the first factor $\hat{P}_G(\mathbf{x})$ is an estimator of the PDF, but this time with the kernel G, instead of K as it was in Equation (3.129). The second factor is a variable bandwidth mean shift vector [93], which considering Equation (3.144) can be written as

$$\mathbf{m} = \mathbf{x}_m - \mathbf{x}, \tag{3.146}$$

thus

$$\mathbf{m} = \frac{\displaystyle\sum_{l=1}^{N} \frac{\mathbf{x}_l}{h_l^{L+2}} g\left(\frac{\|\mathbf{x} - \mathbf{x}_l\|^2}{h_l^2}\right)}{\displaystyle\sum_{l=1}^{N} \frac{1}{h_l^{L+2}} g\left(\frac{\|\mathbf{x} - \mathbf{x}_l\|^2}{h_l^2}\right)} - \mathbf{x}. \tag{3.147}$$

The described variable bandwidth mean shift method reveals many interesting properties, such as very fast convergence, and robustness to outliers. The latter property is due to the very small weights assigned by a kernel to the outliers and in effect their influence is neutralized. Further properties of the variable bandwidth mean shift are discussed in the paper by Comaniciu *et al.* [93]. In the same paper the system for blob segmentation in digital images is presented.

It is based on a five dimensional feature space composed of the three color components, expressed in the orthogonal RGB color space, as well as two spatial coordinates. The mean shift procedure is then applied to this joint spatial–color space. This is somewhat similar to the EMD segmentation outlined in Equation (3.124). For the spatial coordinates the spherical kernel is used, whereas color components are processed with the product kernel. The bandwidths are selected separately for both spaces as well. The blobs are then extracted as groups of pixels with the same connected convergence point. As reported, the system is able to process up to eight frames of 320×240 pixels per second [93].

In practice, however, the method is simplified assuming a fixed bandwidth, i.e. constant value of h for all data points. Under this assumption Equation (3.147) reduces to the following

$$\mathbf{m} = \mathbf{x}_m - \mathbf{x} = \frac{\sum_{l=1}^{N} \mathbf{x}_l g\left(\frac{\|\mathbf{x} - \mathbf{x}_l\|^2}{h^2}\right)}{\sum_{l=1}^{N} g\left(\frac{\|\mathbf{x} - \mathbf{x}_l\|^2}{h^2}\right)} - \mathbf{x}. \tag{3.148}$$

It is interesting to note that $\mathbf{x}$ stands in two components of the difference at the right side of Equation (3.148). However, in the left component $\mathbf{x}$ is used as a reference point to which distances are measured and weighted by the kernel to all the points $\mathbf{x}_l$ in the whole population. If a square window kernel (3.143) is assumed, then $\mathbf{m}$ in Equation (3.148) becomes an average of difference vectors $\mathbf{x}_l$-$\mathbf{x}$ for all points $\mathbf{x}_l$ close enough to $\mathbf{x}$ (i.e. for which the kernel returns 1). All further $\mathbf{x}_l$ are just ignored. For other more complex kernels the differences are further modulated depending on their distance (and possibly other parameters).

Analogously to Equation (3.145) also Equation (3.148) can be rewritten to the following form

$$\nabla P(\mathbf{x}) = \frac{2}{Nh^2} P_G(\mathbf{x}) \mathbf{m}, \tag{3.149}$$

where $P_G(\mathbf{x})$ denotes a nonparametric PDF estimation with a kernel G which profile is given in Equation (3.141), i.e. it is a negative derivative of the profile of the (original) kernel K. This finally yields

$$\mathbf{m} = \frac{Nh^2}{2} \frac{\nabla P(\mathbf{x})}{P_G(\mathbf{x})}. \tag{3.150}$$

The above expression indicates that at location $\mathbf{x}$, the mean shift vector $\mathbf{m}$, which is computed with respect to the kernel G, as in Equation (3.148), is proportional to the estimate of the *density gradient* with the kernel K as in Equation (3.137), additionally normalized by $P_G(\mathbf{x})$. Thanks to this, $\mathbf{m}$ always points toward a direction of maximal increase of the PDF, i.e. to a region dense with data points. At the same time the normalization factor causes large mean shifts in sparse areas since $P_G(\mathbf{x})$ is small in such regions. However, for dense areas the mean shift gets finer, thus the method follows an adaptive gradient ascent. The iterative mean shift procedure for the searching mode of the PDF is shown in Algorithm 3.12.

```
1. Select a starting point x = x₀.

2. Do:

        3. For a point x compute the mean shift vector m from Equation (3.147) for a
        variable bandwidth, or from Equation (3.148) for a fixed bandwidth.

        4. Shift x by m:

                                        x= x + m

while (not convergent)
```

Algorithm 3.12   The iterative mean shift procedure for the searching mode of the probabilistic density function.

In effect, the procedure summarized in Algorithm 3.12 consists of iterative computations of $\mathbf{x}_m$ in Equation (3.144), at a point $\mathbf{x}$ which is $\mathbf{x}_m$ from the previous iteration, until the stop conditions are met. Thus, at first sight, the method can be seen as a version of the k-means clustering method, which is discussed in Section 3.11.1. However, it is actually more robust than k-means since in Equation (3.144) the weighted means are computed. These weights are dependent on the kernel G which follows a derivative of the kernel K. Moreover k-means requires an *a priori* assumption on the number of clusters, as well as being constrained to the spherically symmetric clusters.

Frequently the starting point $\mathbf{x}_0$ in Algorithm 3.12 is set as a one of the data points $\{\mathbf{x}_i\}$ [89]. It was proved that the mean shift procedure is convergent if and only if the kernel K has a convex and monotonically decreasing profile [49]. Convergence can be additionally assured by checking a distance between consecutive iteration steps t and $t + 1$,

$$\|\mathbf{x}_{t+1} - \mathbf{x}_t\| < \varepsilon, \tag{3.151}$$

where ε is a sufficiently small threshold value. As already mentioned, the maximal number of iterations should also be checked to guarantee the stop conditions [53]. Such a strategy follows the paradigms of defensive programming [94].

The problem with the mean shift procedure outlined in Algorithm 3.12 is selection of the bandwidth h_l in (3.147). A practical choice of a fixed value of h should be a trade-off between the bias and variance of the estimator P_K in Equation (3.137) in respect to the true PDF P_T, since it holds that

$$MSE(\mathbf{x}) = E\left[P_K(\mathbf{x}) - P_T(\mathbf{x})\right]^2 = Var\left[P_K(\mathbf{x})\right] + Bias^2\left[P_K(\mathbf{x})\right]^2. \tag{3.152}$$

Certainly we wish to have $MSE(\mathbf{x})$ as small as possible. As pointed out by Comaniciu *et al.* [93], the *Bias* in Equation (3.152) is proportional to h^2, i.e. the smaller the bandwidth, the less biased the estimator. On the other hand, the smaller the bandwidth, the larger the variance *Var* in Equation (3.152), which is proportional to $N^{-1}h^{-L}$. Thus h should be chosen to achieve an optimal compromise. As shown by Comaniciu *et al.*, h can be adapted either with respect to the estimation point $\mathbf{x}$ (a balloon estimator) or to the sample point $\mathbf{x}_l$. The latter has the desirable property of reducing the bias. In [93] two variable bandwidth mean shift methods are proposed. The first assumes a pilot estimate with a fixed bandwidth from which the proportionality constant is derived which is then used for estimation of h_i. The second method is based on an assumption that density in a local neighborhood of pixels is a spherical normal with unknown mean and covariance. The very important property of these methods is that when initialized at a given local position they converge to a nearest point in which the estimator (3.137), with the variable bandwidth h_i, has zero gradient. This local convergence point is just a mode of the PDF.

As pointed out in the paper [89], the pilot density estimate can be obtained from the nearest neighbors method, as follows

$$h_l = \left\| \mathbf{x}_l - \mathbf{x}_l^k \right\|,$$ (3.153)

where $\mathbf{x}_1^k$ is the k-th nearest neighbor of the point $\mathbf{x}_1$. The number of neighbors k should be large enough to ensure sufficient density within the support of majority of the kernels with bandwidths h_l. Usually k increase in accordance with dimensionality L of the data space.

For the mean shift procedure, the task of finding an object, based on a characteristic feature $\mathbf{v}$ of this object, is formulated as finding a discrete location $\mathbf{y}$ in the candidate image for which the associated density $s(\mathbf{y},\mathbf{v})$ is the most similar to the target density $d(\mathbf{v})$. However, for this purpose the described versions of the mean shift rely on *static* distributions of the target and candidates. On the other hand, online selection of the best discriminative features for the mean shift tracking was proposed by Collins *et al.* [95]. The features are chosen to best discriminate between the tracked object and its background. To measure this discriminative power of features the two class variance ratio is computed. Based on this factor the tracking module selects the best two features in each frame. Wang and Yagi propose to incorporate shape–texture features alongside color distributions for reliable tracking [96]. The method updates the target model based on the similarity between the initial and present models which allows adaptation to the changing appearance of the object of interest.

A method for real time tracking of nonrigid objects was developed Comaniciu *et al.* [90]. As reported, the system is able to process up to 30 frames of 352×240 pixels per second. They proposed to reformulate Equation (3.148) to express $\mathbf{x}_m$ as follows

$$\mathbf{x}_m = \frac{\displaystyle\sum_{l=1}^{N} \mathbf{x}_l w_l g \left(\frac{\|\mathbf{x} - \mathbf{x}_l\|^2}{h_l^2} \right)}{\displaystyle\sum_{l=1}^{N} w_l g \left(\frac{\|\mathbf{x} - \mathbf{x}_l\|^2}{h_l^2} \right)}.$$ (3.154)

where w_1 denotes a weight related to the Bhattacharyya distance[7] (3.113) between the model and the target object. In Equation (3.154) the product $w_l g(.)$ can be seen as a factorization of the profile g in Equation (3.148) into the feature and spatial components. The Epanechnikov profile was used in [90] for computation of the histogram and therefore the mean shift was computed with the uniform profile, as presented in Equation (3.143). The target histograms

---

[7] See Table 3.3, p. 240

were computed in the RGB space with $32 \times 32 \times 32$ bins per each channel, which can be implemented as discussed in Section 3.7.3. As reported by Comaniciu *et al.* the tracking system is also robust to partial occlusions, noise, and some distractors [90].

An interesting practical approach for object tracking which utilizes the mean shift approach was proposed by Petrović *et al.* [97]. It relies on application of the naïve Bayesian classifier, as defined in Equation (3.70), operating in a four dimensional feature space consisting of the RGB color channels and rotationally invariant uniform local binary pattern for texture detection.

3.8.2 Continuously Adaptive Mean Shift Method (CamShift)

A mean shift based method for face tracking in video was proposed by Bradski and called a continuously adaptive mean shift (*CamShift*) [98]. Instead of computing g in Equation (3.148) over and over for each data point $\mathbf{x}_l$, in *CamShift* the idea is to tabularize these computations, e.g. with help of the precomputed histogram (3.7.1). In other words, values of a model-target probability function W are assumed to be known for each point $\mathbf{x}_l$. At first a model-histogram of human skin colors was constructed since the method was used in the system for face detection. Then at each pixel position the function W was obtained by looking up a corresponding entry in the model histogram. The same idea can be concluded starting from Equation (3.148), as well as from its generalized version (3.154) with a constant kernel. In the latter case the weighting role is cast into the values of w_l which now can be a more general measure than the Bhattacharyya distance assumed in Equation (3.154). Based on this, the *CamShift* can be expressed as follows

$$\mathbf{m} = \underbrace{\frac{\sum_{l=1}^{N} \mathbf{x}_l W(\mathbf{x}_l)}{\sum_{l=1}^{N} W(\mathbf{x}_l)}}_{\mathbf{x}_m} - \mathbf{x}. \tag{3.155}$$

Thus, after a closer look, the vector $\mathbf{x}_m$ in Equation (3.155) can be interpreted as a centroid of the "mass" expressed by the probability field W, whatever events (features) it represents. Actually W can be any nonnegative signal, as will be shown later. Hence, Equation (3.155) can be rewritten in terms of the statistical moments discussed in Section 2.8

$$\mathbf{x}_m = \left[\frac{m_{10}}{m_{00}}, \frac{m_{01}}{m_{00}} \right]^T, \text{ assuming } m_{00} \neq 0. \tag{3.156}$$

The moments m_{10}, m_{01}, and m_{00} are defined in Equation (2.120).

As alluded to previously, $W(\mathbf{x}_1)$ defines the probability of agreement of a model and target (test) object at the point $\mathbf{x}_1$. In the aforementioned work by Bradski the probability comes directly from the back-projection of the target color histogram. In his work, both the 2D histogram of the normalized red–green components, as well as 1D histogram of the hue component, were tested. The latter was chosen as performing the best.

As shown in Section 2.8, an object can be approximated by a blob of dimensions (l, w) which are proportional to the square roots of the eigenvalues of the inertia tensor, as given in

Equation (2.126). A blob represents a collection of pixels that have coherent spatial, range, and time components. The relative size of an object in the image is expressed by scaling the eigenvalues with the total "mass", given by m_{00}. Moreover, for practical reasons we wish to find and track a rectangle representing the whole object whose dimensions[8] are the same as in Equation (2.125), that is

$$l = 2\sqrt{\frac{\lambda_1}{m_{00}}}, \qquad w = 2\sqrt{\frac{\lambda_2}{m_{00}}}, \tag{3.157}$$

It is interesting to observe that this way obtained rectangles have the same moment **T** as the object [99]. In practice when tracking, we wish to allow the searched windows to slightly grow, so we set

$$l' = \rho l, \qquad w' = \rho w, \tag{3.158}$$

where $1 < \rho < 2$ is a scaling parameter. With the above, the *CamShift* method for a single frame is summarized in Algorithm 3.13.

---

1. Set $\rho = 1.8$;
   Set up stop thresholds τ_x, τ_y, τ_w, and τ_l.
   Set the initial locations and dimensions of the regions of interest R_i.

2. **For each R_i do**:

   $t = 0$

   3. **Do**:

      $t = t + 1$

      Compute the probability distributions in a region R_i.

      Estimate its centroid x_m and dimensions (w',l') from (3.158).

      Align R_i with a new centroid x_m and set its dimensions found in the previous step.

      while $(t < t_{max})$ **and** $(|\Delta x_m| > \tau_x$ **or** $|\Delta y_m| > \tau_x$ **or** $|\Delta w'| > \tau_w$ **or** $|\Delta l'| > \tau_l)$

   4. Trim an exact position of R_i, setting $\rho = 1$ and recomputing (3.158).

Algorithm 3.13   The continuously adaptive mean shift algorithm for multiple regions of interest.

For each frame processed with the *CamShift* the final positions and sizes of the tracked regions R_i are saved. Each new frame is initialized with these data and the above algorithm is repeated. An example of initial localization of regions of interest for tracking was described in the papers by Cyganek [92, 100].

---

[8]Geometry of quadratic forms is discussed in Appendix A.4.

The robustness of the *CamShift* method comes from the fact that the probability values need to be evaluated only in the regions of interest R_i. In the worst case, and usually at the first step of the algorithm, the whole image has to be checked. Such a technique leads to savings in computations. The advantage of the *CamShift* when compared to the basic mean shift method is that the size of the search window is adjusted automatically to the "density" of the pixel distribution within. Thus, size of the final (stable) window tightly encompasses the tracked object, as will be shown in the next sections.

3.8.3 Algorithmic Aspects of the Mean Shift Tracking

The basic formulation of the mean shift and related methods were presented in previous sections. In the simplest approach the method works, though in practice its performance cannot be sufficient. This is caused by many factors, such as the low discriminative power of a simple definition of an object for tracking, as well as noise, geometric deformations, variations in lighting, occlusions, etc. Therefore the basic version of the algorithm can be extended to allow more precise tracking of the objects. Many features of the objects can be combined to improve their discriminative power, i.e. to increase the separation between an object and its background. Sometimes it is also easier to formulate the tracking task in terms of fuzzy logic. Properties of this approach are discussed in the next sections. Tracking can be also formulated in terms of the object/background classification, which in some cases improves the discriminating properties of the method. This is discussed in Section 3.8.3.4. Finally, since appearance of the tracked object, as well as its background, changes from frame to frame, the classifier has to evolve accordingly to adapt to these changes, as will be discussed in Section 3.8.3.5.

3.8.3.1   Tracking of Multiple Features

As already mentioned, an object to be tracked – sometimes called a target – is defined by its model. Frequently it comes in the form of single or multiple histograms of one or multiple dimensions, depending on the dimensionality of discriminative features. Similarly, the target candidate, i.e. a set of pixels which potentially correspond to the target model, is defined by its histograms as discussed in Section 3.7.1.

In the already discussed approach by Comaniciu *et al.* a modification to this scheme was proposed [90]. It consists of weighting entries of the histogram based on the distance of a pixel from the center of the object. Assuming that a set $\{x_l\}$ denotes zero-centered pixel locations of the target object, a function $\xi(x_l)$ can be defined which endows each location with the index of the histogram bin, which in turn corresponds to the feature of a pixel at x_l, as follows

$$\xi : \Re^2 \to \{1 \dots \rho\}. \tag{3.159}$$

In the above ρ denotes the total number of bins in the histogram. The probability of a feature of the target model is additionally weighted with a convex and monotonically decreasing kernel profile k. Thanks to this, pixels which are further from the center of the target add less into histogram, since usually they are cluttered by background or other objects. Thus, the value of a normalized bin of the target histogram h can be computed as proposed in [90]

$$h[i] = C \sum_{l=1}^{N} k\left(\|\mathbf{x}_l\|^2\right)\delta\left[\xi(\mathbf{x}_l) - i\right], \text{ for } 1 \leq i \leq \rho. \tag{3.160}$$

In the formula above δ denotes a Kronecker delta, N stands for a number of pixels in an object, and C is a normalizing constant

$$C = \frac{1}{\sum_{l=1}^{N} k\left(\|\mathbf{x}_l\|^2\right)}. \tag{3.161}$$

In a similar way the histogram of the candidate object is created, although its defining points are not zero-centered since the candidate location is shifted in an image. The two histograms are matched with the Bhattacharyya distance D_B, presented in Equation (3.113).

Another method is the projection of the candidate pixels into the target histograms, as used in the *CamShift* approach (Section 3.8.2). This is usually simpler in realization and faster in execution, as will be discussed in Section 4.6.

Theoretically, for L features (components) which were chosen to define an object for tracking, an L-dimensional histogram should be built. However, such a formulation becomes impractical usually for $L > 3$, even for a low number of bins. Therefore, an independence of features is frequently assumed which allows space reduction from L dimensional to L number of one-dimensional histograms. Nevertheless, it is not difficult to reveal flaws in such a simplification even in the case of color channels. In other words, high distribution of the red component and high distribution of the blue component do not necessarily mean high distribution of the pixels with large red *and* blue components simultaneously. Therefore this assumption, although working in many practical applications, should be chosen with care.

It is also important to indicate the influence of the histogram quantization. Obviously, the wider the bin, the higher the averaging of values of the components. In the two extremes, either each value of a feature falls into a separate bin, or all values can be assigned to one bin, respectively. Therefore there is no general rule for selection of this parameter and it is usually chosen empirically, as discussed in Section 3.7.1.

3.8.3.2 Tracking of Multiple Objects

Frequently there are many objects to be tracked at the same time. If the objects are substantially different, then this problem can be seen as number of independent tracking processes. However, usually this is not the case. For example, in the person tracking system we might be interested in providing the position of each person independently. This obviously complicates the process since objects can interact with each other, cross, occlude themselves, etc. This problem has been addressed in many publications [69, 70, 101, 102, 103, 104]. For instance a simple solution to the tracking of multiple objects of the same class – such as road signs of the same color – was proposed in the paper by Cyganek [92].

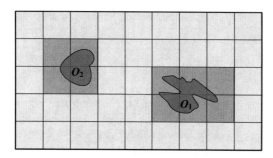

Figure 3.16 Tracking of multiple objects in a single frame.

In this approach a frame is initially divided into a number of tiles, as shown in Figure 3.16. Each tile is assigned a value which indicates whether that tile is active for tracking. Initially all tiles which have values of g in Equation (3.148) greater than 0, in at least one of its pixels, are set as active. Then the tiles are traversed and for each active tile the *CamShift* is launched. On its exit, all tiles which are occupied by a found object (i.e. covered by the returned rectangle) are set as inactive and are given a common tag identifying a found object. If no object is returned, the checked tiles are set as inactive with no object. This way all the tiles are examined. Connected clusters of tiles define occupation regions of the found objects, such as O_1 in Figure 3.16. In the next frame *CamShift* is started in each of the occupation regions. However, independently a process is started which updates the status of all tiles. For example, if a new object moves into the scene then the covered tiles are set as active for tracking, etc. Otherwise, if the objects move into collision then the first tracked object usually obliterates the panes of the second one which in effect disappears. However, this is governed by the parameters of the *CamShift* algorithm which tries to encompass a topologically coherent object (discussed in Section 4.5). Nevertheless, in practice what is problematic is the choice of the size of the tiles. In the road signs tracking system these were set to be 16×16 pixels for the frames with resolution 640×480 [92].

3.8.3.3 Fuzzy Approach to the *CamShift*

Direct computation of the multivariate PDF from point samples, e.g. with the Parzen window or the probabilistic neural network, can become memory and computationally expensive especially for color video of high resolution, whereas direct histogram projection can be impractical. For these reasons, instead of probabilities, application of the fuzzy measures to the *CamShift* in Equation (3.155) can be considered. The fuzzy measures show many desirable features, such as reduced memory and fast computation [105, 44, 106]. The membership functions are frequently derived from the point samples in the form of piecewise-linear functions. Moreover, the fuzzy membership functions are derived from the low pass filtered histograms with a suppression mechanism for values with insufficient magnitude, such as noise or outliers – see also Section 5.2.3.

In the fuzzy approach, for each point of an image an L-dimensional fuzzy measure $\mu_\xi = [\mu_1, \mu_2, \dots, \mu_L]$ is associated for a feature ξ. Then, in step 3 of Algorithm 3.13 a function

$g(\mu_\xi)$ is substituted for the probability in a region R_i, defined as follows:

$$g(\mu_\xi) = \prod_{i=1}^{L} \mu_i. \tag{3.162}$$

As alluded to previously, a feature attributed to a pixel in an image can be color, shape, etc., for which its membership function, either from empirical data, a model, etc. can be computed. Alternatively the following fuzzy function can be used

$$g(\mu_\xi) = \min_{1 \le i \le L} (\mu_i), \tag{3.163}$$

which is simpler in computations. Interestingly, it is not necessary to normalize μ_ξ since normalization is already embedded into Equation (3.155).

It is worth noticing that under this assumption x_m in Equation (3.148) becomes identical to the formula for the center-of-gravity used in the defuzzification stage of fuzzy systems [107]. Thus, the mean shift in terms of fuzzy logic can be seen as *a defuzzification process*. Properties of the *CamShift* method in the fuzzy domain were analyzed in the paper [92].

3.8.3.4   Discrimination with Background Information

To improve tracking, a distribution of the closest background of a tracked object can be taken into consideration. Such approaches were proposed for instance by Collins *et al.* [95], Wang and Yagi [96], and Petrović *et al.* [97]. The two-class Bayesian scheme was used in which class ω_0 denotes a target object, whereas ω_1 its background, as discussed in Section 3.4.5. The classification discriminating function g for a pixel $\mathbf{p}$ directly follows from Equation (3.78)

$$g(\mathbf{p}) = \sum_{k=1}^{L} \log \frac{\max\{P(x_k|\omega_0), \varepsilon\}}{\max\{P(x_k|\omega_1), \varepsilon\}}, \tag{3.164}$$

where x_k denotes one of the L features attributed to $\mathbf{p}$. The probabilities in the above can be estimated from the histograms, for instance with the approximations given in Equation (3.111). Finally, ε is a constant which prevents numerical instabilities when the denominator of the above is close to 0.

A region for collection of background information is built around a region encompassing an object for tracking. For practical reasons these are usually rectangular, as depicted in Figure 3.17. However, frequently the background pixels are more numerous than the ones belonging to the object. Therefore setting the right side of Equation (3.77) to 1 is not justified. On the other hand, it is not possible to consider the whole background due to its great variations, as well as the possible existence of other objects which resemble the one to be tracked. In practice, the same procedure is used to gather features in the area anchored at the point A_0 in Figure 3.17 as it is for all four rectangles defining the background. Each of these is anchored at the points A_1 to A_4, respectively. The question is left on sizes of each of the regions. Some researchers (e.g. Petrović *et al.* [97]) advocate selecting a larger region for the background than the one for the main object. However, the two groups should be more less the same, since

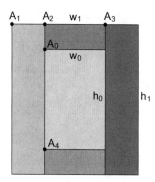

Figure 3.17 Regions defining an object (inner gray) for tracking and its neighboring background (the dashed region).

in the inequality (3.77) its right part denotes just the ratio of probability of occurrence of the two classes.

Thus, proper selection of the scale is necessary here. Therefore, selection of equal areas and the same aspect ratio can be recommended. In this case it is not difficult to show that the dimensions of the rectangles satisfy (see Figure 3.17)

$$w_1 = \sqrt{2}w_0, \text{ and } h_1 = \sqrt{2}h_0. \tag{3.165}$$

However, since an object moves and its distribution can also change, the appearance of the background can exhibit even stronger variations. Therefore the tracking model should be updated more frequently if background is to be taken into consideration.

3.8.3.5 Adaptive Update of the Classifiers

As alluded to previously, the tracked object and its background change from frame to frame. This is caused by movements of objects and changing viewpoints, but also by many other factors such as lighting conditions, noise, occlusions, etc. For these reasons definition of the tracked object should be updated throughout the time of tracking to adapt to changing conditions. However, an abrupt exchange at one frame base could cause false detections since a tracked region is usually approximated with a rectangle, which obliterates some differences between an object and its background. In the paper by Petrović *et al.* [97] a robust method of updating classifiers is presented. Assuming multiple features used for tracking in accordance with the scheme (3.164), one idea is to create a queue of classifiers in the initial number of frames and then update only *one* of the classifiers, i.e. the oldest one is replaced by a new one. By this token F classifiers, each created in each of the F frames, are used to judge the frame number $F + 1$, and so on. This can be written as a direct modification to Equation (3.164)

$$g_t(\mathbf{p}) = \sum_{f=1}^{F} g_{t-fT}(\mathbf{p}) = \sum_{f=1}^{F} \sum_{k=1}^{L} \log \frac{\max\{P_{t-fT}(x_k|\omega_0), \varepsilon\}}{\max\{P_{t-fT}(x_k|\omega_1), \varepsilon\}}, \tag{3.166}$$

where the subscript *t-fT* denotes the time at which the separate classifier was trained, *T* is the distance between training of the two consecutive classifiers. This is a kind of a voting mechanism[9] which makes the whole process more robust.

3.8.4   IMPLEMENTATION of the CamShift Method

Algorithm 3.14 presents a definition of the *TCamShift* class which implements the *CamShift* method of object tracking. As already discussed, its operation is based on computation of the statistical moments. For this purpose the member functions *ComputeMoments* were defined.

```
/////////////////////////////////////////////////////////////
// This class does CamShift tracking
/////////////////////////////////////////////////////////////
class TCamShift
{
   protected:


      // This image is filled with the probabilities assigned
      // to each pixel of an image
      TRealImage * fProbabilityField;

   public:

      enum {   k_m_00 = 0, k_m_10, k_m_01, k_m_02, k_m_20, k_m_11,
               k_max_value, kNumOfMoments };


      typedef double VectorOfMoments[ kNumOfMoments ];

      /////////////////////////////////////////////////////////////
      // This function computes the statistical moments
      // of order: m_00, m_10, m_01, m_02, m_20, m_11
      // used in CamShift.
      /////////////////////////////////////////////////////////////
      //
      // INPUT:
      //                window - a region in which the moments
      //                         will be computed
      //                theVector - the vector of double values
      //                         which upon return contains computed
      //                         moments
      // OUTPUT:
      //                none
      //
      // REMARKS:
      //
      //
```

---

[9]Such voting mechanisms are used in ensembles of classifiers, discussed in Section 5.6.

```
    void ComputeMoments( const RealImageProxy & window, VectorOfMoments &
    theVector );
    // The second version computes the moments directly
    // in the outer coordinate system (i.e. in the space of the input image)
    void ComputeMoments( const TRealImage & inImage, int anchor_x, int
    anchor_y, int pane_width, int pane_height, VectorOfMoments & theVector );

public:

    ////////////////////////////////////////////////////////////////
    // This function crops a subimage from a given one,
    // then computes its histogram (H-S).
    ////////////////////////////////////////////////////////////////
    //
    // INPUT: colorImage - reference to the input color image
    //        left_x, left_y - coords of the top left point of window
    //        widht, height - window width and height
    // OUTPUT:
    //        orphaned histogram object
    //
    // REMARKS:
    //
    //
    LongImage * OrphanHistogramFromWindow( const ColorImage & colorImage,
                    int left_x, int left_y, int width, int height );


    ////////////////////////////////////////////////////////////////
    // This function initializes the probability field
    // from the 2D H-S histogram.
    ////////////////////////////////////////////////////////////////
    //
    // INPUT:
    //        colorImage - reference to the input color image
    //            i.e. the one that needs to be tracked
    //        HS_Histogram - a 2D histogram of the H-S components
    //            of a tracked object.
    //
    // OUTPUT:
    //        true if ok
    //
    // REMARKS:
    //
    //
    bool InitProbabilityField(     const ColorImage & colorImage,
                                   const LongImage & HS_Histogram );

public:

    typedef vector< Region > RegionVector;

    RegionVector    fRegionVector;
```

```
/////////////////////////////////////////////////////////
// This function does CamShift based on the precomputed
// probability field.
/////////////////////////////////////////////////////////
//
// INPUT:
//         r - output parameter that contains found region
//         _anchor_x, _anchor_y, init_w, init_h - initial
//             region for search
//
// OUTPUT:
//      true if returned region is nonempty
//
// REMARKS:
//
//
virtual bool SingleArea_CamShift( Region & r, int _anchor_x,
int _anchor_y, double init_w, double init_h );


/////////////////////////////////////////////////////////
// This function does CamShift of multiple objects in
// the probability field (image).
/////////////////////////////////////////////////////////
//
// INPUT:
//       regionVector - an (empty) vector which upon return
//           contains found regions
//       min_allowed_side_of_found_region - regions which
//           min side is lower than this threshold do not
//           go into the region vector
//
// OUTPUT:
//      none
//
// REMARKS:
//    The probability field should be initialized before that function
//       is called
//
virtual void RegionCamShift( RegionVector & regionVector,
                          int min_allowed_side_of_found_region = 10 );


};
```

Algorithm 3.14 Definition of the *TCamShift* class implementing the *CamShift* method for multiple object tracking in images (the most important members are shown).

The first main operation is launched calling the *InitProbabilityField* function. Its purpose is to initialize the probability field, i.e. a matrix of real positive values in which the tracking takes place. Overriding this function in derived classes allows the implementation of different transformations from features to probability (or fuzzy membership) fields. The next pair of functions, *SingleArea_CamShift* and *RegionCamShift*, carry out *CamShift* tracking of single and multiple objects, respectively. Their details follow Algorithm 3.13. Code detail can be looked up in the attached implementation [6].

3.9 Neural Networks

The idea behind artificial neural networks (ANN) is to create computer models which in some sense mimic the behavior of biological neural networks existing in the brain. In this section we outline three types of classification system which belong to the group of artificial neural networks but which, apart from drawing some ideas from their biological counterpart, realize some of the classification methods discussed so far. These are, respectively, the probabilistic neural network which joins the Bayes classifier with the Parzen method, the Hamming neural network which realizes the one-nearest-neighbor classifier with the Hamming distance orchestrated by the winner-takes-all strategy, as well as the morphological neural network which harnesses mathematical morphology to pattern recognition problems. All of these have found application in different tasks of CV, as shall be discussed in subsequent sections of this book. Nevertheless, the fascinating area of NN is very broad and the interested reader is referred to the ample journal literature on the subject, and also to the books by Haykin [42], Kulkarni [108], as well as Kecman [44] to name a few.

3.9.1 Probabilistic Neural Network

The probabilistic neural network (PNN) follows the classification method based on the Bayes maximum *a posteriori* classification scheme (3.67) and the Parzen kernel PDF estimation (3.129). This network was originally proposed by Specht [86, 87] and is frequently used for the multiple-class classification problems method. Its architecture is composed of four layers as shown in Figure 3.18.

The input layer $\mathbf{X}$ receives input pattern vectors, each of dimension L, which should be normalized. Each vector can belong to one of the C classes. The next layer $\mathbf{W}$ contains the number of weights which store components of the reference patterns. The neurons W_{cn}, each belonging to only one of the ω_c classes ($1 \leq c \leq C$), compute a kernel function of its reference patterns (stored in the weights of $\mathbf{W}$) and the present input, in accordance with the following formula

$$W_{cn}(\mathbf{x}) = K\left(\frac{\mathbf{x} - \mathbf{x}_{cn}}{h_c}\right), \text{ for} 1 \leq c \leq C, \text{ and } 1 \leq l \leq N_c, \tag{3.167}$$

where $\mathbf{x}_{cn}$ denotes an n-th prototype pattern from an c-th class ω_c, h_c is a parameter that controls the effective width of the kernel for that class (i.e. its "zone of influence"), N_c denotes the number of available prototypes (i.e. given data vectors $\mathbf{x}_{cn}$) for that class.

Outputs of the neurons from the layer $\mathbf{W}$ are fed into the summation layer, which is composed of C neurons. Output of each follows Equation (3.129), i.e.

$$g_c(\mathbf{x}) = \alpha_c \sum_{n=1}^{N_c} K\left(\frac{\mathbf{x} - \mathbf{x}_{cn}}{h_c}\right), \tag{3.168}$$

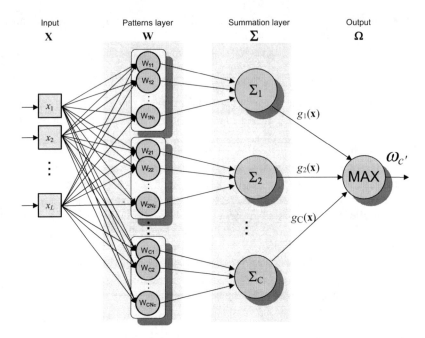

Figure 3.18 Architecture of the probabilistic neural network which consists of four layers: input, patterns, summation, and output layers.

where α_c is a scaling parameter. Usually the Gaussian kernel is selected for K which leads to the following function

$$g_c(\mathbf{x}) = \frac{1}{N_c} \sum_{n=1}^{N_c} e^{-\frac{\|\mathbf{x}-\mathbf{x}_{cn}\|^2}{2\sigma^2}}, \tag{3.169}$$

where the same parameter σ is substituted for h_c for all the classes, and the common multiplicative constant is omitted (in practical implementation factor 2 in the denominator can also be absorbed into the σ). Thus, one neuron of the summation layer with all its preceding neurons composes the kernel PDF estimation path, specific to a given class in accordance with Equation (3.129).

The last layer Ω simply selects one $g_p(\mathbf{x})$ which gives the maximal response in accordance with scheme (3.67). Its index j indicates the class to which the input pattern has been classified by the network

$$\omega_{c'} = \arg \max_{1 \le c \le C} \{g_c(\mathbf{x})\}. \tag{3.170}$$

Thus the whole network realizes the MAP classification rule (3.66) which follows the maximum membership strategy stated in Definition 3.3. This becomes clearly visible when comparing the scheme in Figure 3.18 with Figure 3.1.

The first step of PNN training consists of data normalization which is then used to initialize weights of the **W** layer [87]. However, the second part of the training process concerns determination of the kernel (spread) parameter σ which is not so obvious, as discussed in the previous sections. Many methods have been proposed for this purpose which, for instance, use educated guesses, heuristics [106], the Particle Swarm Optimization [109], or the Dynamic Decay Adjustment [110], to name a few. In the experiments with classification of the Polish road signs the two first methods were tested [111].

Montana proposes a modification to the classic PNN which is based on the utilization of the anisotropic Gaussian in Equation (3.169) and the Atkinson metrics for which the weights are found by the genetic optimization method [112]. This improvement, called the weighted PNN, is invariant to affine changes of data in the populations. However, training process can be time consuming in this case.

PNN still finds broad applications in computer vision problems. For instance Ge *et al.* use PNN in their method for hand gesture recognition and tracking [113]. The input images are first projected to low dimensional space, after which they are used to train the PNN. A new pattern is then classified by PNN, which corresponds to a gesture group to which the input gesture belongs to. A PNN based system for classification of the pictograms of the circular road signs is proposed in the paper [111]. Classification with PNN is based on affine moment invariants which allow moderate affine variations of the input images of pictograms. This technique is further discussed in Sction 5.3.1. Specifically, for each population the standard deviation is computed and their mean value is taken as a common σ. The other possibility here is the already mentioned jackknife method (Section 3.7.4.1) which operates in accordance with the following scheme

```
For each σ in an assumed range of values do:
    For each sample x in the training data set do:
        Exclude a data sample from the training set;
        Test if PNN correctly classifies the just removed sample;
```

Properties of the PNN are summarized in Table 3.5. Nevertheless, the disadvantages listed in Table 3.5 can be mitigated to some extent thanks to the parallel structure of the network. Thus, PNN can be easily implemented e.g. on a graphic card (GPU) achieving much higher throughput than in a serial implementation.

Table 3.5 Summary of the properties of the probabilistic neural network (PNN).

	Advantages	Disadvantages
1	Fast training	Slow response
2	Bayes optimal classifier	High memory requirements
3	Parallel structure	No simple method for σ selection

3.9.2   IMPLEMENTATION – Probabilistic Neural Network

Algorithm 3.15 presents a definition of the *TProbabilisticNeuralNetFor* class which implements the PNN. The class uses the feature implementation presented in Algorithm 3.1. The most important members of that class are *Train_Network* and *Run_Network*. The first does the simple training of PNN which consists of filling up the internal data structure with the normalized features. A little bit more complicated is the second function which operates in accordance with formulas (3.169) and (3.170). Its role is to determine the most probable pattern class for an input feature. Its implementation is shown in Algorithm 3.16.

```cpp
///////////////////////////////////////////////////////////
// This class implements the probabilistic neural network
///////////////////////////////////////////////////////////
template < class T, int FeatureDimension >
class TProbabilisticNeuralNetFor
{
   public:
      enum { kFeatureDimension = FeatureDimension };

      typedef std::auto_ptr< TProbabilisticNeuralNetFor< T, FeatureDimension > >
                                        ProbabilisticNeuralNetFor_AP;

   public:

      typedef typename FeatureSpaceFor< T >                FeatureSpaceFor_PNN;

      typedef typename FeatureSpaceFor_PNN::DataPoint       Feature;

      typedef typename FeatureSpaceFor_PNN::DataRepository  FeatureVector;

      typedef vector< FeatureVector >                       ClassVector;

      enum { kUnknownClass = -1 };

   public:

      int GetNumOfClasses( void ) const { return fClassVector.size(); }

      void ClearAll( void ) { fClassVector.clear(); }

      ///////////////////////////////////////////////////////////
      // This function normalizes a given feature vector
      // dividing each of its elements by the maximal component
      ///////////////////////////////////////////////////////////
      //
      // INPUT:
      //            theFeature - the feature vector to be normalized
      //
      // OUTPUT:
      //
      //
```

```
    // REMARKS:
    //
    //
    void NormalizeSingleFeature( Feature & theFeature );


public:


    ///////////////////////////////////////////////////////////
    // This function computes a SQUARED Euclidian distance
    // between two vectors
    ///////////////////////////////////////////////////////////
    //
    // INPUT:
    //          a, b - two feature vectors
    //
    // OUTPUT:
    //          SUM{ ( a[i] - b[i] )^2 }
    //          i
    //
    // REMARKS:
    //          Other distances can be used in PNN as well.
    //
    double ComputeDistance( const Feature & a, const Feature & b );


    ///////////////////////////////////////////////////////////
    // This function computes the Gaussian kernel for a given
    // distance which should be the sum of squared differences
    // between two vectors (SSD).
    ///////////////////////////////////////////////////////////
    //
    // INPUT:
    //          sum of squared differences between pairs of
    //                corresponding components of the two vectors
    //
    // OUTPUT:
    //          Gaussian Kernel value with sigma parameter
    //                already given in the constructor or set later
    //
    // REMARKS:
    //
    //
    double ComputeKernel( double dist );


public:


    // =======================================================
    // class default constructor
    TProbabilisticNeuralNetFor( double sigma = 1.0 );

    TProbabilisticNeuralNetFor( int numOfClasses, double sigma );
    // =======================================================
```

```
public:

    ////////////////////////////////////////////////////////////
    // This function adds (appends) a new class.
    ////////////////////////////////////////////////////////////
    //
    // INPUT:
    //          none
    //
    // OUTPUT:
    //          none
    //
    // REMARKS:
    //
    //
    void AddClass( void );


    ////////////////////////////////////////////////////////////
    // This function adds a new feature to the given class
    ////////////////////////////////////////////////////////////
    //
    // INPUT:
    //          theFeature - the feature to be added to
    //                  the population of the theClass class
    //          theClass - available klass number
    //
    // OUTPUT:
    //          true on ok
    //          false on error (e.g. wrong class number)
    //
    // REMARKS:
    //
    //
    bool AddFeature( const Feature & theFeature, int theClass );


    ////////////////////////////////////////////////////////////
    // This function trains the whole network. The training
    // consists of data normalization. If data is already
    // normalized then you don't need to call this before run.
    ////////////////////////////////////////////////////////////
    //
    // INPUT:
    //          none
    //
    // OUTPUT:
    //          total number of features in the network
    //
    // REMARKS:
    //
    //
    virtual int Train_Network( void );
```

```
/////////////////////////////////////////////////////////
// This function runs the network
/////////////////////////////////////////////////////////
//
// INPUT:
//          theInputFeature - the input feature to
//             be classified
//
// OUTPUT:
//          class number or
//          kUnknownClass if classification not possible
//
// REMARKS:
//
//
virtual int Run_Network( const Feature & theInputFeature );

};
```

Algorithm 3.15 Definition of the *TProbabilisticNeuralNetFor* class implementing the probabilistic neural network for pattern classification (the most important members are shown).

The first loop <14–36> in function *Run_Network*, shown in Algorithm 3.16, traverses all pattern classes. Then in lines <24–26> the sum from Equation (3.169) is computed which, after being normalized by the number of data for that pattern class in line <28>, is checked for the highest response. The best found value and class index are saved in lines <31–35>. Finally, the function returns the index of the pattern class with the highest probability.

```
1    template < class T, int FeatureDimension >
2    int TProbabilisticNeuralNetFor< T, FeatureDimension >::Run_Network(
3                                        const Feature & theInputFeature )
4    {
5       int theRetClass = kUnknownClass;
6
7       const int kNumOfClasses = fClassVector.size();
8
9       register int i, j;
10
11      double theBestResponseAsSoFar = -1.0;
12
13      // For each population (i.e. for all classes)
14      for( i = 0; i < kNumOfClasses; ++ i )
15      {
16          FeatureVector & theFeatureVector = fClassVector[ i ];
17
18          if( kNumOfFeatures == 0 )
19             continue;          // skip that empty class
20
21          double outputVal = 0.0;
```

```
22
23          // Access each feature from a single class
24          for( j = 0; j < kNumOfFeatures; ++ j )
25            outputVal += ComputeKernel(
26                ComputeDistance( theInputFeature, theFeatureVector[ j ] ) );
27
28          outputVal /= (double)kNumOfFeatures;
29
30          // Save the best response
31          if( outputVal > theBestResponseAsSoFar )
32          {
33              theBestResponseAsSoFar = outputVal;
34              theRetClass = i;
35          }
36      }
37
38      return theRetClass;
39  }
```

Algorithm 3.16 Implementation of the *Run_Network* function (member of the *TProbabilisticNeural-NetFor* class) which finds the most probable pattern class for a given input feature.

The auxiliary functions *ComputeDistance* and *ComputeKernel* are actually used to compute the exponent of a squared distance of two vectors.

3.9.3 Hamming Neural Network

The Hamming neural network (HNN), wose structure is depicted in Figure 3.19, allows the classification of patterns whose features can be measured with the Hamming distance. This

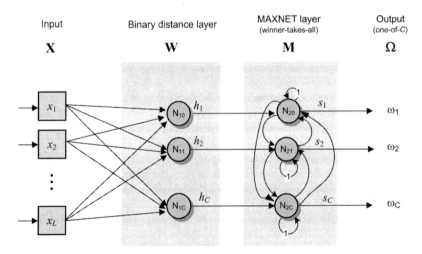

Figure 3.19 Structure of the Hamming neural network for classification of binary patterns.

network was originally proposed by Lippman [114], and directly realizes the one-nearest-neighbor classification rule discussed in Section 3.7.4.2.

The auto-associative version of HNN contains four, and the hetero-associative five, layers of neurons. It belongs to the group of recursive NNs [42], similarly to the Hopfield network. However it has a much larger pattern capacity than the latter. It is also much faster during training and recognition stages [115]. Thanks to this HNN finds many applications in pattern recognition. However, since the network accepts binary patterns, the input patterns should be coded as binary values. Therefore when processing images, for example, it is preferable not to use intensity values directly and instead use some transformed signals. For this purpose the nonparametric Census or Rank transformations are good candidates, as shown in many experiments [53, 116, 117, 118]. An example of application of HNN is the system for road sign recognition that will be discussed in Section 5.7. A number of HNNs is used to built an ensemble of expert classifiers which are able to select a test pattern in a group of deformable prototypes.

HNN classifies a test pattern, represented by a feature vector $\mathbf{x}$, into one of the C classes. These are encoded into the network by means of one prototype pattern per class. Therefore, HNN follows the one-nearest-neighbor classification strategy (Section 3.7.4.2), i.e. in the recognition stage a single nearest vector to the input pattern $\mathbf{x}$ is found (in the sense of the Hamming distance) and its class c_i is returned. The layers of the HNN can be characterized as follows:

1. The input layer $\mathbf{X}$ accepts binary vectors, i.e. with components from the set $\{-1, +1\}$. Modification to this can include the new "neutral" state with value 0 which can be used to indicate the "unknown" state [115].
2. The binary distance layer computes the Hamming distance between the input pattern and each prototype pattern stored during the training stage in the weights of the matrix $\mathbf{W}$.
3. The MAXNET layer selects the winning neuron in the recursive winner-takes-all process. Operation of this layer is governed by the trained matrix $\mathbf{M}$ containing weights among neurons of this layer. A characteristic feature is that all these weights are negative except for the self-connection of a neuron to itself which contains the weight with value 1 (this is not present e.g. in the Hopfield network).
4. Output layer Ω, after reaching a stable state, contains at most one neuron – a winner – whose output is different from 0. It indicates the class c_i that the input vector $\mathbf{x}$ has been classified to.
5. The decoder layer (optional, not shown in Figure 3.19), presents only in the hetero-associative version of HNN. It is responsible for coding an output class into an output pattern (a vector of dimension M).

The training stage of HNN is fast and simple, especially when compared with other neural networks [42]. It consists of setting up rows of the matrix $\mathbf{W}$ with the prototype patterns

$$\mathbf{w}_i = \mathbf{x}_i, \tag{3.171}$$

where $1 \leq i \leq C$ is a number of a prototype pattern $\mathbf{x}_i$, each of length L, $\mathbf{w}_i$ is the i-th row of the matrix $\mathbf{W}$, which is of dimensions $C \times L$. The computation time is linear with the size and number of input patterns C.

The recursive layer MAXNET is responsible for selection of a winning neuron which indicates a class of the input pattern. Neurons in the MAXNET layer are characterized by a self-connection, i.e. the weight $m_{ii} = 1$ for all $1 \leq i \leq C$. All other weights are kept negative. In effect a neuron amplifies its own response whereas all other neurons try to extinguish its excitation by an amount which is proportional to the connecting weight and values (strength) of other neurons in that layer. Initialization of the MAXNET layer consists of assigning negative values to the square matrix $\mathbf{M}$ (of dimensions $C \times C$) except of the main diagonal (which is 1.0 since it corresponds to the self-excitation of a neuron). In the original formulation to initialize $\mathbf{M}$ Lippman proposes using the following values [114]

$$m_{kl} = \begin{cases} -\dfrac{1}{C-1} + \xi_{kl}, & for \quad k \neq l \\ 1, & for \quad k = l \end{cases} \tag{3.172}$$

where $1 \leq k, l \leq C, C > 1$, ξ is a sufficiently small random value for which $|\xi| \ll 1/(C-1)$. A modification to the above was proposed by Floréen [115], which consists of assigning the same value ε_k to all m_{kl} for $k \neq l$

$$m_{kl} = m_k = \begin{cases} -\dfrac{1}{C-1}\left(1 - \dfrac{1}{L}\right)^{\frac{k-1}{C-1}}, & for \quad k \neq l \\ 1, & for \quad k = l. \end{cases} \tag{3.173}$$

Computed this way values of $\mathbf{M}$ are near-optimal in terms of the network convergence [115]. The computational complexity of this scheme is of order $O(C \log(LC))$ in the general case, and $O(C \log(L))$ if there exists a unique prototype vector which is the nearest to the input pattern. The other advantage of using Equation (3.173) is significant reduction in memory since the square matrix $\mathbf{M}$ reduces to a vector of length C. However, it appears that the most efficient and still convergent solution consists of setting equal weights for all neurons in the MAXNET layer, as follows

$$m_{kl} = m_k = \begin{cases} -\dfrac{1}{C-t}, & for \quad k \neq l \\ 1, & for \quad k = l, \end{cases} \tag{3.174}$$

where t is a time step in the *classification* stage (see formula (3.178)). However, values of m_k in (3.174) need to be modified at each step during the classification stage. In this case the convergence is achieved in $p-1-r$ steps, where $r > 1$ stands for a number of nearest prototypes stored in $\mathbf{W}$. Choosing different training models (3.172)–(3.174) influences the speed of convergence of the classification stage, depending on the type of stored prototypes. Nevertheless, for all of them convergence was proved [115].

In the classification (run-time) stage neurons of the distance layer compute the binary Hamming distance between the input pattern $\mathbf{x}$ and the prototype patterns already stored in $\mathbf{W}$, as follows

$$h_i(\mathbf{x}, \mathbf{W}) = 1 - \frac{1}{L} D_H(\mathbf{x}, \mathbf{w}_i), \tag{3.175}$$

where $1 \leq i \leq C$, $b_i \in [0,1]$ is a value of an i-th neuron in this layer, $D_H(\mathbf{x},\mathbf{w}_i) \in \{0,1,\dots,L\}$ is a Hamming distance (i.e. the amount of mismatched positions of the two vectors) between the input patterns $\mathbf{x}$ and the i-th stored prototype pattern $\mathbf{w}_i$ [116, 117]. Usually all feature vectors are assumed to have coefficients from the set $\{-1,+1\}$, then (3.175) reduces to

$$h_i (\mathbf{x}, \mathbf{W}) = \frac{1}{2} \left(\frac{\mathbf{w}_i \mathbf{x}}{L} + 1 \right) = \frac{1}{2} \left(\frac{1}{L} \sum_{r=1}^{L} w_{ir} x_r + 1 \right). \tag{3.176}$$

For the input vectors with values from the set $\{0,1\}$, Equation (3.175) can be written as follows

$$h_i (\mathbf{x}, \mathbf{W}) = 1 - \frac{\mathbf{w}_i \mathbf{x}}{L} = 1 - \frac{1}{L} \sum_{r=1}^{L} w_{ir} x_r. \tag{3.177}$$

As alluded to previously, Floréen proposed extending the allowable values from the binary set $\{-1,+1\}$ to the ternary $\{-1,0,+1\}$ where value 0 indicates an additional state whose role is to express uncertain conditions (i.e. "*don't know*" state).

During classification the MAXNET layer performs recursive computations to select a winner neuron in accordance with the following scheme

$$s_i [t + 1] = \theta \left(\sum_{j=1}^{L} m_{ij} s_j [t] \right) = \theta \left(s_i [t] + \sum_{j=1,i \neq j}^{L} m_{ij} s_j [t] \right), \tag{3.178}$$

where $s_i[t]$ is an output of the i-th neuron in MAXNET at the iteration step t, while θ denotes a function which suppresses all negative values to 0, as follows

$$\theta (x) = \begin{cases} x, & x > 0 \\ 0, & x \leq 0. \end{cases} \tag{3.179}$$

This means that if the output of a neuron from this layer goes beyond 0 its signal s_i is then also set to 0. In consequence such a neuron has no further influence on the process of selecting a winner (it is simply off). The goal of the iterative process (3.178) is to proceed until only one neuron has a value different than 0. This neuron is the winner of the process and its index indicates the determined class of the input pattern. Table 3.6 lists features of the HNN.

It is also worth noticing that the just described process of selecting a winner in the MAXNET layer can be used in other types of networks.

Table 3.6 Summary of the properties of the Hamming Neural Network (HNN).

	Advantages	Disadvantages
1	Fast training	Binary input
2	Fast response	Only binary distance (Hamming)
3	High capacity	Iterative response

3.9.4   IMPLEMENTATION of the Hamming Neural Network

Algorithm 3.17 presents a definition of the *TExHammingNeuralNet* class. It implements the
base version of the Hamming neural network, in accordance with the method described in
the previous section. The main data structure is the binary matrix **W** to store the prototype
patterns, as outlined in Equation (3.171). The binary structure in the presented code is just a
binary image of class *BinaryImage* which is optimized to store and access single bits [53]. It
is copied from the input provided to the constructor *TExHammingNeuralNet* of this class. The
second data structure, *fHammingComputer*, is a real valued vector which contains Hamming
distance $h_1, \ldots, h_C$, as shown in Figure 3.19. Since the input values are 0/1, the Hamming is
computed in accordance with Equation (3.177).

```
//////////////////////////////////////////////////////////////////
// This class implements the Hamming neural network.
//
//
// The running algorithm has been chosen according to:
// Floreen P.: The Convergence of Hamming memory networks.
// IEEE Trans. Neural Networks, 1991, Vol.2, s. 449-457
//
// The recent modification assume the BinaryImage
// stores the input patterns, as well as that the input test
// vector is in the form of a BinaryImage. This saves
// memory significantly.
// Number of columns in the image corresponds to the
// length of a single pattern. Number of rows is the
// same as the number of input patterns.
//
//////////////////////////////////////////////////////////////////
class TExHammingNeuralNet
{
   public:

      enum { kWrongPatIndex = -1 };

      typedef vector< double >    DoubleVector;

   protected:

      // Class inherent variables
      // EACH ROW CONTAINS A SINGLE PATTERN !!!
      BinaryImage * fInputPatterns; // W1 - effectively stores input vectors

      DoubleVector * fHammingComputer;   // output of the first neuron layer -
                                         // computes the Hamming distance

   protected:

      // This activation function implements the WTA algorithm.
      // The thresh value plays an important role in a dynamics of the net.
      // Usually it is set to 0.
```

```
double MAXNET_Activation_Fun( double val, double thresh = 0.0 )
{ return val >= thresh ? val : 0.0; }

public:


    // ==========================================================
    //////////////////////////////////////////////////////////////
    // Class constructor
    //////////////////////////////////////////////////////////////
    //
    // INPUT:
    //
    //              orphaned_patterns - orphaned binary image
    //                 containing the reference patterns in
    //                 the form of image rows. The number of
    //                 patterns corresponds to the number of
    //                 columns.
    //              Must not be 0.
    //
    // OUTPUT:
    //              none
    //
    // REMARKS:
    //              The input image is orphaned by a caller
    //                 on behalf of this class.
    //
    TExHammingNeuralNet( BinaryImage * orphanedInPatterns );
    // ==========================================================


    //////////////////////////////////////////////////////////////
    // This function runs the network to find the winner
    //////////////////////////////////////////////////////////////
    //
    // INPUT:
    //              singleInputPattern - a single row of a binary
    //                 pattern (size must match the already stored
    //                 reference patterns)
    //              theScore - optional parameter that receives
    //                 score achieved by the winning neuron
    //
    // OUTPUT:
    //          number of the winner (i.e. the class)
    //
    // REMARKS:
    //
    //
    virtual int Run( const BinaryImage & singleInputPattern, double *
        theScore = 0 );
};
```

Algorithm 3.17 Definition of the *TExHammingNeuralNet* class implementing the basic version of the Hamming neural network (the most characteristic members are shown). Code available from [6].

In the response time this version of HNN follows the iterative process (3.178) with the weights computed in accordance with Equation (3.174). This means that the weights of the MAXNET layer do not need to be stored since they are computed online. The protected member function *MAXNET_Activation_Fun* is a threshold function given in Equation (3.179). The described iterative response process is initiated calling the virtual function *Run* from the *TExHammingNeuralNet* class. Its only input parameter is the just classified pattern **x**. On output the function returns its class index *c* to which **x** was classified, as well as optionally a value of the winning neuron which can indicate strength of the response (however, if this is only slightly higher than 0.0, then the classification process cannot be seen as reliable). The simple code of the *Run* function can be looked up in the attached software [6].

```
class TRandomWeights_ExHammingNeuralNet : public TExHammingNeuralNet
{
   protected:

      int    fNumOfIterations;      // number of iterations for the first-type-net

      TRealImage * fMAXNETWeights;    // This is the matrix of the MAXNET layer M
      // It will be intialized with some random numbers.
      // However, it is square matrix of size: inPatterns x inPatterns :(

   public:

      // ============================================================
      //////////////////////////////////////////////////////////////
      // Class constructor
      //////////////////////////////////////////////////////////////
      //
      // INPUT:
      //            orphaned_patterns - orphaned binary image
      //               containing the reference patterns in
      //               a form of image rows. The number of
      //               patterns corresponds to the number of
      //               columns.
      //               Must not be 0.
      //            maxNumOfIter - maximal number of allowed
      //               iteration for the MAXNET layer
      //
      // OUTPUT:
      //            none
      //
      // REMARKS:
      //            The input image is orphaned by a caller
      //               on behalf of this class.
      //            This version is slower and requires more memory
      //               than the base class. However,
      //               it has a little different dynamics.
      //
```

```
        TRandomWeights_ExHammingNeuralNet( BinaryImage * orphaned_inPatterns,
                                           int maxNumOfIter = 100 );
    // =========================================================

protected:


    ////////////////////////////////////////////////////////////
    // This function initializes the MAXNET square matrix
    ////////////////////////////////////////////////////////////
    //
    // INPUT:
    //          kInputPatterns - new number of class patterns
    //             the size of the MAXNET matrix is:
    //             kInputPatterns x kInputPatterns
    //
    // OUTPUT:
    //          none
    //
    // REMARKS:
    //          Old class settings are disposed.
    //
    void InitializeMAXNETWeights( int kInputPatterns );

public:


    ////////////////////////////////////////////////////////////
    // This function runs the network to find the winner
    ////////////////////////////////////////////////////////////
    //
    // INPUT:
    //          singleInputPattern - a single row of a binary
    //             pattern (size must match the already stored
    //             reference patterns)
    //          theScore - optional parameter that receives
    //             score achieved by the winning neuron
    //
    // OUTPUT:
    //          number of the winner (i.e. the class)
    //
    // REMARKS:
    //
    //
    virtual int Run( const BinaryImage & singleInputPattern, double *
        theScore = 0 );
};
```

Algorithm 3.18 Definition of the *TRandomWeights_ExHammingNeuralNet* class implementing a specialized version of the Hamming neural network with randomly initialized weights.

Algorithm 3.18 presents the definition of the *TRandomWeights_ExHammingNeuralNet* class, derived from the *TExHammingNeuralNet*. Its main difference compared to the base class is a different method for computation of the weights **M** of the MAXNET layer. In this case, these are computed by the protected member *InitializeMAXNETWeights* in accordance with Equation (3.172). Its parameter *kInputPatterns* stands for *C* in Equation (3.172). In many applications this method results in faster convergence (i.e. selection of the winner) during the response of the network, which is due to the stochastic factor ξ_{kl} present in the weights. However, the price to pay is the memory necessary to store the *fMAXNETWeights* square matrix of dimensions $C \times C$ containing the weights.

Objects of the *TRandomWeights_ExHammingNeuralNet* class were used in the version of the system for road sign recognition for classification of the binary pictograms, as will be discussed in Section 5.7.

3.9.5   Morphological Neural Network

Morphological neural networks (MNN) constitute a very interesting and sometimes even surprising group of neural networks. They exhibit many desirable properties such as high pattern capacity, resistance to the erosive and dilative type of noise, as well as the fact that the response of MNN is obtained in just one step. All these features make them very attractive to the pattern classification community, especially in the context of real-time systems.

MNN were originally proposed by Ritter *et al.* as a response to the requirements of very fast neural solutions [119], and since then have found broad interest among groups of researchers. MNN are based on mathematical lattice algebra, which properties will be discussed. From their beginning MNN were used to build associative pattern memories resistant to morphological distortions (such as e.g. salt-and-pepper noise, etc.). In this respect the works by Ritter *et al.* [120], Ritter and Iancu [121], Raducanu *et al.* [122], as well as Sussner [123] are recommended reading. An extension from the binary to gray scale memories was proposed by Sussner and Valle [124]. A different application was proposed by Villaverde *et al.* [125] who proposed using MNN for simultaneous localization and mapping (SLAM), which is a key problem in robotics [67]. They solve the indoor nonmetric SLAM problem with the help of visual information obtained from the morphologically independent images. These, in turn, correspond to the vertices of the convex hull encompassing data points in a high dimensional space. MNN were also applied to classification of binary pictograms in the road sign recognition system proposed by Cyganek [100].

Let us now present the basic concepts of MNN which reside on mathematical lattice theory. Figure 3.20 depicts a model of a *morphological neuron* which operates in accordance with the following equation [119]

$$y_k = \theta \left(p_k \bigvee_{i=1}^{L} r_{ik}(x_i + w_{ik}) \right), \tag{3.180}$$

where r_{ik} is a pre-synaptic response which transfers excitatory ($r_{ik} = +1$), or inhibitory ($r_{ik} = -1$), incitation of an *i-th* neuron, p_k is the post-synaptic response of a *k-th* neuron to the total input signal, $\vee$ denotes a *max* product, and finally θ is a saturation function defined in

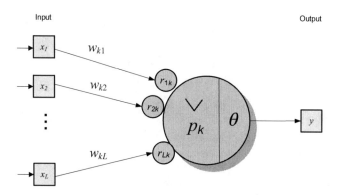

Input

Output

Figure 3.20 Structure of a morphological neuron.

Equation (3.179). More often than not values of r_{ik} and p_k are positive. The *max* product $\vee$ for two matrices $\mathbf{A}_{pq}$ and $\mathbf{B}_{qr}$ is a matrix $\mathbf{C}_{pr}$, with elements c_{ij} is defined as follows [119]

$$c_{ij} = \bigvee_{k=1}^{q} \left(a_{ik} + b_{kj} \right).$$ (3.181)

In an analogous way the *min* operator $\wedge$ is defined

$$c_{ij} = \bigwedge_{k=1}^{q} \left(a_{ik} + b_{kj} \right).$$ (3.182)

Thus, comparing the well known model of the perceptron[10] with the morphological neuron we can easily see the exchange of multiplication of inputs x_i with the synaptic weights into their summation, as well as summation of these products into their max value.

Figure 3.21 depicts the structure of the MNN used for classification of binary pictograms which was used in the system for road sign recognition [100]. It is a version of morphological associative memory in which a set of N input/output pairs is given as $(\mathbf{x}_1, \mathbf{y}_1), \ldots, (\mathbf{x}_N, \mathbf{y}_N)$. The pattern $\mathbf{x}$ is a linear version of an image of a sign and $\mathbf{y}$ is binary version of a pattern's class. Thus, the input layer $\mathbf{X}$ contains a binarized pictogram of a tracked sign. In experiments described in [100] its dimensions were warped to 256×256 (i.e. 8 kB).

Binarization and sampling was carried out in accordance with the method described in [126]. Vectors $\mathbf{y}$ decode (in the *one-of-N* code) classes of pictograms. From pairs $(\mathbf{x}_i, \mathbf{y}_i)$ the matrices $\mathbf{X} = (\mathbf{x}_1, \ldots, \mathbf{x}_N)$ and $\mathbf{Y} = (\mathbf{y}_1, \ldots, \mathbf{y}_N)$ are created. In the simplest formulation with one layer, a matrix $\mathbf{W}$ or $\mathbf{M}$ of weights is determined from $\mathbf{X}$ and $\mathbf{Y}$ as follows [120]

$$\mathbf{W}_{\mathbf{XY}} = \bigwedge_{i=1}^{N} \left(\mathbf{y}_i \times (-\mathbf{x}_i)^T \right),$$ (3.183)

$$\mathbf{M}_{\mathbf{XY}} = \bigvee_{i=1}^{N} \left(\mathbf{y}_i \times (-\mathbf{x}_i)^T \right),$$ (3.184)

---

[10]The perceptron computes: $y = b + \sum_{i=1}^{L} x_i w_i$, where x_i is the input signal at the i-th synapse, w_i is a synaptic weight, and b denotes a bias [42].

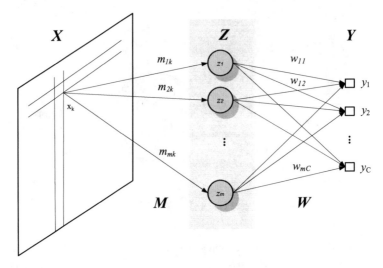

Figure 3.21 Architecture of the morphological neural network set for image processing (based on [100]).

where $\times$ denotes the *morphological outer product* of vectors

$$\mathbf{y} \times \mathbf{x}^T = \begin{bmatrix} y_1 + x_1 & \cdots & y_1 + x_n \\ \vdots & \ddots & \vdots \\ y_m + x_1 & \cdots & y_m + x_n \end{bmatrix}, \tag{3.185}$$

which is analogous to the outer product of vectors, in which addition is substituted for multiplication, however. Let us also notice that for any real number a its *additive conjugate* is given as follows

$$a^* = -a. \tag{3.186}$$

Thus, for all $a,b \in \mathfrak{R}$ the following holds

$$a \wedge b = \left(a^* \vee b^*\right)^*. \tag{3.187}$$

On the other hand, T in Equation (3.185) denotes a "simple" transposition of a vector.
   With $\mathbf{W}$ and $\mathbf{M}$ defined in Equations (3.183)–(3.184) the following holds

$$\mathbf{W_{XY}} \vee \mathbf{x}_i = \mathbf{y}_i, \tag{3.188}$$

$$\mathbf{M_{XY}} \wedge \mathbf{x}_i = \mathbf{y}_i, \tag{3.189}$$

even for *erosively or dilatively distorted* versions $\tilde{\mathbf{x}}_i$ of $\mathbf{x}_i$, for $\mathbf{W}$ and $\mathbf{M}$ respectively. Proofs of these statements can be found in [120, 122, 123]. In other words, a perfect recall is guaranteed if there are some pixels in a prototype image $\mathbf{x}$ which values are greater for $\mathbf{W}$ (or lower for $\mathbf{M}$)

Table 3.7   Summary of the properties of the morphological neural network (MNN).

Advantages	Disadvantages
1 Fast training	Lattice based (requires order relation on data)
2 Fast response	
3 High capacity	
4 Associative memory resistant to the morphological noise	

in this image from the maximum of all the remaining patterns x_i stored in that network. For binary images this means a unique pixel with value "1" in each of the prototypes x_i. This in turn is connected to the concept of morphological independence and strong independence [120]. It also provides a methodology for construction of the MNNs.

However, to make MNN robust to random, i.e. dilative, *and* erosive noise, Ritter *et al.* proposed a kernel version of MNN. The idea is to replace the associative memory **W** or **M** with *a series* of two memories **M'** and **W'** which are connected by a specially formed intermediate pattern **Z**, called *a kernel* [120]. The structure of this network is depicted in Figure 3.21. It operates in the scheme: input-**M'**$_{ZZ}$-**W'**$_{ZY}$-output. This can be written as follows

$$\mathbf{W}_{ZY}^T \vee \underbrace{\left(\mathbf{M}_{ZZ}^T \wedge \tilde{\mathbf{x}}_i\right)}_{\mathbf{z}_i} = \mathbf{y}_i, \tag{3.190}$$

where $\tilde{\mathbf{x}}_i$ is a randomly corrupted input pattern. $\mathbf{M}^T{}_{ZZ}$ and $\mathbf{W}^T{}_{ZY}$ in Equation (3.190) are found in accordance with the formula in (3.183), however for different matrices and after finding a kernel **Z**. Construction of the kernel **Z** is outlined by the theorems given in [119, 123]. Table 3.7 presents main features of the MNN.

Morphological scale space (MSS) was proposed by Raducanu *et al.* to define a version of the morphological neural network [122]. Their heteroassociative morphological memories exhibit robust recall in the presence of random (i.e. erosive and dilative) noise. This follows from the continuity properties of the morphological scale-space (discussed in Appendix A.2). That is, if the input patterns are eroded or dilated with the same structural element, then their local extremes will be preserved. Therefore if the morphological memory is constructed with the patterns dilated with the same structural element and scale σ, then the recognition of the original patterns will be preserved if the dilative pseudo kernel of the patterns contains local maxima of intrinsic scale σ or higher. More details can be found in the paper by Raducanu *et al.* [122].

3.9.5.1   IMPLEMENTATION of the Morphological Neural Network

Algorithm 3.19 presents a definition of the auxiliary *TMorphoOperationsFor* class which implements basic operations of lattice algebra, among them those described by Equations

(3.181), (3.182), (3.185), and (3.186). The last group of members (*Is_Equal*, etc.) of the *TMorphoOperationsFor* class implement the comparisons which are defined for lattices.

```
/////////////////////////////////////////////////////////////
// This class implements the basic operations of
// the morphological algebra.
/////////////////////////////////////////////////////////////
template < typename T >
class TMorphoOperationsFor
{
   public:

      typedef TImageFor< T >    MorphMatrix;

   public:

      /////////////////////////////////////////////////////////
      // This function returns a morpho conjugate matrix
      /////////////////////////////////////////////////////////
      //
      // INPUT:
      //        a - reference to the input n x m matrix A
      //
      // OUTPUT:
      //        An orphaned conjugate matrix A* of A
      //
      // REMARKS:
      //        For a real number R, its additive conjugate R*
      //        is defined as:
      //            R* = - R
      //
      //        It holds also that:
      //            R** = R
      //            R ^ S = ( R* v S* )*
      //
      //        For an n x m matrix A its conjugate A* is
      //          of dimensions m x n, where
      //            a*[i,j] = -a[j,i]
      //
      MorphMatrix * Orphan_Conjugate( const MorphMatrix & a ) const;


      /////////////////////////////////////////////////////////
      // This function computes a morphological maximum
      // of the two matrices.
      /////////////////////////////////////////////////////////
      //
      // INPUT:
      //        a - a reference to the p x q matrix
      //        b - a reference to the p x q matrix
      //
```

```
// OUTPUT:
//       an orphaned p x q matrix of the max
//
// REMARKS:
//       The two matrices A and B have to be of the
//         same size.
//
//       C = A V B       <==>
//       c[i,j] = max ( a[i,j], b[i,j] )
//
MorphMatrix * Orphan_Max( const MorphMatrix & a,
                          const MorphMatrix & b ) const;


////////////////////////////////////////////////////////////
// This function computes a morphological minimum
// of the two matrices.
////////////////////////////////////////////////////////////
//
// INPUT:
//       a - a reference to the p x q matrix
//       b - a reference to the p x q matrix
//
// OUTPUT:
//       an orphaned p x q matrix of the max
//
// REMARKS:
//       The two matrices A and B have to be of the
//         same size.
//
//       C = A ^ B       <==>
//       c[i,j] = min ( a[i,j], b[i,j] )
//
MorphMatrix * Orphan_Min( const MorphMatrix & a,
                          const MorphMatrix & b ) const;


////////////////////////////////////////////////////////////
// This function computes a morphological max product
// of the two matrices.
////////////////////////////////////////////////////////////
//
// INPUT:
//       a - a reference to the p x q matrix
//       b - a reference to the q x r matrix
//
// OUTPUT:
//       an orphaned p x r matrix of the max product
//
// REMARKS:
//       C = A V B       <==>
//       c[i,j] = V ( a[i,k] + b[k,j] )
//                k
```

```
//
MorphMatrix * Orphan_Max_Product( const MorphMatrix & a,
                                  const MorphMatrix & b ) const;



///////////////////////////////////////////////////////////
// This function computes a morphological min product
// of the two matrices.
///////////////////////////////////////////////////////////
//
// INPUT:
//      a - a reference to the p x q matrix
//      b - a reference to the q x r matrix
//
// OUTPUT:
//      an orphaned p x r matrix of the max product
//
// REMARKS:
//      C = A ^ B       <==>
//      c[i,j] = ^ ( a[i,k] + b[k,j] )
//               k
//
MorphMatrix * Orphan_Min_Product( const MorphMatrix & a,
                                  const MorphMatrix & b ) const;



///////////////////////////////////////////////////////////
// This function returns a matrix which is an outer product
// of the two vectors, for which we assume that they are column
// vectors.
///////////////////////////////////////////////////////////
//
// INPUT:
//      a - a reference to the n x 1 vector
//      b - a reference to the m x 1 vector
//
// OUTPUT:
//      an orphaned n x m cross product
//
// REMARKS:
//      Since the two vectors are column vectors, then
//      the second one has to be transposed for this
//      operation, i.e. we take B'.
//
//       C = A x B'
//
//      It holds also that:
//       C = A v B' = A ^ B'
//
MorphMatrix * Orphan_Cross_Product( const MorphMatrix & a,
                                    const MorphMatrix & b ) const;


///////////////////////////////////////////////////////////
```

```
// The function minimum-updates a matrix c by
// a cross-product a x b'.
//
// For each element at [i,j]
//
//     c[i,j] = min { c[i,j], ( a x b' )[i,j] }
//
////////////////////////////////////////////////////////////
//
// INPUT:
//       c - a reference to the matrix c which will
//           be updated by a minimum of c and a x b'
//       a - a reference to the n x 1 vector
//       b - a reference to the m x 1 vector
//
// OUTPUT:
//       true if operation successful
//       false otherwise
//
// REMARKS:
//       It is assumed that a and b are column vectors.
//       The matrix c has to be of size a_rows by b_rows.
//
bool Min_Update_Cross_Product( MorphMatrix & c, const MorphMatrix & a,
                               const MorphMatrix & b ) const;


////////////////////////////////////////////////////////////
// The function maximum-updates a matrix c by
// a cross-product a x b'.
//
// For each element at [i,j]
//
//     c[i,j] = min { c[i,j], ( a x b' )[i,j] }
//
////////////////////////////////////////////////////////////
//
// INPUT:
//       c - a reference to the matrix c which will
//           be updated by a minimum of c and a x b'
//       a - a reference to the n x 1 vector
//       b - a reference to the m x 1 vector
//
// OUTPUT:
//       true if operation successful
//       false otherwise
//
// REMARKS:
//       It is assumed that a and b are column vectors.
//       The matrix c has to be of size a_rows by b_rows.
//
bool Max_Update_Cross_Product( MorphMatrix & c, const MorphMatrix & a,
                               const MorphMatrix & b ) const;
```

```
public:

    // Operations
    bool Is_Equal              ( const MorphMatrix & a,
                                 const MorphMatrix & b ) const;
    bool Is_LessThan           ( const MorphMatrix & a,
                                 const MorphMatrix & b ) const;
    bool Is_LessOrEqualThan    ( const MorphMatrix & a,
                                 const MorphMatrix & b ) const;
    bool Is_GreaterThan        ( const MorphMatrix & a,
                                 const MorphMatrix & b ) const;
    bool Is_GreaterOrEqualThan( const MorphMatrix & a,
                                 const MorphMatrix & b ) const;
};
```

Algorithm 3.19 Definition of the *TMorphoOperationsFor* template class implementing basic lattice based morphological operations on matrices.

Operations of the *TMorphoOperationsFor* class are used in implementation of the MNN, realized by the *TMorphoNeuralNet* class. Its definition is listed in Algorithm 3.20. The class follows operations outlined in Equations (3.188)–(3.190). For this purpose its stores the matrices **W**, **M** respectively (protected part of the class).

```
///////////////////////////////////////////////////////////
// This class implements a simple version
// of the morphological neural network
///////////////////////////////////////////////////////////
class TMorphoNeuralNet
{
    protected:

        TRealImage *    fMatrix_W;
        TRealImage *    fMatrix_M;

    public:


        ///////////////////////////////////////////////////////////
        // This function adds a pair of patterns <x,y> to the network
        ///////////////////////////////////////////////////////////
        //
        // INPUT:
        //        x - reference to the first pattern
        //        y - reference to the pattern associated with x
        //
        // OUTPUT:
        //        true if ok
        //        false otherwise
        //
```

```
// REMARKS:
//
//       Images x and y are converted into vector representation
//       with '1' denoting an object and '0' for a background
//
bool AddPatternsPair( const MonochromeImage & x,
                      const MonochromeImage & y );



////////////////////////////////////////////////////////////////
// This function returns a pattern (associative memory)
////////////////////////////////////////////////////////////////
//
// INPUT:
//       x - input pattern
//       y_cols - number of columns of output pattern
//       y_rows - number of columns of output pattern
//
// OUTPUT:
//       associated (memorized) pattern
//
// REMARKS:
//       y_cols x y_rows need to be provided since
//       patterns are stored as vectors
//
MonochromeImage * RecallPattern(   const MonochromeImage & x,
                                   int y_cols, int y_rows );
};
```

Algorithm 3.20   Definition of the *TMorphoNeuralNet* implementing the morphological neural network.

The main functions of the *TMorphoOperationsFor* class are *AddPatternsPair* and *Recall-Pattern*. The former is used to enter a new associative pair of patterns **x** and **y** into the networks, which can be seen as a training process. The latter, is called to retrieve an associated pattern to the given one, i.e. this is a recall mode. The auxiliary parameters for the *RecallPattern* function are dimensions of the image which will be reconstructed from its vectorized version (although only one is necessary, two parameters are requested to assure consistency of data). Implementations of the two classes are contained in the attached code library [6].

3.10   Kernels in Vision Pattern Recognition

Application of kernels contributed to development of new data classification methods which in many areas outperform the already known solutions, but it also allowed transformation of the known methods into their kernel counterparts which also show many advantages over their original formulations. The main two ideas behind the kernel methods are as follows:

1. They allow data classification in so called feature space which usually is *higher dimensional* than the input data space.
2. However, direct transformation of data into the feature space is not required since all computations are expressed solely in terms of the *inner product* (dot product) of vectors belonging to the feature space, thus avoiding cumbersome (or sometimes not feasible) direct computations.

At first transformation of data into a higher dimension can sound ridiculous, so why bother anyway? The reward comes in the form of easier separation of data classes which in the feature space can become linearly separable. More formally the answer is provided by the Cover theorem which states that the probability of data being linearly separable increases if they are nonlinearly mapped to a higher dimensional feature space [127]. This can be shown considering N points in the L dimensional space. A linear dichotomy is any partitioning of a data set by a linear function into two nonoverlapping subsets. For example a set of 2D points, i.e. those which lie on a plane, can be separated into a number of linear dichotomies by the number of lines which are 1D geometrical structures. We can generalize this example going into higher dimensions and states after Cover such that in a general case of an L dimensional space containing N points a total number of possible linear dichotomies is given by [127]

$$O(N, L) = 2 \sum_{i=0}^{L} \binom{N-1}{i},$$

where (.) denotes a Newton symbol.[11] The geometric structures that separate the dichotomies are formed by hyperplanes of dimension $L - 1$. However, a total number of groupings of any size out of a set of N elements is 2^N. Thus, the probability of the possibility of a partition of N points in L dimensional space into two linearly separable sets is given as follows [5, 42, 127]

$$P(N, L) = \frac{O(N, L)}{2^N} = \begin{cases} \frac{1}{2^{N-1}} \sum_{i=0}^{L} \binom{N-1}{i} & for \quad N > L+1 \\ 1 & otherwise. \end{cases} \quad (3.191)$$

We see that if the number of points is very small compared to their dimension, then the points are always linearly separable. However, the more realistic is the first branch of Equation (3.191). In this case, considering a constant number of points N, the only way of increasing $P(N,L)$ is to increase L, i.e. the number of dimensions.

---

[11]Let us remember that $\binom{N}{k} \overset{df}{=} \frac{N!}{k!(N-k)!}$ is the number of ways one can choose a set of k out of N elements.

After this introduction we can take a look into some more formal definitions, followed by examples of commonly used kernels.

Definition 3.4

$\mathcal{X}$ is a unitary vector space (a pre-Hilbert space) if for each pair of points $\mathbf{x}_i$ and $\mathbf{x}_j$ belonging to $\mathcal{X}$ an inner product is defined as the following mapping

$$\langle \mathbf{x}_i, \mathbf{x}_j \rangle : \mathcal{X} \times \mathcal{X} \to \mathcal{R}, \tag{3.192}$$

which fulfills the following conditions:

1. Symmetry -

$$\langle \mathbf{x}_i, \mathbf{x}_j \rangle = \langle \mathbf{x}_j, \mathbf{x}_i \rangle. \tag{3.193}$$

2. Positive definiteness -

$$\langle \mathbf{x}, \mathbf{x} \rangle \geq 0, \text{ and if } \langle \mathbf{x}, \mathbf{x} \rangle = 0 \text{ then } \mathbf{x} = \mathbf{0}. \tag{3.194}$$

3. Associativity -

$$\langle c\mathbf{x}_i, \mathbf{x}_j \rangle = c \langle \mathbf{x}_i, \mathbf{x}_j \rangle, \text{ for any scalar } c \in \mathcal{R}, \text{ and} \tag{3.195}$$

4. Distributivity -

$$\langle \mathbf{x}_i + \mathbf{x}_j, \mathbf{x}_k \rangle = \langle \mathbf{x}_i, \mathbf{x}_k \rangle + \langle \mathbf{x}_j, \mathbf{x}_k \rangle. \quad \square \tag{3.196}$$

For completeness, let us remember that a special case is an inner product of a vector with itself which is called a squared norm (or a quadratic metric). It is given as follows

$$\|\mathbf{x}\|^2 = \langle \mathbf{x}, \mathbf{x} \rangle = \mathbf{x}^T \mathbf{x} = \sum_{k=1}^{L} x_k^2, \tag{3.197}$$

where L is the dimensionality of the vector space (see also the definition of the tensor inner product Equation (2.17) on p. 19)

Usually a Hilbert space is defined as a pre-Hilbert space which is complete and infinite-dimensional [128].

Definition 3.5

A symmetric continuous vector function

$$K : \mathcal{X} \times \mathcal{X} \to \mathcal{R}, \tag{3.198}$$

which for any set of points $x_i \in X$ satisfies the following Mercer's condition

$$\sum_{i=1}^{N}\sum_{j=1}^{N} k_i k_j K\left(x_i, x_j\right) \geq 0, \qquad (3.199)$$

where N is a positive integer, and k_i and k_j are real numbers, is called the *positive semidefinite kernel* [8, 9, 42, 129].   □

The last equation also means that for any set of points $x_i \in X$ the symmetric matrix of their kernel values

$$\mathbf{K} = \begin{bmatrix} K\left(x_1, x_1\right) & \cdots & K\left(x_1, x_N\right) \\ \vdots & \ddots & \vdots \\ K\left(x_N, x_1\right) & \cdots & K\left(x_N, x_N\right) \end{bmatrix}, \qquad (3.200)$$

is positive semidefinite, i.e. all determinants of its minors are positive [128, 130]. $\mathbf{K}$ is called a kernel matrix.

Some further properties of such defined kernels are given by the following.

Theorem 3.1 K is a positive semidefinite *kernel* if and only if there exists a vector function [131]

$$\Phi : \mathcal{X} \rightarrow \mathcal{F}, \qquad (3.201)$$

which maps a point x_i, belonging to the input space $\mathcal{X}$, into the $\Phi(x_i)$, which belongs to the inner product Hilbert space $\mathcal{F}$, and for which

$$K(x_i, x_j) = \langle \Phi\left(x_i\right), \Phi\left(x_j\right)\rangle = \Phi^T\left(x_i\right)\Phi\left(x_j\right). \quad \square \qquad (3.202)$$

Summarizing, the kernel methods operate as follows:

- Map all data points from the input space into the higher dimensional feature space $\mathcal{F}$.
- Perform classification, regression, etc. in the feature space.

Let us observe application of the kernel method for classification of the radially distributed data classes, as shown in Figure 3.22(a) [132].

Let us now choose a polar transformation of the input coordinates, that is the mapping function from Equation (3.201) becomes

$$\Phi : \begin{bmatrix} x_1 \\ x_2 \end{bmatrix} \rightarrow \begin{bmatrix} r \\ \varphi \end{bmatrix} = \begin{bmatrix} \sqrt{x_1^2 + x_2^2} \\ \arctan\left(x_2/x_1\right) \end{bmatrix}. \qquad (3.203)$$

After transformation of the points from Figure 3.22(a) they become easily separable by a linear function (and also based on the value of only one coordinate), as shown in Figure 3.22(b).

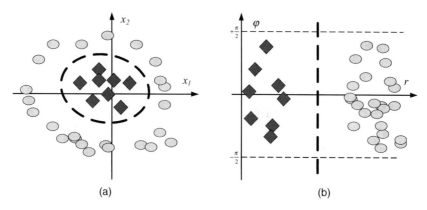

Figure 3.22 Two classes of objects. Nonlinearly separable (a). Linearly separable after polar mapping (b).

However, there are a few problems with the above scheme. The first is how to find a function Φ for our task, which is usually not as simple as the one depicted in Figure 3.22(a). Even if we had that function somehow, the second problem is computation of all the $\Phi(\mathbf{x}_i)$ mappings which may be cumbersome due to the nonlinearity and high dimensionality of $\mathcal{F}$. However, as already mentioned at the beginning of this section, a direct computation of $\Phi(\mathbf{x}_i)$ can be avoided if we can express all operations in terms of the inner product. That is, we wish to express all our computations exclusively in the form of products $\Phi^T(\mathbf{x}_i)\Phi(\mathbf{x}_j)$ which from Equation (3.202) are kernel values that are simple scalars. This is called a *kernel trick*. Its applications we shall see in the subsequent section.

When dealing with kernels, sometimes the opposite problem arises, that is after having obtained a solution in the feature space we wish to find a corresponding point in the input space. This is a so called *pre-image problem*, encountered for instance in data denoising or data compression with kernels. Nevertheless, it can be shown that in some cases such a pre-image point does not exist and the only solution is to find an approximate pre-image [8, 133]. This situation is depicted in Figure 3.23. More precisely, a solution $\mathbf{Z}$ to a kernel method is usually

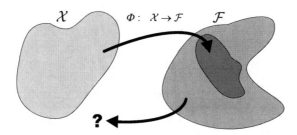

Figure 3.23 Mapping from the input space $\mathcal{X}$ into the feature space $\mathcal{F}$. However, the opposite direction is not always possible since the points in the span of Φ are not always images of some input points. This is the pre-image problem.

obtained in the form of an expansion of the N Φ-mapped input data points, such as

$$\mathcal{Z} = \sum_{i=1}^{N} \alpha_i \Phi(\mathbf{x}_i). \tag{3.204}$$

The above comes from the so called *representer theorem*. The values of α_i and Φ in Equation (3.204) depend on a kernel algorithm. Notwithstanding, the above can be computationally demanding since the sum spans through all input points. However there are methods which allow shorter representation of the solution

$$\mathcal{Z} = \sum_{i=1}^{M} \beta_i \Phi(\mathbf{y}_i), \tag{3.205}$$

with an assumption that $M < N$. The parameters β_i in Equation (3.205) are obtained by means of optimization methods, whereas $\mathbf{y}_i$ can be found either by selecting them among the $\mathbf{x}_i$ or by iteratively constructing the synthetic patterns. Both methods are described with details in the book by Schölkopf and Smola [8].

3.10.1 Kernel Functions

Let us now consider some common kernel functions. The first and the most intuitive is the inner product kernel, given as follows

$$K_L(\mathbf{x}_i, \mathbf{x}_j) = \langle \mathbf{x}_i, \mathbf{x}_j \rangle = \mathbf{x}_i^T \mathbf{x}_j. \tag{3.206}$$

The second one is the polynomial kernel

$$K_P(\mathbf{x}_i, \mathbf{x}_j) = (\langle \mathbf{x}_i, \mathbf{x}_j \rangle + R)^d, \tag{3.207}$$

where d and R are parameters (frequently R is set to 1, and d to 2). It can be easily generalized to incorporate any other kernels K_X

$$K_P(\mathbf{x}_i, \mathbf{x}_j) = (K_X(\mathbf{x}_i, \mathbf{x}_j) + R)^d, \tag{3.208}$$

Interestingly, the following fuzzy like function defined for two scalars is also a kernel

$$K_I(x_i, x_j) = \min(x_i, x_j). \tag{3.209}$$

It is called an intersection kernel [134]. However, the most frequently used is the Gaussian kernel[12] K_G, defined as follows [9, 129, 135]

$$K_G(\mathbf{x}_i, \mathbf{x}_j) = e^{-\frac{D_E^2(\mathbf{x}_i, \mathbf{x}_j)}{2\sigma^2}} = e^{-\frac{\|\mathbf{x}_i - \mathbf{x}_j\|^2}{2\sigma^2}}, \tag{3.210}$$

[12] Also called the Radial Basis Function (RBF) kernel.

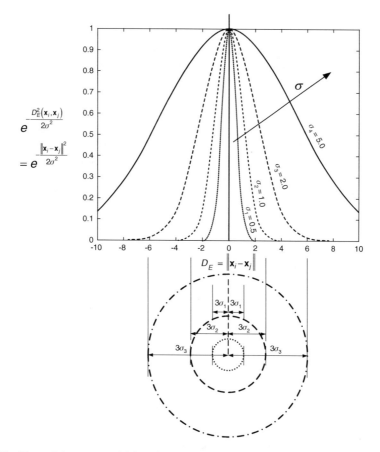

$$e^{-\frac{D_E^2(\mathbf{x}_i,\mathbf{x}_j)}{2\sigma^2}}$$

$$= e^{-\frac{\|\mathbf{x}_i-\mathbf{x}_j\|^2}{2\sigma^2}}$$

Figure 3.24 Plots of the exponential function for different values of the parameter σ. Nonzero values are practically for distances d not exceeding 3σ.

where $D_E(\mathbf{x}_i,\mathbf{x}_j)$ denotes the Euclidean distance between $\mathbf{x}_i$ and $\mathbf{x}_j$, and σ is a parameter which controls the spread of the kernel.

To understand the properties and operation of the Gaussian kernel (3.210) we need to analyze its values in respect to a distance d between two data points $\mathbf{x}_i$ and $\mathbf{x}_j$, given a scaling parameter σ. Values of K_G for different distances at four values of σ are depicted in Figure 3.24. This is a well known plot from all statistical books, but for us the most important conclusion is that the value of K_G decays quite rapidly and nonlinearly[13] toward 0, as the distance between the points tends towards 3σ and further. So, for example if $\sigma = 2$, then points distant by 6 and more are treated by the K_G in the same way, i.e. its value approaches 0. Thus, the influence region spans 3σ around a chosen reference point, say $\mathbf{x}_i$, as depicted in Figure 3.24. In other words, only the points inside the ring of radius 3σ will be attributed by K_G as having some kind of "neighborhood" with respect to the point $\mathbf{x}_i$. All others are treated the same, i.e. as far and not

---

[13]The rate of this decay is a derivative of the exponent function, which is also the exponent function.

influencing points. On the other hand, when σ grows then the Gaussian function is getting wider and more "flat" and in extremes it takes a constant value. In such a case all distances among the points would be the same. These considerations give us some insight into a role of the parameter σ, though in practice its choice depends on a task K_G that it is used for. We saw that too small values of σ lead to situations where each distance of a pair of points is zero, and conversely at the other extreme when σ is excessively large, all distances will tend to one. Thus, when choosing σ for a given task we need to rely on some rough estimate of σ, which is then finely tuned e.g. by means of grid search [136], etc. The primary estimate of σ can be based on some knowledge of statistics of the input data set, such as its variance. These issues are discussed in the next sections devoted to kernel segmentation (3.11.3) and support vector machines (3.12).

Choice of a proper kernel for a classification task, as well as its specific parameters, is a challenge and usually requires some experimentation, heavily dependent on the type of data. Here, once again, the no-free-lunch theorem manifests itself. However, having a toolset of available kernels, construction of new ones is also possible with one or a combination of the following rules.

Lemma 3.1   If K is a kernel then the following are also kernels:

1. A generalization of the Gaussian kernel (3.210) -

$$K_{GG}(\mathbf{x}_i, \mathbf{x}_j) = e^{-\gamma D^2(\mathbf{x}_i, \mathbf{x}_j)}, \tag{3.211}$$

where $D(\mathbf{x}_i,\mathbf{x}_j)$ is a metric and γ is a scaling parameter [129];
2. Linear combination of kernels

$$K(\mathbf{x}_i, \mathbf{x}_j) = \alpha K_1(\mathbf{x}_i, \mathbf{x}_j) + \beta K_2(\mathbf{x}_i, \mathbf{x}_j), \tag{3.212}$$

where $\alpha, \beta \in \mathcal{R}^+$ and K_1 and K_2 are kernels;
3. Product of the kernels

$$K(\mathbf{x}_i, \mathbf{x}_j) = K_1(\mathbf{x}_i, \mathbf{x}_j) K_2(\mathbf{x}_i, \mathbf{x}_j), \tag{3.213}$$

where K_1 and K_2 are kernels;
4. Product of function values

$$K(\mathbf{x}_i, \mathbf{x}_j) = f(\mathbf{x}_i) f(\mathbf{x}_j), \tag{3.214}$$

where f is a real-valued function;
5. Product

$$K(\mathbf{x}_i, \mathbf{x}_j) = \mathbf{x}_i \mathbf{P} \mathbf{x}_j, \tag{3.215}$$

where $\mathbf{P}$ is a symmetric positive semidefinite matrix;
6. Normalized kernel

$$\hat{K}(\mathbf{x}_i, \mathbf{x}_j) = \frac{K(\mathbf{x}_i, \mathbf{x}_j)}{\sqrt{K(\mathbf{x}_i, \mathbf{x}_i) K(\mathbf{x}_j, \mathbf{x}_j)}} \qquad \square \tag{3.216}$$

Obviously the parameters in (3.210) and (3.211) are related as follows

$$\gamma = \frac{1}{2\sigma^2}.$$
(3.217)

If $D(\mathbf{x}_i,\mathbf{x}_j)$ is a metric, then K_{GG} is a proper kernel that fulfills the Mercer condition (3.199).

A special case of Equation (3.211) is the Mahalanobis kernel which accounts for the distribution of given data set [2]. It is expressed by the following formula [13]

$$K_M(\mathbf{x}_i, \mathbf{x}_j) = e^{-(\mathbf{x}_i-\mathbf{x}_j)^T \Sigma_{\{x\}}^{-1}(\mathbf{x}_i-\mathbf{x}_j)},$$
(3.218)

where $\Sigma_{\{x\}}$ is the covariance matrix, which is computed for a given population of data points $\{\mathbf{x}_i\}$ used to build the classifier. For instance in the direct pixel classification problem, these data can be color values that come from the representative objects of interest (such as human faces, road signs, etc.). The covariance matrix can be computed in accordance with Equation (3.24), presented while discussing PCA in Section 3.3.1.1. However, in Equation (3.218) an inverse of $\Sigma_{\{x\}}$ is required. In the case of low dimensional data (such as color pixels) this does not pose a significant problem. Moreover, $\Sigma_{\{x\}}^{-1}$ has to be computed once during training, thus not delaying the classification stage.

Finally, let us observe that for the discussed kernels (3.210), (3.211), and (3.218) it holds that

$$K(\mathbf{x}, \mathbf{x}) = 1.$$
(3.219)

Ah-Pine proposes a different normalization of kernels which extends the well known cosine normalization (3.216). In this context let us observe that Equation (3.216) can be written as [137]

$$\hat{K}^0(\mathbf{x}_i, \mathbf{x}_j) = \frac{\langle \Phi(\mathbf{x}_i), \Phi(\mathbf{x}_j) \rangle}{\|\Phi(\mathbf{x}_i)\|\|\Phi(\mathbf{x}_j)\|} = \left\langle \frac{\Phi(\mathbf{x}_i)}{\|\Phi(\mathbf{x}_i)\|}, \frac{\Phi(\mathbf{x}_j)}{\|\Phi(\mathbf{x}_j)\|} \right\rangle = \cos \sphericalangle(\Phi(\mathbf{x}_i), \Phi(\mathbf{x}_j))$$
(3.220)

From the above we obtain immediately that for all $\mathbf{x} \in \mathcal{X}$

$$\hat{K}(\mathbf{x}, \mathbf{x}) = 1,$$
(3.221)

which means that objects in the feature space are projected onto a unit hypersphere. In this light the main idea of the method by Ah-Pine is to use a normalization factor based on a ratio of vector norms in the feature space. More precisely, the method is based on the generalized mean, which for a sequence of L values x_i is defined as follows

$$\mathcal{M}^t(x_1, \ldots, x_L) = \left[\frac{1}{L} \sum_{i=1}^{L} x_i^t \right]^{\frac{1}{t}},$$
(3.222)

where t is a parameter of the normalization. Based on Equation (3.216) and (3.222) the normalized kernel of order t is defined as [137]

$$\hat{K}^t(\mathbf{x}_i, \mathbf{x}_j) = \frac{K(\mathbf{x}_i, \mathbf{x}_j)}{\mathcal{M}^t[K(\mathbf{x}_i, \mathbf{x}_i), K(\mathbf{x}_j, \mathbf{x}_j)]}. \tag{3.223}$$

Similarly to Equation (3.216) the normalization (3.223) is equivalent to projecting data from the feature space to a unit hypersphere. However, as we will soon see, it allows the extension of Equation (3.216) to account for differences in norms of the vectors in the feature space. Now, introducing Equation (3.202) into (3.223) and considering Equation (3.195) as well as (3.220) yields

$$
\begin{aligned}
\hat{K}^t(\mathbf{x}_i, \mathbf{x}_j) &= \frac{\langle \Phi(\mathbf{x}_i), \Phi(\mathbf{x}_j) \rangle}{\mathcal{M}^t[\langle \Phi(\mathbf{x}_i), \Phi(\mathbf{x}_i) \rangle, \langle \Phi(\mathbf{x}_j), \Phi(\mathbf{x}_j) \rangle]} = \\[2mm]
&= \frac{\| \Phi(\mathbf{x}_i) \| \, \| \Phi(\mathbf{x}_j) \| \left\langle \dfrac{\Phi(\mathbf{x}_i)}{\| \Phi(\mathbf{x}_i) \|}, \dfrac{\Phi(\mathbf{x}_j)}{\| \Phi(\mathbf{x}_j) \|} \right\rangle}{\mathcal{M}^t[\| \Phi(\mathbf{x}_i) \|^2, \| \Phi(\mathbf{x}_j) \|^2]} \\[2mm]
&= \frac{\cos \sphericalangle(\Phi(\mathbf{x}_i), \Phi(\mathbf{x}_j))}{\dfrac{\mathcal{M}^t[\| \Phi(\mathbf{x}_i) \|^2, \| \Phi(\mathbf{x}_j) \|^2]}{\| \Phi(\mathbf{x}_i) \| \, \| \Phi(\mathbf{x}_j) \|}} = \frac{\cos(\Phi(\mathbf{x}_i), \Phi(\mathbf{x}_j))}{\mathcal{M}^t\left[\dfrac{\| \Phi(\mathbf{x}_i) \|}{\| \Phi(\mathbf{x}_j) \|}, \dfrac{\| \Phi(\mathbf{x}_j) \|}{\| \Phi(\mathbf{x}_i) \|} \right]}.
\end{aligned}
\tag{3.224}
$$

The above can be written in the following form

$$
\begin{aligned}
\hat{K}^t(\mathbf{x}_i, \mathbf{x}_j) &= \frac{\cos \sphericalangle(\Phi(\mathbf{x}_i), \Phi(\mathbf{x}_j))}{\mathcal{M}^t[\gamma, \gamma^{-1}]} \\[2mm]
&= \frac{\cos \sphericalangle(\Phi(\mathbf{x}_i), \Phi(\mathbf{x}_j))}{\left[\dfrac{1}{2}(\gamma + \gamma^{-1}) \right]^{\frac{1}{t}}} = \underbrace{\left[\left(\frac{2}{\gamma^{2t}+1} \right)^{\frac{1}{t}} \gamma \right]}_{\hat{\gamma}} \cdot \cos(\Phi(\mathbf{x}_i), \Phi(\mathbf{x}_j)),
\end{aligned}
\tag{3.225}
$$

where the parameter

$$\gamma = \max \left[\frac{\| \Phi(\mathbf{x}_i) \|}{\| \Phi(\mathbf{x}_j) \|}, \frac{\| \Phi(\mathbf{x}_j) \|}{\| \Phi(\mathbf{x}_i) \|} \right] = \frac{\max(\| \Phi(\mathbf{x}_i) \|, \| \Phi(\mathbf{x}_j) \|)}{\min(\| \Phi(\mathbf{x}_i) \|, \| \Phi(\mathbf{x}_j) \|)} \tag{3.226}$$

expresses *the degree of difference between the two norms* $\| \Phi(\mathbf{x}_i) \|$ and $\| \Phi(\mathbf{x}_j) \|$. If these are equal then $\gamma = 1$. However, the higher the difference between these two norms, the larger the value of γ which can rise toward $+\infty$. It was shown that for $t > 0$ (3.225) is also a kernel and a distance $\sqrt{2(1 - \hat{K}^t(\mathbf{x}_i, \mathbf{x}_j))}$ fulfills the metric properties [137]. From Equation (3.225) it follows easily that $\hat{K}^t(\mathbf{x}_i, \mathbf{x}_j)$ extends the cosine kernel $\hat{K}^0(\mathbf{x}_i, \mathbf{x}_j)$, given by Equation (3.220), taking into account the difference between the norms of the two vectors in the feature space. The new information is conveyed by the multiplicative factor $\hat{\gamma}$ in Equation (3.225) which depends on the chosen value of the parameter t (which should be nonnegative). $\hat{K}^0(\mathbf{x}_i, \mathbf{x}_j)$ is a special case of $\hat{K}^t(\mathbf{x}_i, \mathbf{x}_j)$ for $t = 0$, whereas for $t \to \infty$ $\hat{K}^t(\mathbf{x}_i, \mathbf{x}_j) = \hat{K}^0(\mathbf{x}_i, \mathbf{x}_j)/\gamma$. The method was tested by Ah-Pine with some of the well known reference data sets from the UCI repository [138]. The k-means clustering[14] initialized by the kernel-PCA with the $\hat{K}^t(\mathbf{x}_i, \mathbf{x}_j)$ kernel showed superior results for $t > 0$ when compared to the Gaussian and cosine kernels. Detailed results can be found in [137].

An interesting research area is the development of new kernels suitable for specific classification tasks. In particular, kernels for image recognition were investigated by Barla *et al.* [139] as well as by Odone *et al.* [140]. In both papers it was shown that the histogram intersection D_i, discussed in Section 3.5 on p. 240, follows Mercer's conditions. Thus, for two histograms $\mathbf{x}_i$ and $\mathbf{x}_j$ with L bins, the histogram intersection kernel is defined as

$$K_H\left(\mathbf{x}_i, \mathbf{x}_j\right) = \sum_{k=1}^{L} \min\left(x_{ik}, x_{jk}\right),\tag{3.227}$$

where $\mathbf{x}_{ik}$ and $\mathbf{x}_{jk}$ are the k-th bins of the two histograms, respectively. It was also shown that the Hausdorff distance, which will be discussed in Section 5.4.2 on p. 428, if modified, can also be a proper Mercer kernel. The latter was then used for such vision tasks as novelty detection and view-based identification systems [140]. Kernel methods have found broad applications in many areas of signal processing. For instance Arenas-García *et al.* discuss kernel multivariate analysis in remote sensing and feature extraction [141].

3.10.2   IMPLEMENTATION – Kernels

Implementation of the kernels is straightforward and follows the mathematical notation used in Section 3.10. Nevertheless, we need to remember that kernel functions are called in different contexts and frequently from nested loops. The other variable parameter is the length of data vectors used in functions such as Equations (3.206)–(3.211). Therefore each kernel will be endowed with two members: a data parameter denoting length of data and a kernel function which returns a scalar value. For all operations we assume a computer approximation of real data and use a long floating point representation *double*. Figure 3.25 depicts a class hierarchy that implements the already discussed kernels. *TKernel* is a base class that's role is to define a common interface to the whole hierarchy. For this purpose *TKernel* defines two pure virtual functions. The first is a function operator which computes the value of a kernel for given input parameters. The role of the second, named *Clone()*, is to reproduce (spawn) a given kernel. This operation is used in some contexts in which one wishes to create a copy of a kernel given a kernel object, however, usually accessed only by a pointer to the base class. The technique is called a prototype design pattern [142, 143]. From *TKernel* two families of kernels are derived. The first contains the two classes, *RBF_Kernel* and *Mahalanobis_Kernel*, which operate in accordance with Equations (3.210) and (3.218), respectively. However, in the case of the Gaussian kernel we use a slightly different representation which utilizes only one multiplicative parameter γ instead of dividing by $2\sigma^2$. Thus, its constructor expects two parameters: positive data size and real gamma. This is similar to the representation (3.211) and proves to be more efficient for computations. Nevertheless, thanks to the helper functions

[14]Discussed in Section 3.11.1.

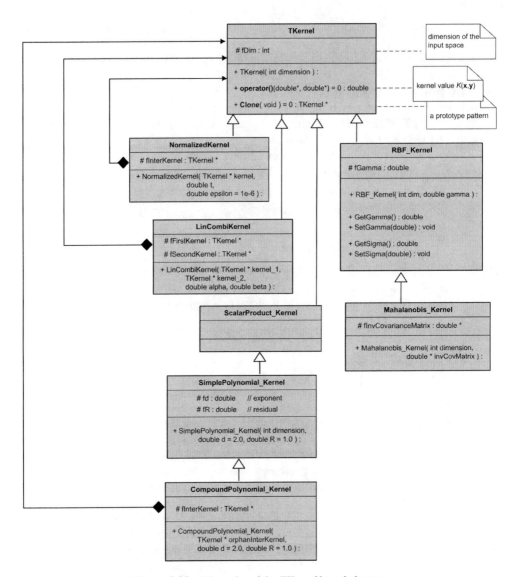

Figure 3.25   Hierarchy of the *TKernel* based classes.

the kernel can be provided with either parameter γ or σ, whichever is more convenient. The *Mahalanobis_Kernel* is derived from the *RBF_Kernel*. In the constructor it requires the inverse of the matrix **A**, such as in Equation (3.218). The *Mahalanobis_Kernel* reduces to the *RBF_Kernel* if $\mathbf{A}^{-1}$ is a diagonal with parameters γ. Thus, although it would be theoretically possible to reverse the relationship between these two classes, such an implementation would be inefficient for the *RBF_Kernel*.

As an example computation of the Gaussian kernel is shown in Algorithm 3.21.

```
virtual double operator() ( double * x, double * y ) const
{
    // At first compute the squared Euclidean distance between x and y
    register double tmp = 0.0;
    register double sum = 0.0;
    for( register int d = 0; d < fDim; ++ d )    // fDim member has a
                                                 //    number of components
    {
        tmp = * x ++ - * y ++;
        sum += tmp * tmp;
    }

    return exp( - fGamma * sum );
}
```

Algorithm 3.21 Implementation of the function operator for the *RBF_Kernel.*

In the computation of the squared distance between the two input vectors we rely only on data pointers, whereas their dimension is assumed to be greater or equal to the preset *fDim* member. This speeds up computations, but also allows a change of dimensions in the run time. In some applications the loop in Algorithm 3.21 can be sped up, e.g. by the loop unfolding technique [94] or making the code operate concurrently, as discussed in Appendix A.6. Nevertheless, at first such an improvement should be measured to compare performance and used for rather large lengths of vectors (such as hundreds or more), which in practice is rarely the case.

The second family of derived classes in Figure 3.25 starts from the the *ScalarProduct_Kernel*, followed by the *SimplePolynomial_Kernel* and ends with the *CompoundPolynomial_Kernel*. The first one computes a kernel K_L from Equation (3.206). The second computes Equation (3.208) with K_X being K_L which is taken from its base class. Finally, the *CompoundPolynomial_Kernel* computes Equation (3.207) for any provided kernel. Therefore it is implemented as a composite pattern,[15] i.e. it is a kernel and it also contains a kernel. It is interesting to note that data size assumed in the *CompoundPolynomial_Kernel* is taken to be the same as in the supplied to its constructor (orphaned[16]) internal kernel *orphanInterKernel* (see Figure 3.25). A similar idea of composite pattern is used in the other two subclasses, namely the *LinCombiKernel* and the *NormalizedKernel*. The former implements the idea of the normalizes kernels expressed in Equation (3.212), whereas the latter Equation (3.216).

Application of any kernel is very simple. As a prerequisite it is sufficient to provide two vectors of real values. An example is shown in the listing Algorithm 3.22.

---

[15]In Figure 3.25 we incorporate the so called strong whole-part model denoted by the black diamond. It means that a container object contains a containee object and if the former is deleted this incurs deletion of the latter. A more in-depth discussion of this, as well as of the weak whole-part model (usually depicted with a white diamond), can be found in the literature [144, 145, 146, 147].

[16]An orphaned object in C++ means that a receiving component is responsible for its deletion.

```
1   /////////////////////////////////////////////////////////////
2   // A function to test the Gaussian (RBF) kernel and to
3   // show how it can be used in a context.
4   /////////////////////////////////////////////////////////////
5   //
6   // INPUT:
7   //       none
8   //
9   // OUTPUT:
10  //       none
11  //
12  // REMARKS:
13  //
14  //
15  void GaussianKernel_TestFunction( void )
16  {
17    int num_of_elems = 16;
      // Number of components in each feature point can be dynamic
18
19    // The two vectors: a = [ 0, 0, ..., 0 ]T, b = [ 1, 1, ..., 1 ]T
20    FeatureSpaceForDouble::DataPoint a( num_of_elems, 0.0 ),
                                       b( num_of_elems, 1.0 );
21    // This is the same as vector< double > a, b
22
23    // Creating the Gaussian kernel we use gamma
24    RBF_Kernel   theGaussianKernel( num_of_elems, 1.0 );
25
26    // But we can also set the sigma as well
27    double vec_distance_1 = sqrt( Compute_SquaredEuclidDistance_For( a, b ) );
28    double sigma_1 = vec_distance_1 / 2.0;
29    // set to half a distance of a and b (i.e. 4.0)
30    theGaussianKernel.SetSigma( sigma_1 );
31
32    double k_0 = theGaussianKernel( & a[ 0 ], & b[ 0 ] );
33
34    // But we can also set different length of data vectors
35    theGaussianKernel.SetDim( 4.0 );   // new dimension for the kernel is 4
36    double vec_distance_2 = 2.0;   // so a distance between a and be
                                        becomes 2.0
37
38    // and compute again the Gaussian kernel
39    double k_1 = theGaussianKernel( & a[ 0 ], & b[ 0 ] );
40
41    // Let's change sigma to fit into a new dimensioin
42    double sigma_2 = vec_distance_2 / 2.0;
43    theGaussianKernel.SetSigma( sigma_2 );
44
45    // and compute new kernel value
46    double k_2 = theGaussianKernel( & a[ 0 ], & b[ 0 ] );
47  }
```

Algorithm 3.22   An example of using the *RBF_Kernel* and setting its sigma parameter.

In the above functions we rely on the flexible definition of features provided by the *FeatureSpaceFor* template class, in which each feature is defined as a STL vector. This, in turn, allows variable lengths and data types of features, whatever are chosen. Definitions of kernels follow this idea and also allow easy changes in data lengths and other parameters in the run-time. In the function *GaussianKernel_TestFunction* we choose to define two vectors, one zero, and the second with all components set to one, all of length 16 (defined in line <17>). Thus, a distance among these vectors is 4. Then the *RBF_Kernel* is created in line <24>, providing a γ parameter to its constructor. However, we can easily change the σ parameter in the *RBF_Kernel*, as in line <30>, and then also in <43>. A value of the Gaussian kernel is computed in line <32>. After that we can easily change the length of data, the σ parameter, and again a value of the kernel is computed in lines <39> and <46>, for different parameter σ, however. These comply with the values read from the plot in Figure 3.24. Let us also note that in all cases the distance between the two vectors does not exceed 3σ, beyond which we would always get 0.

Algorithm 3.23 contains an example of using the *CompoundPolynomial_Kernel* and its parameters.

```
1    //////////////////////////////////////////////////////////////
2    // A function to test the CompoundPolynomial_Kernel kernel
3    // and to show how it can be used in a context.
4    //////////////////////////////////////////////////////////////
5    //
6    // INPUT:
7    //        none
8    //
9    // OUTPUT:
10   //        none
11   //
12   // REMARKS:
13   //
14   //
15   void CompoundPolynomialKernel_TestFunction( void )
16   {
17       int num_of_elems = 4;// Number of components in each feature point
         can be dynamic
18
19       // The two vectors: a = [ 0, 0, ..., 0 ]T, b = [ 1, 1, ..., 1 ]T
20       FeatureSpaceForDouble::DataPoint   a( num_of_elems, 0.0 ),
       b( num_of_elems, 1.0 ); // This is the same as vector< double > a, b
21
22       double vec_distance = sqrt( Compute_SquaredEuclidDistance_For( a, b ) );
23
24       double gamma = 1.0;         // a parameter for the internal kernel
25
26       double d = 2.0, R = 1.0;    // parameters of the polynomial
27
28       CompoundPolynomial_Kernel    theCompKernel(
                              new RBF_Kernel(num_of_elems,gamma),d, R );
29
30
```

```
31      reinterpret_cast< RBF_Kernel * >( theCompKernel.GetInterKernel() )
        ->SetSigma( vec_distance / 3.0 );
                        // let's change sigma  (3*sigma=vector_distance)
32
33      double k_0 = theCompoundKernel( & a[ 0 ], & b[ 0 ] );
34
35      reinterpret_cast< RBF_Kernel * >( theCompKernel.GetInterKernel() )
        ->SetSigma( vec_distance / 2.0 );
                        // let's change sigma (2*sigma=vector_distance)
36
37      double k_1 = theCompKernel ( & a[ 0 ], & b[ 0 ] );
38    }
```

Algorithm 3.23   An example of using the *CompoundPolynomial_Kernel* and its parameters.

An interesting feature is that actually two kernels are created, in accordance with Equation (3.207), as placed in line <28>. Nevertheless, it is still possible to control parameters of both kernels, such as in lines <31> and <35>.

3.11   Data Clustering

The process of data clustering is aimed at discovering internal structures and relations between data. In other words, the role of clustering is to divide input data into a number of data sets, called *data partitions*, which exhibit some common property within each partition. On the other hand, we usually wish the partitions to be sufficiently separated from each other, so the discovered categories among data are distinctive. This means that data clustering relies heavily on similarity measures which can be applied to a chosen set of features representing a given type of data.

Many methods have been developed for data clustering, since it is no surprise that data clustering belongs to the central problems of computer vision, pattern recognition, data mining, and many other areas. Therefore the problem has found broad interest from many groups of researchers and the literature on the subject is ample. A reader interested in an overview of existing methods is referred to the books by Gan *et al.* [148], Xu and Wunsch [149], Theodoridis and Koutroumbas [5], as well as Duda *et al.* [2], to name a few.

In this section we only discuss a small, but very important, subgroup of these methods which have their roots in the so called *k*-means clustering method. Its basic idea is, at the first step, to assume a number of cluster centers in a data space and then repeatedly assign all remaining data points to the closest of these centers. In the next step the center positions are refined to become centers of mass for each group of data. The whole process gets repeated until no changes in centers is observed. This way fixed groups of points (partitions) are obtained. Alternatively, the described process can be also seen as the creation of separating hypersurfaces among clusters. Recommended introductory texts on this group of methods are the paper by Filipone [150], which provides a survey on kernel and spectral methods, as well as the book by de Oliveira and Pedrycz [151], which deals with the modified *k*-means called fuzzy clustering, as will be discussed in Section 3.11.2. Further references concerning detailed problems will also be provided in the following sections.

When constraining our interest to visual signals, data clustering is frequently used for image segmentation, i.e. the process of finding distinct groups of similar regions or objects in images. Thus, the problem of data clustering can be seen as an object detection method. In this case, we usually say that an image is partitioned into objects (foreground) and the rest (background). Simple examples of pixel clustering based on Euclidean similarity in the RGB color space are presented in Figure 3.26.

Figure 3.26 depicts two color images and the corresponding scatter plots of their colors in the space spanned by the red–green–blue axis. The first image, shown in Figure 3.26(a), is an

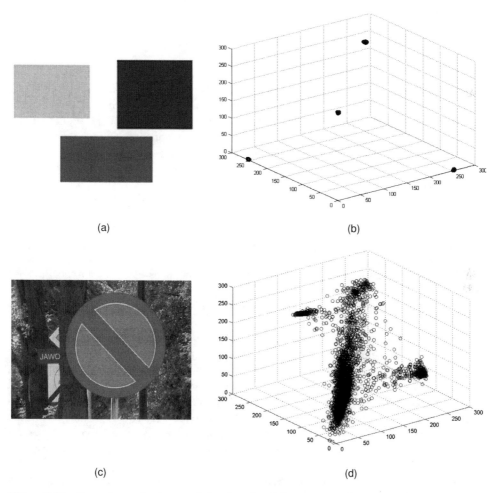

(a) (b)

(c) (d)

Figure 3.26 Exemplary color images (left column) and the corresponding scatterplots of their colors in the RGB color space (right column). (For a color version of this figure, please see the color plate section.)

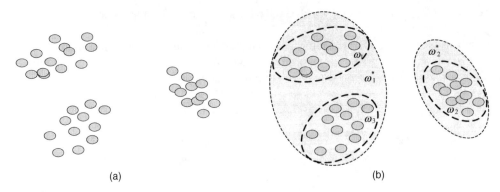

Figure 3.27 Data points which form clusters (a). Possible partitions of data points into two or three clusters (b).

artificially created image which contains three squares, red, green, and blue, respectively, on a white background. Thus, the corresponding distribution of colors, shown in Figure 3.26(b), contains four well separated clusters of characteristic colors of that image. On the other hand, Figure 3.26(c) and Figure 3.26(d) show a real color image with the scatterplot of its colors, respectively. It is evident that in the last case clustering is not that obvious, though a few point concentration areas can be detected from the plot. In the next sections we discuss the basic theory, as well as implementation aspects, of the most popular groups of data clustering methods, starting with the *k*-means algorithm.

3.11.1   The k-Means Algorithm

In many practical situations real data tend to have different "densities of occurrence," i.e. for chosen attributes data frequently are not spread uniformly across the attribute space. Instead, some areas of data concentration can be observed. A clustering method allows identification of data clusters, i.e. usually separate partitions (subsets) of data which, in some chosen sense, are *more similar* to each other within that partition than with data from other partitions. Thus, depending on the chosen similarity criteria, it is assumed that data tend to groups itself into clusters. However, the problem is how to predict the number of clusters or whether we can determine that from data as well. For example let us consider some data points depicted in Figure 3.27a. It is clear that they form three clusters, say, as depicted in Figure 3.27(b). However, we can also assume that there are only two clusters and partition data into two classes ω^*_1 and ω^*_2. At the other extreme, we can say that each point represents a separate cluster. Thus, in general case the nonsupervised data clustering problem is not unique.

There are many algorithms of data clustering, one of which is the *k*-means method [2, 5]. For a given set of N points $\{\mathbf{x}_i\}$ the algorithm requires an *a priori* selection of a number M of clusters. Let us denote them as D_m, where $1 \leq m \leq M$. Then, for each partition and initial mean $\boldsymbol{\mu}_m$ is selected. A simple and frequently used method is random selection of M data from the training set. After initialization the method proceeds iteratively to solve the following optimization problem

$$Q = \min \left(\sum_{m=1}^{M} \sum_{i=1}^{N} w_{mi} \underbrace{\left\| \mathbf{x}_i - \boldsymbol{\mu}_m \right\|_L^2}_{d_{im}} \right), \tag{3.228}$$

for w_{mi} and $\boldsymbol{\mu}_m$. In the above d_{im} denotes a chosen distance between the data points, which most commonly is the Euclidean metric, M denotes number of clusters, N is the total number of points, and finally

$$w_{mi} \in \{0, 1\} \tag{3.229}$$

is a binary discrete weight value which takes on 1 only if a point $\mathbf{x}_i$ belongs to the m-th cluster, otherwise it is 0. The above minimization problem is solved sequentially by assigning each point $\mathbf{x}_i$ to such partition D_m which mean $\boldsymbol{\mu}_m$ is *the closest* to $\mathbf{x}_i$. This way data clusters are formed. More formally, for each $m \in \{1,2,\ldots,M\}$, a cluster D_m consists of all such points from $\{\mathbf{x}_i\}$ which are the closest to $\boldsymbol{\mu}_m$ in terms of the L-norm. That is, these are the points which fulfill the condition

$$D_m = \left\{ \mathbf{x}_i : \underset{\substack{1 \leq l \leq N \\ 1 \leq q \leq M, q \neq m}}{\forall} \left\| \mathbf{x}_i - \boldsymbol{\mu}_m \right\|_L < \left\| \mathbf{x}_i - \boldsymbol{\mu}_q \right\|_L \right\}. \tag{3.230}$$

To find a solution to Equation (3.228) let us try to find out for which values of $\boldsymbol{\mu}_m$ the following condition is met

$$\frac{\partial Q}{\partial \mathbf{x}_i} = \sum_{m=1}^{M} \sum_{i=1}^{N} w_{mi} \frac{\partial d_{im}^2}{\partial \mathbf{x}_i} = 0. \tag{3.231}$$

A derivative of d_{im} can be easily obtained by first expanding it to the following form

$$d_{im}^2 = \left\| \mathbf{x}_i - \boldsymbol{\mu}_m \right\|_E^2 = \langle (\mathbf{x}_i - \boldsymbol{\mu}_m), (\mathbf{x}_i - \boldsymbol{\mu}_m) \rangle = \langle \mathbf{x}_i, \mathbf{x}_i \rangle - 2\langle \mathbf{x}_i, \boldsymbol{\mu}_m \rangle + \langle \boldsymbol{\mu}_m, \boldsymbol{\mu}_m \rangle$$
$$= \mathbf{x}_i^T \mathbf{x}_i - 2\mathbf{x}_i^T \boldsymbol{\mu}_m + \boldsymbol{\mu}_m^T \boldsymbol{\mu}_m. \tag{3.232}$$

Then

$$\frac{\partial d_{im}^2}{\partial \mathbf{x}_i} = 2\left(\mathbf{x}_i - \boldsymbol{\mu}_m \right). \tag{3.233}$$

Inserting Equation (3.233) into Equation (3.231) and expanding the latter yields

$$\sum_{i=1}^{N} w_{1i}\left(\mathbf{x}_i - \boldsymbol{\mu}_1\right) + \sum_{i=1}^{N} w_{2i}\left(\mathbf{x}_i - \boldsymbol{\mu}_2\right) + \cdots + \sum_{i=1}^{N} w_{Mi}\left(\mathbf{x}_i - \boldsymbol{\mu}_M\right) = 0. \quad (3.234)$$

Since a point $\mathbf{x}_i$ can only belong exclusively to one cluster, therefore its corresponding weight $w_{\cdot i}$ is 1 only for one cluster. Therefore to meet the condition (3.234) for all points $\mathbf{x}_i$, each $\boldsymbol{\mu}_m$ needs to fulfill the following condition

$$\sum_{i=1}^{N} \left(w_{1i}\mathbf{x}_i - w_{1i}\boldsymbol{\mu}_1\right) = 0,$$

$$\sum_{i=1}^{N} w_{mi}\mathbf{x}_i - \boldsymbol{\mu}_m \sum_{i=1}^{N} w_{mi} = 0,$$

which finally leads to the following expression

$$\boldsymbol{\mu}_m = \frac{1}{N_m} \sum_{\mathbf{x}_i \in D_m} \mathbf{x}_i = \frac{\displaystyle\sum_{i=1}^{N} w_{im}\mathbf{x}_i}{\displaystyle\sum_{i=1}^{N} w_{im}}, \quad (3.235)$$

where N_m denotes a number of points belonging to the m-th cluster D_m, which equals the sum of all weights w_{im} for that cluster. The second derivative of Equation (3.234) with respect to $\mathbf{x}_i$ is positive due to the property (3.229), so Equation (3.228) gets minimized with Equation (3.235).

The above two steps (3.230) and (3.235) are repeated until a convergence state is reached, i.e. there are no new re-assignments of points to the clusters. This way means $\boldsymbol{\mu}_m$ do not change between two consecutive iterations k and $k + 1$, respectively. That is

$$\mathop{\forall}_{1 \le m \le M} : \left\| \boldsymbol{\mu}_m^{(k)} - \boldsymbol{\mu}_m^{(k+1)} \right\| < \varepsilon, \quad (3.236)$$

for some positive value of a threshold ε. Let us also observe that due to Equation (3.235) the means $\boldsymbol{\mu}_m$, except for the initial setup, do not necessarily belong to the input data set $\{\mathbf{x}_i\}$. In other words, these are usually "new" data points. For some types of data this might be problematic.

In practice the necessity of the initial selection of the number of clusters is the most troublesome step since usually there are no clear criteria for them. This means also that the method can proceed differently for each initialization since the initial means are selected stochastically. This is the reason why during experiments application of the same clustering method frequently leads to different results. To somewhat remedy this, such experiments need to be run a number of times and the results analyzed collectively, providing their statistical properties in a form of some measures on cluster quality, such as the ones expressed by formulas (3.257) or (3.258), as will be discussed in Section 3.11.4.

It is worth noticing that the discussed k-means is a tool for information processing, especially for (lossy) signal compression by the so called *vector quantization*, as already pointed out in Section 2.12.9. At first, the existence of specific features is assumed in a signal. In color images, these can be color codes by themselves. In monochrome images a practical approach is to select blocks of 2×2 or 3×3 pixels which form 4D or 9D features, respectively. Then k-means is run on the signal data with an *a priori* chosen number of k centers. In effect, k possibly new centers are obtained with partitioning information for each data point, i.e. it is an index which uniquely assigns data to its closest centroid. This process is called signal *encoding*, while the centers found are stored in the so called *codebook*. Now the real compression step commences. It consists of replacing each data with a number (a code or index) of the partition it belongs to. Practically, this number requires much fewer bits than the original feature vector. For instance, if an 8-bit index can be substituted for each 4D vector, then we easily receive 4 times compression. Nevertheless, we need also to store the codebook as well. In practice there are many factors, such as the number and initial values of the centroids, coding method, etc., that need to be properly chosen for a method to result in subjectively best image quality. More information can be found in the literature [3, 42, 152].

Finally, let us mention that there are close relations between the k-means and the graph based clustering methods, called spectral clustering, as presented e.g. in the paper by Filipone [150]. There is also a close relation of these to the NMF method [153], discussed in Section 2.12.10, as well as the vector quantization already outlined in Section 2.12.9 in the discussion on subspace data representation. This, in turn, leads to another interesting group of methods called *self organizing maps* (SOM), originally proposed by Kohonen [154].

3.11.2 Fuzzy c-Means

Bezdek proposed a fuzzy version of the k-means algorithm described in the previous section and called it the fuzzy c-means (FCM) [155]. The main idea here is to allow points to belong simultaneously to many clusters, with different membership values, however. Such a strategy frequently pays off in practice, since in many cases points are not tightly grouped around the means. Under this formulation, the optimization problem Q in Equation (3.228) becomes modified as follows

$$Q = \min \left(\sum_{m=1}^{M} \sum_{i=1}^{N} w_{mi}^{\gamma} \underbrace{\left\| \mathbf{x}_i - \boldsymbol{\mu}_m \right\|_L^2}_{d_{im}} \right), \tag{3.237}$$

where γ is a parameter that controls the fuzziness of the solution. It should fulfill the condition $\gamma \geq 1$, usually also γ is kept lower than 10. When γ approaches 1, the described algorithm becomes the crisp version of the k-means method described in the previous section. A derivation of this property can be found e.g. in Frigui [156]. Also the weights w_{im} can now take *any* value from the range from 0 to 1, i.e.

$$w_{mi} \in [0, 1]. \tag{3.238}$$

Thus, w_{mi} describes a fuzzy membership of a point $\mathbf{x}_i$ into an m-th cluster, which in this context can also be seen as a probability measure, since for any i it is assumed that

$$\sum_{m=1}^{M} w_{mi} = 1. \tag{3.239}$$

It holds also that

$$0 < \sum_{i=1}^{N} w_{mi} < N. \tag{3.240}$$

A solution to the fuzzy c-means is given by Equation (3.228) with the constraints (3.238) and (3.239) is given as follows. Looking for a solution to the optimization problem (3.228) we can partially rely on the already computed optimal conditions (3.231) with respect to $\mathbf{x}_i$. However, in this case we need to consider that the weight w is raised to the power γ and that w is a continuous value in the range [0,1]. Therefore we need to find out the optimality conditions with respect to w_{mi}, with the constraint (3.239), however. For this purpose the following Lagrange optimization functional[17] is created [155, 158]

$$L(w_{mi}) = \underbrace{\sum_{m=1}^{M} \sum_{i=1}^{N} w_{mi}^{\gamma} d_{im}^2}_{a} - \underbrace{\lambda \left(\sum_{m=1}^{M} w_{mi} - 1 \right)}_{b}. \tag{3.241}$$

Now, taking into consideration that a distance d_{im} between the points is independent of the weights, the two derivatives of L with respect to w_{mi} and λ can be evaluated as follows

$$\frac{\partial L(w_{mi}, \lambda)}{\partial w_{mi}} = \gamma w_{mi}^{\gamma-1} d_{mi}^2 - \lambda, \text{ and} \tag{3.242}$$

$$\frac{\partial L(w_{mi}, \lambda)}{\partial \lambda} = \sum_{m=1}^{M} w_{mi} - 1. \tag{3.243}$$

Equating Equation (3.242) to 0 yields

$$w_{mi} = \left(\frac{\lambda}{\gamma d_{mi}^2} \right)^{\frac{1}{\gamma-1}}. \tag{3.244}$$

---

[17]The theorem of Lagrange multipliers provides sufficient conditions for a function $a(\mathbf{x})$ to reach its extreme, given certain constraint $b(\mathbf{x}) = 0$. For this purpose the functional

$$L(\mathbf{x}) = a(\mathbf{x}) - \lambda b(\mathbf{x}),$$

is constructed, where λ denotes a so called Lagrange multiplier [31, 81, 159]. The sufficient conditions for existence of an extreme of $a(\mathbf{x})$ require vanishing of the following partial derivatives

$$\frac{\partial L(\mathbf{x}, \lambda)}{\partial \mathbf{x}} = 0 \text{ and } \frac{\partial L(\mathbf{x}, \lambda)}{\partial \lambda} = 0.$$

By the same token, equating Equation (3.243) to 0 leads to the following condition

$$\sum_{m=1}^{M} w_{mi} = 1,$$

(3.245)

which complies with assumption (3.239). Now, inserting Equation (3.244) into Equation (3.245) yields

$$\sum_{m=1}^{M} \left(\frac{\lambda}{\gamma d_{mi}^2} \right)^{\frac{1}{\gamma-1}} = 1,$$

(3.246)

which, after some rearrangements, provides

$$\frac{\lambda}{\gamma} = \left[\sum_{m=1}^{M} \left(\frac{1}{d_{mi}^2} \right)^{\frac{1}{\gamma-1}} \right]^{\gamma-1}.$$

(3.247)

Finally, inserting Equation (3.247) into Equation (3.244) yields

$$w_{mi} = \frac{\left(d_{mi}^2 \right)^{\frac{1}{1-\gamma}}}{\sum_{p=1}^{M} \left(d_{pi}^2 \right)^{\frac{1}{1-\gamma}}}.$$

(3.248)

The denominator of the above equation contains a summation that spans through distances of a single i-th point $\mathbf{x}_i$ to all possible mean points $\boldsymbol{\mu}_m$. Therefore we have changed the summation index to p to avoid confusion. If $d_{mi} = 0$ for some indices m and i, then $w_{mi} = 1$ and all other weights for that cluster are set to 0. For the weights given by Equation (3.248), the means are computed again in accordance with Equation (3.235). Additionally, the algorithm proceeds iteratively until no change in w_{mi} or $\boldsymbol{\mu}_m$ is detected.

Finally, it is worth noticing that for some problems the constraint (3.239) is too restrictive, i.e. only a relative distance of the points to the means is taken into consideration when assigning membership values of the points to the clusters. By dropping Equation (3.239) the fuzzy c-means becomes a *possibilistic* c-means, as proposed by Krishnapuram and Keller [160, 161]. In this version, the membership w_{mi} denotes a possibility of a point $\mathbf{x}_i$ being a member of the m-th cluster.

3.11.3   Kernel Fuzzy c-Means

A kernel version of the fuzzy c-means assumes mapping of data into the feature space, as discussed in Section 3.10, and then applying fuzzy c-means clustering. The method was first proposed by Wu *et al.* [162]. Again, the main expectation is to obtain better partitioning of data in the feature space. Similarly to the fuzzy c-means, after initialization of the mean values $\boldsymbol{\mu}_i$ and choice of the number of clusters M, the method proceeds iteratively by assigning each point $\mathbf{x}_i$ to the closest mean $\mathbf{\mu}_m$ in M clusters, this time in the feature space however. The functional (3.228) now becomes

$$Q = \min \left(\sum_{m=1}^{M} \sum_{i=1}^{N} w_{mi}^{\gamma} \underbrace{\| \Phi(\mathbf{x}_i) - \Phi(\mathbf{\mu}_m) \|^2}_{D_{im}^{2\Phi}} \right), \tag{3.249}$$

where N again denotes a number of data points and γ is a fuzziness parameter. In the above $\Phi(\mathbf{x}_i)$ denotes a function that maps the data point $\mathbf{x}_i$ into the higher dimensional feature space. However, as already pointed out in Section 3.10, due to the kernel trick we never need to directly compute values of $\Phi(\mathbf{x}_i)$, since in all steps of the algorithm we always compute mutual distances of points in the feature space.

Let us now expand a square distance $D_{im}^{2\Phi}$ between the points $\mathbf{x}_i$ and $\mathbf{x}_r$ in the *feature* space. This will appear useful in further derivations and will give us an insight into the method of distance computation in the feature space. $D_{im}^{2\Phi}$ evaluates as follows

$$D_{im}^{2\Phi}(\mathbf{x}_m, \mathbf{x}_i) = \| \Phi(\mathbf{x}_m) - \Phi(\mathbf{x}_i) \|^2 = \langle (\Phi(\mathbf{x}_m) - \Phi(\mathbf{x}_i)), (\Phi(\mathbf{x}_m) - \Phi(\mathbf{x}_i)) \rangle =$$

$$\langle \Phi(\mathbf{x}_m), (\Phi(\mathbf{x}_m) - \Phi(\mathbf{x}_i)) \rangle - \langle \Phi(\mathbf{x}_i), (\Phi(\mathbf{x}_m) - \Phi(\mathbf{x}_i)) \rangle =$$

$$\langle \Phi(\mathbf{x}_m), \Phi(\mathbf{x}_m) \rangle - \langle \Phi(\mathbf{x}_m), \Phi(\mathbf{x}_i) \rangle - \langle \Phi(\mathbf{x}_i), \Phi(\mathbf{x}_m) \rangle + \langle \Phi(\mathbf{x}_i), \Phi(\mathbf{x}_i) \rangle = \tag{3.250}$$

$$K(\mathbf{x}_m, \mathbf{x}_m) - 2K(\mathbf{x}_m, \mathbf{x}_i) + K(\mathbf{x}_i, \mathbf{x}_i)$$

In the above we used respectively a definition of a vector norm (3.197), the distributive property of the inner product (3.196), as well as the kernel property (3.202). Equation (3.250) can be used to measure a squared distance in the feature space between two points if their positions are known in the data space. However, this is not always the case.

Again and as in the previous versions of the k-means algorithms we also start from the choice of the initial mean values $\mathbf{\mu}_m$. These can be randomly chosen from the data points. Then the initial distances can be easily computed based on Equation (3.250). However, in the case of the kernel c-means formulation of the mean values is usually indirect, i.e. we do not have an explicit formulation of $\mathbf{\mu}_m$ in the feature space and usually we cannot map them back to the data space. In general, we cannot even assume the existence of $\mathbf{\mu}_m$ in the data space. This is called the pre-image problem, discussed in Section 3.10. Nevertheless, each center in the feature space can be expressed as a weighted sum of all the data points. That is, based on Equation (3.204) a center $\mathbf{\mu}_m^{\Phi}$ can be expressed indirectly in the feature space as

$$\mathbf{\mu}_m^{\Phi} = \sum_{i=1}^{N} a_{mi} \Phi(\mathbf{x}_i), \tag{3.251}$$

where a_{mi} denote scalar weights. Under this assumption the formula (3.250) on distances between data and the mean point becomes

$$D_{mi}^{2\Phi}(\mathbf{x}_i, \mathbf{\mu}_m) = K(\mathbf{x}_i, \mathbf{x}_i) - 2 \sum_{s=1}^{N} a_{ms} K(\mathbf{x}_i, \mathbf{x}_s) + \sum_{s=1}^{N} \sum_{t=1}^{N} a_{ms} a_{mt} K(\mathbf{x}_s, \mathbf{x}_t). \tag{3.252}$$

With this assumption minimization of Q in Equation (3.249) leads to the following update equations for the values of the weights w_{mi} in Equation (3.230) and a_{mi} in Equation (3.251), as follows [162]

$$w_{mi} = \frac{\left(D_{mi}^{2\Phi}\right)^{1/(1-\gamma)}}{\sum_{k=1}^{M} \left(D_{ki}^{2\Phi}\right)^{1/(1-\gamma)}}, \tag{3.253}$$

$$a_{mi} = \frac{w_{mi}^{\gamma}}{\sum_{s=1}^{N} w_{ms}^{\gamma}}. \tag{3.254}$$

If for some indices m, i it holds that $D_{mi} = 0$, then $w_{mi} = 1$ and all other weights for that cluster become 0. Finally, the new mean values change in accordance with Equation (3.251), but they do not need to be directly computed. However, their change entails new distances between the points, which need to be recomputed from Equation (3.252).

The algorithm stops when a change of each weight w_{mi} in Equation (3.253) in two consecutive iterations is less than a chosen threshold value ε, i.e.

$$\left| w_{mi}^{(k)} - w_{mi}^{(k+1)} \right| < \varepsilon, \tag{3.255}$$

where k and $k+1$ denote two consecutive steps of the algorithm. In practice the convergence is very fast. However, for iterative procedures it is good practice to check an iterative loop for both, the convergence conditions (3.255), and an excessive number of runs, and stop the loop whatever is met first. Such a strategy was also implemented in the attached implementation of the clustering methods [6].

It is worth noticing that for some, especially large, sets of data the kernel fuzzy c-means can be simplified, as proposed by Zhang and Chen [163]. In their approach the main idea is to assume that cluster centers can always be looked for in the input data space, rather than indirectly in the feature space as described by Equation (3.251). Also, in this approach only the Gaussian kernel (3.210) is assumed. Although not in every instance, this approach may be well justified, in practice it greatly shortens computation time. A speed up in kernel computations can also be obtained with some more advanced methods, such as the compression proposed by Arif and Vela [164]. Their compression method assumes replacement of data projections from the input to the feature space with a nonlinear approximation by a neural network.

3.11.4 Measures of Cluster Quality

Means of assessment of the quality of segmentation are very important, especially if there is no prior information on the possible number of clusters in the data. Cluster validity measures, also called cluster validity indices, have found broad interest in the research community. However, depending on the type of data, the quality of clusters can be understand differently, since it depends on a chosen measure of a distance among data. As the simplest measure the ratio between cluster compactness and cluster separability can be evaluated [2]. As already pointed out, the idea of k-means clustering is to partition the input set of data into M clusters, such that the points inside each cluster are the closest to the mean of that cluster, whereas the cluster centers are maximally separated from each other. Thus, a qualitative insight into the k-means clustering can be gained by analyzing the total sums of distances, defined as follows

$$S_m = \sum_{x \in D_m} \| x - \mu_m \|^2,$$
(3.256)

$$S_t = \sum_{m=1}^{M} S_m.$$
(3.257)

These should be as minimal as possible. However, the k-means algorithm does not guarantee the global optimum in terms of S_m and S_t, though in practice the convergence condition(3.236) and (3.255) are met very fast. In other words, different starting conditions (i.e. number and positions of the means) can provide different solutions. Moreover, for any given solution it is difficult to tell whether it is optimal.

In the case of fuzzy versions of the k-means clustering, entropy of the membership values (3.253) can serve as an indicator of the quality of the clustering process. It can be computed from the following formula [155]

$$E(M) = - \sum_{m=1}^{M} \sum_{i=1}^{N} w_{mi} \log w_{mi}.$$
(3.258)

Thus, E depends on the type of data and the number of clusters M. If a number of clusters M is not known in advance, then clustering can be performed for a different number of clusters, and the number M^* with the minimal entropy (3.258) can be chosen to build an ensemble, that is

$$M^* = \arg \min_{M} (E(M)).$$
(3.259)

In practice such a strategy can be used to provide a hint to the number of means, which is usually data dependent.

Figure 3.28 depicts results of computation of the entropy in Equation (3.259) for five data sets *Iris, Ionosphere*, and *Wisconsing Breast Cancer* (*WBC*) from the Berkeley repository [165, 166], as well as for the data set of the red and blue samples used to segment an image in Figure 4.14 (p. 363), and the last being a set of manually gathered color points, as depicted in Figure 4.15 (p. 364), respectively. Based on Equation (3.259) for each data set the preferable number of clusters is two, except for the *Iris* set which has three classes.

Another cluster validity measure was proposed by Xie and Beni [167]

$$S = \frac{\sum_{m=1}^{M} \sum_{i=1}^{N} w_{mi}^2 \| x_i - \mu_m \|^2}{N \min_{i,m(i \neq m)} \left\{ \| \mu_i - \mu_m \|^2 \right\}},$$
(3.260)

that's smaller values indicate better partitioning by a fuzzy clustering algorithm. Such behavior is easily to read out from (3.260) since, as already pointed out, a principle of any c-mean

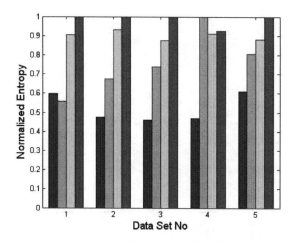

Figure 3.28 Normalized entropy for five data sets. Each data set was clustered a number of times for $M = 2 \ldots 5$ clusters. For each clustering, entropy was computed in accordance with Equation (3.258). Used data sets: 1-*Iris*, 2-*Ionosphere*, 3-*Wisconsin Breast Cancer* (*WBC*), 4-manually gathered sample points of red and blue colors of Polish road signs, 5- manually gathered point samples, as shown in Figure 4.15 (from [168]).

algorithm is to minimize the nominator of Equation (3.260). On the other hand, its denominator is maximized if the means are maximally spread apart from each other. In literature it is reported as the one of the most indicative performance measures for clustering [169, 170].

An interesting proposition of a cluster validity index that aims to find an optimal number of clusters in one of the variants of the fuzzy c-means algorithm was recently proposed by Le Capitaine and Frélicot [171]. This paper also contains an interesting overview of other existing cluster validity indices.

In this section we cited the measures which in some form were used in the methods discussed in this book. Nevertheless, the literature on the subject is ample. Some other interesting papers in the field, which contain further references, are [169, 170, 172]. The book by Theodoridis and Koutroumbas contains a chapter on cluster validity [5]. This list is certainly not complete.

3.11.5   IMPLEMENTATION Issues

Figure 3.29 depicts a flowchart of the family of k-means algorithms which have a number of common steps which differ only in the formulas used for their computations, as discussed in Sections 3.11.1 to 3.11.3, respectively. This property is used in object based implementation, defining a hierarchy of classes with virtual members which implement the abovementioned specific versions of these common steps.

Figure 3.30 depicts a hierarchy of the k-means class hierarchy. Actually there are two paths of classes, all originated from the base *k_Means* class. It contains all data structures and member functions used in all clustering activities. The most pronounced are two containers,

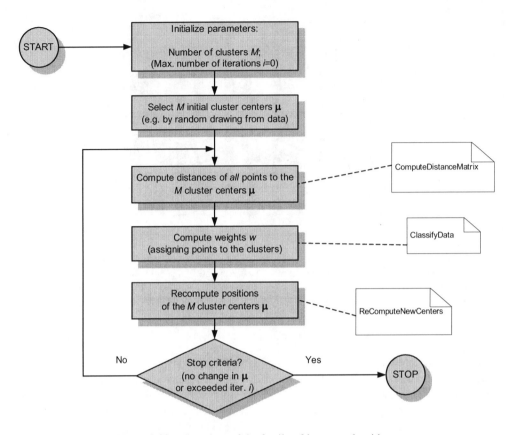

Figure 3.29 Flowchart of the family of *k*-means algorithms.

fDataRepository for the whole set of input (not segmented) points and *fMeans* which is a repository to store the *k*-mean points. The two most important operations are the overloaded function operators, depicted in Figure 3.30. When called they start the segmentation process. The only difference between them is that in the first a number of mean points is provided, which starting values are then randomly selected from the data repository. The second function allows the starting mean points to be provided by the caller (in the case of image segmentation these can be, for instance, the known start color values, etc.). The other members of the classes can be looked up in the attached code [6].

The derived *Fuzzy_k_Means* class implements the fuzzy *k*-means algorithm. Thus, it requires two new data members which are the exponent *fPowerFactor*, denoted as γ in Equation (3.249), as well as the *fProbability* matrix which stores membership values for each point with respect to each cluster, denoted as *w* in Equation (3.249). The next derived class is the *FuzzyAnnealing_k_Means* which constitutes a modification to its predecessor *Fuzzy_k_Means* and allows a change of the *fPowerFactor* on each iteration of the algorithm. For this purpose it contains three members which allow such iteration (*fPowerFactor_From*, *fPowerFactor_To*, and *fPowerFactor_Step*). Such a strategy helps in some situations to obtain better partitioning

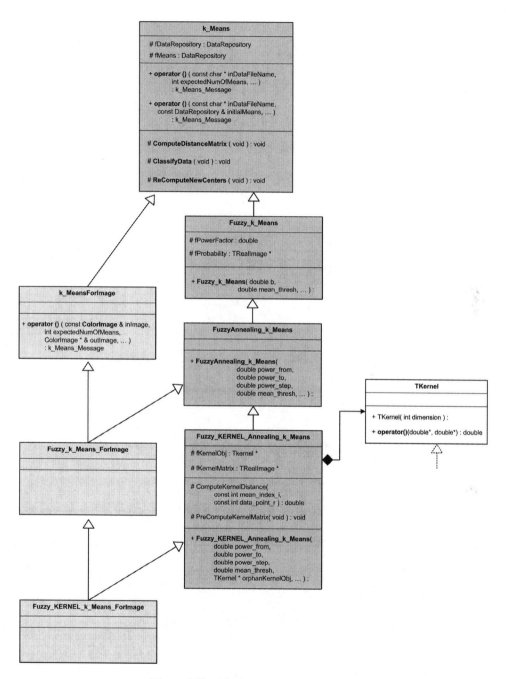

Figure 3.30   The *k_means* class hierarchy.

of data points. The last in the main path is the *Fuzzy_KERNEL_Annealing_k_Means* class which carries out *k*-means in the feature space thanks to the provided kernel object from the *TKernel* hierarchy (depicted in Figure 3.25, p. 302), as described in Section 3.11.3.

The addition to the main path is the path of *k*-means classes which specialize in clustering of color images. These are the *k_MeansForImage*, *Fuzzy_k_Means_ForImage*, and *Fuzzy_KERNEL_k_Means_ForImage* classes for the crisp, fuzzy, and kernel fuzzy *k*-means for color images, respectively. These are specialized and read data directly from the image rather than from an arbitrary data file. The other functions bear the same semantics as the case of the main path in the hierarchy.

Algorithm 3.24 presents a definition of the base class *k_Means*, implementing the base for the family of *k*-means methods. The main members implement the stages outlined in the flowchart in Figure 3.29.

```
/////////////////////////////////////////////////////////////////////////
// Codes to report specific actions in the k-means hierarchy
enum k_Means_Message {
    kOk,
    kCannotOpenFileForRead,
    kCannotOpenFileForWrite,

    // if number of chosen k-means exceeds number of data points
    kInsufficientNumberOfData,

    // if no convergence after set number of iterations of the k-means
    kExceededMaxNumOfIterations,

    // if e.g. data points in the input file are of different lengths
    kInconsistendDataFile,

    // if fDistanceTable is 0 or has wrong dimensions (means wrong order
    //    of function call)
    kDistanceTableNotInitialized,

    // if fProbability in the fuzzy k-means not created
    kProbFieldNotCreated_Error,

    // returned if there is a mismatch between internal
    // data structures and calling parameters
    kCallDimensionsMismatch_Error

    // add new messages below
    // ...
};
/////////////////////////////////////////////////////////////////////////


/////////////////////////////////////////////////////////////////
// This class implements the k-means algorithm.
/////////////////////////////////////////////////////////////////
```

```
class k_Means
{
    public:

        enum { kInvalidClass = -1 };

        typedef double ElemFeatureType;

        typedef FeatureSpaceFor< ElemFeatureType >::FeatureType
                                                        FeatureType;

        typedef FeatureSpaceFor< ElemFeatureType >::DataPoint
                                                        DataPoint;

        typedef FeatureSpaceFor< ElemFeatureType >::DataRepository
                                                        DataRepository;

    protected:

        DataRepository      fDataRepository;    // storage of all data points
        DataRepository      fMeans;             // storage of the mean centers

        // matrix of distances from the points to the mean-centers
        TRealImage *    fDistanceTable;

        typedef vector< int > IndexVector;

        // each data has an entry in this vector which contains index of
            a closest center
        IndexVector         fDataClasses;

        typedef vector< bool > ChangeVector;

        // contains flags which tell whether a corresponding center has
            changed
        ChangeVector        fCentreChangeIndicator;

    public:

        // =========================================================
        ///////////////////////////////////////////////////////////
        // Class default constructor
        ///////////////////////////////////////////////////////////
        //
        // INPUT:
        //      normalizeDataBeforeSegmentation - if true then
        //          initial data is normalized to the range:
        //          minNormalVal - maxNormalVal
        //
        // OUTPUT:
        //
        //
        // REMARKS:
```

```
    //
    //
    k_Means(      bool normalizeDataBeforeSegmentation = false,
                  ElemFeatureType minNormalVal = -1,
                  ElemFeatureType maxNormalVal = +1 );
    // =========================================================

public:

    //////////////////////////////////////////////////////////////
    // This function selects an initial set of the mean centers
    // in accordance with a chosen heuristics.
    //////////////////////////////////////////////////////////////
    //
    // INPUT:
    //    num_of_means_to_select - a number of mean centers
    //
    // OUTPUT:
    //    status of the operation
    //
    // REMARKS:
    //
    //
    virtual k_Means_Message SelectTheInitialMeans(
                                    int num_of_means_to_select );


    //////////////////////////////////////////////////////////////
    // This function copies the externally supplied means.
    //////////////////////////////////////////////////////////////
    //
    // INPUT:
    //      initialMeans - a repository with the initial mean values
    //
    // OUTPUT:
    //      status value
    //
    // REMARKS:
    //
    virtual k_Means_Message SelectTheInitialMeans(
                                    const DataRepository & initialMeans );

protected:

    //////////////////////////////////////////////////////////////
    // This function computes distances of all points to all
    // means
    //////////////////////////////////////////////////////////////
    //
    // INPUT:
    //      none
    //
    // OUTPUT:
    //      none
    //
```

```
// REMARKS:
//        fDataClasses is also filled - each data
//          is assigned kInvalidClass
//
virtual void ComputeDistanceMatrix( void );


//////////////////////////////////////////////////////////////
// This function assigns data to the classes (computes
// the weights w)
//////////////////////////////////////////////////////////////
//
// INPUT:
//     none
//
// OUTPUT:
//     none
//
// REMARKS:
//
//
virtual void ClassifyData( void );


//////////////////////////////////////////////////////////////
// This function re-computes new mean values
//////////////////////////////////////////////////////////////
//
// INPUT:
//       none
//
// OUTPUT:
//       number of means that have been updated
//       (i.e. number of changed values)
//
// REMARKS:
//       Internal class members are changed.
//
virtual int ReComputeNewCenters( void );


//////////////////////////////////////////////////////////////
// This function computes a squared Euclidean distance.
//////////////////////////////////////////////////////////////
//
// INPUT:
//       a, b - two data points for which a distance
//        is computed
//
// OUTPUT:
//       ** squared ** Euclidean distance
//
// REMARKS:
//
//
virtual double ComputeDistance( const DataPoint & a,
                                const DataPoint & b );
```

```
/////////////////////////////////////////////////////////////
// This function scales data to the required range.
// Each feature (scalar) is scaled independently.
/////////////////////////////////////////////////////////////
//
// INPUT:
//      kMinVal - a point which defines a minimal limit
//         for EACH component independently
//      kMaxVal - a point which defines a maximal limit
//         for EACH component independently
//
// OUTPUT:
//      true if ok
//
// REMARKS:
//
//
virtual bool NormalizeInputData(
  const FeatureSpaceFor< ElemFeatureType >::DataPoint & kMinVal,
  const FeatureSpaceFor< ElemFeatureType >::DataPoint & kMaxVal,
  FeatureSpaceFor< double >::DataPoint * minDataLimit = 0,
  FeatureSpaceFor< double >::DataPoint * maxDataLimit = 0 );

public:

/////////////////////////////////////////////////////////////
 // This function restores original data range
/////////////////////////////////////////////////////////////
//
// INPUT:
//      none
//
// OUTPUT:
//      none
//
// REMARKS:
//      fDataRepository is restored to the
//         original dynamic range
//      This function can be called after
//         operator ()
//
virtual void DeNormalizeInputData( void );

public:

/////////////////////////////////////////////////////////////
// This function does k-means on the input data.
/////////////////////////////////////////////////////////////
//
// INPUT:
//    inDataFileName - a full path of a data file; Each
//       data entry should occupy a single line (i.e. up to '\n').
//       Single features are separated by blanks (space, tab)
```

```
//      expectedNumOfMeans - number of k-means (i.e. "k")
//      outMeanCenters - optional repository for output of final
//        k-means
//      maxAllowableIteration - a fuse for max number of iterations
//      madeIterations - optional parameter for actual number
//        of iterations
//
// OUTPUT:
//      status of the operation (see k_Means_Message)
//
// REMARKS:
//
//
virtual k_Means_Message operator () ( const char * inDataFileName,
  int expectedNumOfMeans, DataRepository * outMeanCenters,
  int maxAllowableIteration = 1000, int * madeIterations = 0 );


//////////////////////////////////////////////////////////////
// This function does k-means on the input data.
//////////////////////////////////////////////////////////////
//
// INPUT:
//      inDataFileName - a full path of a data file; Each
//        data entry should occupy a single line (i.e. up to '\n').
//        Single features are separated by blanks (space, tab)
//      initialMeans - external table of the initial means
//      outMeanCenters - optional repository for output of final
//        k-means
//      maxAllowableIteration - a fuse for max number of iterations
//      madeIterations - optional parameter for actual number
//        of iterations
//
// OUTPUT:
//      status of the operation (see k_Means_Message)
//
// REMARKS:
//
//
virtual k_Means_Message operator () ( const char * inDataFileName,
    const DataRepository & initialMeans, DataRepository *
    outMeanCenters,
    int maxAllowableIteration = 1000, int * madeIterations = 0 );


//////////////////////////////////////////////////////////////
// This function does k-means on the input data.
//////////////////////////////////////////////////////////////
//
// INPUT:
//      outMeanCenters - optional repository for output of final
//        k-means
//      maxAllowableIteration - a fuse for max number of iterations
//      madeIterations - optional parameter for actual number
//        of iterations
//
```

```
// OUTPUT:
//      status of the operation (see k_Means_Message)
//
// REMARKS:
//
//      Here we assume that data has been loaded ok
//      to the member fDataRepository
//      and the means have been selected in the member fMeans
//      Also other member must be initialized.
//      These are: fDistanceTable and fDataClasses.
//
virtual k_Means_Message operator () ( DataRepository *
   outMeanCenters, int maxAllowableIteration, int * madeIterations );


public:

   ///////////////////////////////////////////////////////////
   // This function computes the squared error distances Ji
   // as well as the total error Je.
   ///////////////////////////////////////////////////////////
   //
   // INPUT:
   //
   //      Ji_matrix - A matrix of square distances (notation
   //         as in Duda et al., pg. 548)
   //      cluster_counter - Contains number of elements of each cluster
   //      Je - contains sum of all Ji
   //
   // OUTPUT:
   //
   //      kOk or error code
   //
   // REMARKS:
   //         It should be launched after classifiction.
   //
   k_Means_Message ComputeStatistics( vector< double > & Ji_matrix,
          vector< int > & cluster_counter, double & Je ) const;
};
```

Algorithm 3.24 Definition of the *k_Means* class (the most important members are shown).

   The class *k_Means* starts with a number of *typdef* definitions which facilitate further naming of important structures of the class. The role of the *SelectTheInitialMeans* members is selection of the specified number of initial means, which is the main input parameter that controls clustering. The protected virtual function *ComputeDistanceMatrix* computes the Euclidean distances between each point and the selected mean points. These are stored in the matrix data member *fDistanceTable*. An auxiliary function is the *ComputeDistance* which computes a distance between just two points. Also, there is a function *NormalizeInputData* which does the (optional) data normalization. The member *ClassifyData* assigns the points to the partitions represented by each mean data point. The next member, *ReComputeNewCenters*, recomputes new centers of the data partitions, which follows the corresponding step shown in the diagram in Figure 3.29. The whole k-means clustering is orchestrated by the set of *operator ()* functions which implement the whole action diagram in Figure 3.29. The presented members can be overridden in the derived classes. For instance, if data points are provided from other source than a file, then the *fDataRepository* repository can be directly loaded from that buffer. These and other code details can be looked up in the attached code [6]. The last member shown in Algorithm 3.24 is the *ComputeStatistics* function which computes the measure of cluster quality (3.256) and (3.257).

3.12   Support Vector Domain Description

The support vector machine (SVM) is a relatively new type of a classifier originally invented by Vapnik and first published in 1995 in the paper by Cortes and Vapnik [173]. Characteristic to SVM is transformation of data into so called feature space, in which the classification can hopefully be achieved with a linear hypersurface. However, amazingly at a first glance, the new space is of a *higher* dimension than the original. The transformation is carried out with a kernel, discussed in Section 3.10. However, as already discussed, direct computation of the high dimensional representation in the feature space is not necessary due to the kernel trick, which consists simply in computation of scalar products at all stages of classification. Recommended reading on architecture and many aspects of classification with SVM for different data can be found in the books by Vapnik [43], Haykin [42], and Abe [135], to name a few. In the rest of this section we concentrate on a relatively less known classification version of SVM in which only one class of patterns is available. Therefore this type of classifier is called a *one-class SVM* and the method *SVM domain description*. The method has proved to be very useful in image processing and computer vision, as will be shown. The one-class SVM was introduced and examined by Tax and Duin [174, 175], Ben-Hur *et al.* [176], as well as by Schölkopf and Smola [8] and Camastra and Verri in the context of data segmentation [177].

The idea of data clustering with SVM is simple and consists of computing a closed boundary around the n-dimensional input data in the form of an n-dimensional hypersphere, as shown in Figure 3.31. The hypersphere is characterized with its center $\mathbf{a}$ and radius r [174]. Points entirely enclosed inside the hypersphere are called inliers (the black dots in Figure 3.31). Points on the border are *support vectors* (SV), whereas outliers are outside the hypersphere (gray points). During operation a test point is classified as a belonging to the class if it falls inside this hypersphere. Otherwise it is rejected as an outlier.

When constructing the hypersphere its volume, which is proportional to r^n, should be minimal to tightly encompass the training points. Hence, to find parameters of the hypersphere the following minimization functional Θ can be created [174, 175]

$$\Theta\left(\mathbf{a}, r\right) = r^2 \tag{3.261}$$

with the constraint

$$\underset{1 \leq i \leq N}{\forall} : \|\mathbf{x}_i - \mathbf{a}\|^2 \leq r^2, \tag{3.262}$$

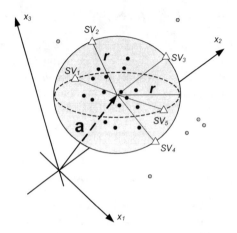

Figure 3.31   The idea of boundary classification methods is to build a hypersphere with a center **a** and radius r around a distribution of training points. Points entirely enclosed inside the hypersphere are called inliers (black dots). Points on the border of the hypersphere (triangles) are called support vectors (SV). Outliers are outside the hypersphere (gray points).

where $\mathbf{x}_i$ denotes a training data, N stands for a number of training data points, $\mathbf{a}$ is the center of the hypershpere and r its radius. It is worth noticing that in the above a minimization with respect to r^2 implies minimization with respect to r^n, simplifying further derivations though.

Additionally to allow some outliers in the training set and to make the classifier more robust, some of the training points are allowed to have their distances to the center of the hypersphere larger than r. For this purpose the slack variables ξ_i are entered into Equation (3.261) which represent an additional penalty factor. However, in the work by Bicego and Figueiredo [178] this idea was further extended by augmenting data points with some weights w_i which can represent the importance measure of each of the data points. For instance, the weights w_i can be the values w_{mi} assigned in the c-means clustering process after fixing the index m (see Equations (3.248) and (3.253)). Such an idea was proposed in the paper by Cyganek to build an ensemble of classifiers after the initial partitioning of the data space with unsupervised clustering [168]. This idea is further discussed in Section 4.2.4.

To account for the weights, these can simply multiply the slack variables ξ_i [178]. Hence, if with a point $\mathbf{x}_i$ a small weight w_i is associated, then its corresponding slack variable ξ_i will become larger, allowing $\mathbf{x}_i$ to lie further from the center $\mathbf{a}$ of the hypersphere. This, in effect, lowers an impact of $\mathbf{x}_i$ on the center and radius of the hypersphere. Incorporating these ideas into Equation (3.261) yields

$$\Theta\left(\mathbf{a}, r\right) = r^2 + C \sum_{i=1}^{N} w_i \xi_i \tag{3.263}$$

with the following constraints to ensure that almost all objects are within the hypersphere

$$\forall_i : \|\mathbf{x}_i - \mathbf{a}\|^2 \le r^2 + \xi_i, \quad \xi_i \ge 0, \quad 0 \le w_i \le 1, \tag{3.264}$$

where C is a parameter that controls the optimization process. The larger C, the fewer outliers are possible at the larger volume of the hypersphere.

Given a set of training points $\{\mathbf{x}_i\}$, solution to the Equations (3.263) and (3.264), can be obtained by means of the Lagrange multipliers [174, 175]

$$Q_L(r, \mathbf{a}, \alpha_i, \beta_i, \xi_i) = r^2 - \sum_i \alpha_i \left[r^2 + \xi_i - \|\mathbf{x}_i - \mathbf{a}\|^2 \right] + C \sum_i w_i \xi_i - \sum_i \beta_i \xi_i, \quad (3.265)$$

in which

$$\alpha_i \geq 0 \text{ and } \beta_i \geq 0, \quad (3.266)$$

are so called Lagrange multipliers for the constraints (3.264) [159, 179]. The functional Q_L has to be minimized with respect to r, $\mathbf{a}$, and ξ_i, and simultaneously maximized with respect to α_i and β_i. To solve this problem partial derivatives of Q_L with respect to the mentioned variables are computed, and after that their roots need to be found, as follows

$$\frac{\partial Q_L}{\partial r} = 0: \quad \frac{\partial Q_L}{\partial r} = 2r \left(1 - \sum_i \alpha_i \right), \quad (3.267)$$

$$\frac{\partial Q_L}{\partial \mathbf{a}} = 0: \quad \frac{\partial Q_L}{\partial \mathbf{a}} = 2 \left(\mathbf{a} \sum_i \alpha_i - \sum_i \alpha_i \mathbf{x}_i \right) = 0, \quad (3.268)$$

$$\frac{\partial Q_L}{\partial \xi_i} = 0: \quad \frac{\partial Q_L}{\partial \xi_i} = Cw_i - \alpha_i - \beta_i = 0. \quad (3.269)$$

Equation (3.267) immediately yields

$$\sum_i \alpha_i = 1. \quad (3.270)$$

The above and Equation (3.268) yield

$$\mathbf{a} = \sum_i \alpha_i \mathbf{x}_i. \quad (3.271)$$

This means that the center $\mathbf{a}$ can be uniquely expressed as a linear combination of data points for which $\alpha_i \neq 0$. Finally, from Equation (3.269) the following condition is obtained

$$\alpha_i = Cw_i - \beta_i, \quad (3.272)$$

which can be written as $\beta_i = Cw_i - \alpha_i$. In conjunction with Equation (3.266) this indicates that if α_i is further constrained to the range

$$0 \leq \alpha_i \leq Cw_i, \quad (3.273)$$

then Equation (3.266) is always fulfilled, and this way β_i can be replaced with $(Cw_i - \alpha_i)$.

Now by introducing Equations (3.270)–(3.272) into Equation (3.265) the so called Wolfe dual form Q_W is obtained [159, 179]

$$Q_W = r^2 + C \sum_i w_i \xi_i - r^2 \sum_i \alpha_i - \sum_i \alpha_i \xi_i + \sum_i \alpha_i \|\mathbf{x}_i\|^2 - 2\mathbf{a} \sum_i \alpha_i \mathbf{x}_i$$
$$+ \|\mathbf{a}\|^2 \sum_i \alpha_i - \sum_i (Cw_i - \alpha_i) \xi_i \qquad (3.274)$$

which reduces to

$$Q_W = \sum_{i=1}^{N} \alpha_i K(\mathbf{x}_i, \mathbf{x}_i) - \sum_{j=1}^{N} \alpha_j \sum_{i=1}^{N} \alpha_i K(\mathbf{x}_i, \mathbf{x}_j). \qquad (3.275)$$

In this case $K(\mathbf{x}_i, \mathbf{x}_j)$ denotes a scalar product between its arguments. However, it can be replaced with any other kernel, such as the very popular Gaussian kernel K_G in Equation (3.210), or others already discussed in Section 3.10. The set of α_i that maximizes Equation (3.275) can be computed taking into consideration the constraint provided in Equation (3.273).

The Kuhn–Tucker optimality conditions indicate that at the optimal point of the functional Equation (3.265) the following conditions need to hold [31]

$$\xi_i \beta_i = 0, \qquad (3.276)$$
$$\alpha_i \left[r^2 + \xi_i - \|\mathbf{x}_i - \mathbf{a}\|^2 \right] = 0. \qquad (3.277)$$

for all i. Hence from Equation (3.277) we conclude that a training point $\mathbf{x}_i$ for which $\alpha_i > 0$ and $\xi_i > 0$ lies outside the hypersphere. For such points from Equation (3.276) it follows that $\beta_i = 0$. Similarly from Equation (3.273) it follows that $\alpha_i = Cw_i$. Such points are called *bounded support vectors* (BSV). On the other hand, if for a point $\mathbf{x}_i$ its corresponding $\xi_i = 0$, then such a point falls inside the hypersphere or it lies exactly on its border. To differentiate this case it is easy to observe that if for a given point $\mathbf{x}_i$ it holds that $0 < \alpha_i < Cw_i$, then in consequence such a point falls exactly on the border of the hypersphere. These points become the *support vectors* (SV). All other points are inliers for which, to reinforce Equation (3.277), the corresponding $\alpha_i = 0$. These three different groups of points are exemplified in Figure 3.31. They are also summarized in Table 3.8.

Let us observe that at least two support vector points are necessary to define a hypersphere (since they lie on its border, whereas the BSVs are excluded at the same time). In light of

Table 3.8 Summary of the groups of points in relation to the hypersphere.

Point category	Relation to the hypersphere	Conditions on α_i, β_i, and ξ_i	
Inliers	$\|\mathbf{x}_i - \mathbf{a}\|^2 < r^2$	$\alpha_i = 0$, $\beta_i = C$, and $\xi_i = 0$	(3.278)
Support vectors (SV)	$\|\mathbf{x}_i - \mathbf{a}\|^2 = r^2$	$0 < \alpha_i < Cw_i$, $\beta_i > 0$, and $\xi_i = 0$	(3.279)
Bounded support vectors (BSV)	$\|\mathbf{x}_i - \mathbf{a}\|^2 > r^2$	$\alpha_i = Cw_i$, $\beta_i = 0$, and $\xi_i > 0$	(3.280)

Equation (3.270), we notice that if $C \geq 1$ then Equation (3.269) cannot be fulfilled for any i. Thus, in this case there are not any outliers, i.e. there are not any BSVs.

Based on the above derivations, a closed formula of a distance d from the center $\mathbf{a}$ of the hypersphere to a test point $\mathbf{x}_x$ can be derived. That is, if $d \leq r$, then

$$d^2(\mathbf{x}_x, \mathbf{a}) \leq r^2 \tag{3.281}$$

and we classify $\mathbf{x}_x$ as belonging to the class enclosed by this hypersphere. Otherwise, it is an outlier. From Equations (3.250) and (3.271) d^2 can be expressed as follows

$$d^2(\mathbf{x}_x, \mathbf{a}) = K(\mathbf{x}_x, \mathbf{x}_x) - 2 \sum_{i \in Idx(SV)} \alpha_i K(\mathbf{x}_x, \mathbf{x}_i) + \sum_{j \in Idx(SV)} \alpha_j \sum_{i \in Idx(SV)} \alpha_i K(\mathbf{x}_i, \mathbf{x}_j), \tag{3.282}$$

where $Idx(SV)$ denotes a set of indices of all support vectors found for this problem. The summation in the above takes on only such $\mathbf{x}_i$ which are SVs, since for the inliers we have $\alpha_i = 0$, whereas BSVs do not fulfill the required optimization criteria.

The SVs are placed on the boundary of the hypersphere, and thus are equidistant to $\mathbf{a}$. Therefore the following holds

$$r^2 = d^2(\mathbf{x}_s, \mathbf{a}) = K(\mathbf{x}_s, \mathbf{x}_s) - 2 \sum_{i \in Idx(SV)} \alpha_i K(\mathbf{x}_s, \mathbf{x}_i) + \sum_{j \in Idx(SV)} \alpha_j \sum_{i \in Idx(SV)} \alpha_i K(\mathbf{x}_i, \mathbf{x}_j), \tag{3.283}$$

where $\mathbf{x}_s$ denotes any of the support vectors. Introducing Equations (3.282) and (3.283) into Equation (3.281) yields

$$K(\mathbf{x}_x, \mathbf{x}_x) - 2 \sum_{i \in Idx(SV)} \alpha_i K(\mathbf{x}_x, \mathbf{x}_i) \leq K(\mathbf{x}_s, \mathbf{x}_s) - 2 \sum_{i \in Idx(SV)} \alpha_i K(\mathbf{x}_s, \mathbf{x}_i), \tag{3.284}$$

which for some kernels satisfying

$$K(\mathbf{x}_r, \mathbf{x}_r) = const, \tag{3.285}$$

for all r, e.g. the Gaussian kernel, simplifies to computation of the following parameter

$$\delta = \sum_{i \in Idx(SV)} \alpha_i K(\mathbf{x}_x, \mathbf{x}_i) - \underbrace{\sum_{i \in Idx(SV)} \alpha_i K(\mathbf{x}_s, \mathbf{x}_i)}_{\tau} \tag{3.286}$$

and finally checking if

$$\delta \geq 0. \tag{3.287}$$

The second term in Equation (3.286) is constant in the recognition stage, thus it can be precomputed to a value τ which denotes a cumulative kernel-distance of a chosen SV to all other SVs. Thus, the training process of the OC-SVM needs to compute the set of support vectors as well as their corresponding parameters α_i. On the other hand, classification of a test pattern $\mathbf{x}_x$ follows Equation (3.287). It simply checks if $\mathbf{x}_x$ belongs to a class represented by a set of SVs or not.

Finally, let us observe that after inserting Equation (3.273) into (3.270) it can be concluded that the weights should also be constrained with respect to the training parameter C, as follows

$$\sum_{i=1}^{N} w_i \geq \frac{1}{C}. \tag{3.288}$$

As shown by Tax in [180], comparing the SVM classification rule (3.286) and Gaussian kernel K_G (3.210) with the mixture of Gaussian equation (3.83), it is easy to note that such a SVM is actually a MoG classifier with α_i being weighting parameters. As alluded to in Section 3.5, for the latter it is required to sum up to 1, which condition is also fulfilled for α_i thanks to the property (3.270). However, in contrast to the parametric formulation of MoG, the SVM-RBF classifier does not require *a priori* selection of the number of the Gaussians J in Equation (3.83). These are selected automatically when solving the SVM quadratic maximizing problem (3.265), which is a great advantage. Moreover, for sufficiently large values of the parameter $\gamma = 1/\sigma^2$ in Equation (3.210), i.e. values larger than an average nearest neighbor distance among data, all kernel values of the K_G tend toward zero, since the following holds

$$\lim_{\gamma > \gamma_{\min}, i \neq j} e^{-\gamma \|\mathbf{x}_i - \mathbf{x}_j\|^2} = 0, \tag{3.289}$$

where $\gamma_{\min} = 1/\min^2 \|\mathbf{x}_i\text{-}\mathbf{x}_j\|$ (for all pairs of i, j such that $i \neq j$) is a minimal squared distance found among a pair of training data points. In this case, considering the constraint (3.270), it follows that the Lagrangian (3.265) is minimized when all data objects become support vectors with equal weights $\alpha_i = 1/N$. Thus, each data point supports a locally limited and equally weighted Gaussian function spread around that point. In effect in this case SVM is identical to the nonparametric Parzen density estimation discussed in Section 3.7.4. At the other extreme, for sufficiently small values of γ, i.e. $\gamma < \gamma_{\max} = 1/\max^2 \|\mathbf{x}_i\text{-}\mathbf{x}_j\|$, it can be shown that SVM approximates a spherically shaped solution, i.e. the same as with the polynomial kernel with a degree 1 [175, 180].

Followed the paper by Tax and Duin [175] let us also note that under some assumptions the formulation presented in this section of the one-class SVM which is based on hypersphere model is equivalent to the hypersurface based classification presented by Schölkopf and Smola [8].

SVM classifiers have gained great attention in the pattern recognition community due to their superior results mostly with respect to improved generalization and classification. This was verified in different classification tasks and with variable data sets. As alluded to previously, this is an outcome of the two new concepts in SVMs of margin maximization and mapping into a feature domain. The latter is accomplished with the kernels. Therefore a new idea came about of augmenting other classification methods with the maximizing margins and/or application of kernels. Following this concept new types of kernel neural networks, kernel fuzzy classifiers, kernel-PCA, kernel-k-means, etc. have emerged [9, 135]. These constitute an interesting "upgrade" for all of the already established systems with "classical" classifiers, now with new properties, however.

The problems of feature selection and application of domain knowledge into the process of model building with SVM are analyzed in the paper by Barzilay and Brailovsky [181]. The authors strongly advocate using a knowledge domain to lower the number of features used to train SVM (and other classifiers as well). This remains in accordance with the methodology and practical systems presented in this book. An *a priori* knowledge domain allows a diminishing influence of noisy parameters in the high dimensional feature space, which usually results in the lower number of support vectors and better recognition accuracy. As an example, in [181] the *a priori* knowledge was used for construction of the kernel function. For the particular problem of binary texture classification a kernel was proposed which takes into consideration neighboring pixels around a chosen point in an image. As was demonstrated experimentally, this kernel outperformed other regular kernel functions (such as the polynomial, neural network, or RBF kernels), both in terms of the lower number of support vectors and the best recognition accuracy. The search for new methods of robust feature selection and kernel construction for the SVM is a very interesting and promising direction of further research in computer vision.

As alluded to previously, the original formulation of SVM assumes binary classification, for which exemplar prototype data for each class are available [173]. However, it is also assumed that the two groups contain roughly the same number of data points. The problem becomes complicated if the number of data in one of the classes is different by a few orders of magnitude. This happens for many computer vision tasks for which exemplars of that class are comparatively less numerous than all other objects. For instance the color segmentation problem falls into this category. A method for such a problem that employs SVM was proposed by Zou *et al.* [182], as well as by Pan *et al.* [183]. However, in their works the segmentation task was stated as a two-class problem which leads to the problematic choice of "background" color data. As a consequence an unbalanced training data set is obtained which necessitates a larger number of support vectors, as well as leading to a slower execution time. In such cases the discussed one-class version of SVM can be very useful as shown in the paper by Cyganek devoted to the real-time color segmentation for road sign selection [184]. This will be discussed further in Section 4.2.3.2.

Another interesting application of the one-class SVM is data filtering. For instance outlier denoising was proposed by Park *et al.* [185]. Their idea consists of projection of the data seen as outliers, i.e. data to be filtered, into a hypersphere which defines a set of pure data, and then finding a nearest neighbor to this projection. Such a neighbor can mean a denoised version of the outlier.

Finally, Jin *et al.* proposed a method for face recognition which is based on a color model with a mixture of Gaussians. However, classification is then performed with the help of the one-class SVM [186].

3.12.1   *Implementation of Support Vector Machines*

Since their invention, many methods of SVM training have been developed and implemented. The majority of implementations are based on the Sequential Minimal Optimization (SMO) method which efficiently solves the minimization problem as given in Equations (3.263) and (3.264). There are many available software packages for different mutations of the SVM classifiers. These and many other useful resources can be accessed from the website devoted to SVMs [187]. Well known implementations are provided in the *mySVM* package by Rüping [188], as well as *SVM^{light}* by Joachims [189]. Another, slightly simplified approach is discussed in the book by Press *et al.* [17]. However, the most popular and frequently used is the LIBSVM library provided by Chang and Lin [190]. The latter also contains implementation of the WOC-SVM, discussed in the previous section. This version with a suitable adapter interface was also used in the presented experiments.

3.12.2 Architecture of the Ensemble of One-Class Classifiers

The presented one-class classifiers are useful in all situations in which only exemplars of one class are available. For instance, sometimes it is possible to obtain only the parameters of a machine operating in its "normal" conditions. Then, to classify "normal" from "abnormal" situations a one-class classifier can be trained with the available normal cases. Then, this way a trained classifier can tell if a classified data point belongs to the "normal" cases or not, which can be interpreted as malfunction conditions for that machine. Such a situation frequently arises in the classification of objects encountered in digital images. In this case it is possible to provide an appearance model of an object for detection or tracking, whereas all the others for which it can be difficult to provide definitions can be treated by the classifier as unknown (or background).

However, in the case of OC-SVM classifiers and different classification data, which will be used in the examples presented in this section, parameters of the obtained classifier can be greatly extended by prior clustering of the input data into disjoined sets which, however, are less spread. Then, each partition is used independently to train one OC-SVM classifier. Thanks to the data clustering, each classifier operates with a significantly reduced and potentially more condensed partition of data. In other words, it is now easier to build a tighter hypersphere around data, as was presented in Figure 3.31. The outlined idea of unsupervised clustering and training of OC-SVM is depicted in Figure 3.32. Available data can be partitioned with any of the clustering methods which were discussed in Section 3.11. In the presented experiments the family of k-mean algorithms were used [168]. The main clustering parameter which needs to be selected is the number of expected partitions. However, as previously discussed, this can be chosen by experimentally performing clustering for different numbers of initial clusters, then choosing such a configuration which results in the best clustering quality, as discussed in Section 3.11.4. Choice of the fuzzy version of the k-means provides further information on data partitioning in a form of fuzzy membership values (3.248) which represents a degree of membership of a data point to a given partition, as discussed in Section 3.11.2. These values can be treated as weights for the weighted version of the OC-SVM classifiers. This idea, as well as application of the kernel version of k-means to partition the input data space, is further discussed in the paper [168].

During operation, each member classifier of the ensemble responds independently. Then, all responses are gathered and analyzed in the arbitration unit, as shown in Figure 3.32. Its role is to provide a single response of the ensemble. In the case of one-class training this is whether a tested object belongs to the represented class or not.

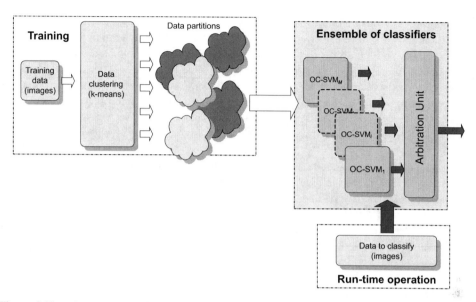

Figure 3.32 Block diagram of the ensemble of one-class classifiers. Each of the classifiers is trained with a data partition obtained after the unsupervised *k*-means clustering of the input data (white arrows). During run-time operation all classifiers independently classify a test pattern. Their answers are gathered in the arbitration unit which provides a final response (dark arrows).

Prior clustering of the input data space introduces *diversity* among the classifiers in the ensemble. Thanks to this the obtained accuracy of the ensemble is usually better than for a single classifier. The other way to obtain diversity is to use different classifiers or to apply boosting or bagging in the data set. The latter will be further discussed in Section 5.8.3.

k-means clustering of training data was previously proposed by Kuncheva, however with a different approach to classification [191]. Another similar system was proposed by Rahman and Verma. Their idea was to create ensembles of classifiers based on the layered clustering of the input data set [192]. A layer in this concept is an independent run of the *k*-means clustering method which, due to different settings, usually results in dissimilar data partitioning. In effect, the same pattern can belong to different clusters in different layers. However, in this way a diversity of classifiers can be achieved. In other words, a pattern can be used many times to train different base classifiers at different layers. In the experiments reported by Rahman and Verma the backpropagation neural networks were used (though it would be interesting to see other classifiers in this role). During operation for a test pattern, first an appropriate cluster at each layer is found by searching for a minimal distance between this test pattern and consecutively all centers of all clusters. This way at each layer a responding base classifier is found which is allowed to answer a class of the test pattern. The final response of the ensemble is obtained by means of the majority voting fusion method. The number of clusters for the *k*-means was chosen from 1 to 10, though in this method a higher number of clusters is beneficial. The method showed very good classification results on 20 benchmark data sets from the UCI Machine Learning Repository [138]. This was acknowledged by conducting a two-tailed Wilcox on signed rank tests [4].

Yang *et al.* [170] also propose a novel approach to the training of the *binary* SVM classifiers. Their main idea is first to perform prior clustering of data of the two training classes with the fuzzy c-means with Gaussian kernels. This yields data partitions and, additionally, thanks to the fuzzy clustering, each data is endowed with a value denoting its degree of membership to that cluster. Usually outliers and noise are assigned lower weights. Then, in this way the obtained partitions are checked for the farthest pair of clusters from different classes which are finally used to train the fuzzy version of the binary SVM. The method was tested on a number of benchmark and artificially generated data sets and showed better performance than the "classical" approach to the binary SVM. However, two things must be remembered when using this method. The first is that the k-means clustering algorithm depends on random initialization. Therefore usually the method is run a number of times and the best clustering – due to certain criteria – is retained [167]. The second thing about the method by Yang *et al.* is that the reported difference of accuracy between the fuzzy clustering FCM and its kernel version KFCM for many test benchmarks is unremarkable. In addition the latter requires more computational effort. Similar results are reported in the paper by Graves and Pedrycz [193]. These indicate that FCM and KFCM still deserve further research.

Finally, let us note that further discussion and examples on ensembles of classifiers in the context of pattern recognition follow in Section 5.6.

3.13 Appendix – MATLAB® and other Packages for Pattern Classification

A MATLAB® toolbox for pattern recognition, called *PRTools4*, was developed by Duin *et al.* [194]. It contains a few hundred MATLAB® procedures for different pattern recognition methods, such as linear and higher degree polynomial classifiers, normal density based classification, nonlinear classification, feature selection, regression, and many others for data set and image manipulation, as well as their plots.

For readers of the classical book by Duda *et al.* [2] there is a MATLAB® package of the pattern classification routines available [195]. It contains all the algorithms described in the book, as well as some additions. Also useful is a short introduction to MATLAB® and to its graphical user interface.

Also interesting and very useful is the Statistical Pattern Recognition Toolbox developed at the Czech Technical University [196]. It contains a vast number of MATLAB® functions for linear discriminating function, feature extraction, density estimation and clustering, SVM, visualization, and regression.

3.14 Closure

3.14.1 Chapter Summary

In this chapter a number of important classification techniques were presented. We started with formulation of the probabilistic classification framework. Then the group of subspace classification methods were discussed. One of the most important methods from this category is the principal component analysis and both its theoretical derivation as well as two different implementations are provided with details. PCA has found many applications in computer vision. We also present an application of PCA to the processing of multichannel images, as well as to the problem of background subtraction. Both exemplify two different implementations of PCA. However, subspace and PCA-like methodology is also employed in the tensor based classifier, discussed in Section (5.8.1). The basics of PCA lie in statistics, so not surprisingly just after the discussion on PCA we discuss statistical methods of object recognition. This section started with a short introduction to the probabilistic framework with the most noticeable Bayes decision rule as well as the maximum *a posteriori* classification method. Then the parametric and nonparametric methods were discussed. In the former we provided details of the mixture of Gaussians, especially the underlying importance of the feature subspace and its complement to proper formulation of the decision function. The next section dealt with the Kalman filter, one of the most important methods of object tracking. On the other hand, the section on nonparametric methods started from a discussion on a practical group of methods based on histogram analysis. Implementation of the 3D histograms was also provided. Then the Parzen and its derived nearest-neighbor methods were discussed. These were followed by a section on the very effective mean shift method that's main idea is to trace the mode of a distribution. As shown this can be applied to broad group of computer vision methods, such as filtering, segmentation, as well as object tracking. With respect to the latter, the effective *CamShift* approach, as well as its implementation, are provided. Neural networks were next in the group of discussed methods. The probabilistic, Hamming, and morphological networks were discussed. Then a section on important kernel methods followed. Their main idea is to transform features into a higher dimensional space in which classification can be linear. This idea was first applied to the support vector machines, and then incorporated into other classification methods. With these ideas in mind, we discussed a group of data clustering methods, derived from the k-means base algorithm. Finally, special one-class support vector machines were discussed. This was followed by a specification of the ensemble which joins data clustering and a set of cooperating one-class SVMs. The majority of the presented techniques were underpinned with working code examples and references to literature.

3.14.2 Further Reading

Good reading on linear spaces and projections is the book by Moon and Stirling on mathematical methods in signal processing [31]. To study details of linear algebra or just for revising some of the topics, the books by Meyer [130] and by Trefeten and Bau [197] can be recommended.

Kernel methods have become more and more popular due to their outstanding properties in pattern recognition. These were originally applied to SVM classifiers and were then employed in other, already known, classification methods such as neural networks, PCA, Fisher discriminating, or Bayes frameworks. These are very well described in the book by Schölkopf and Smola [8]. Also recommended reading on kernels is the book by Shawe-Taylor and Cristianini [9]. There is a special Internet web page devoted to the kernel methods in pattern recognition where one can find further links to the books, papers, software, etc. devoted to the kernel machines [198].

Recommended reading on architecture and many aspects of classification with SVM for different data can be found in the books by Vapnik [43], Haykin [42], and Abe [135],

A very comprehensive introduction and discussion on different segmentation methods is presented in the book by Theodoridis and Koutroumbas [5]. This is also recommended as a first stop for people looking for a good introductory text in pattern recognition. An in-depth treatment of different fuzzy clustering methods, as well as their applications, can be found in the excellent book edited by Oliveira and Pedrycz [151]. One of the first chapters of this book, written by Kruse *et al.*, provides a good overview on this group of clustering methods [199]. A competitive agglomeration version of the c-means, which allows an automatic determination of a number of cluster centers in the optimization process, can be found in the paper by Frigui and Krishnapuram [200]. A little older but also useful reading shedding light on the fuzzy c-varieties, c-shells, c-spherical, c-ellipsoidal, etc., is the book by Höppner *et al.* [158]. The paper by Filippone *et al.* contains a good overview on kernel and spectral clustering methods [150]. Finally, in-depth information on the varieties of clustering methods can be found in the books by Xu and Wunsch [149], as well as in that by Gan *et al.* [148]. A method of contour detection and hierarchical image segmentation is described in the paper by Arbeláez *et al.* [201]. This paper contains further references to the recent articles on the subject.

Methods and algorithms on systems (ensembles) of cooperating classifiers are addressed in the book by Kuncheva [4]. Many interesting issues on using ensemble methods in pattern classification can be looked up in the book by Rokach [202]. Finally, a nice introduction to ensemble based systems in decision making is the paper by Polikar [203].

Recommended reading on the tracking of multiple objects are the publications [69, 70, 101, 102, 103, 104]. Finally, a nice treatment on the design and analysis of machine learning experiments can be accessed in the book by Alpaydin [1].

Problems and Exercises

1. Show the formula (3.24) based on definitions of the mean and covariance matrix given in Equations (3.7) and (3.11), respectively.
2. Check the two implementation methods of PCA. Measure their computation time.
3. Using the method of PCA for multichannel images in Section 3.3.1.2, decompose a color image into three eigenchannels. Then median filter only the most important one and compose the channels back. Analyze the results. Repeat with median filtering of all image channels.
4. Based on Equations (3.39) and (3.44) show that $\tilde{\mathbf{x}}$ and $\mathbf{x}'$ are perpendicular, i.e. $\tilde{\mathbf{x}} \perp \mathbf{x}'$.
5. Build a PCA subspace classification system for handwritten digit recognition. Check the recognition results using the complementary subspace $\bar{S}$, as discussed in Section 3.3.2.
6. Check the basic idea of the linear Kalman filter, discussed in Section 3.6, and show that a product of two Gaussian pdfs is also a Gaussian. Find its parameters.
7. Implement a method of pattern classification, such as handwritten digits, using the k-nearest neighbour approach discussed in Section 3.7.4.2. Analyze its properties with respect to the chosen k.
8. Show that the Hamming neural network operates as the one-nearest-neighbor classifier.
9. Check operation of the morphological neural network on occluded image patterns.
10. Run different versions of k-means data clustering with color images in different color spaces. Try to find a number of segmentation groups from analysis of the peaks in the 3D histogram with a properly chosen bin width.
11. Compare the cluster validation measures (3.258) with (3.260). What are their similarities and differences? Can these two be joined together?

12. Run the LIBSVM with the UCI test data sets [138]. Check its operation with different kernels and data normalization methods. Apply grid search method to find optimal kernel settings.

References

[1] Alpaydin E.: Introduction to Machine Learning. 2nd edition, MIT Press, 2010.

[2] Duda R.O., Hart P.E., Stork D.G.: Pattern Classification. Wiley, 2001.

[3] Hastie T., Tibshirani R., Friedman J.: The Elements of Statistical Learning. Springer, 2nd edition, 2009.

[4] Kuncheva L.I.: Combining Pattern Classifiers. Methods and Algorithms. Wiley Interscience, 2005.

[5] Theodoridis S., Koutroumbas K.: Pattern Recognition. Fourth Edition. Academic Press, 2009.

[6] http://www.wiley.com/go/cyganekobject

[7] Jolliffe I.T.: Principal component analysis. Springer, 2002.

[8] Schölkopf B., Smola A.J.: Learning with Kernels, MIT Press, 2002.

[9] Shawe-Taylor J., Cristianini N.: Kernel Methods for Pattern Analysis. Cambridge University Press, 2004.

[10] Gu Y.: KPCA algorithm for hyperspectral target/anomaly detection. In Kernel Methods for Remote Sensing and Data Analysis edited by Camps-Valls G. and Bruzzone L., Wiley, pp. 353–373, 2009.

[11] Liu Y., Liu Y., Chan K.C.C.: Tensor Distance Based Multilinear Locality-Preserved Maximum Information Embedding. IEEE Transactions on Neural Networks, Vol. 21, No. 11, pp. 1848–1854, 2010.

[12] Lee J.A., Verleysen M.: Nonlinear Dimensionality Reduction. Springer, 2007.

[13] Abdi H., Williams L.J.: Principal Component Analysis. Wiley Interdisciplinary Reviews: Computational Statistics, Vol. 2, No. 4, pp. 433–459, 2010.

[14] Gonzalez R.C, Woods R.E.: Digital Image Processing, 2nd Edition, Prentice Hall, 2002.

[15] Tuzel O., Porikli F., Meer P.: Pedestrian Detection via Classification on Riemannian Manifolds. IEEE Transactions on Pattern Analysis and Machine Intelligence, Vol. 30, No. 10, pp. 1713–1727, 2008.

[16] Vidal R., Ma Y., Sastry S.: Generalized Principal Component Analysis (GPCA). IEEE Transactions on Pattern Analysis and Machine Intelligence, Vol. 27, No. 12, pp. 1945–1959, 2005.

[17] Press W.H., Teukolsky S.A., Vetterling W.T., Flannery B.P.: Numerical Recipes in C. The Art of Scientific Computing. Third Edition. Cambridge University Press, 2007.

[18] Tuncer Y, Tanik M.M., Allison D.B.: An overview of statistical decomposition techniques applied to complex systems. Computational Statistics & Data Analysis, Vol. 52, No. 5, pp. 2292–2310, 2008.

[19] http://www.wiley.com/legacy/wileychi/cyganek3dcomputer/supp/HIL_Manual_01.pdf

[20] Turk M.A., Pentland A.P.: Face recognition using eigenfaces. IEEE Conference on Computer Vision and Pattern Recognition, pp. 586–590, 1991.

[21] Dhillon I.S., Inderjit S.; Parlett, Beresford N., Vömel C.: The Design and Implementation of the MRRR Algorithm, ACM Transactions on Mathematical Software, Vol. 32, No 4, pp. 533–560, 2006.

[22] Lang B.: Direct Solvers for Symmetric Eigenvalue Problems. Modern Methods and Algorithms of Quantum Chemistry. Proceedings, Second Edition, J. Grotendorst (Ed.), John von Neumann Institute for Computing, Jülich, NIC Series, Vol. 3, pp. 231–259, 2000.

[23] Soille P.: Morphological Image Analysis. Principles and Applications. Springer, 2003.

[24] Ritter G., Wilson J.: Handbook of Computer Vision Algorithms in Image Algebra, CRC Press, 2001.

[25] Piccardi M.: Background subtraction techniques: a review. 2004 IEEE International Conference on Systems, Man and Cybernetics, Vol. 4, pp. 3099–3104, 2004.

[26] Hu W., Li X., Zhang X., Shi X., Maybank S., Zhang Z.: Incremental Tensor Subspace Learning and Its Applications to Foreground Segmentation and Tracking. International Journal of Computer Vision, Springer, Vol. 91, No. 3, pp 303–327, 2011.

[27] Caseiro R., Martins P., Henriques J.F., Batista J.: A nonparametric Riemannian framework on tensor field with application to foreground segmentation. Pattern Recognition, Vol. 45, pp. 3997–4017, 2012.

[28] Oliver N.M., Rosario B., Pentland A.P.: A Bayesian Computer Vision System for Modeling Human Interactions. IEEE Transactions on Pattern Analysis and Machine Intelligence, Vol. 22, No. 8, pp. 831–843, 2000.

[29] http://i21www.ira.uka.de/image_sequences/

[30] Wold S.: Pattern recognition by means of disjoint principal components models. Pattern Recognition, Vol. 8, No. 3, pp. 127–139, 1976.

[31] Moon T.K., Stirling W.C.: Mathematical Methods and Algorithms for Signal Processing. Prentice-Hall 2000.

[32] Xu L., Krzyzak A., Oja E.: Neural Nets for Dual Subspace Pattern Recognition Method. International Journal of Neural Systems, Vol. 2, No. 3, pp. 169–184, 1991.

[33] Prakash M., Murty M.N.: Extended subspace methods of pattern recognition. Pattern Recognition Letters, Vol. 17, No. 11 pp. 1131–1139, 1996.

[34] Moghaddam B., Pentland A.: Probabilistic Visual Learning for Object Representation. IEEE Transactions on Pattern Analysis and Machine Intelligence, Vol. 19, No. 7, pp. 696–710, 1997.

[35] Branden K.V., Hubert M.: Robust Classification in High Dimensions Based on the SIMCA Method. Chemometrics and Intelligent Laboratory Systems, Vol. 79, pp. 10–21, 2005.

[36] Bischof H., Leonardis A., Pezzei F.: A robust subspace classifier. 14th International Conference on Pattern Recognition, Vol. 1, pp. 114–116, 1998.

[37] Ritter G., Gallegos M.: Outliers in statistical pattern recognition and an application to automatic chromosome classification. Pattern Recognition Letters Vol. 18, No. 6, pp. 525–539, 1997.

[38] Shakhnarovich G., Moghaddam B.: Face Recognition in Subspaces. Mitsubishi Electric Research Laboratories (http://www.merl.com), Technical Report TR2004-041, 2004.

[39] Raoa A., Noushath S.: Survey. Subspace methods for face recognition. Computer Science Review, Vol. 4, pp. 1–17, 2010.

[40] Bishop C.M.: Pattern Recognition and Machine Learning. Springer 2006.

[41] Webb A.: Statistical Pattern Recognition. Wiley, 2002.

[42] Haykin S.: Neural Networks and Learning Machines. 3rd Edition, Pearson Education, 2009.

[43] Vapnik V.N.: The Nature of Statistical Learning Theory, Springer, 2000.

[44] Kecman V.: Learning and Soft Computing: Support Vector Machines, Neural Networks, and Fuzzy Logic Models. MIT Press, 2001.

[45] Cyganek B.: A Real-Time Vision System for Traffic Signs Recognition Invariant to Translation, Rotation and Scale. ACIVS 2008, France, Lecture Notes in Computer Science, Vol. 5259, Springer, pp. 278–289, 2008.

[46] Cheng Y.: Mean shift, mode seeking, and clustering. IEEE Transactions on Pattern Analysis and Machine Intelligence, Vol. 15, No. 6, 1993, pp. 602–605.

[47] Fukunaga K., Hostetler L.D.: The estimation of the gradient of a density function, with application in pattern recognition. IEEE Transactions on Information Theory, Vol. 21, No. 1, pp. 32–40, 1975.

[48] Chellappa R., Roy-Chowdhury A.K., Zhou S.K.: Recognition of Humans and their Activities Using Video. Morgan & Claypool Publishers, 2005.

[49] Comaniciu D., Meer P.: Mean Shift: A Robust Approach Toward Feature Space Analysis. IEEE Transactions on Pattern Analysis And Machine Intelligence, Vol. 24, No. 5, pp. 603–619, 2002.

[50] Papoulis A.: Probability, Random Variables, and Stochastic Processes. Third Edition, McGraw-Hill, 1991.

[51] Maesschalck De R., Jouan-Rimbaud D., Massart D.L.: Tutorial. The Mahalanobis distance. Chemometrics and Intelligent Laboratory Systems, Vol. 50, pp. 1–18, 2000.

[52] Sung K., Poggio T.: Example-Based Learning for View-Based Human Face Detection. Proc. Image Understanding Workshop, Monterey, California USA, 1994.

[53] Cyganek B., Siebert J.P.: An Introduction to 3D Computer Vision Techniques and Algorithms, Wiley, 2009.

[54] Jones M.J., Rehg J.M.: Statistical Color Models with Application to Skin Detection. International Journal of Computer Vision, Vol. 46, No. 1, pp. 81–96, 2002.

[55] Yang M-H, Ahuja N.: Gaussian mixture model for human skin color and its application in image and video databases, Proceedings of SPIE'99, 1999.

[56] Vezhnevets V., Sazonov V., Andreeva A.: A survey on pixel-based skin color detection techniques, Proc. Graphicon-2003, 2003.

[57] Phung S.L., Bouzerdoum A., Chai D.: Skin Segmentation Using Color Pixel Classification: Analysis and Comparison. IEEE Transactions on Pattern Analysis and Machine Intelligence, Vol. 27, No. 1, pp. 148–154, 2005.

[58] Orazio De T., Leo M., Guaragnella C., Distante A.: A visual approach for driver inattention detection. Pattern Recognition 40, pp. 2341–2355, 2007.

[59] Shih P., Liu C.: Face detection using discriminating feature analysis and Support Vector Machine. Pattern Recognition, Vol. 39, No. 2, pp. 260–276, 2006.

[60] Liu C., Wechsler H.: Robust Coding Schemes for Indexing and Retrieval from Large Face Databases, IEEE Transactions on Image Processing, Vol. 9, No. 1, pp. 132–137, 2000.

[61] Rasmussen C.E., Williams C.K.I.: Gaussian Processes for Machine Learning. MIT Press, 2006.

[62] Kalman R.E.: A new approach to linear filtering, prediction problems. Transactions of the ASME Journal of Basic Engineering, pp. 35–45, 1960.

[63] Welch G., Bishop G.: An Introduction to the Kalman Filter. Technical Report TR 95-041. University of North Carolina at Chapel Hill, 2006.

[64] Faragher R.: Understanding the Basis of the Kalman Filter Via a Simple and Intuitive Derivation. IEEE Signal Processing Magazine, Vol. 29, No. 5, pp. 128–132, 2012.

[65] Bradski G., Kaehler A.: Learning OpenCV. Computer Vision with the OpenCV Library. O'Reilly, 2008.

[66] Piovoso M., Laplante P.A.: Kalman filter recipes for real-time image processing. Real-Time Imaging, Vol. 9, No. 6, pp. 433–439, 2003.

[67] Thrun S., Burgard W. Fox D.: Probabilistic Robotics. MIT Press, 2006.

[68] Ristic B., Arulampalam S., Gordon N.: Beyond the Kalman Filter. Particle Filters for Tracking Applications. Artech House, 2004.

[69] Khan Z., Balch T., Dellaert F.: MCMC-based particle filtering for tracking a variable number of interacting targets. IEEE Transactions on Pattern Analysis and Machine Intelligence, Vol. 27, No. 11, pp. 1805–1819, 2005.

[70] Okuma K., Taleghani A., de Freitas N., Little J.J., Lowe D.G.: A boosted particle filter: multitarget detection and tracking. European Conference on Computer Vision ECCV'04, Lecture Notes in Computer Science, No. 3021, Springer, pp. 28–39, 2004.

[71] Pratt W.K.: Digital Image Processing. Third Edition. Wiley, 2001.

[72] Myler H.R., Weeks A.: The Pocket Handbook of Image Processing Algorithms in C. Prentice-Hall 1993.

[73] Scott D.W.: Averaged shifted histogram. WIREs Computational Statistics, Wiley, Vol. 2, pp. 160–164, March/April 2010.

[74] Bhattacharyya A.: On a Measure of Divergence Between Two Statistical Populations Defined by their Probability Distributions. Bull. Calcutta Mathematic Society, Vol. 35, pp. 99–110, 1943.

[75] Cover T.M., Thomas J.A.: Elements of Information Theory. 2nd Edition. Wiley, 2006.

[76] Aherne F.J., Thacker N.A., Rockett P.I.: The Bhattacharyya Metric as an Absolute Similarity Measure for Frequency Coded Data. Kybernetika, Vol. 34, No. 4, pp. 363–368, 1998.

[77] Hudson J.E.: Signal Processing Using Mutual Information. IEEE Signal Processing Magazine, Vol. 23, No. 6, pp. 50–58, 2006.

[78] Escolano F., Suau P., Bonev B.: Information Theory in Computer Vision and Pattern Recognition. Springer, 2009.

[79] Rubner Y., Tomasi C., Guibas L.J.: The Earth Mover's Distance as a Metric for Image Retrieval. International Journal of Computer Vision, Vol. 40, No. 2, pp. 99–121, 2000.

[80] Bazaraa M.S., Jarvis J.J., Sherali H.D.: Linear Programming and Network Flows. 3rd edition, Wiley, 2005.

[81] Fletcher R.: Practical Methods of Optimization. 2nd edition, Wiley, 2003.

[82] Grauman K., Darrell T.: Contour Matching Using Approximate Earth Mover's Distance, in [83], 2005.

[83] Shakhnarovich G., Darrell T., Indyk P.: Nearest-Neighbor Methods in Learning and Vision. Theory and Practice. MIT Press, 2005.

[84] Cormen T.H., Leiserson C.E., Rivest R.L., Stein C.: Introduction to Algorithms. MIT Press, 2001.

[85] Comaniciu D., Ramesh V., Meer P.: Kernel-Based Object Tracking. IEEE Transactions on Pattern Analysis And Machine Intelligence, Vol. 25, No. 5, pp. 564–577, 2003.

[86] Specht D.F.: Probabilistic neural networks, Neural Networks, 1(3), pp. 109–118, 1990.

[87] Specht D.F.: Enhancements to Probabilistic Neural Networks, International Joint Conference on Neural Networks, Vol. I, pp. 761–768, 1992.

[88] Fukunaga K.: Introduction to Statistical Pattern Recognition. 2nd edition. Academic Press, 1990.

[89] Shimshoni I., Georgescu B., Meer P.: Adaptive Mean Shift Based Clustering in High Dimensions, in [83], pp. 203–220, 2005.

[90] Comaniciu D., Ramesh V., Meer P.: Real-time tracking of non-rigid objects using mean shift. IEEE Conference on Computer Vision and Pattern Recognition CVPR '00, Vol. 2, pp. 142–149, 2000.

[91] Kim K.I., Jung K., Kim J.H.: Texture-Based Approach for Text Detection in Images Using Support Vector Machines. IEEE Transactions on Pattern Analysis and Machine Intelligence, Vol. 25, No. 12, pp. 1631–1639, 2003.

[92] Cyganek B.: Real-Time Road Signs Tracking with the Fuzzy Continuously Adaptive Mean Shift Algorithm. Lecture Notes in Computer Science, Vol. 5097, Springer, pp. 217–228, 2008.

[93] Comaniciu D., Ramesh V., Meer P.: The variable bandwidth mean shift and data-driven scale selection. Proceedings of the eighth IEEE Int. Conference on Computer Vision, 2001. ICCV 2001. Vol. 1, pp. 438–445, 2001.

[94] McConnell S.: Code Complete. 2nd edition. Microsoft Press, 2004.

[95] Collins R.T., Liu Y., Leordeanu M.: Online Selection of Discriminative Tracking Features. IEEE Transactions on Pattern Analysis And Machine Intelligence, Vol. 27, No. 10, pp. 1631–1643, 2005.

[96] Wang J., Yagi Y.: Integrating Color and Shape-Texture Features for Adaptive Real-Time Object Tracking. IEEE Transactions on Image Processing, Vol. 17, No. 2, 235–240, 2008.

[97] Petrović N., Jovanov L., Pižurica A., Philips W.: Object Tracking Using Naïve Bayesian Classifiers. Lecture Notes in Computer Science, Vol. 5259, pp. 775–784, 2008.

[98] Bradski G.R.: Computer Vision Face Tracking for Use in a Perceptual User Interface, Intel Technical Report, 1998.

[99] Klette R., Rosenfeld A.: Digital Geometry. Morgan-Kaufman, 2004.

[100] Cyganek B.: Neuro-Fuzzy System for Road Signs Recognition. Lecture Notes in Computer Science, Vol. 5163, Springer, pp. 503–512, 2008.

[101] Argyros A.A., Lourakis M.I.A.: Real-Time Tracking of Multiple Skin-Colored Objects with a Possibly Moving Camera T. Pajdla and J. Matas (Eds.): ECCV 2004, Lecture Notes in Computer Science, Vol. 3023, pp. 368–379, 2004.

[102] Argyros A., Papadourakis V.: Multiple objects tracking in the presence of long-term occlusions. Computer Vision and Image Understanding, No. 114, No. 7, pp. 835–846, 2010.

[103] Wu B., Nevatia R.: Detection and Tracking of Multiple, Partially Occluded Humans by Bayesian Combination of Edgelet Based Part Detectors. International Journal of Computer Vision, Vol. 75, No. 2, pp. 247–266, 2007.

[104] Taj M., Cavallaro A.: Distributed and Decentralized Multicamera Tracking. IEEE Signal Processing Magazine, Vol. 28, No. 3, pp. 46–58, 2011.

[105] Zadeh L.A.: Fuzzy sets. Information and Control, Vol. 8, pp. 338–353, 1965.

[106] Masters T.: Practical Neural Network Recipes in C++, Academic Press, 1993.

[107] Driankov D., Hellendoorn H., Reinfrank M.: An Introduction to Fuzzy Control. 2nd edition, Springer, 1996.

[108] Kulkarni A.D.: Computer Vision and Fuzzy-Neural Systems. Prentice-Hall, 2001.

[109] Georgiou V.L., Pavlidis N.G., Parsopoulos K.E., Alevizos P.D., Vrahatis M.N.: Optimizing the Performance of Probabilistic Neural Networks in a Bionformatics Task. Proceedings of the EUNITE 2004 Conference, 2004.

[110] Bertold M., Diamond R.: Constructive Training of Probabilistic Neural Networks. Neurocomputing, Vol. 19, No. 1-3, pp. 167–183, 1998.

[111] Cyganek B.: Circular Road Signs Recognition with Affine Moment Invariants and the Probabilistic Neural Classifier. Lecture Notes in Computer Science, Vol. 4432, Springer, pp. 508–516, 2007.

[112] Montana D.: A Weighted Probabilistic Neural Network. Technical Report, 1999.

[113] Ge S.S., Yang Y., Lee T.H.: Hand gesture recognition and tracking based on distributed locally linear embedding. Image and Vision Computing 26, pp. 1607–1620, 2008.

[114] Lippman R.: An introduction to computing with neural nets. IEEE Transactions on Acoustic, Speech, and Signal Processing, v.ASSP-4, pp. 4–22, 1987.

[115] Floréen P.: Computational Complexity Problems in Neural Associative Memories. PhD Thesis, University of Helsinki, Department of Computer Science, Finland, 1992.

[116] Cyganek B.: Comparison of Nonparametric Transformations and Bit Vector Matching for Stereo Correlation. Springer, Lecture Notes in Computer Science, Vol. 3322, pp. 534–547, 2004.

[117] Cyganek B.: Matching of the Multi-channel Images with Improved Nonparametric Transformations and Weighted Binary Distance Measures, Lecture Notes in Computer Science, Vol. 4040, Springer, pp. 74–88, 2006.

[118] Zabih R., Woodfill J.: Non-parametric Local Transforms for Computing Visual Correspondence. Computer Science Department, Cornell University, Ithaca, 1998.

[119] Ritter G., X., Sussner P., Diaz J., L.: Morphological Associative Memories, IEEE Transactions on Neural Networks, Vol. 9, No. 2, pp. 281–293, 1998.

[120] Ritter G.,X., Urcid G., Iancu L.: Reconstruction of Patterns from Noisy Inputs Using Morphological Associative Memories. Journal of Mathematical Imaging and Vision, Vol. 19, No. 2, pp. 95–111, 2003.

[121] Ritter G.,X., Iancu L.: A Lattice Algebraic Approach to Neural Computation, in Handbook of Geometric Computing. Applications in Pattern Recognition, Computer Vision, Neuralcomputing, and Robotics, edited by Corrochano E.B., Springer, pp. 97–127, 2005.

[122] Raducanu B., Graña M., Albizuri F.X.: Morphological Scale Spaces and Associative Morphological Memories: Results on Robustness and Practical Applications. Journal of Mathematical Imaging and Vision, Vol. 19, No. 2, pp. 113–131, 2003.

[123] Sussner P.: Observations on Morphological Associative Memories and the Kernel Method. Neurocomputing, Vol. 31, No. 1-4, Elsevier Science, pp. 167–183, 2000.

[124] Sussner P., Valle M.E.: Gray-Scale Morphological Associative Memories. IEEE Transactions on Neural Networks, Vol. 17, No. 3, pp. 559–570, 2006.

[125] Villaverde I., Graña M., d'Anjou A.: Morphological neural networks and vision based simultaneous localization and mapping. Integrated Computer-Aided Engineering, Vol. 14, No. 4, pp. 355–363, 2007.

[126] Cyganek B.: Circular Road Signs Recognition with Soft Classifiers. Integrated Computer-Aided Engineering, IOS Press, Vol. 14, No. 4, pp. 323–343, 2007.

[127] Cover T.M: Geometrical and statistical properties of systems of linear inequalities with appllications in pattern recognition. IEEE Transactions on Electronic Computers, Vol. EC-14, pp. 326–334, 1965.

[128] Korn G.A., Korn T.M.: Mathematical Handbook for Scientists and Engineers. Dover Publications, 2000.

[129] Cristianini N., Shawe-Taylor J.: An introduction to Support Vector Machines and other kernel-based learning methods. Cambridge University Press, 2002.

[130] Meyer C.D.: Matrix Analysis and Applied Linear Algebra. SIAM, 2000.

[131] Aronszajn N.: Theory of reproducing kernels. Transactions American Mathematical Society, No. 68, pp. 337–404, 1950.

[132] Lampert C.H.: Kernel Methods in Computer Vision, Foundations and Trends® in Computer Graphics and Vision: Vol. 4: No 3, pp 193–285, 2009.

[133] Honeine P., Richard C.: Preimage Problem in Kernel-Based Machine Learning. IEEE Signal Processing Magazine, Vol. 28, No. 2, pp. 77–88, 2011.

[134] Maji S., Berg A., Malik J.: Classification Using Intersection Kernel SVMs Is Efficient. IEEE Conference on Computer Vision and Pattern Recognition, pp. 1–8, 2008.

[135] Abe S.: Support Vector Machines for Pattern Classification. Springer, 2005.

[136] Hsu C-W., Chang C-C., Lin C-J.: A Practical Guide to Support Vector Classification. Department of Computer Science and Information Engineering, National Taiwan University, (www.csie.ntu.edu.tw/~cjlin/papers/guide/guide.pdf), 2003.

[137] Ah-Pine J.: Normalized Kernels as Similarity Indices. M.J. Zaki et al. (Eds.): PAKDD 2010, Part II, Lecture Notes in Artificial Intelligence, Vol. 6119, Springer, pp. 362–373, 2010.

[138] Frank A., Asuncion A.: UCI Machine Learning Repository (http://archive.ics.uci.edu/ml). Irvine, CA: University of California, School of Information and Computer Science, 2010.

[139] Barla A., Franceschi E., Odone F., Verri A.: Image Kernels. Lecture Notes in Computer Science, Vol. 2388, pp. 83–96, 2002.

[140] Odone F., Barla A., Verri A.: Building Kernels from Binary Strings for Image Matching. IEEE Transactions on Image Processing Vol. 14, No. 2, pp 169–180, 2005.

[141] Arenas-García J., Petersen K.B.: Kernel multivariate analysis in remote sensing feature extraction. In Kernel Methods for Remote Sensing and Data Analysis edited by Camps-Valls G. and Bruzzone L., Wiley, pp. 329–352, 2009.

[142] Gamma E., Helm R., Johnson R., Vlissides J.: Design Patterns. Elements of Reusable Object-Oriented Software. Addison-Weseley 1995.

[143] Vlissides J.: Pattern Hatching. Design Patterns Applied. Addison Wesley, 1998.

[144] Bruel J-M, Henderson-Sellers B., Barbier F, Le Parc A., France R.B.: Improving the UML Metamodel to Rigorously Specify Aggregation and Composition. Proceedings of the OOIS'2001, 7th Int. Conference on Object Oriented Information Systems, Calgary, Canada. Springer, pp. 5–14, 2001.

[145] Henderson-Sellers B., Barbier F.: Black and White Diamonds, UML'99, Lecture Notes in Computer Science, Vol. 1723, Springer, pp. 550–565, 1999.

[146] Henderson-Sellers B.: UML - the Good, the Bad or the Ugly? Perspectives from a panel of experts. Software and System Modeling, Vol. 4, No. 1, pp. 4–13, 2005.

[147] http://www.sparxsystems.com/resources/uml2_tutorial/uml2_classdiagram.html

[148] Gan G., Ma C., Wu J.: Data Clustering. Theory, Algorithms, and Applications. SIAM Press, 2007.

[149] Xu R., Wunsch II D.C.: Clustering. IEEE Press, 2009.

[150] Filippone M., Camastra F., Masullia F., Rovetta S.: A survey of kernel and spectral methods for clustering. Pattern Recognition, Vol. 41, No. 1, pp. 176–190, 2008.

[151] Oliveira de J.V., Pedrycz W.: Advances in Fuzzy Clustering and its Applications, Wiley, 2007.

[152] Gersho A., Gray R.M.: Vector Quantization and Signal Compression. The Springer International Series in Engineering and Computer Science, Springer, 1991.

[153] Ding C., He X., Simon H.D.: On the Equivalence of Nonnegative Matrix Factorization and Spectral Clustering. SIAM International Conference on Data Mining, pp. 606–610, 2005.

[154] Kohonen T.: Self-Organizing Maps. Springer, 2000.

[155] Bezdek J.: Pattern Recognition With Fuzzy Objective Function Algorithms. Plenum Press, New York, 1981.

[156] Frigui H.: Simultaneous Clustering and Feature Discrimination with Applications, in Advances in Fuzzy Clustering and its Applications by J.V. de Oliveira and W. Pedrycz [151], Wiley, pp. 285–312, 2007.

[157] Duda R.O. and Hart P.E.: Use of the Hough Transformation to Detect Lines and Curves in Pictures. Communications ACM, Vol. 15, pp. 11–15, 1972.

[158] Höppner F., Klawonn F., Kruse R., Runkler T.: Fuzzy Cluster Analysis. Methods for Classification, Data Analysis and Image Recognition. Wiley, 1999.

[159] Bertsekas D.P.: Constraint Optimization and Lagrange Multiplier Methods. Athena Scientific, 1996.

[160] Krishnapuram R. and Keller J.: A possibilistic approach to clustering. IEEE Transactions on Fuzzy Systems Vol. 1, No. 2, pp. 98–110, 1993.

[161] Krishnapuram R., Keller J.: The possibilistic c-means algorithm: insights and recommendations. IEEE Transactions on Fuzzy Systems, Vol. 4, No. 3, pp. 385–393, 1996.

[162] Wu Z., Xie W., Yu J.: Fuzzy C-Means Clustering Algorithm based on Kernel Method. Fifth International Conference on Computational Intelligence and Multimedia Applications (ICCIMA'03), pp. 1–6, 2003.

[163] Zhang D., Chen S.: Clustering incomplete data using kernel-based fuzzy c-means algorithm. Neural Processing Letters, Vol. 18, No. 3, pp. 155–162, 2003.

[164] Arif O., Vela P.A.: Kernel Map Compression for Speeding the Execution of Kernel-Based Methods. IEEE Transactions on Neural Networks, Vol. 22, No. 6, pp. 870–879, 2011.

[165] University of California, Database (ftp://ftp.ics.uci.edu/pub/machine-learning-databases/), 2011.

[166] Blake C., Keogh E., Merz C.: UCI repository of machine learning databases. University of California, Irvine, Department of Information and Computer Science (www.ics.uci.edu/~mlearn/MLRepository.html). 1998.

[167] Xie X.L., Beni G.: A Validity Measure for Fuzzy Clustering. IEEE Transactions on Pattern Analysis and Machine Intelligence, Vol. 13, No. 8, pp. 841–847, 1991.

[168] Cyganek B.: One-Class Support Vector Ensembles for Image Segmentation and Classification. Journal of Mathematical Imaging & Vision, Vol. 42, No. 2-3, Springer, pp. 103–117, 2012.

[169] Pal N.R., Bezdek J.C.: On Cluster Validity for the Fuzzy c-Means Model. IEEE Transactions on Fuzzy Systems, Vol. 3, No. 3, pp. 370–380, 1995.

[170] Yang X., Zhang G., Lu J., Ma J.: A Kernel Fuzzy c-Means Clustering-Based Fuzzy Support Vector Machine Algorithm for Classification Problems With Outlier or Noises. IEEE Transactions on Fuzzy Systems, Vol. 19, No. 1, 105–115, 2011.

[171] Capitaine Le H., Frélicot C.: A Cluster-Validity Index Combining an Overlap Measure and a Separation Measure Based on Fuzy-Aggregation Operators. IEEE Transactions on Fuzzy Systems, Vol. 19, No. 3, 580–587, 2011.

[172] Rezaee M.R., Lelieveldt B.P.F., Reibe J.H.C.: A new cluster validity index for the fuzzy c-mean. Pattern Recognition Letters, Vol. 19, No. 3-4, pp. 237–246, 1998.

[173] Cortes C., Vapnik V.: Support vector machines. Machine Learning, Vol. 20, pp. 273–297, 1995.

[174] Tax D.M.J., Duin R.P.W.: Support vector domain description. Pattern Recognition Letters, Vol. 20, pp. 1191–1199, 1999.

[175] Tax D.M.J., Duin R.P.W.: Support Vector Data Description, Machine Learning 54, pp. 45–66, 2004.

[176] Ben-Hur A., Horn D., Siegelmann H.T., Vapnik V.: Support Vector Clustering. Journal of Machine Learning Research Vol. 2, pp. 125–137, 2001.

[177] Camastra F., Verri A.: A novel kernel method for clustering, IEEE Transactions on Pattern Analysis and Machine Intelligence, Vol. 27, No. 5, pp. 801–805, 2005.

[178] Bicego M., Figueiredo M.A.T.: Soft clustering using weighted one-class support vector machines. Pattern Recognition 42, pp. 27–32, 2009.

[179] Bertsekas D.P.: Convex Analysis and Optimization. Athena Scientific, 2003.

[180] Tax D.M.J.: One-class classification. PhD thesis, TU Delft University, 2001.

[181] Barzilay O., Brailovsky V.L.: On domain knowledge and feature selection using a support vector machine. Pattern Recognition Letters, Vol. 20, No. 5, pp. 475–484, 1999.

[182] Zou A-M., Hou Z-G., Tan M.: Support Vector Machines (SVM) for Color Image Segmentation with Applications to Mobile Robot Localization Problems, Lecture Notes in Computer Science Vol. 3645, pp. 443–452, 2005.

[183] Pan C., Yan X-G., Zheng C-X.: Fast Training of SVM for color-based image segmentation. The 3rd International Conference on Machine Learning and Cybernetics, pp. 3820–3825, 2004.

[184] Cyganek B.: Color Image Segmentation With Support Vector Machines: Applications To Road Signs Detection. International Journal of Neural Systems, Vol. 18, No. 4, World Scientific Publishing Company, pp. 339–345, 2008.

[185] Park J., Kang D., Kim J., Kwok J.T., Tsang I.W.: Pattern De-Noising Based on Support Vector Data Description. Proceedings of the International Joint Conference on Neural Networks, Montreal, Canada, pp. 949–953, 2005.

[186] Jin H., Liu Q., Lu H., Tong X.: Face Detection Using One-Class SVM in Color Images. 7th International Conf. on Signal Processing ICSP '04, pp. 1431–1434, 2004.

[187] http://www.support-vector-machines.org

[188] Rüping S.: mySVM - Manual. AI Unit University of Dortmund, Computer Science Departmnet, 2000.

[189] http://svmlight.joachims.org/

[190] Chang C-C., Lin C-J.: LIBSVM, a library for support vector machines (2001). Software available at http://www.csie.ntu.edu.tw/~cjlin/libsvm

[191] Kuncheva L.I.: Cluster-and-selection method for classifier combination, Proc. 4th International Conference on Knowledge-Based Intelligent Engineering Systems & Allied Technologies (KES'2000), Brighton, UK, pp. 185–188, 2000.

[192] Rahman A., Verma B.: Novel Layered Clustering-Based Approach for Generating Ensemble of Classifiers. IEEE Transactions on Neural Networks, Vol. 22, No. 5, pp. 781–792, 2011.

[193] Graves D., Pedrycz W.: Kernel-based fuzzy clustering and fuzzy clustering: A comparative experimental study. Fuzzy Sets and Systems 161, pp. 522–543, 2010.

[194] Duin R.P.W., Juszczak P., Paclik P., Pekalska E., de Ridder D., Tax D.M.J., Verzakov S.: PRTools4: A Matlab® Toolbox for Pattern Recognition, Delft University of Technology, 2007.

[195] Stork D.G., Yom-Tov E.: Computer Manual in MATLAB® to accompany Pattern Classification. Second Edition, Wiley-Interscience, 2004.

[196] Statistical Pattern Recognition Toolbox (http://cmp.felk.cvut.cz/cmp/software/stprtool/), 2012.

[197] Trefethen L.N., Bau D.: Numerical Linear Algebra. SIAM 1997.

[198] http://www.kernel-machines.org

[199] Kruse R., Döring C., Lesot M-J.: Fundamentals of Fuzzy Clustering, in Advances in Fuzzy Clustering and its Applications, ed. de Oliveira J.,V., Pedrycz W. [151], Wiley, pp. 3–30, 2007.

[200] Frigui H., Krishnapuram R.: Clustering by competitive agglomeration. Pattern Recognition, Vol. 3, No. 7, pp. 1109–1119, 1997.

[201] Arbeláez P., Maire M., Fowlkes C., Malik J.: Contour Detection and Hierarchical Image Segmentation. IEEE Transactions on Pattern Analysis and Machine Intelligence, Vol. 33, No. 5, pp. 898–916, 2011.

[202] Rokach L.: Pattern Classification Using Ensemble Methods. Series in Machine Perception and Artificial Intelligence, Vol. 75, World Scientific, 2010.

[203] Polikar R.: Ensemble Based Systems in Decision Making. IEEE Circuits and Systems Magazine, Vol. 6, No. 3, pp. 21–45, 2006.

4

Object Detection and Tracking

4.1 Introduction

This section is devoted to selected problems in object detection and tracking. Objects in this context are characterized by their salient features, such as color, shape, texture, or other traits. Then the problem is telling whether an image contains a defined object and, if so, then indicating its position in an image. If instead of a single image a video sequence is processed, then the task can be to track, or follow, the position and size of an object seen in the previous frame and so on. This assumes high correlation between consecutive frames in the sequence, which usually is the case. Eventually, an object will disappear from the sequence and the detection task can be started again.

Detection can be viewed as a classification problem in which the task is to tell the presence or absence of a specific object in an image. If it is present, then the position of the object should be provided. Classification within a group of already detected objects is usually stated separately, however. In this case the question is formulated about what particular object is observed. Although the two groups are similar, recognition methods are left to the next chapter. Thus, examples of object detection in images are, for instance, detection of human faces, hand gestures, cars, and road signs in traffic scenes, or just ellipses in images. On the other hand, if we were to spot a particular person or a road sign, etc. we would call this recognition. Since detection relies heavily on classification, as already mentioned, one of the methods discussed in the previous section can be used for this task. However, not least important is the proper selection of features that define an object. The main goal here is to choose features that are the most characteristic of a searched object or, in other words, that are highly discriminative, thus allowing an accurate response of a classifier. Finally, computational complexity of the methods is also essential due to the usually high dimensions of the feature and search spaces. All these issues are addressed in this section with a special stress on automotive applications.

4.2 Direct Pixel Classification

Color conveys important information about the contents of an environment. A very appealing natural example is a coral reef. Dozens of species adapt the colors of their skin so as to be as indistinguishable from the background as possible to gain protection from predators.

Object Detection and Recognition in Digital Images: Theory and Practice, First Edition. Bogusław Cyganek.
© 2013 John Wiley & Sons, Ltd. Published 2013 by John Wiley & Sons, Ltd.

The latter do the same to outwit their prey, and so on. Thus, objects can be segmented out from a scene based exclusively on their characteristic colors. This can be achieved with direct pixel classification into one of the two classes: objects and background. An object, or pixels potentially belonging to an object, are defined providing a set or range of their allowable colors. A background, on the other hand, is either also defined explicitly or can be understood as "all other values." Such a method is usually applied first in a chain on the computer vision system to sieve out the pixels of one object from all the others. For example Phung *et al.* proposed a method for skin segmentation using direct color pixel classification [1]. Road signs are detected by direct pixel segmentation in the system proposed by Cyganek [2]. Features other than color can also be used. For instance Viola and Jones propose using Haar wavelets in a chain of simple classifiers to select from background pixels which can belong to human faces [3].

Although not perfect, the methods in this group have an immense property of dimensionality reduction. Last but not least, many of them allow very fast image pre-processing.

4.2.1 Ground-Truth Data Collection

Ground-truth data allow verification of performance of the machine learning methods. However, the process of its acquisition is tedious and time consuming, because of the high quality requirements of this type of data.

Acquisition of ground-truth data can be facilitated by an application built for this purpose [4, 5]. It allows different modes of point selection, such as individual point positions, as well as rectangle and polynomial outlines of visible objects, as shown in Figure 4.1.

An example of its operation for points marked inside the border of a road sign is depicted in Figure 4.2. Only the positions of the points are saved as meta-data to the original image. These can then be processed to obtain the requested image features, i.e. in this case it is color in the chosen color space. This tool was used to gather point samples for the pixel-based classification for human skin selection and road sign recognition, as will be discussed in the next sections.

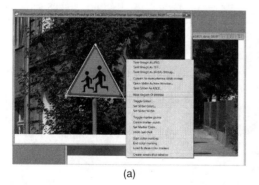 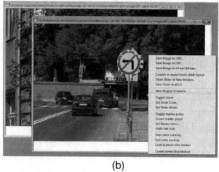

(a)   (b)

Figure 4.1   A road sign manually outlined by a polygon defined by the points marked by an operator. This allows selection of simple (a) and more complicated shapes (b). Selected points are saved as meta-data to an image with the help of a context menu. Color versions of this and subsequent images are available at www.wiley.com/go/cyganekobject.

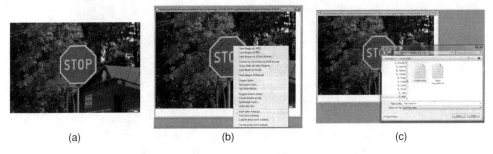

(a)   (b)   (c)

Figure 4.2 View of the application for manual point marking in images. Only the positions of the selected points are saved in the form of meta-data to the original image. These can be used to obtain image features, such as color, in the indicated places.

4.2.2   CASE STUDY – Human Skin Detection

Human skin detection gets much attention in computer vision due to its numerous applications. The most obvious is detection of human faces for their further recognition, human hands for gesture recognition,[1] or naked bodies for parental control systems [6, 7], for instance.

Detection of human skin regions in images requires the definition of characteristic parameters such as color and texture, as well as the choice of proper methods of analysis, such as used color space, classifiers, etc. There is still ongoing research in this respect. As already discussed, a method for human skin segmentation based on a mixture of Gaussians was proposed by Jones and Rehg [8]. Their model contains $J = 16$ Gaussians which were trained from almost one billion labeled pixels from the RGB images gathered mostly from the Internet. The reported detection rate is 80% with about 9% of false positives. A similar method based on MoG was undertaken by Yang and Ahuja in [9].

On the other hand, Jayaram *et al.* [10] report that the best results are obtained with histogram methods rather than using the Gaussian models. They also pointed out that different color spaces improve the performance but not consistently. However, a fair trade-off in this respect is the direct use of the RGB space. A final observation is that in all color spaces directly partitioned into achromatic and chromatic components, performance was significantly better if the luminance component was employed in detection. Similar results, which indicate the positive influence of the illumination component and the poor performance of the Gaussian modeling, were reported by Phung *et al.* [1]. They also found that the Bayesian classifier with the histogram technique, as well as the multilayer perceptron, performs the best. The Bayes classifier operates in accordance with Equation (3.77), in which $\mathbf{x}$ is a color vector, ω_0 denotes a "skin," whereas ω_1 is a "nonskin" classes, as described in Section 3.4.5. However, the Bayes classifier requires much more memory than, for example, a mixture of Gaussians. Therefore there is no unique "winner" and application of a specific detector can be driven by other factors such as the computational capabilities of target platforms.

With respect to the color space, some authors advocate using perceptually uniform color spaces for object detection based on pixel classification. Such an approach was undertaken by Wu *et al.* [11] in their fuzzy face detection method. The front end of their detection constitutes

---

[1] A method for gesture recognition is presented in Section 5.2.

Table 4.1 Fuzzy rules for skin detection in sun lighting.

Rule no	Rule description
R_1:	*Range of skin color components in daily conditions found in experiments* **IF** $R > 95$ **AND** $G > 40$ AND $B > 20$ **THEN** $T_0 = high$;
R_2:	*Sufficient separation of the RGB components; Elimination of gray areas* **IF** $max(R,G,B)\text{-}min(R,G,B) > 15$ **THEN** $T_1 = high$;
R_3:	*R, G should not be close together* **IF** $|R\text{-}G| > 15$ **THEN** $T_2 = high$;
R_4:	*R must be the greatest component* **IF** $R > G$ **AND** $R > B$ **THEN** $T_3 = high$;

skin segmentation operating in the Farnsworth color space. A perceptual uniformity of this color space makes the classification process resemble subjective classification made by humans due to similar sensitivity to changes of color.

Surveys on pixel based skin detection are provided in the papers by Vezhnevets *et al.* [12], by Phung *et al.* [1], or the recent one by Khan *et al.* [13]. Conclusions reported in the latter publication indicate that the best results were obtained with the cylindrical color spaces and with the tree based classifier (Random forest, J48). Khan *et al.* also indicate the importance of the luminance component in feature data, which stays in agreement with the results of Jayaram *et al.* [10] and Phung *et al.* [1].

In this section a fuzzy based approach is presented with explicit formulation of the human skin color model, as proposed by Peer *et al.* [14]. Although simple, the conversion of the histogram to the membership function greatly reduces memory requirements, while fuzzy inference rules allow real-time inference. A similar approach was also undertaken to road sign detection based on characteristic colors, which is discussed in the next section (4.2.3).

The method consists of a series of the fuzzy *IF . . . THEN* rules presented in Table 4.1 for daylight conditions and in Table 4.2 for artificial lighting, respectively. These were designed based on expert knowledge from data provided in the paper by Peer *et al.* [14], although other models or modifications can be easily adapted.

The combined (aggregated) fuzzy rule for human skin detection directly in the RGB space is as follows

$$R_{HS}: \quad \textbf{IF } T_{0-3} \textit{ are high } \textbf{OR } T_{4-6} \textit{ are high } \textbf{THEN } H = high; \tag{4.1}$$

Table 4.2 Fuzzy rules for flash lighting.

Rule no	Rule description
R_5:	*Skin color values for flash illumination* **IF** $R > 220$ **AND** $G > 210$ **AND** $B > 170$ **THEN** $T_4 = high$;
R_6:	*R and G components should be close enough* **IF** $|R\text{-}G| \leq 15$ **THEN** $T_5 = high$;
R_7:	*B component has to be the smallest one* **IF** $B < R$ **AND** $B < G$ **THEN** $T_6 = high$;

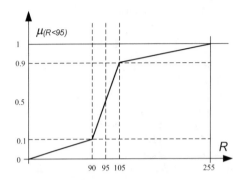

Figure 4.3   A possible membership function for the relation R > 95.

The advantage of the fuzzy formulation (4.1) over its crisp version is that the influence of each particular rule can be controlled separately. Also, new rules can be easily added if necessary. For instance in the rule R_1 when checking the condition for the component R being greater than 95 this can be assigned different values than simple "true" or "false" in the classical formulation. Thus, in this case knowing a linear membership function presented in Figure 4.3, the relation $R < 95$ can be evaluated differently (in the range from 0 to 1) depending on a value of R. Certainly, a type of membership function can be chosen with additional "expert" knowledge. Here, we assume a margin of noise in the measurement of R which in this example spans from 90–105. Apart from this region we reach two extremes for R "significantly lower" with the membership function spanning 0–0.1 and for R "significantly greater" with a corresponding membership function from 0.9–1. Such fuzzy formulation has been shown to offer much more control over a crisp formulation. Therefore it can be recommended for tasks which are based on some empirical or heuristic observations. A similar methodology was undertaken in fuzzy image matching, discussed in the book by Cyganek and Siebert [15], or in the task of figure detection, discussed in Section 4.4. The fuzzy AND operation can be defined with the multiplication or the minimum rule of the membership functions [16], as it was already formulated in Equations (3.162) and (3.163), respectively.

On the other hand, for the fuzzy implication reasoning the two common methods of Mamdani and Larsen,

$$\mu_{P \Rightarrow C}(x, y) = \min(\mu_P(x), \mu_C(y))$$
$$\mu_{P \Rightarrow C}(x, y) = \mu_P(x)\mu_C(y)$$

(4.2)

can be used [17, 18]. In practice the Mamdani rule is usually preferred since it avoids multiplication. It is worth noting that the above inference rules are conceptually different from the definition of implication in the traditional logic. Rules (4.2) convey the intuitive idea that the truth value of the conclusion C should not be larger than that of the premise P.

In the traditional implication if P is *false* and C is *true*, then $P \Rightarrow C$ is defined also to be *true*. Thus, assuming about 5% transient region as in Figure 4.3, the rule R_1 in Table 4.1 for exemplary values $R = 94$, $G = 50$, and $B = 55$ would evaluate to $(0.4 \times 0.95 \times 0.96) \times 1 \approx 0.36$, in accordance with the Larsen rule in (4.2). For Mamdani this would be 0.4. On the other hand, the logical AND the traditional formulation would produce *false*. However, the result of the implication would be *true*, since *false* $\Rightarrow$ *true* evaluates to *true*. Thus, neither crisp *false*, nor *true*, reflect the insight into the nature of the real phenomenon or expert knowledge (in our case these are the heuristic values found empirically by Peer *et al.* [14] and used in Equation (4.1)).

The rule R_{HS} in (4.1) is an aggregation of the rules R_1–R_6. The common method of fuzzy aggregation is the maximum rule, i.e. the maximum of *the output* membership functions of the rules which "fired." Thus, having output fuzzy sets for the rules the aggregated response can be inferred as

$$\mu_H = \max\left(\mu_{P\Rightarrow C}^1, \ldots, \mu_{P\Rightarrow C}^n\right), \tag{4.3}$$

where $\mu_{P\Rightarrow C}$ are obtained from (4.2). Finally, from μ_H the "crisp" answer can be obtained after defuzzification. In our case the simplest method for this purpose is also the maximum rule (we need a "false" or "true" output), although in practice the centroid method is very popular.

The presented fuzzy rules were then incorporated into the system for automatic human face detection and tracking in video sequences. For face detection the abovementioned method by Viola and Jones was applied [3]. For tests the OpenCV implementation was used [19, 20]. However, in many practical examples it showed high rate of false positives. These can be suppressed however at the cost of the recall factor. Therefore, to improve the former without sacrificing the latter, the method was augmented with a human skin segmentation module to take advantage if color images are available. Faces found this way can be tracked, for example, with the method discussed in Section 4.6. The system is a simple cascade of a prefilter, which partitions a color image into areas-of-interest (i.e. areas with human skin), and a cascade for face detection in monochrome images, as developed by Viola and Jones. Thus, the prefilter realizes the already mentioned dimensionality reduction, improving speed of execution and increasing accuracy. This shows the great power of a cascade of simple classifiers which can be recommended in many tasks in computer vision. The technique can be seen as an ensemble of cooperating classifiers which can be arranged in a series, parallel, or a mixed fashion. These issues are further discussed in Section 5.6. The system is depicted in Figure 4.4.

In a cascade simple classifiers are usually employed, for which speed is preferred over accuracy. Therefore one of the requirements is that the preceding classifier should have a high

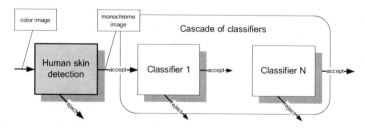

Figure 4.4 A cascade of classifiers for human face detection. The first classifier does dimensionality reduction selecting only pixels-of-interest based on a model of a color for human skin based on fuzzy rules.

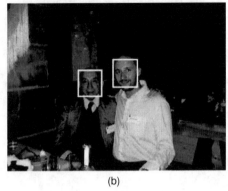

(a)   (b)

Figure 4.5   Results of face detection with the system in Figure 4.4. A color skin map (a). A test image with outlined face areas (b).

recall factor. In other words, it is better if such a classification module has a high ratio of false positives rather than too high false negative, i.e. it passes even possibly wrong answers to the next classification stage rather than rejecting too many. If this is done, then there is still hope that the next expert module in the chain will have a chance to correctly classify an object, and so on. Thus, in the system in Figure 4.4 the human skin detector operates in accordance with the fuzzy method (4.1). For all relations in the particular rules of (4.1) the fuzzy margin of 5% was set as presented in Figure 4.3. Summarizing, this method was chosen for three reasons. Firstly, as found by comparative experiments, it has the desirable property of a high recall factor, for the discussed reasons, at the cost of slightly lower precision when compared with other methods. Secondly, it does not require any training and it is very fast, allowing run-time operation. Thirdly, it is simple to implement.

Figure 4.5(a) depicts results of face detection in a test color image carried out in the system presented in Figure 4.4. Results of human skin detection computed in accordance with (4.1) are shown in Figure 4.5(a). The advantage of this approach is a reduction in the computations which depend on the contents of an image, since classifiers which are further along in the chain exclusively process pixels passed by the preceding classifiers. This reduction reached up to 62% in the experiments with different images downloaded from the Internet from the links provided in the paper by Hsu [21].

4.2.3   CASE STUDY – Pixel Based Road Signs Detection

In this application the task was to segment out image regions which could belong to road signs. Although shapes and basic colors are well defined for these object, in real situations there can be high variations of the observed colors due to many factors, such as materials and paint used in manufacturing the signs, their wear, lighting and weather conditions, and many others. Two methods were developed which are based on manually collected samples from a few dozen images from real traffic scenes. In the first approach a fuzzy classifier was built from the color histograms. In the second, the one-class SVM method, discussed in Section 3.8.4, was employed. These are discussed in the following sections.

4.2.3.1 Fuzzy Approach

For each of the characteristic colors for each group of signs their color histograms were created based on a few thousand samples gathered. An example of the red component in the HSV color space and for the two groups of signs is presented in Figure 4.6. Histograms allow assessment of the distributions of different colors of road signs and different color spaces. Secondly, they allow derivation of the border values for segmentation based onsimple thresholding. Although not perfect, this method is very fast and can be considered in many other machine vision tasks (e.g. due to its simple implementation) [22].

Based on the histograms it was observed that the threshold values could be derived in the HSV space which give an insight into the color representation. However, it usually requires prior conversion from the RGB space.

From these histograms the empirical range values for the H and S channels were determined for all characteristic colors encountered in Polish road signs from each group [23]. These are given in Table 4.3. In the simplest approach they can be used as threshold values for segmentation. However, for many applications the accuracy of such a method is not satisfactory.

The main problem with crisp threshold based segmentation is usually the high rate of false positives, which can lower the recognition rate of the whole system. However, the method is one of the fastest ones.

Better adapted to the actual shape of the histograms are the piecewise linear fuzzy membership functions. At the same time they do not require storage of the whole histogram which can be a desirable feature especially for the higher dimensional histograms, such as 2D or 3D. Table 4.4 presents the piecewise linear membership functions for the blue and yellow colors of the Polish road signs obtained from the empirical histograms of Figure 4.7. Due to specific Polish conditions it was found that detection of warning signs (group "A" of signs) is more reliable based on their yellow background rather than their red border, which is thin and usually greatly deteriorated.

Experimental results of segmentation of real traffic scenes with the different signs are presented in Figure 4.8 and Figure 4.9. In this case, the fuzzy membership functions from Table 4.4 were used. In comparison to the crisp thresholding method, the fuzzy approach allows more flexibility in classification of a pixel to one of the classes. In the presented experiments such a threshold was set experimentally to 0.25. Thus, if for instance for a pixel p, if $min(\mu_{HR}(p), \mu_{SR}(p)) \geq 0.25$, it is classified as possibly the red rim of a sign.

It is worth noticing that direct application of the Bayes classification rule (3.77) requires evaluation of the class probabilities. Its estimation using, for instance, 3D histograms can even occupy a matrix of up to $255 \times 255 \times 255$ entries (which makes 16 MB of memory assuming only 1 byte per counter). This could be reduced to 3×255 if channel independence is assumed. However, this does not seem to be justified especially for the RGB color space, and usually leads to a higher false positive rate. On the other hand, the parametric methods which evaluate the PDF with MoG do not fit well to some recognition tasks what results in poor accuracy, as frequently reported in the literature [10, 1].

4.2.3.2 SVM Based Approach

Problems with direct application of the Bayes method, as well as the sometimes insufficient precision of the fuzzy approach presented in the previous section, has encouraged the search

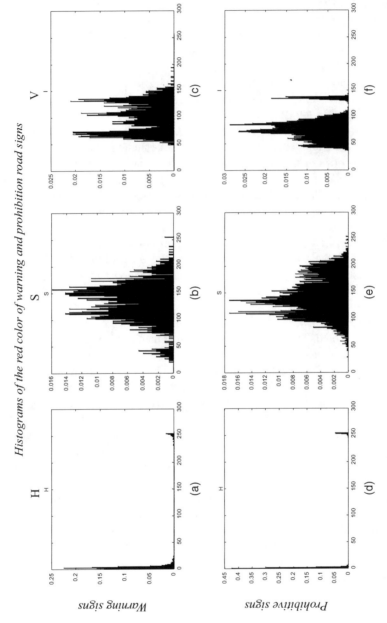

Figure 4.6 HSV histograms of the red color of warning signs (a,b,c) and the prohibitive signs in daylight conditions (d,e,f).

Table 4.3 Empirical crisp threshold values for different colors encountered in Polish road signs. The values refer to the normalized [0–255] HSV space.

	H	S
Red	[0–10] ∪ [223–255]	[50–255]
Blue	[120–165]	[80–175]
Yellow	[15–43]	[95–255]
Green	[13–110]	[60–170]

for other classifiers. In this case the idea is to use the one-class SVM, operating in one of the color spaces, to select pixels with the characteristic colors of the road signs. Once again, the main objectives of the method are accuracy and speed, since it is intended to be applied in real-time applications. Operation of the OC-SVMs is discussed in Section 3.12. In this section we outline the main properties and extensions of this method [2].

The idea is to train the OC-SVM with color values taken from examples of pixels encountered in images of real road signs. This seems to fit well to the OC-SVM since significantly large amounts of low dimensional data from one class are available. Thus, a small number of SVs is usually sufficient to outline the boundaries of the data clusters. A small amount of SVs means faster computation of the decision function which is one of the preconditions for automotive applications. For this purpose and to avoid conversion the RGB color space is used. During operation each pixel of a test image is checked to see if it belongs to the class or not with the help of formulas (3.286) and (3.287). The Gaussian kernel (3.211) was found to provide the best results.

A single OC-SVM was trained in a 10-fold fashion. Then its accuracy was measured in terms of the ROC curves, discussed in Appendix A.5. However, speed of execution – which is a second of the important parameters in this system – is directly related to the number of support vectors which define a hypersphere encompassing data and are used in classification of a test point, as discussed in Section 3.12. These, in turn, are related to the parameter γ of the Gaussian kernel (3.211), as depicted in Figure 4.10. For $\gamma \leq 10$ processing time in the software implementation is in the order of 15–25 ms per frame of resolution 320×240

Table 4.4 Piecewise linear membership functions for the red, blue, and yellow colors of Polish road signs.

Attribute	Piecewise-linear membership functions – coordinates (x,y)
Red H (H_R)	(0, 1) - (7, 0.9) - (8, 0) - (245, 0) - (249, 0.5) - (255, 0)
Red S (S_R)	(75, 0.4) - (80, 1) - (180, 1) - (183, 0)
Blue H (H_B)	(125, 0) - (145, 1) - (150, 1) - (160, 0)
Blue S (S_B)	(100, 0) - (145, 1) - (152, 1) - (180, 0)
Yellow H (H_Y)	(20, 0) - (23, 1) - (33, 1) - (39, 0)
Yellow S (S_Y)	(80, 0) - (95, 0.22) - (115, 0.22) - (125, 0) - (128, 0) - (150, 0.48) - (155, 0.48) - (175, 0.18) - (200, 0.22) - (225, 1) - (249, 0.95) - (251, 0)

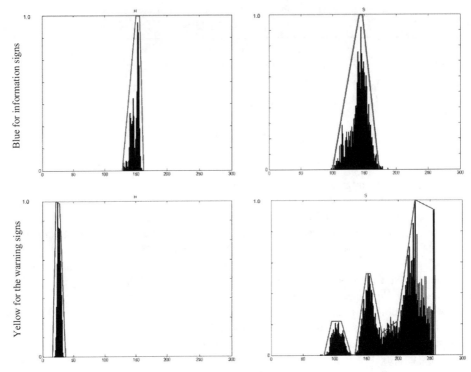

Figure 4.7 Piecewise linear membership function created from the histograms of color values for selected road signs (from [22]).

Figure 4.8 Results of fuzzy segmentation for different road signs. Color versions of this and subsequent images are available at www.wiley.com/go/cyganekobject.

Figure 4.9   Results of image segmentation with the fuzzy method for different road signs.

which is an acceptable result for automotive applications. Thus, in the training stage the two parameters of OC-SVM need to be disovered which fulfill the requirements

Other kernels, such as the Mahalanobis (3.218) or a polynomial gave worse results. For the former this caused the much higher number of support vectors necessary for the task, leading to much slower classification. The latter resulted in the worst accuracy.

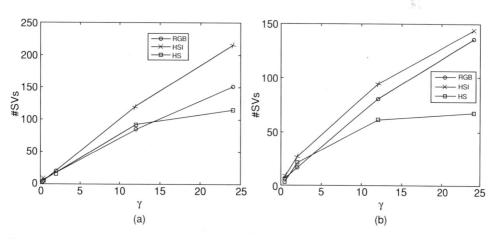

Figure 4.10   Number of support vectors with respect to the parameter γ of the Gaussian kernel for the blue (a) and red (b) sample points and in different color spaces for $C = 1$ (from [2]).

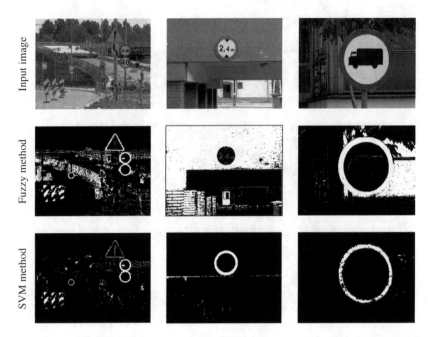

Figure 4.11   Comparison of image segmentation with the fuzzy method (middle row) and the one-class SVM with RBF kernel (lower row) (from [2]). (For a color version of this figure, please see the color plate section.)

Segmentation with the presented method proved to be especially useful for objects which are placed against a similar background, as shown Figure 4.11. In this respect it allows more precise response as compared with the already discussed fuzzy approach, in which only two color components are used in classification. It can be seen that the fuzzy method is characteristic of lower precision which manifests with many false positives (middle row in Figure 4.11). This leads to incorrect figure detections and system response which will be discussed in the next sections.

On the other hand, the SVM based solutions can suffer from overfitting in which their generalization properties diminish. This often happens in configurations which require comparatively large numbers of support vectors. Such behavior was observed for the Mahalanobis kernel (3.218), and also for the Gaussian kernel (3.211) for large values of the parameter γ. However, the RBF kernel operates well for the majority of scenes from the verification group, i.e. those which were not used for training, such as those presented in Figure 4.11. However, to find the best operating parameters, as well as to test and compare performance of the OC-SVM classifier with different settings, a special methodology was undertaken which is described next. Thanks to its properties the method is quite versatile. Specifically it can be used to segment out pixels of an object from the background, especially if the number of samples in the training set is much smaller than the expected number of all other pixels (background).

The used f-fold cross-validation method consists of dividing a training data set into f partitions of the same size. Then, sequentially, $f - 1$ partitions are used to train a classifier, while the remaining data is used for testing. The procedure follows sequentially until all partitions have been used for testing. In implementation the LIBSVM library was employed,

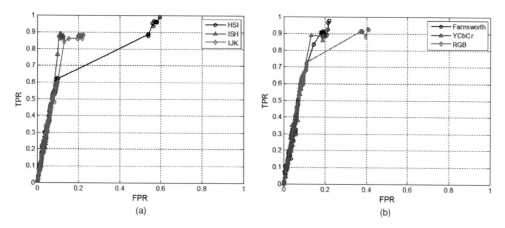

Figure 4.12 ROC curves of the OC-SVM classifier trained with the red color of Polish prohibition road signs in different color spaces: HIS, ISH, and IJK (a), Farnsworth, YCbCr, and RGB (b). Color versions of the plots are available at www.wiley.com/go/cyganekobject.

also discussed in Section 3.12.1.1. In this library, instead of the control parameter C, the parameter $\upsilon = 1/(CN)$ is assumed [24]. Therefore training can be stated as a search for the best pair of parameters (γ, υ) using the described cross-validation and the grid search procedure [24]. Parameters of the search grid are preselected to a specific range which show promising results. In the presented experiments the search space spanned the range $0.0005 \leq \gamma \leq 56$ and $0.00005 \leq \upsilon \leq 0.001$, respectively.

Figure 4.12 depicts ROC curves for the OC-SVM classifier tested in the 10-fold cross-validation fashion for the red color encountered in prohibition road signs. Thus, to compute a single point in the ROC curve an entire f-fold cycle has to be completed. In other words, if in this experiment there are 10 sets of the training data, then for a single ROC point the classifier has to be trained and checked 10 times (i.e. each time with $10 - 1 = 9$ sets used to build and 1 left for performance checking). The *FPR* and *TPR* values of a single point are then the arithmetic average of all 10 build-check runs. Six color spaces were tested. These are HIS, ISH, and IJK, shown in Figure 4.12(a) and RGB, YCbCr, and Farnsworth in Figure 4.12(b). The best results were obtained for the perceptually uniform Farnsworth color space (black in Figure 4.12(b)). Apart from the Farnsworth color space the YCbCr space gave very good results with the lowest *FPR*. It is interesting since computation of the latter from the original RGB space is much easier. Differences among other color spaces are not so significant. These and other color spaces are discussed in Section 4.6. In this context the worst performance was obtained for the HSI color space. As shown in Table 4.5, the comparably high number of support vectors means that in this case the OC-SVM with the RBF kernel was not able to closely encompass this data set.

Figure 4.13 shows two ROC curves of the OC-SVM classifier trained on blue color samples which were collected from the obligation and information groups of road signs (in Polish regulations regarding road signs these are called groups C and D, respectively [23]). In this experiment a common blue color was assumed for the two groups of signs. The same 10-fold cross-validation procedure and the same color spaces were used as in the case of the red

Table 4.5 Best parameters found for the OC-SVM based on the f-fold cross-validation method for the red and blue color signs. The grid search method was applied with the range $0.0005 \leq \gamma \leq 56$ and $0.00005 \leq \upsilon \leq 0.001$.

Color space	Red			Blue		
	γ	υ	#SVs	γ	υ	#SVs
HSI	5.6009	0.000921	52	0.0010	0.0002400	3
ISH	0.0010	0.000287	2	0.0010	0.0003825	4
IJK	0.0020	0.000050	2	0.0020	0.0009525	8
Farnsworth	**0.2727**	**0.001000**	**8**	**0.0910**	**0.0002875**	**8**
YCbCr	0.0010	0.000050	2	0.0010	0.0003350	4
RGB	1.2009	0.000020	10	2.8009	0.0001450	25

color data. The best performance once again was obtained for the Farnsworth color space, though other color spaces do almost as well, including the HSI space. The RGB space shows the highest FPR rate, though at relatively large TPR values. Such characteristics can be also verified by analyzing the number of support vectors contained in Table 4.5. In this case it is the largest one (25) which shows the worst adaptation of the hypersphere to the blue color data.

Only in one case does the number of support vectors (#SVs) exceed ten. For the best performing Farnsworth color space #SVs is 5 for red and 3 for blue colors, respectively. A small number of SVs indicates sufficient boundary fit to the training data and fast run time performance of the classifier. This, together with the small number of control parameters, gives a significant advantage of the OC-SVD solution. For instance a comparison of OC-SVM with other classifiers was reported by Tax [25]. In this report the best performance on many

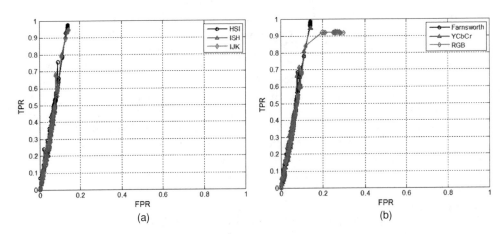

Figure 4.13 ROC curves of the OC-SVM classifier trained with the blue color of Polish information and obligation road signs in different color spaces: HIS, ISH, and IJK (a), Farnsworth, YCbCr, and RGB (b). Color versions of the plots are available at www.wiley.com/go/cyganekobject.

test data was achieved by the Parzen classifier (Section 3.7.4). However, this required a large number of prototype patterns which resulted in a run-time response that was much longer than for other classifiers. On the other hand, the classical two-class SVM with many test data sets requires a much larger number of SVs.

4.2.4   Pixel Based Image Segmentation with Ensemble of Classifiers

For more complicated data sets than discussed in the previous section, for example those showing specific distribution, segmentation with only one OC-SVM cannot be sufficient. In such cases, presented in Section 3.12.2 the idea of prior data clustering and building of an ensemble operating in data partitions can be of help. In this section we discuss the operation of this approach for pixel-based image clustering. Let us recall that operation of the method can be outlined as follows:

1. Obtain sample points characteristic to the objects of interest (e.g. color samples);
2. Perform clustering of the point samples (e.g. with a version of the k-means method); for best performance this process can be repeated a number of times, each time checking the quality of the obtained clustering; after the clustering each point is endowed with a weight indicating strength of membership of that point in a partition;
3. Form an ensemble consisting of the WOC-SVM classifiers, each trained with points from different data partitions alongside with their membership weights.

Thus, to run the method, a number of parameters need to be preset both for the clustering and for the training stages, respectively. In the former the most important is the number of expected clusters M, as well as parameters of the kernel, if the kernel version of the k-means is used (Section 3.11.3). On the other hand, for each of the WOC-SVM member classifiers two parameters need to be determined, as discussed in the previous sections. These are the optimization constant C (or its equivalent $\nu = 1/(NC)$), given in Equation (3.263), as well as the σ parameter if the Gaussian kernel is chosen (other kernels can require different parameters, as discussed in Section 3.10). However, the two parameters can be discovered by a grid search, i.e. at first a coarse range of the parameters can be checked, and then a more detailed search around the best values can be performed [24]. As already mentioned, the points in each partition are assigned weights. However, for a given cluster $1 \leq m \leq M$ the weights have to fulfill the summation condition (3.240), i.e.

$$0 < \sum_{i=1}^{N_m} w_{mi} < N_m. \tag{4.4}$$

Therefore, combining condition (3.288) with (4.4) the following is obtained

$$1 < \frac{1}{C} \leq \sum_{i=1}^{N_m} w_{mi} \leq N_m. \tag{4.5}$$

Thus, for a given partition and its weights the training parameter C should be chosen in accordance with the following condition

$$\frac{1}{\sum\limits_{i=1}^{N_m} w_{mi}} \leq C < 1. \tag{4.6}$$

In practice, a range of C and σ values is chosen and then for each the 10-fold cross-validation is run. That is, the available training set is randomly split into 10 parts, from which 9 are used for training, and 1 left for testing. The procedure is run number of times and the parameters for which the best accuracy was obtained are stored.

The described method assumes twofold transformation of data to the two different feature spaces. The first mapping is carried out during the fuzzy segmentation. The second is obtained when training the WOC-SVM classifiers. Hence, by using different kernels or different sets of features for clustering and training, specific properties of the ensemble can be obtained.

The efficacy of the system can be measured by the number of support vectors per number of data in partitions, which should be the minimum possible for the required accuracy. Thus, efficacy of an ensemble of WOC-SVMs can be measured as follows [26]

$$\rho = \sum\limits_{i=1}^{M} \rho_i = \sum\limits_{i=1}^{M} \frac{\#SV_i}{\#D_i}, \tag{4.7}$$

where $\#SV_i$ is the number of support vectors for a data set D_i, M denotes the number of members of the ensemble, and $\#D_i$ is a cardinality of the i-th data subset. A rule of thumb is to keep ρ_i below 0.1–0.2. Higher values usually indicate data overfitting and as a consequence worse generalization properties of the classifier. The value of Equation (4.7) can be used in the training stage to indicate a set of parameters with sufficiently small ρ. On the other hand, if only one subset i shows excessive value of its ρ_i, then new clustering of this specific subset can be considered. In other cases, the clustering process can be repeated with different initial numbers of clusters M.

During operation a pixel is assigned as belonging to the class only if accepted by exactly one of the member classifiers of that ensemble. Nevertheless, depending on the problem this arbitration rule can be relaxed, e.g. a point can be also assigned if accepted by more than one classifier, etc. The classification proceeds in accordance to Equations (3.286) and (3.287). Thus, its computation time depends on a number $\#SV$, as used in (4.7). Nevertheless, for a properly trained system $\#SV$ is much lower than the total number of original data. Therefore, the method is very fast and this is what makes it attractive to real-time applications.

In the following two experimental results exemplifying the properties of the proposed ensemble of classifiers for pixel based image segmentation are presented [26]. In the first experiment, a number of samples of the red and blue colors occurring in the prohibitive and information road signs, respectively, were collected. Then these two data sets were mixed and used to train different versions of the ensembles of WOC-SVMs, presented in the previous sections. Then the system was tested for an image in Figure 4.14(a). Results of the red-and-blue color segmentation are presented in Figure 4.14(b–d), for $M = 1$, 2, and 5 clusters, respectively. We see a high number of false positives in the case of one classifier, i.e. for $M = 1$. However, the situation is significantly improved if only two clusters are used. In the case of

(a)

(b)

(c)

(d)

Figure 4.14 Red-and-blue color image segmentation with the ensemble of WOC-SVMs trained with the manually selected color samples. An original 640 × 480 color image of a traffic scene (a). Segmentation results for $M = 1$ (b), $M = 2$ (c), and $M = 5$ (d) (from [26]). (For a color version of this figure, please see the color plate section.)

five clusters, $M = 5$, we notice an even lower number of false positives. However, the red rim of the prohibition sign is barely visible indicating lowered generalization properties (i.e. tight fit to the training data).

In this experiment the kernel c-means with Gaussian kernels was used. Deterministic annealing was also employed. That is, the parameter γ in (3.253) starts from 3 and is then gradually lowered to the value 1.2.

The second experiment was run with an image shown in Figure 4.15(a) from the Berkeley Segmentation Database [27]. This database contains manually outlined objects, as shown in Figure 4.15(b). From the input image a number of color samples of bear fur were manually gathered, as shown in Figure 4.15(c).

The image in Figure 4.16(a) depicts manually filled animals in an image, based on their outline in Figure 4.15(b). Figure 4.16(b–c) show results of image segmentation with the ensemble composed of 1 and 7 members, respectively. As can be seen, an increase in the number of members in the ensemble leads to fewer false positives. Thanks to the ground-truth data in Figure 4.16(a) these can be measured quantitatively, as precision and recall (see Section A.5). These are presented in Table 4.6.

Figure 4.15 A 481 × 321 test image (a) and manually segmented areas of image from the Berkeley Segmentation Database [27] (b). Manually selected 923 points from which the RGB color values were used for system training and segmentation (c), from [26]. Color versions of the images are available at the book web page [28]. (For a color version of this figure, please see the color plate section.)

The optimal number of clusters was obtained with the entropy criterion (3.259). Its values for color samples used to segment images in Figure 4.14(a) and Figure 4.15(a) are shown in Figure 3.28 with the groups of bars for the 4th and 5th data set.

From the results presented in Table 4.6 we can easily see that highest improvements in accuracy are obtained by introducing a second classifier. This is due to the best entropy parameter for the two classes in this case, as shown in Figure 3.28. Then accuracy improves with increasing numbers of classifiers in the ensemble, reaching a plateau. Also, kernel based clustering allows slightly better precision of response as compared with the crisp version. Further details of this method, also applied to data sets other than images, can be found in paper [26].

4.3   Detection of Basic Shapes

Detection of basic shapes such as lines, circles, ellipses, etc. belongs to one of the fundamental low-level tasks of computer vision. In this context the basic shapes are those that can be described parametrically by means of a certain mathematical model. For their detection the most popular is the method by Hough [29], devised over half a century ago as a voting method

Figure 4.16 Results of image segmentation based on chosen color samples from Figure 4.15(c). Manually segmented objects from Figure 4.15(b–c) used as a reference for comparison. Segmentation results with the ensemble of WOC-SVMs for only one classifier, $M = 1$ (b) and for $M = 7$ classifiers (b). Gaussian kernel used with parameter $\sigma = 0.7$ (from [26]).

Table 4.6 Accuracy parameters precision P vs. recall R of the pixel based image segmentation from Figure 4.15 with results shown in Figure 4.16 (from [26]).

M	P	R	P	R
	Crisp k-means		Kernel c-means (Gaussian kernel)	
1	0.63	0.97	0.63	0.97
2	0.63	0.95	0.76	0.90
5	0.77	0.92	0.79	0.92
7	0.76	0.91	**0.81**	**0.91**

for recognition of lines, it was then introduced to the computer vision community by Duda and Hart [30], and further extended to detection of arbitrary shapes by Ballard [31]. However, in the case of general shapes the method is computationally extensive.

A good overview on the Hough method and its numerous variations can be found for instance in the book by Davies [32]. However, what is less known is that application of the structural tensor, discussed in Section 2.7, can greatly facilitate detection of basic shapes. Especially fast and accurate information can be obtained by analyzing the local phase φ of the tensor (2.94), as well as its coherence (2.97). Such a method, called *orientation-based Hough transform*, was proposed by Jähne [33]. The method does not require any prior image segmentation. Instead, for each point the structural tensor is computed which provides three pieces of information, that is, whether a point belongs to an edge and, if so, what is its local phase and what is the type of the local structure.

Then, only *one* parameter is left to be determined, the distance p_0 of a line segment to the origin of the coordinate system. The relations are as follows (see Figure 4.17).

$$\frac{x_2 - x_2^0}{x_1^0 - x_1} = \operatorname{ctg} \varphi, \quad x_2^0 = p_0 \sin \varphi, \quad x_1^0 = p_0 \cos \varphi, \tag{4.8}$$

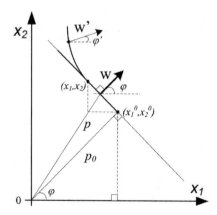

Figure 4.17 Orientation-based Hough transform and the *UpWrite* method.

(there are the lower and upper indices, not the powers) which after rearranging yield

$$\underbrace{\begin{bmatrix} \cos\varphi & \sin\varphi \end{bmatrix}}_{\mathbf{w}} \begin{bmatrix} x_1 \\ x_2 \end{bmatrix} = p_0. \tag{4.9}$$

In the above $\mathbf{w} = [\cos\varphi, \sin\varphi]^{\mathsf{T}}$ is a normal vector to the sought line and p_0 is a distance of a line segment to the center of the image coordinate system.

It is interesting to observe that such an orientation-based approach is related to the idea called the *UpWrite* method, originally proposed for detection of lines, circles, and ellipses by McLaughlin and Alder [34]. Their method assumes computation of local orientations as the phase of the dominant eigenvector of the covariance matrix of the image data. Then, a curve is found as a set of points passing through *consecutive* mean points m of local pixel blobs with local orientations that follow, or track, the assumed curvature (or its variations). In other words, the inertia tensor (or statistical moments) of pixel intensities are employed to extract a curve – these were discussed in Section 2.8. Finally, the points found can be fitted to the model by means of the least-squares method.

The two approaches can be connected into the method for shape detection in multichannel and multiscale signals[2] based on the structural tensor [35]. The method joins the ideas of the orientation-based Hough transform and the *UpWrite* technique. However, in the former case the ST was extended to operate in multichannel and multiscale images. Then the basic shapes are found in accordance with the additional rules. On the other hand, it differs from the *UpWrite* method mainly by application of the ST which operates on signal gradients rather than statistical moments used in the *UpWrite*. The two approaches are discussed in the next sections. Implementation details can be found in the papers [35, 36].

4.3.1 Detection of Line Segments

Detection of compound shapes which can be described in terms of line segments can be done with trees or with simple grammar rules [35, 37, 38]. In this section the latter approach is discussed. The productions describe expected local structure configurations that could contain a shape of interest. For example the S_A and $S_{D,E,F,T}$ productions help find silhouettes of shapes for the different road signs (these groups are named "A" and "D", "E", "F", "T"). They are formed by concatenations of simple line segment L_i. The rules are as follows

$$S_A \rightarrow L_1 L_2 L_3, \quad S_{D,E,F,T} \rightarrow L_3 L_4. \tag{4.10}$$

The line segments L_i are defined by the following productions

$$L_i \rightarrow L(\eta_i \pi, p_i, \kappa_i), L \rightarrow L_H | L_U, \tag{4.11}$$

where L_i defines a local structure segment with a slope $\pi/\eta_i \pm p_i$ which is returned by the detector L controlled by a set of specific parameters κ_i. The segment detector L, described by

[2]These can be *any* signals, so the method is not restricted to operating only with intensity values of the pixels.

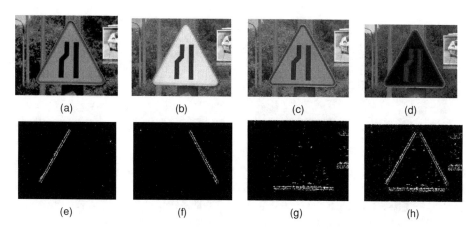

Figure 4.18 Shape detection with the S_A grammar rule. For detection the oriented Hough transform, computed from the structural tensor operating in the color image at one scale, was used. Color version of the image is available at www.wiley.com/go/cyganekobject.

the second production in Equation (4.11), can be either the orientation-based Hough transform L_H from the multichannel and multiscale ST [35], or the *UpWrite* L_U.

If all L_i of a production are parsed, i.e. they respond with a nonempty set of pixels (in practice above a given threshold), then the whole production is also fulfilled. However, since images are multidimensional structures, these simple productions lack spatial relations. In other words, a production defines only necessary, but not sufficient, conditions. Therefore further rules of figure verification are needed. These are discussed in Section 4.4.

Figure 4.18 depicts the results of detection of triangular shapes with the presented technique. The input is a color image shown in Figure 4.18a. Its three color channels R, G, and B, presented in Figure 4.18(b–d), are used directly to compute the ST, as defined in Equation (2.107) on p. 53. The weights are the same $c_k = {}^1/_3$ for all channels. The parameters η_i in (4.11) are $\eta_1 = 1/3$, $\eta_2 = 2/3$, and $\eta_3 = 0$. Parameter p_i which controls slope variation is $p_i = 3\%$, i.e. it is the same for all component detectors. Results of the L_1, L_2, and L_3 productions, as well as their combined output, are depicted in Figure 4.18(e–h), respectively. The shape that is found can be further processed to find parameters of its model, e.g. with the Hough transform. However, in many applications explicit knowledge of such parameters is not necessary. Therefore in many of them a detected shape can be tracked, as discussed in Section 3.8, or it can be processed with the adaptive window technique, discussed in Section 4.4.3.

4.3.2 UpWrite *Detection of Convex Shapes*

As alluded to previously, components of the ST provide information on areas with high local structure together with their local phases, as discussed in Section 2.7. The former can be used to initially segment an image into areas with strong local structures (such as curves, for instance), then the latter provides their local curvatures. These, in turn, can be tracked as long as they do not differ significantly, or in other words, to assure curvature continuity. This forms a foundation to the version of the *UpWrite* method presented here which is based on the structural tensor.

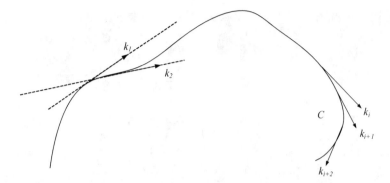

Figure 4.19 Curve detection with the *UpWrite* tensor method. Only places with a distinct structure are considered, for which their local phase is checked. If a change of phase fits into the predefined range, then the point is included into the curve and the algorithm follows.

The condition on strong local structure can be formulated in terms of the eigenvalues of the structural tensor (2.92), as follows

$$\lambda_1 > \tau, \quad \lambda_2 \approx 0, \tag{4.12}$$

where τ is a constant threshold. In other words, phases of local structures will be computed *only* in the areas for which there is one dominating eigenvalue. A classification of types of local areas based on the eigenvalues of the ST can be found in publications such as [39, 40, 15]. A similar technique for object recognition with local histograms computed from the ST is discussed in Section 5.2.

Figure 4.19 depicts the process of following local phases of a curve. A requirement of curve following from a point to point is that their local phases do not differ more than by an assumed threshold. Hence, a constraint on the gradient of curvature is introduced

$$\Delta \varphi = \varphi_k - \varphi_{k+1} < \kappa, \tag{4.13}$$

where κ is a positive threshold. Such a formulation allows detection of convex shapes, however. Thus, choice of the allowable phase change κ can be facilitated providing the degree of a polygon approximating the curve. The method is depicted in Figure 4.20.

In this way $\Delta \varphi$ from Equation (4.13) can be stated in terms of a degree N of a polygon, rather than a threshold κ, as follows

$$\Delta \varphi_{max} = \frac{2\pi}{N}, \text{ and } \Delta \varphi = \varphi_k - \varphi_{k+1} < \frac{2\pi}{N}. \tag{4.14}$$

In practice it is also possible to set some threshold on the maximum allowable distance between pairs of consecutive points of a curve. This allows detection of curves in real *discrete* images in which it often happens that the points are not locally connected mostly due to image distortions and noise.

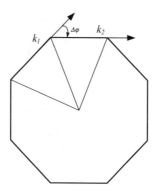

Figure 4.20 The allowable phase change in each step of the method can be determined providing a degree of the approximating polygon.

Figure 4.21 presents results of detection of the circular objects in real traffic scenes. Detected points for the allowable phase change, set with a polygon of degree $N = 180$, are visualized in Figure 4.21b. The maximum separation between consecutive points was set to not exceed 4 pixels.

Figure 4.22 also shows detection of oval road signs. In this case, however, the finer phase change was allowed, setting $N = 400$. The same minimal distance was used as in the previous example.

The method is fast enough for many applications. In the C++ implementation this requires about 0.4 s on average to process an image of 640 × 480 pixels. At first some time is consumed for computation of the ST, as discussed in Section 2.7.4.1. Although subsequent phase computations are carried out exclusively in areas with strong structure, some computations are necessary to follow a curve with backtracking. That is, the algorithm assumes to find the longest possible chain of segments of a curve. A minimal allowable length of a segment is set as a parameter. If this is not possible then it backtracks to the previous position and starts in other direction, if there are such possibilities. Nevertheless, memory requirements are

(a) (b)

Figure 4.21 Detection of ovals in a real image (a). Detected points with the *UpWrite* tensor method for the allowable phase change as in a polygon of degree $N = 180$ (b). Color versions of the images are available at www.wiley.com/go/cyganekobject.

(a)   (b)

Figure 4.22   Detection of ovals in a real image (a). Detected points with the method for the allowable phase change set to $N = 400$ (b). (For a color version of this figure, please see the color plate section.)

moderate, i.e. some storage is necessary for ST as well as to save the positions of the already processed pixels. Such requirements are convenient when compared with other algorithms, such as circle detection with the Hough method.

The next processing steps depend on the application. If parameters of a curve need to be determined, then the points can be fitted to the model by the voting technique as in the Hough transform. Otherwise, the least-squares method can be employed to fit a model to data [41, 42]. However, such a method should be able to cope with outliers, i.e. the points which do not belong to a curve at all and which are results of noise. In this respect the so called RANSAC method could be recommended [43, 44]. It has found broad application in other areas of computer vision, such as determination of the fundamental matrix [15, 45, 46]. Nevertheless, in many practical applications the parameters of a model are irrelevant or a model is not known. For example in the system for road sign recognition, presented in Section 5.7, such information would be redundant. A found object needs to be cropped from its background and then, depending on the classifier, it is usually registered to a predefined viewpoint and size. For this purpose a method for the tight encompassing of a found set of points is more important. This can be approached with the adaptive window growing method, discussed in Section 4.4.3. The mean shift method can also be used (Section 3.8).

4.4   Figure Detection

Many objects can be found based on detection of their characteristic points. The problem belongs to the dynamically changing domain of sparse image coding. The main idea is to detect characteristic points belonging to an object which are as much as possible invariant to potential geometrical transformation of the view of that object, as well as to noise and other distortions. The most well known point descriptors are SIFT [47], HOG [48], DAISY [49], SURF [50], as well as many of their variants, such as PCA-SIFT proposed by Ke and Sukthankar [51], OpponentSIFT [52], etc. A comparison of sparse descriptors can be found in the paper by Mikolajczyk and Schmid [53]. They also propose an improvement called the *gradient location and orientation histogram* descriptor (GLOH), which as reported outperforms SIFT in many cases. These results were further verified and augmented in the paper by Winder and Brown [54]. Their aim was to automatically learn parameters of the local descriptors based on a set of patches from the multi-image 3D reconstruction with well known ground-truth matches. Interestingly, their conclusion is that the best descriptors are those with log polar histogramming regions and feature vectors composed from rectified outputs of the steerable quadrature filters. The paper by Sande *et al.* also presents an interesting overview of efficient methods of objects' category recognition with different sparse descriptors, tested on the PASCAL VOC 2007 database [55] with an indication of the OpponentSIFT for its best performance [52]. Description of objects with covariance matrices is proposed in the paper by Tuzel *et al.* [56]. However, the covariance matrices do not form a vector space, so their space needs to be represented as a connected Riemannian manifold, as presented in the paper [56].

Finally, let us note that rather than by their direct appearance model, objects sometimes can be detected indirectly, i.e. by their some other characteristic features. For instance, a face can be inferred if two eyes are detected. Similarly a warning road sign, which depending on a country is a white or a yellow triangle with a red rim, can be found by detecting its three corners, etc. Nevertheless, such characteristic points do not necessarily mean the sought object exists. In other words, these are usually necessary but not sufficient conditions of existence of an object in an image. Thus, after detecting characteristic points, further verification steps are required. Such an approach undertaken to detect different shapes or road signs is discussed in the subsequent sections. Nevertheless, the presented methods can be used in all applications requiring detection of objects defined in a similar way, that is either by their characteristic points or with specification of a "mass" function describing their presence in an image, as will be discussed.

4.4.1   Detection of Regular Shapes from Characteristic Points

Many regular shapes can be detected if the positions of their salient points are known. These are the points for which an *a priori* knowledge is provided. For instance, for triangular, rectangular, diamond like, etc. shapes these can naturally be their corners.

In the proposed approach each point and its neighborhood are examined to check if a point fulfills conditions of a salient (characteristic) point. This is accomplished with the proposed *salient point detector* (SPD). It can operate directly with the intensity signals or in a transformed space. However, the method can be greatly simplified if, prior to detection, an image is segmented to a binary space, as discussed in Section 4.2. Such an approach was proposed for detection of triangular and rectangular road signs [57].

Figure 4.23a presents the general structure of the SPD. For each pixel **P** its neighborhood is selected which is then divided into a predefined number of parts. In each of these parts a distribution of selected features is then computed. Thus, a point **P** is characterized by its position in an image and N distributions. These, in turn, can be compared with each other or matched with a predefined model [15]. In practice a square neighborhood divided into eight parts, as depicted in Figure 4.23b, proved to be sufficient for detection of basic distributions of features. In this section we constrain our discussion to such a configuration operating in binary images.

Practical realization of the SPD, depicted in Figure 4.23(b), needs to account for a discrete grid of pixels. Therefore the symmetrical SPD was created – see Figure 4.24(a) – which is composed of four subsquares which are further divided into three areas, as shown in Figure 4.24(b).

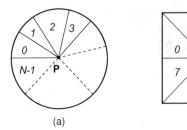

(a)   (b)

Figure 4.23 Detection of salient points with the SPD detector. A neighborhood of a pixel **P** is divided into N parts. In each of them a distribution of selected features is checked and compared with a model (a). In practice a square neighborhood is examined which is divided into eight parts (b) (from [22]).

Each of the subsquares S_i is of size $h_i \times v_i$ pixels, which are further divided into three regions, such as L_0, D_0 (diagonal), and U_0 in Figure 4.24(b). Usually D_0 is joined with U_0 and counts as one region DU_0. For example, there are 81 pixels in a 9×9 subsquare; From these 36 belong to L_0 and $36 + 9 = 45$ to DU_0. These two, i.e. L_0 and DU_0 are called further subregions R_i which can be numbered as e.g. in Figure 4.23(b).

Since a binary signal is assumed, detection is achieved by counting the number of bits attributed to an object in each of the regions R_i. Hence, for each point **P** of an image a series of eight counters c_i is provided. These counters are then compared with the model. If a match is found then **P** is classified as a salient point of a given type. Thus the whole process can also be interpreted as pixel labeling.

Figure 4.25 shows results of the SPD used for detection of triangular, rectangular, and diamond shaped road signs. If, for instance, subregions no. 5 and 6 are almost entirely filled while all others are empty, then possibly a point can be the top corner of a triangle. Similarly, if the panes 0 and 1 are filled, whereas the others are empty, then a bottom-right corner of a rectangle can be assumed. The method proved to be very accurate and fast in real applications [57]. It only requires definitions of the models which are expressed as ratios of counters for

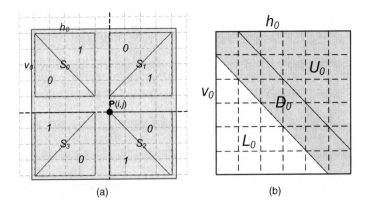

(a)   (b)

Figure 4.24 Symmetrical SPD detector on a discrete grid around a pixel **P** at location (i, j) divided into four subsquares S_i (a). Each subsquare is further divided into three areas (b) (from [22]).

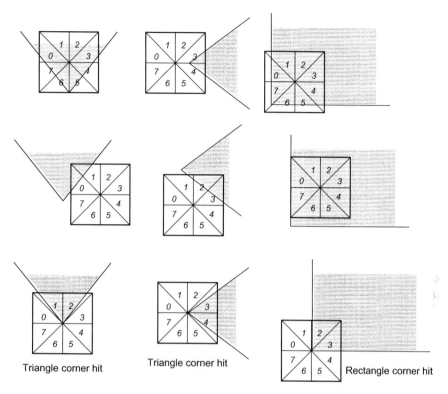

Triangle corner hit   Triangle corner hit   Rectangle corner hit

Figure 4.25 Detection of salient points with the SDLBF detector. A central point is classified based on the fill ratios of each of the eight subregions (from [22]).

each of the eight subregions. This can be accomplished with the definition of flexible fuzzy rules, as will be shown.

Figure 4.26 shows definitions of the salient points for detection of triangles, rectangles, and diamonds which are warning and information road sign shapes.

The necessary fill ratios for each subsquare are controlled by the fuzzy membership function, depicted in Figure 4.27. Three functions {*low, medium, high*} of the fill ratio f are defined. Their membership values depend on the ratio (expressed in %) of a number of set pixels to the total capacity of a subregion. A set of fuzzy rules (4.15) was defined for detection of different

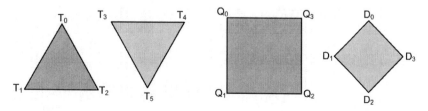

Figure 4.26 Salient points for detection of basic shapes – triangles, rectangles, and diamonds – of warning and information road sign shapes (from [22]).

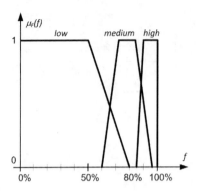

Figure 4.27 Fuzzy membership functions for the fill ratio f of the subregions (from [22]).

signs based on their salient points shown in Figure 4.26. The fuzzy output indicates if a given pixel is a characteristic point or not. In these the Mamdani inference rule (4.2) was employed. In the above fuzzy rules R_i is a fuzzy fill ratio of an *i-th* sub-region, in ordering as defined in Figure 4.23b, R_{other} denotes the fuzzy fill ratio of all other subregions than explicitly used in a rule, and finally, T_i, Q_i, and D_i denote fuzzy membership of the salient points as defined in Figure 4.26. The rules (4.15) take also into account that the two types of subregions, L_0 and DU_0 in Figure 4.24(b), are not symmetrical.

The last parameters that need to be set are the size and number of subregions in the SPD. These have to be tailored to the expected shape and size of the detected objects. These parameters also depend on the resolution of the input images. However, it was observed that eight subregions is a good trade-off between accuracy and speed of computations. Similarly, the size of the SPD does not greatly affect the accuracy of the method.

R_1:   **IF** $R_5 = medium$ **AND** $R_6 = medium$ **AND** $R_{other} = low$ **THEN** $T_0 = high$;
R_2:   **IF** $R_2 = medium$ **AND** $R_3 = high$ **AND** $R_{other} = low$ **THEN** $T_1 = high$;
R_3:   **IF** $R_0 = high$ **AND** $R_1 = medium$ **AND** $R_{other} = low$ **THEN** $T_2 = high$;

R_4:   **IF** $R_1 = medium$ **AND** $R_2 = medium$ **AND** $R_{other} = low$ **THEN** $T_5 = high$;
R_5:   **IF** $R_6 = medium$ **AND** $R_7 = high$ **AND** $R_{other} = low$ **THEN** $T_4 = high$;
R_6:   **IF** $R_4 = high$ **AND** $R_5 = medium$ **AND** $R_{other} = low$ **THEN** $T_3 = high$;

R_7:   **IF** $R_4 = high$ **AND** $R_5 = high$ **AND** $R_{other} = low$ **THEN** $Q_0 = high$;   (4.15)
R_8:   **IF** $R_2 = high$ **AND** $R_3 = high$ **AND** $R_{other} = low$ **THEN** $Q_1 = high$;
R_9:   **IF** $R_0 = high$ **AND** $R_1 = high$ **AND** $R_{other} = low$ **THEN** $Q_2 = high$;
R_{10}:   **IF** $R_6 = high$ **AND** $R_7 = high$ **AND** $R_{other} = low$ **THEN** $Q_3 = high$;

R_{11}:   **IF** $R_5 = high$ **AND** $R_6 = high$ **AND** $R_{other} = low$ **THEN** $D_0 = high$;
R_{13}:   **IF** $R_3 = high$ **AND** $R_4 = high$ **AND** $R_{other} = low$ **THEN** $D_1 = high$;
R_{14}:   **IF** $R_1 = high$ **AND** $R_2 = high$ **AND** $R_{other} = low$ **THEN** $D_2 = high$;
R_{15}:   **IF** $R_0 = high$ **AND** $R_7 = high$ **AND** $R_{other} = low$ **THEN** $D_3 = high$;

Last but not least, it is worth noticing that the method allows detection of rotated or slightly deformed or occluded shapes. This is a very useful feature of the proposed technique, especially when applied to detection of objects in real images. Further details are provided in [57]. Some real examples obtained with the presented technique are also discussed in Section 4.4.5.

4.4.2 Clustering of the Salient Points

To cope with shape deformations and noise the number of subregions in the SPD is reduced. For instance, based on experiments in many applications up to eight subregions is sufficient. Also the classification rules allow some degree of variation in the ratios of the filled and empty subregions (see the fuzzy rules (4.15)). As a result SPD usually reports a number of points that fulfill a predefined rule rather than a single location (i.e. the returned points tend to create local "cliques"). However, usually we are interested in having just one point representing such a group of close points. Thus, the next step of processing consists of locating such clusters of points and their replacement with a single location at the center of gravity of such a cluster.

For clustering, let us assume that SPD returned a set S_P of the points

$$S_P = \{ \mathbf{P}_0, \mathbf{P}_1, \ldots, \mathbf{P}_N \} = \{ (x_0, y_0), (x_1, y_1), \ldots, (x_N, y_N) \} . \tag{4.16}$$

The clusters K_i are defined as subsets of S_P for which it is assumed that a maximal distance of any pair of points in a cluster does not exceed a certain threshold value which, at the same time, is much smaller than a minimal distance between centers of any two other clusters. Thus, to determine K_i the distances between the points need to be determined. However, a number M of the clusters is not known either.

The set of all clusters $C(S_p)$ is denoted as follows:

$$C (S_P) = \{ K_1, K_2, \ldots, K_M \} . \tag{4.17}$$

Then, for each cluster its center of gravity is found, which is finally represents the whole cluster. This process results with the set M of m points

$$H = \{ \bar{K}_1, \ldots, \bar{K}_M \} = \{ (\bar{x}_1, \bar{y}_1), \ldots, (\bar{x}_m, \bar{y}_m) \} , \tag{4.18}$$

where

$$\bar{x}_p = \frac{1}{\#K_p} \sum_{x_p \in K_p} x_p, \quad \text{and} \quad \bar{y}_p = \frac{1}{\#K_p} \sum_{y_p \in K_p} y_p. \tag{4.19}$$

The clustering algorithm is controlled by the maximal distance d_τ between any two points above which the points are classified as belonging to different clusters. This means that if for two points $\mathbf{P}_i$ and $\mathbf{P}_j$ it holds that

$$D (P_i, P_j) \le d_\tau, \tag{4.20}$$

where D denotes the Euclidean distance, then these points belong to one cluster.

For a set S_P, containing n points, the process of its clustering starts with building the distance matrix $\mathbf{D}$ which contains distances for each pair drawn from the set S_P. There is $n(n-1)/2$ of such pairs. Thus, $\mathbf{D}$ is a triangular matrix with a zero diagonal.

```
1. Set the cluster counter        j = 0.
   Set a distance threshold       d_r.
   Construct the distance matrix   D.

2. Do:

       3. Take the first not clustered point P_i from the set S_P.

       4. Create a cluster K_j which contains P_i.

       5. Mark P_i as already clustered .

       6. For all not clustered points P_i from S_P do:

              7. If in K_j there is a close neighbor to P_i, i.e. inequality (4.20) holds:

                     8. Add P_i to K_j.

                     9. Set P_i as clustered.

       j = j + 1

   while there are not non-clustered points in S_P
```

Algorithm 4.1 Clustering of salient points.

The clustering Algorithm 4.1 finds the longest distinctive chains of points in the S_P. A chain contains at least two points. Hence, for each point in a chain there is at least one other point which is not further than d_τ. However, the clusters can contain one or more points. In our experiments d_τ takes values from 1 to 5 pixels. This is a version of the nearest-neighbor clustering algorithm in which the number of clusters is determined by the threshold d_τ. This is more convenient than for instance the k-means method discussed in Section 3.11.1, in which a number of clusters need to be known *a priori*.

4.4.3 Adaptive Window Growing Method

The detection technique with salient points SDP cannot be used to detect other shapes, for which a definition of a few characteristic points is difficult or impossible, e.g. ellipses. Thus, to solve this problem the idea is to first segment an image based on characteristic features of such objects (color, texture, etc.), and then find areas which contain dense clusters of such points. This can be achieved with the mean shift procedure discussed in Section 3.8. However, a simpler and faster technique is called *the adaptive window growing* method (AWG) which

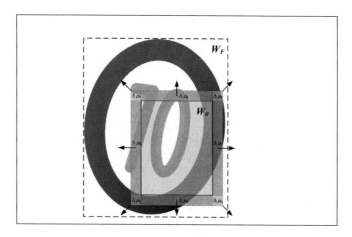

Figure 4.28 Adaptive region growing technique for fast shape detection. An initial window W_0 grows in all eight directions until a stopping criteria is reached. The final size of the region is W_F.

has some resemblance to the connected components method [58]. A rectangular window W is expanded in all eight directions around a place with high evidence of existence of an object. An example of the operation of this method is depicted in Figure 4.28. The only requirement is that the outlined region is described by a nonnegative "mass" function μ, for which it is assumed that the higher its value, the strongest the belief that a pixel belongs to an object. Thus μ can be PDF or a fuzzy membership function. Hence, the versatility of the method.

Expansion in a given direction is possible if there is a nonnegative increase in density of a new region. This is computed as a ratio of the object "mass" function divided by a number of pixels in a growing area. In effect, starting from an initial size W_0 a final window W_F is reached which tightly encompasses an object, as shown in Figure 4.28. In practice it is sufficient to set a minimal thresholds on a "mass" update in all directions of expansion to avoid divisions. Hence, a stop criteria in a direction k becomes

$$\Delta \mu_k < \tau_k, \tag{4.21}$$

where τ_k denotes an expansion threshold. In our experiments, in which μ was a fuzzy membership function conveying a degree of color match, τ_k is in the order of *0.1–10*.

The algorithm is guaranteed to stop either when condition (4.21) is fulfilled for all k, i.e. in all directions, or when the borders of an image are reached. Further details are provided in the paper [59].

The topological properties of a found shape are controlled by the expansion factor. If the window is allowed to grow at most by one pixel each step, then the neighbor-connected shapes are detected. Otherwise, the sparse versions can be obtained.

Once a shape is detected, it is usually cropped from the image, simply marking all pixels of the found shape as background. In the used HIL platform this is easily achieved with objects of the *TMaskedImageFor* class. Then the algorithm proceeds to find possibly other objects, until all pixels in the image have been visited [15].

4.4.4   Figure Verification

As alluded to in the previous sections, detected salient points provide valuable information on the possible vertices of the sought shapes. Moreover, all of them are additionally annotated, i.e. it is known whether a salient point can be a lower left or upper right corner of a rectangle, etc. However, the existence of separated salient points does not necessarily mean that they are vertices of a sought figure, e.g. a single triangle. Thus, there is a necessity for subsequent verification which relies on checking *all possible configurations* of the salient points. Certainly this requires further assumptions on what is actually sought. For instance whether we are looking for equilateral triangles, shapes of a certain size, or whether the figures can occlude each other, etc.

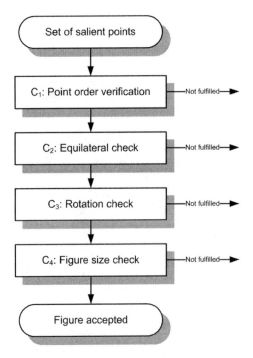

Algorithm 4.2   Rules for verification of a triangle given its salient points.

This information can be provided e.g. in the form of fuzzy rules. Such an approach was undertaken in the system for road sign recognition, which we discuss in this section [22, 60, 61]. For instance, for triangles it is checked whether a candidate figure is equilateral and whether its base is not excessively tilted. The former condition is checked by measuring lengths of sides while the latter is checked by measuring the slope of the base side. Similar conditions are set and checked for other shapes as well. On the other hand, application of the fuzzy rules gives enough flexibility to express expert knowledge. Although in the presented system these were hard coded, a straightforward implementation of an imperative language would allow simple formulation and dynamic processing of such rules [62, 63].

Algorithm 4.2 presents a flow chart of rules which allow detection of only those triangles whose dimensions and/or positions fulfill these rules. The rules in Algorithm 4.2 were composed for the road sign detection system, though their order and the rules by themselves can be easily changed. For instance the rule C_1 verifies the order of the salient points. That is, a triangle is assumed to be defined by three salient points T_0, T_1, and T_2 which are attributed to the corresponding vertices of a triangle, as depicted in Figure 4.26. However, if T_1 is a left vertex whereas T_2 is a right one, then if the order of actually detected points is reversed (which can be checked comparing their horizontal coordinates), such a configuration will be invalid. Once C_1 is passed, the other rules are checked in a similar manner. It is worth noting that if possible the rules should be set and then checked in order of the higher probability of rejecting the points. For instance, if we expect that more frequently figure size check rule (C_4 in Algorithm 4.2) is not fulfilled than the rotation check C_3, then their order should be changed. Thanks to this we can save on computations.

However, the rules can also be given some freedom of uncertainty or, in other words, they can be fuzzified. Thus, each rule can output a value of a membership function rather than a crisp result as "true" or "false." Then the whole verification rule could be as follows

$$\text{V}_1\colon \textbf{IF } C_1 = high \textbf{ AND } C_2 = high \textbf{ AND } C_3 = high \textbf{ AND } C_4 = high \textbf{ THEN } F = high;$$

$$(4.22)$$

A similar approach can be used to verify other shapes, which are not necessarily based on salient points. For instance, an oval returned by the adaptive region growing method (4.4.3) can be checked to fill some geometrical constraints. For example the smallest regions can be rejected since usually they do not provide sufficient information for the classification process. For example, in the aforementioned road sign recognition system these were the regions that were less than 10% of the average size $(N + M)/2$ of the input images of $N \times M$ pixels.

For verification of circles all four squares, anchored at the corners of a rectangle circumscribed on that circle, are checked, as shown in Figure 4.29. In each corner of the square circumscribed around the circle we enter the square $\square ABCD$ with the already found side x, as follows

$$x = \left(1 - \frac{1}{\sqrt{2}}\right) a \approx 0.3a, \tag{4.23}$$

where a denotes a side of the square encompassing the circle. Then the fill ratio of the set pixels in the triangle $\triangle ABD$ is checked. If this ratio exceeded about 15% then the conditions for the proper circular signs are assumed as not fulfilled.

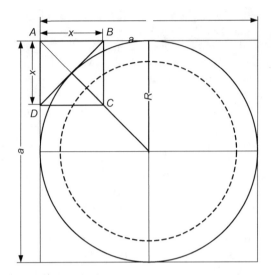

Figure 4.29   Verification rules for circles (from [59]).

4.4.5   CASE STUDY – Road Signs Detection System

The already presented methods of image segmentation, detection, and figure verification have found application in the vision system for recognition of Polish road signs [57] [59]. The applications of such a system are ample [61]. Figure 4.30 shows the architecture of the front-end of this system that's role is the detection of the signs and then construction of the feature

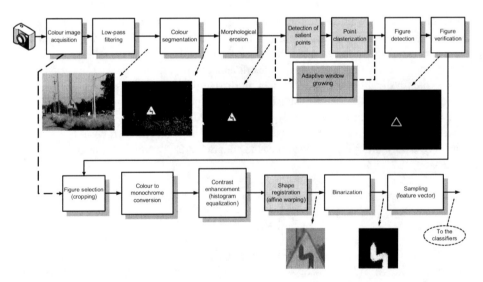

Figure 4.30   Architecture of the front-end detection used in road sign recognition systems (from [22]). Color versions of this and subsequent images are available at www.wiley.com/go/cyganekobject.

vector from their pictograms for further classification. This front-end is also used in the road signs recognition system discussed in Section 5.7.

The first module carries out color image acquisition and its filtering. The purpose of the filtering module is not only filtering of noise but also adjusting the image resolution. Sufficient resolution for simultaneously reliable and fast recognition of signs was found to be in the range 320×240 up to 640×480. Images with higher resolution cause excessive computations and therefore they are transformed to the preferred dimensions with the binomial interpolation in the RGB space [15]. The next step is image segmentation for which two methods were tested: the fuzzy one and the SVM, discussed in Sections 4.2.3.1 and 4.2.3.2, respectively.

From segmentation a binary image is obtained. This contains some noise which is removed with the morphological erosion filter. Usually the square structural element of 3×3 or 5×5 pixels is sufficient. Then shape detection takes place. There are two alternative methods for this depending on the type of a shape for detection. That is, triangles, rectangles, or diamonds can be detected with the algorithm based on salient points, discussed in Section 4.4.1. This has an additional advantage since the salient points can be used directly to register the found object to the predefined frame. This technique was discussed in the papers [59, 64], as well as in the book [15]. On the other hand, the adaptive window growing method, presented in Section 4.4.3, is more general since it allows detection of *any* connected shape. However, this method does not allow such easy registration since it does not provide a set of matched points of the shape, although corners of the encompassing window are available and can be used for this task. Because of this, circular signs (i.e. the prohibition and information signs) require special classifiers which can cope with the internal rotation and perspective deformation of an object. These issues are discussed in the next chapter.

Figure detection and verification constitute the next stages of processing, discussed in Section 4.4.4, from which a final answer is provided on the type and position of the found figures.

Then a detected object is cropped from the original color image and transformed into a monochrome version since the features for classification are binary. Finally, if possible the object is registered to a predefined frame to adjust its view to the size and viewpoint of the prototypes used for classification. As alluded to previously, such registration is possible if positions of the salient points are known. Three such points are sufficient to determine an inverse homography, which is then used to drive the image warping to change the viewpoint of the detected object [15, 59]. Objects detected with the adaptive window growing method are not registered, however. Finally, the object is binarized, as described in [59], from which a feature vector is created for classification, as discussed in Section 5.7.2.

Figure 4.31 depicts results of detection of warning signs from a real traffic scene. Fuzzy color segmentation was used that's output is visible in Figure 4.31(b). This is then morphologically eroded, from which the salient points are detected, as shown in Figure 4.31(d). In the next step, detected figures are cropped from the image and registered. These are shown in Figure 4.31(e–f), respectively. Detection of the warning signs from another traffic scene are shown in Figure 4.32. It is interesting to note the existence of many other yellow objects present in that image. However, thanks to the figure verification rules only the triangles that comply with the formal definition of the warning signs are output.

Figure 4.33 depicts stages of detection of information (rectangular) and warning signs (inverted triangle) in a real traffic scene. Salient points are computed from two different segmentation fields, which are blue and yellow color areas, respectively. The points for these

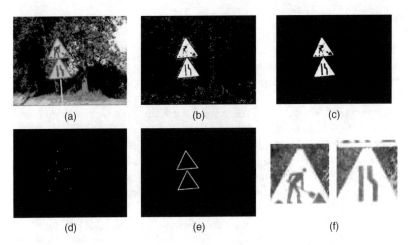

Figure 4.31 Detection of warning signs. The original image (a), fuzzy color segmentation (b), eroded (c), salient points (d), detected figures (e), cropped and registered signs (f) (from [22]).

two groups are visible in Figure 4.33(b,e). Finally, the detected shapes are shown in Figure 4.33(c) and Figure 4.33(f), respectively.

Figure 4.34 shows the stages of detection of the diamond like shapes of information signs. These are also detected after yellow segmentation. Interestingly, detection in this case is also recognition since there is only one such a sign in the image.

Finally, Figure 4.35 depicts the stages of detection of circular prohibition signs. Segmentation in this case is performed for red color characteristic for this group of signs. However, rather than with salient points, shapes are detected with help of the adaptive growing method,

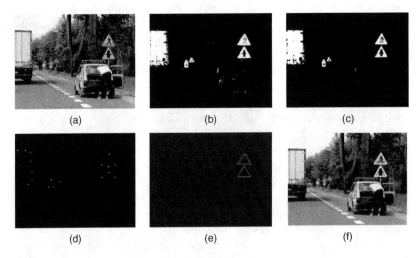

Figure 4.32 Detection of the warning signs in the real traffic scene (a), the color segmented map (b), after erosion (c), salient points (d), detected figures (e,f) (from [22]).

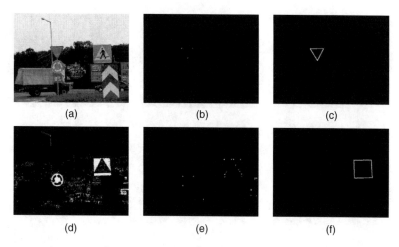

Figure 4.33 Detection of road signs in a real traffic scene (a). Salient points after yellow segmentation (b). Detected inverted triangle (c). The scene after blue segmentation (d) with salient points (e). A detected and verified rectangle of a sign (f) (from [22]).

presented in Section 4.4.3. At some step of processing two possible objects are returned, as shown in Figure 4.35(d). From these only the one that fulfills shape and size requirements is returned, based on the figure verification rules presented in Section 4.4.4.

Since classification requires only binary features, a detected shape is converted to the monochrome version, from which binary features are extracted. Actually conversion from color to the monochrome representation is done by taking only one channel from the RGB representation, rather than averaging the three. This was found to be superior in processing different groups of signs.

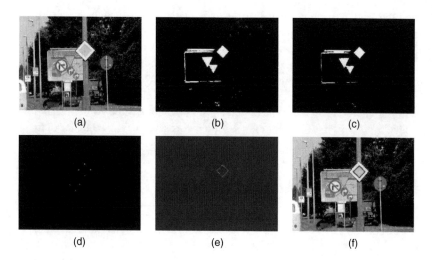

Figure 4.34 Stages of detection of diamond shapes (information signs) (from [22]).

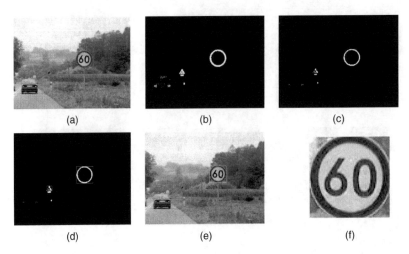

Figure 4.35 Detection of the prohibition sign from the real scene (a), fuzzy color segmentation (b), results of morphological erosion (c), two regions obtained with the adaptive window method (d), one figure fulfilling requirements of a circular road sign (e), the cropped and registered sign (f) (from [22]).

Accuracy of detection was tested on a database containing a few hundred real traffic scenes in daily conditions. Table 4.7 presents the measured accuracy in terms of the precision (P) vs. recall (R) parameters (see Section A.5). The measurement were made under the control of a human operator since ground-truth data is not available. A detected sign was qualified either as correctly or incorrectly detected. This was judged based on visual inspection of the operator. Small variations of a few pixels were accepted as positive responses since the classification modules can easily tolerate that. In general accuracy is above 91%, though it can be noticed that in Table 4.7 there are two different groups which follow two different detection methods, i.e. with salient points and adaptive window growing. For all groups the R parameter was lower than the P. This follows rather strict rules of detection, which results in some signs not being detected but with a small number of false positives at the same time. After closer inspection it appears that more often than not the problems were caused by the segmentation module incorrectly classifying pixels. In this field the SVM based pixel segmentation method performs better, however with some computational penalty (Section 4.2.3.2). The next module that usually causes errors in final classification is the morphological filtering which sometimes removes important areas for subsequent detection of salient points. However, more often this was preceded by very sparse segmentation. In other words, there is no evidence of inappropriate

Table 4.7 Accuracy of detection for different types of geometrical figures in daily conditions (last column AWG).

Type of a figure:	△	▽	□	◇	○
Precision	0.97	0.99	0.97	0.96	0.91
Recall	0.96	0.99	0.92	0.95	0.95

operation of the morphological module when supplied with a good segmentation field. Some problems are also encountered if a sign is partially occluded, especially if the occluded area is one of the salient points.

The lowest recall was noticed for the group of rectangular signs. This seems to be specific to the testing set of images which contain information signs taken in rather bad conditions. For all groups, except the inverted triangles and diamonds, the number of tested images was similar ($\sim$50 for each category). The two mentioned groups are simply less numerous in the traffic scenes (only 20 examples for each). Precision for the salient point detectors reached 0.97–0.99 for the triangles. Such high precision results from the stringent process of finding salient points and then the multistage figure verification process (Section 4.4.4). On the other hand, it should be pointed out that the database contains traffic scenes taken only in daylight conditions (sunny and rainy conditions). Nevertheless, the method shows good results in rainy conditions, and also for deblurred images. Tests with the night images show much worse results mostly due to insufficient lighting conditions which almost always lead to incorrect segmentation (the signs are simply not detected). Such conditions require different acquisition and processing methods.

The last column in Table 4.7 provides P and R factors for the circular shapes which were detected with the AWG method (Section 4.4.3). Accuracy here is about 5% lower compared to the salient points method. This is mostly caused by the lack of the point verification step. Hence, it sometimes happens that AWG returns on an object which is not a sign.

Software implementation of the presented road sign detection system allows real-time processing of a video stream of resolution 320 × 240. The fastest execution shows the AWG method than detection based on salient points, since in the latter each point has to be checked by the SPD detector. This suggests that for time critical applications the AWG detection can be used for all types of objects as this is a faster method. However, its accuracy is slightly worse, as has already been pointed out.

4.5 CASE STUDY – Road Signs Tracking and Recognition

As already mentioned, object detection means finding the position of an object in an image, and certainty that it is present. On the other hand, tracking of an object means finding the positions of this particular object in a sequence of images. In this process we take an indirect assumption that there is a correlation among subsequent images. Therefore for an image detected in one frame, it is highly probable that it will also appear in the next one, and so on. Obviously, its position and appearance can change from frame to frame. An object to be tracked is defined in the same way as for detection. More information on tracking can be found in the literature, e.g. in the books by Forsyth and Ponce [65] or by Thrun *et al.* [66].

In this section we present a system for road sign recognition in color video [67]. Processing consists of two stages: tracking with a fuzzy version of the *CamShift* method (Section 3.8.3.3) and then classification with the morphological neural networks MNN (Section 3.9.4). Detection of the signs is based on their specific colors. Fuzzy rules operating in the HSV color space allow reliable detection of the borders of the signs observed in daily conditions. The fuzzy map is then used by the *CamShift* method to track a sign in consecutive frames. The inner part of the tracked region, i.e. its pictogram, is cropped from the image and binarized, as described in Section (4.4.5). A pictogram is then fed to the MNN classifier. Because the pictograms of the signs can be represented as binary patterns it became possible to build efficient kernels for their classification with MNN. Such an approach allows recognition of pictograms that are contaminated with a random noise (erosive or dilative) or which are slightly deformed due to the properties of the acquisition system. Thanks to this the system allows fast (HMM converge in one step) and sufficiently accurate processing of real traffic scenes.

As alluded to previously, the pictograms are binary images. Thus, a simple (i.e. weak) independence among patterns is a sufficient condition for the existence of a kernel [68, 69]. The independence among patterns means that each pattern has at least one component which is greater than a maximum value computed over all other patterns. Desirable (but not absolutely necessary) is the existence of a minimal representation which implies at most one nonzero entry in each kernel vector $\mathbf{z}^\alpha$ in Equation (3.190).

It is assumed that deformations of the detected pictograms, which are due to geometric transformation of the acquisitions system as well as noise, can to some degree be accommodated by the MNN. As already mentioned, the prototype pictograms are taken from their formal specification [23]. Thanks to this, the prototypes can be easily changed to allow detection of other road signs or even other objects, such as letters, cars, etc. The prototypes constitute the reference binary patterns $\mathbf{x}^\alpha$ when creating the kernel $\mathbf{Z}$ in Equation (3.190). A bit "1" is assigned to a pictogram symbols, and "0" to the background. The original idea is then to find skeletons of the prototype patterns and use them to train the MNN. However, these have to be verified to fulfill the independence conditions.

Figure 4.36 depicts some prototypes of Polish road signs from the database presented in Figure 5.28(a) (p. 446) with their skeletons. These are then used to train the MNN. The main idea behind using skeletons in training of the MNN is to allow some small vertical and horizontal shifts in the test patterns. In other words, even if a pattern is shifted by few pixels then the majority of skeleton pixels still are inside the shifted version. Certainly, each skeleton of the shifted patterns is different, but in this method only one skeleton, computed from the reference (i.e. not shifted) pattern, is used. Skeletonization is carried out with the iterative morphological algorithm with a square structural element of 3×3 pixels [70, 71].

Verification for pattern independence is done by checking one-against-all for all skeletons. If the independence cannot be fulfilled then correction of the checked patterns is attempted by

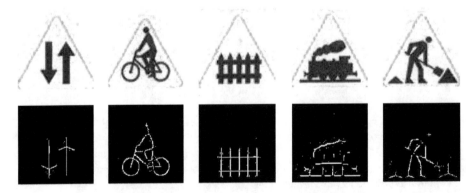

Figure 4.36 Some prototype patterns from the formal database (upper row) and their skeletons (lower row) used to train the morphological neural network.

removing pixels which cause failure of the independence conditions. Alternatively a trouble-some pattern can be shifted and the procedure repeated. Further details are described in [67]. In our experiments only some patterns were modified. However, there is still ongoing research on incorporating deformable patterns into the described training procedure. This would make the classifiers more robust to some geometric transformations of the observed patterns. Basically the novel procedure is finding a common core pattern for all allowable deformations of each prototype pattern (for binary patterns this is the logical AND function). Then the already described training method can be employed.

For the MNN with binary patterns $\mathbf{x}$ their kernels $\mathbf{z}$ should comply with the following conditions [72]

1. For all α: $\mathbf{z}^{\alpha} \leq \mathbf{x}^{\alpha}$ as well as
2. $\mathbf{z}^{\alpha} \wedge \mathbf{z}^{\beta} = \mathbf{0}$ and $\mathbf{z}^{\alpha} \not\leq \mathbf{x}^{\beta}$ (i.e. there is at least one component of $\mathbf{z}^{\alpha}$ which is greater from the corresponding component of $\mathbf{x}^{\beta}$) whenever $\alpha \neq \beta$.

Further properties of the MNN are discussed in Section 3.9.5. Considering the above, the corrected patterns are used to create the kernel vectors $\mathbf{z}^{\alpha}$ as described.

Figure 4.37 depicts the results of tracking the A-17 warning sign ("Children"). In this case the membership functions are constructed from the H and S channels of a subset of red color values pertinent to the signs of this group, with help of the method presented in Section 4.2.3.1.

Figure 4.37 Tracking of the A-17 road sign in a real traffic situation. Input video frames from a moving scene (top row). Fuzzy membership functions computed from the min function (3.162) (middle row). Objects detected and tracked by the fuzzy *CamShift* (bottom) (from [67]). Color versions of the images are available at www.wiley.com/go/cyganekobject.

The fuzzy membership function (3.163) is used in this case which requires fewer computations than Equation (3.162). The images in the first row of Figure 4.37 show three consecutive frames of the video stream acquired from a moving car. The middle row visualizes the membership functions which then are used in the fuzzy *CamShift* algorithm, discussed in Section 3.8. These have a direct influence on the detection results which are shown in the third row of Figure 4.37. It can be observed that due to reflections of the border of the windshield the spectra of color of the right frame in Figure 4.37 no longer fall into the specification obtained from the training color samples. In consequence the nonzero membership field is strongly reduced which leads to a much smaller tracking region. In this case it does not contain the whole object and the pictogram cannot be recognized correctly. However, the previous two responses were correct which is sufficient for the system to verify a sign. We can assume an even larger group of unanimous responses, at the cost of time however. The problem with proper segmentation is the most severe in this system. Thus usually only daylight conditions without shine and reflection produce acceptable results. More reliable color segmentation algorithms, as well as shape detection, can improve the performance.

Figure 4.38 presents the results of tracking of the B-33 sign ("40 km/h speed limit"), acquired in daylight conditions. For testing the fuzzy membership function given in Equation (3.162) was chosen.

Qualitative parameters of the system were measured in terms of precision (P) vs. recall (R) parameters. Results of these measurements are presented in Table 4.8. The system is able to process 320×240 color frames in real time, i.e. 15–20 frames per second which is sufficient

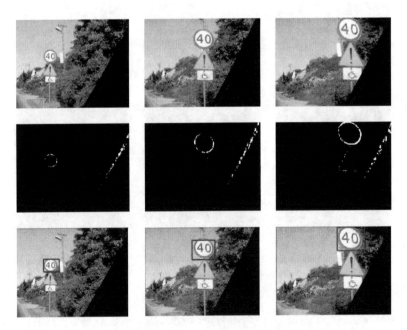

Figure 4.38 Tracking the B-33 prohibition sign in a real traffic situation. Input color frames acquired in a moving vehicle (top). Fuzzy membership functions (3.163) (middle). Objects detected and tracked by the fuzzy *CamShift* (bottom) (from [67]).

Table 4.8   Accuracy of recognition for different groups of signs: warning signs of group "A" and prohibition signs of group "B". Precision (P) versus recall (R).

Warning signs (group "A")		Prohibition signs (group "B")	
P	R	P	R
0.87	0.89	0.80	0.88

for automotive applications. It is also resistant to noise, small lighting variations, and small geometrical deformations of the signs due to the projective distortions of the acquisition system [67].

The values of the recall parameter are affected mostly by performance of the detectors. Precision is lower for the circular signs. This is mostly due to small pictograms which do not provide enough information to the MNN. This can be improved with other acquisition systems. However, the parameters depend on specific settings as well as the type and quality of the input images. Only results for daytime conditions are provided. At night or under worse weather conditions the recall parameter falls down due to large distortions, as discussed in Section 4.4.5.

4.6   CASE STUDY – Framework for Object Tracking

In this section a fuzzy *CamShift* version, discussed in Section 3.8.3.3, is presented as applied to the problem of object tracking, which are selected by a user in the images processed by the system for movie restoration. In this case the size of the input frames which are in the full HD standard[3] (this is 1k × 2k, 3 × 10 bits per color channel) can be problematic.

In this system an object is defined by a user through selection of a rectangular window in a reference frame. Then features from this region-of-interest are collected and processed for further tracking. The most natural features are colors and textures. Regarding color, the most popular is the RGB space. However, during experiments other color spaces were proven to provide more descriptive information about regions-of-interest. This is the case, for instance, for the *orthogonal* color space derived from the RGB, as shown by the following discussion which is based on the papers by Ohta *et al.* [73] and Pătraşcu [74].

The inner product in the RGB space for two color vectors $\mathbf{c}_1 = [R_1,G_1,B_1]^T$ and $\mathbf{c}_2 = [R_2,G_2,B_2]^T$ can be defined as

$$\langle \mathbf{c}_1, \mathbf{c}_2 \rangle = R_1 R_2 + G_1 G_2 + B_1 B_2. \tag{4.24}$$

Ohta *et al.* analyzed segmentation of the color images with respect to the best features for that purpose [73]. As shown, the most simple application of the bare R,G, and B signals does not lead to the best results in this respect. Instead, attempts were made to systematically derive other features from these basic three. The aim is to find such features which are the

---

[3]The HDTV standard began in the NHK Japan Broadcasting Corporation. It assumes resolutions of up to 1920x1080 lines and an aspect ratio of 16:9. HDTV assumes a Y'$C_B C_R$ color coding scheme. Thus the pixel rate is about 60 MPixels per second [75].

most discriminative. In other words, a variance of a feature inside an object should be low, whereas its variance across different objects should be large. For this purpose, test images were segmented with the recursive thresholding method. Then, in each of the segmented regions, the PCA analysis was performed on the RGB values of the pixels contained there. In effect, the principal components are obtained which have the highest variance, and thus high discriminative power. As discussed in Section (3.3.1), for the $L = 3$ components PCA requires computation of the 3×3 covariance matrix Σ_x which eigendecomposition results in three eigenvectors $\mathbf{t}_i$ ordered in accordance with the three eigenvalues λ_i. In the next step, RGB values in each segmented region are transformed to the uncorrelated features based on Equation (3.14), that is

$$\mathbf{y}_i = \mathbf{T}\,(\mathbf{c}_i - \mathbf{m}_\mathbf{c})\,, \tag{4.25}$$

where $\mathbf{c}_i = [R_i,G_i,B_i]^T$ represents a color vector for an i-th pixel in a region and $\mathbf{m}_\mathbf{c}$ denotes a mean of all color vectors in that region. As shown by Ohta et al., segmentation based on these $\mathbf{y}_i$ leads to better segmentation results. However, for different images and different segmentation regions this approach requires dynamic computation of the PCA decomposition to obtain their specific set of features $\mathbf{y}_i$. In practice this is computationally costly. Instead, a uniform base of three eigenvectors is sought which could be common to the PCA decomposition of a large group of color images. If found, this could be used to segment new images as well. Depending on experiments different "common bases" can be assumed, such as the following one [76]

$$\mathbf{t}_1 = [1, 1, 1]^T, \quad \mathbf{t}_2 = [1, 0, 0]^T, \quad \mathbf{t}_3 = [0, 1, -1]^T. \tag{4.26}$$

However, vectors (4.26) are not orthogonal. Their orthogonalization can be achieved with the Gram–Schmidt method [77], as shown in the following steps

$$\mathbf{a}_1 = \frac{\mathbf{t}_1}{\|\mathbf{t}_1\|} = \frac{1}{\sqrt{3}}[1, 1, 1]^T,$$

$$\tilde{\mathbf{a}}_2 = \mathbf{t}_2 - \langle \mathbf{a}_1, \mathbf{t}_2 \rangle\, \mathbf{a}_1 = \frac{1}{3}[2, -1, -1]^T, \text{ then } \mathbf{a}_2 = \frac{\tilde{\mathbf{a}}_2}{\|\tilde{\mathbf{a}}_2\|} = \frac{1}{\sqrt{6}}[2, -1, -1]^T, \tag{4.27}$$

$$\tilde{\mathbf{a}}_3 = \mathbf{t}_3 - \langle \mathbf{a}_1, \mathbf{t}_3 \rangle\, \mathbf{a}_1 - \langle \mathbf{a}_2, \mathbf{t}_3 \rangle\, \mathbf{a}_2 = [0, 1, -1]^T, \text{ then } \mathbf{a}_3 = \frac{\tilde{\mathbf{a}}_3}{\|\tilde{\mathbf{a}}_3\|} = \frac{1}{\sqrt{2}}[0, 1, -1]^T,$$

From (4.27) the orthogonal color space IJK is obtained by the projection of the RGB values onto the new base $\mathbf{a}_1, \mathbf{a}_2,$ and $\mathbf{a}_3$, as follows

$$I = \langle \mathbf{c}, \mathbf{a}_1 \rangle = [R, G, B] \cdot \frac{1}{\sqrt{3}}[1, 1, 1]^T = \frac{R + G + B}{\sqrt{3}},$$

$$J = \langle \mathbf{c}, \mathbf{a}_2 \rangle = [R, G, B] \cdot \frac{1}{\sqrt{6}}[2, -1, -1]^T = \frac{2R - G - B}{\sqrt{6}}, \tag{4.28}$$

$$K = \langle \mathbf{c}, \mathbf{a}_3 \rangle = [R, G, B] \cdot \frac{1}{\sqrt{2}}[0, 1, -1]^T = \frac{G - B}{\sqrt{2}},$$

The first component I in (4.28) is achromatic, whereas J and K denote chromatic components. For instance, the orthonormal color space IJK was used by Pătraşcu for fuzzy segmentation of images based on triangular functions [74]. Good results were also obtained in the segmentation

(a)   (b)

Figure 4.39   Color test image "Camerman" (a) and its monochrome version (b). Color version of the image is available at www.wiley.com/go/cyganekobject.

exclusively in the subspace JK, with respect to the segmentation in the "full" RGB space. This inspired application of the IJK space to tracking of color images, as will be discussed.

ISH is a perceptual color system derived from the IJK space [74]. The achromatic component is the same I, whereas the chromatic components are expressed in terms of saturation S and hue H

$$S = \sqrt{J^2 + K^2}, \quad H = atan2\,(K, J), \tag{4.29}$$

where the function *atan2* is explained in the description of Equation (2.94) on p. 48. Figure 4.39 presents a test image "*Camerman*" in RGB and its monochrome version for comparison with other color spaces which are presented in Figure 4.40.

For all color spaces except RGB, there is one achromatic and two chromatic channels. Although the former are the same, the chromatic channels differ. Conversely, the Farnsworth color space is a perceptually uniform color space [11]. The main goal of such color spaces is to achieve a proportional change in visual importance, as perceived by humans, on a change of a color value. In other words, small changes of color values should not produce rapid changes in perception of colors. Such property of a color space usually leads to more visually plausible results of computations with color components, such as filtering, especially when using the Euclidean distance between color values.

However, in many practical cases color components are not discriminative enough. Therefore when composing a fuzzy membership function describing an object, some other features need to be incorporated into Equations (3.162) and (3.163). In Section 2.7 the structural tensor was presented which, for each pixel of an image, either color or monochrome, associates three components which denote types of local structures. Further, each component is treated independently which is justified especially in the orthogonal color spaces (4.28) and (4.29). These are then quantized in a series of 1D histograms, as described in Section 3.5, from which the fuzzy membership functions are created.

Table 4.9 contains description of the most important parameters that control the behavior of the tracking system built upon the aforementioned assumptions. These are partitioned into three main groups: features, *CamShift* related, and other.

Figure 4.41 shows tracking results of a boat in four frames of a movie test sequence. Each of the presented frames is separated by three frames in the video sequence. Fuzzy field is initialized once in the first frame (upper left) from the rectangular window which defines an object to track (i.e. a boat). Only the IJK color space is used to build the fuzzy field as described in Section 3.8.3.3. Each color channel is quantized and stored in a separate 1D histogram. The

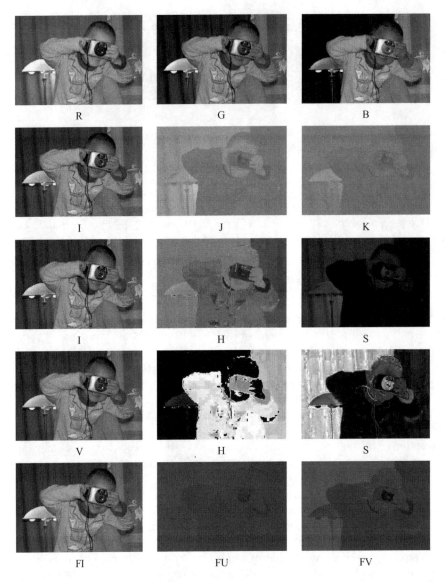

Figure 4.40   Three channels of the test image in different color spaces (from top to bottom): RGB, IJK (4.28), IHS (4.29), VHS, Farnsworth (FI, FU, FV).

bin widths of the histograms in the experiment presented in Figure 4.41 were 2, 4, 2 for the I, J, K channels, respectively. The system assumes description of the tracked objects exclusively with four corner points, that is, an object can be represented only by a quadrilateral. This proved to be a fair compromise in precision and speed of computation as compared to more accurate outlines. Also, shown quadrilaterals are slightly larger to show their interior.

More insight into the process can be gained by analyzing the fuzzy fields of different sets of features used for tracking. These are depicted in Figure 4.43 for the scene in Figure 4.41. Each

Table 4.9 Parameters that control the tracking system.

Group	Parameter	Possible values	Remarks
Features for tracking	Color space.	RGB, YCrCb, IJK, HSI, HVS, Farnsworth	Input images come in the RGB format. All other spaces need additional conversion.
	Quantization value for each color channel.	1–255	All channels are normalized to span 0–255.
	Structural tensor (optional).	on/off	If "on" then structural tensor will be used as image features for tracking
	Quantization value for each component of the structural tensor.	1–255	Tensor components normalized to span 0–255.
CamShift	Iteration resize factor ρ in (3.158).		This parameter controls how expanding the algorithm is.
	Minimum density threshold.	0–100%	A threshold of the density of the tracked object. If it falls down below this threshold CamShift stops.
Other	Change video size.	New size of video (the same aspect ratio)	This parameters allows tracking in a warped video.
	Start tracking in the set rectangle or use the whole image.	on/off	If "on" then the object is tracked based on its position as found in the previous frame; otherwise the whole image is used.
	Reinitialize fuzzy field on each frame.	on/off	If "on" then the fuzzy field is recomputed from the current position of the tracked object; otherwise it is computed only once for the initial frame.

row of Figure 4.43 depicts the RGB, IJK, ISH, as well as for IJK operating with the structural tensor (ST). In the latter case each channel of the ST was gathered in the histograms with bins of widths 4, 8, 4 for the magnitude, coherence, and angle, respectively.

Based on Figure 4.43, it can be seen that color spaces show different parts of an object of interest. Unfortunately, this can change from image to image, so their proper choice needs some experimentation. However, the situation improves with more discriminative color spaces, such as IJK or ISH. In this example, the most focused region of interest (ROI) for the requested object (a boat in this case) is obtained with the IJK color space and structural tensor (last row of Figure 4.43). This proved to be an improvement over other methods that rely exclusively on color information, especially when comparing tracking results and speed of computations (IJK is obtained from the RGB by a linear combination of RGB as shown in Equation (4.28), also ST only requires convolution and multiplication, as discussed in [15]). Results of tracking with IJK and the structural tensor are shown in Figure 4.42. It can be observed that such

Figure 4.41   Tracking of a boat in a color video sequence. Fuzzy field initialized once in the first frame (upper left) which defines an object to track. The IJK space provides features to track. (For a color version of this figure, please see the color plate section.)

compositions of features allow more precise tracking since the feature space is not excessively influenced by other objects and noise. Nevertheless, a proper choice of the widths of bins in the histograms can be problematic in some cases.

Figure 4.44 shows a comparison of tracking of a small red car in a video from the cameras placed in another moving car. In this case the IJK and ST components were used together. Fuzzy field resulting from only color components is depicted in Figure 4.45a. On the other hand, fuzzy field of the IJK and ST is visible in Figure 4.45b. The latter contains less false positives.

The tracked objects presented in Figure 4.44 are skewed since the rectangles placed by the *CamShift* procedure are oriented alongside the major axis of the underlying blob in the fuzzy field as given in Equation (2.125). Tracking in a frame starts from the position of an object inherited from the previous frame, and so on. However, it is also possible to search the whole image for an object (Table 4.9). This option is useful if an initial position of an object is not known (e.g. a model comes from a different image) or an object has been lost during tracking.

The computations were done with the help of the software framework [28]. However, methods like this are highly parallel and can benefit from parallel implementations on multicore systems, e.g. with the help of the OpenMP library discussed in Appendix A.6, or graphic cards

Figure 4.42 Boat tracking with the IJK and structural tensor components. More precise tracking of an object due to more descriptive features. (For a color version of this figure, please see the color plate section.)

GPU. As already pointed out, the set of 1D histograms does not provide the best model for object tracking. Better results can be obtained for example with the OC-SVM, as presented in [78]. However, there are many methods which greatly improve these simplest schemes, as reported in the literature [79, 80, 81, 82, 83, 84].

4.7 Pedestrian Detection

Due to its practical importance, the problem of automatic pedestrian detection by a visual system mounted in a moving car has found broad interest among research groups. If operating reliably, such systems could become part of the DAS systems which will be mounted in cars. Exemplary traffic scenes with people crossing streets are depicted in the upper row of Figure 4.46. Manually outlined silhouettes are shown in the lower row. These can be used to train classification systems. There are many pedestrian databases with ground-truth data. One of the best known is the Caltech Pedestrian Dataset which consists of almost ten hours of video taken from a car traveling in daytime urban conditions [85]. This received tremendous work of annotating contained there few thousands of pedestrians with bounding boxes. The annotation includes temporal correspondence between bounding boxes as well as labeled occluding areas. This and a few other databases are discussed in the recent paper by Dollár *et al.* evaluating the

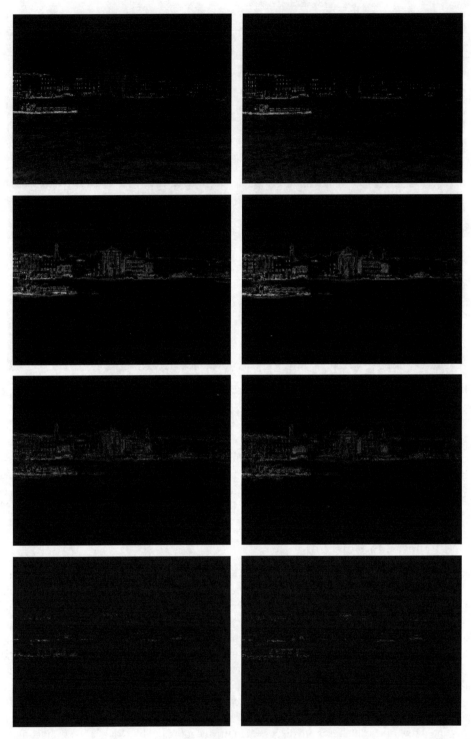

Figure 4.43 Visualization of the fuzzy fields acquired with different tracking parameters used in the tracking of an object in Figure 4.41 .

Figure 4.44 Tracking results of a red car in a traffic sequence with the IJK and ST components. User drawn white rectangle defines an object to track (left image). (For a color version of this figure, please see the color plate section.)

state of the art in pedestrian detection [86]. The authors evaluated the performance of sixteen state-of-the-art detectors using six data sets.

A very influential survey on pedestrian detection in the context of DAS is presented in the paper by Geronimo *et al.* [87]. More than a 100 recent works on pedestrian detection are discussed, however in order of processing stage rather than issue date. Their conclusions are

Figure 4.45 Comparison of the membership fields in the tracking frames in the RGB (left) and IJK with structural tensor components (right).

Figure 4.46   Real traffic scenes with pedestrians (upper). Outlined silhouettes of pedestrians (lower). Color versions of the images are available at www.wiley.com/go/cyganekobject.

that the performance of available methods are still far from the point of being mounting in a serial car. Also, the importance of systems operating in both daytime and NIR spectra is indicated as an expected way of developing new pedestrian detection systems [88, 89]. Figure 4.47 presents the common stages of the majority of pedestrian detection systems discussed in the literature [87].

First the preprocessing stage is concerned with low-level image adjustments, such as control of the exposure or the dynamic range. High dynamic range images (HDR) which usually cover wide spectrum from the visible up to NIR range are especially useful at this stage. The latter greatly facilitates night operations or those in poor visibility conditions. In some cases cameras can require calibration, especially if a stereo-camera setup is envisioned. For the latter Labayrade *et al.* introduced the V-disparity [90] and U-V by Hu *et al.* [91, 92]. The main goal of finding the vertical(V)-disparity is to separate obstacles from the road surface. In this domain, the longitudinal profile of the road plane can be described as a piecewise linear curve. Obstacles, on the other hand, will be projected as vertical lines. The U-V-disparity extends this concept and allows the detection of surfaces other than those only perpendicular to the vertical plane of the stereo-rig mounted in a car.

Foreground segmentation is responsible for the extraction of regions of interest (ROI) which potentially contain person silhouettes, which are then used for further processing. In

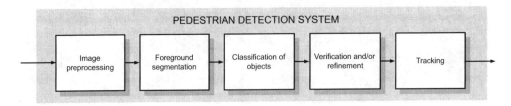

Figure 4.47   Common processing stages in pedestrian detection systems.

this respect the primary role is played by the stereo systems which provide depth information [79, 93]. Motion analysis is also useful at this stage. Pedestrian detection systems which rely on stereo processing are discussed in another part of this section.

Regions selected in the previous stage are fed into the object classification module. Its role is to tell if the observed object is likely to contain a view of a pedestrian and with what probability. Many methods have been proposed in this respect, starting from holistic ones, through sparse, up to implicit shape models, which shall be discussed below.

Many of the contemporary pedestrian detection methods rely on the seminal works on object recognition with sliding window detectors and SVM classifiers by Papageorgiou and Poggio [94], as well as the frequently mentioned work by Viola and Jones who joined Haar wavelets, computed in the integral image, with the AdaBoost cascade of classifiers [3]. Another inspiring discovery was the application of the already mentioned histograms of oriented gradients (HOG) by Dalal and Triggs [48].

The other branch of research relies on using different shape features. In this respect to represent shapes locally Wu and Nevatia use a large set of short line and curve segments, which they call *edgelet* features [80]. A similar approach with *shapelets* was proposed by Sabzmeydani and Mori [95]. These are computed from the gradients in local patches of an image. Combinations of different features and classifiers are discussed in the paper by Walk *et al.* [96]. Very effective object descriptors appear to be the covariance matrices computed from vectors of different image features, as proposed by Tuzel *et al.* [56]. These were also successively applied to pedestrian detection.

The implicit shape model approach was proposed by Leibe *et al.* [97]. First the keypoints, such as SIFT or Harris corners, are found in an image, around which some descriptors are collected. Then, these are clustered to construct a codebook. From this codebook an implicit shape model (ISM) is built which specifies the positions of the codebook on an object. By this technique, allowed shapes are defined for different views of a class of objects. In practice the approach has been shown to be flexible and allows the recognition of different classes of objects seen from different views. In the recognition stage the keypoints of a tested object are computed and then matched to the codebook. These then vote for an object hypothesis using the Generalized Hough Transform (GHT). Consistent hypotheses are then selected from the local maxima of the continuous (not discretized) voting space with help of the mean-shift method.

At this point it is worthy noting after Geronimo *et al.* that one of the problems in object recognition systems is high pose variability [87]. In this respect holistic classifiers are not sufficient. More appropriate ways are parts-based hybrid systems joining different cooperating classifiers operating with combinations of features.

The role of the verification/refinement step in the diagram in Figure 4.47 is to filter out false positives, as well as to perform some tuning of the segmented objects to provide a more accurate outline of an object. Actually, this step can be seen as a second and more refined classification stage. Such an approach is undertaken in many comuter vision systems, such as road signs (Section 5.7) or driver's eye recognition systems (Section 5.9), respectively, as discussed in this book. Validation can be done by exploring other sources of information than those used in the classification stage. These can be person gait analysis, the silhouette of the head and shoulders, inward motion, as well as information on a (relative) distance and cross-validation in the case of stereo setups [79].

The last step in Figure 4.47 concerns object tracking which facilitates detection of objects of interest in consecutive frames, thus decreasing the number of false positives. It also provides useful information for human motion analysis which can be used for further verification of human behavior. Kalman filter and mean-shift methods are discussed in Sections 3.6 and 3.8, respectively. An interesting approach for pedestrian tracking with tracklets and detection residuals was presented by Singh et al. [81]. First a robust detector is employed to find high confidence partial track segments (called *tracklets*). Then, these are fed to the global optimization framework. The originality of this method follows from the idea of using unassociated low confidence detections between tracklets (called *residuals*) to improve overall tracking performance. Due to lighting coditions, tracking in the FIR (thermal[4]) spectrum can be more reliable. Such an approach was undertaken by Goubet et al. for pedestrian tracking using thermal infrared imaging [82], as well as by Li and Gong who developed a real-time version of the FIR tracker [83].

Although in this section we have discussed some recent achievements in pedestrian detection, it is worth noting that the discussed algorithms are far more general and have found application in other vision tasks, such as surveillance, visual search engines, as well as other automotive systems, such as road sign recognition, driver fatigue monitoring, etc. At this point let us present some conclusions on the pedestrian detection systems we have discussed.

Gavrila and Mundi presented the PROTECTOR system for pedestrian detection [79]. Their system is organized as a cascade of modules, each successively narrowing the search space. The system integrates two, sparse and dense, stereovision systems. The first is used to generate a region of interest. Pedestrian detection is obtained by exemplar-based shape analysis, and a weighted combination of texture based classifiers, each responsible for detection of a particular body pose, as well as depth information from the stereo modules. Shapes are detected based on the template matching with the chamfer distance transformation. However, to account for the variability of appearance model, a few thousand exemplars were collected to define pedestrian shape distribution. To allow efficient matching a hierarchical approach is undertaken. Dense stereo is used to verify parts of the region from the template matching module for contamination with background pixels. Finally, a simple α-β tracking[5] is applied to overcome gaps in detection. The paper by Enzweiler and Gavrila discussing advances in monocular pedestrian detection methods [100] could also be of interest.

An interesting proposition for a NIR stereo system for detection of pedestrians at road intersections was recently proposed by the Migma® company, as described in [98]. Their system is composed of two high-resolution color CCD cameras and a hundred IR LED diodes operating at 850 nm. It is able to detect pedestrians in the range of about 25 m in dark conditions. During daylight, the visible spectrum is used. However, no details on the algorithms that were used is provided.

Sappa et al. propose an efficient approach to onboard stereovision system pose estimation especially designed for urban operations [93]. Three-dimensional raw data points are used, from which their 2D representation is computed. Then the RANSAC least-squares method is used to fit a plane to the observed road. Finally, the height and pitch of the stereo rig are computed relative to the fitted road plane.

Assheton and Hunter propose a shape-based voting method for pedestrian detection by shape and tracking [84]. Their method – a mixture of uniform and Gaussian Hough transform (MOUGH) – is based on an observation that edgels of a rigid object at a certain

---

[4]In some publications this is called TIR – thermal infrared.
[5]This is a version of the second order Kalman filter (3.6) with fixed gain [99].

Table 4.10 Summary of the most common features for pedestrian detection.

1	Local features (combination of sliding window, HOG, histograms of flow (HOF), Haar wavelets, shapelets, edgelets, etc.) [48, 80, 95, 96, 97]
2	Motion analysis (optical flow) [81, 96], thermal spectrum [82, 83]
3	Stereovision [79, 93, 98]

orientation are roughly distributed according to a mixture of Gaussians. The Hough space is built by voting using a mixture of Gaussians optimized by the Expectation-Maximization principle (3.5). Thanks to this, the method is able to search for images with slightly deformed shapes.

As alluded to previously, the two best performing classifiers reported in literature are SVM and AdaBoost. Wang and Ma recently proposed a Multi-Pose Learning Boosted Integrable Features Pool (MPL-Boosted IFP) for pedestrian detection [101]. As shown, the method exhibits a good balance between accuracy and performance. In their work different groups of features that can be obtained quickly thanks to the integral image were tested. These were then used to build a large number of candidate weak classifiers by using linear SVM. Then, the MPL-Boost method allows selection of the best of the weak classifiers. As reported, apart from the high speed up ratio, the precision of detection achieved is higher than in the HOG based methods. Table 4.10 and Table 4.11 summarize reported the most commonly used features and best performing classifiers for pedestrian detection.

Detailed analysis of the performance of the aforementioned methods tested on different pedestrian data sets is available from the paper by Dollár *et al.* [86], as well as from the Caltech pedestrian web page [85]. A breakdown of available data sets for evaluation of pedestrian detection systems is contained in Table 4.12. As indicated by Dollár *et al.* the best two methods in terms of speed of operation and accuracy are:

1. The Fastest Pedestrian Detector in the West (FPDW) which utilizes Haar features computed over multiple channels of color, grayscale, gradient of magnitude [107];
2. Multifeature methods combined with motion features, as proposed by Walk *et al.* [96].

A GPU based implementation of multisensor pedestrian detection was presented by Weimer *et al.* [108]. Their system joins hypothesis generation from a laser scanner and HOG features from thermal IR video signals which are then fed into the Kalman filter and SVM with Gaussian kernel.

As concluded by Dollár *et al.*, even the best methods perform poorly in cases of low resolution or partial occlusions. This is in agreement with our experiments on road sign recognition in which good results are mainly obtained for the good daylight conditions. Thus, it seems that the development of methods of object detection and recognition in poor conditions

Table 4.11 Summary of the best classifiers for pedestrian detection.

1	SVM [48], with modification of intersection kernels [102]
2	AdaBoost [3] with further modification into MPLBoost [101]

Table 4.12   Available data sets for evaluation of pedestrian detection systems,

1	Caltech Pedestrian Training and Testing Dataset [85]
2	INRIA Pedestrian Test Dataset [103]
3	ETH Pedestrian Dataset [104]
4	TUD-Brussels Pedestrian Dataset [105]
5	Daimler Pedestrian Dataset [106]

caused by acquisition parameters, weather, etc. deserve special research attention. This is also an open area on the development of hybrid systems joining new types of sensors, such as NIR, FIR cameras for instance, and new processing methods, also in the area of computer vision. Finally, we would like to stress the importance of further development of stereo systems for automotive applications, potentially operating with the HDR images.

4.8   Closure

4.8.1   Chapter Summary

In this chapter various methods of object detection and tracking were discussed. These, in turn, were based on the basic classification methods discussed in previous chapters. We started with methods of image segmentation based on low-level pixel classification. Fuzzy and one-class SVM based approaches were discussed. Their applications to human skin and road sign segmentation were discussed, also in the context of different color spaces. The main point of the presented methods is fast operation, but also high accuracy. In this respect the superior performance of ensembles of classifiers was presented. Then we dealt with the problem of basic shape detection, especially in the context of the real-time road sign recognition. Many of the presented methods from this group were based on the structural tensor (2.7). Then figure detection and validation were presented, based on fuzzy logic rules as well as clustering methods. A case study of a system for road sign detection was presented. Then tracking systems were discussed. The first concerned real-time tracking of road signs, whereas the second one showed a general framework for color and texture based object tracking. Both systems are based on the mean-shift method, discussed in Section 3.8. The chapter ended with an overview of recent vision systems for pedestrian detection.

4.8.2   Further Reading

Vision based object detection and tracking belongs to the dynamically changing domain of computer vision. Tracing up-to-date advances in this area is possible with the help of relevant journals, such as *IEEE Transactions on Pattern Analysis, Image Processing*, and *Neural Networks*, as well as *International Journal of Computer Vision* and *Pattern Recognition*, to name but a few.

A recent overview of the most important methods in computer vision is contained in the book by Szeliski [109]. This is also recommended reading with its well balanced description of the main ideas, as well as practical information. The book also has ample links to important publications, repositories with test data, as well as code packages.

An introduction to 3D computer vision with a discussion on low level image feature detectors is provided in the book by Cyganek and Siebert [15].

A highly recommended source of practical methods connected with in-depth theory on some important aspects of feature extraction in digital images is provided in the book by Nixon and Aguado [42].

The algorithmic aspect of processing color images is discussed in the book by Koschan and Abidi [110]. The other recommended reading on color image processing is the book by Lee [76]. A classical position in this area is the book by Wyszecki and Stiles [111].

Finally, the book by Yoo deals with many interesting topics in CV, such as statistical pattern recognition, segmentation, registration, and many more [112]. Especially well addressed are problems of medical image processing.

Problems and Exercises

1. Download some images from the Internet and collect regions with human skin. Use them to train the OC-SVM classifier. Create new features which are composed of a color values, as before, and new components of a color variance in small pixel neighborhoods, such as 3×3. Train OC-SVM and compare the results.
2. Use the database from the previous task and train the probabilistic neural network, discussed in Section 3.9.1. Compare results of skin segmentation with OC-SVM and the neural network. Compare execution time.
3. Collect a set of color images. Perform their color segmentation with the k-means method (Section 3.11.1). Using the methodology discussed in Section 4.6 find a set of common color base vectors that best describe the segmented regions. Compare them with the bases (4.26). Build orthogonal color space from your vectors.
4. Train the morphological neural network with images of simple icons. Check the accuracy of the network with icons contaminated by noise. Check the network with occluded icons. How discriminative are occluded icons to the network?

References

[1] Phung S.L., Bouzerdoum A., Chai D.: Skin Segmentation Using Color Pixel Classification: Analysis and Comparison. IEEE Transactions on Pattern Analysis and Machine Intelligence, Vol. 27, No. 1, pp. 148–154, 2005.

[2] Cyganek B.: Color Image Segmentation With Support Vector Machines: Applications To Road Signs Detection. International Journal of Neural Systems, Vol. 18, No. 4, World Scientific Publishing Company, pp. 339–345, 2008.

[3] Viola P., Jones M.: Robust real-time face detection, International Journal of Computer Vision, Vol. 57, No. 2, pp. 137–154, 2004.

[4] Cyganek B., Socha K.: A multi-tool for ground-truth stereo correspondence, object outlining and points-of-interest selection. VIGTA '12 Proceedings of the 1st International Workshop on Visual Interfaces for Ground Truth Collection in Computer Vision Applications, Article No. 4, 2012.

[5] Cyganek B.: Software Tool for Colour Data Acquisition, *http://home.agh.edu.pl/~cyganek/RCA.zip*, 2007.

[6] Arentz W.A., Olstad B.: Classifying offensive sites based on image content. Computer Vision and Image Understanding, Vol. 94, Issue 1-3, pp. 295–310, 2004.

[7] Lee J-S., Kuo Y-M., Chung P-C., Chen E-L.: Naked image detection based on adaptive and extensible skin color model. Pattern Recognition, Vol. 40, No. 8, pp. 2261–2270, 2007.

[8] Jones M.J., Rehg J.M.: Statistical Color Models with Application to Skin Detection. International Journal of Computer Vision, Vol. 46, No. 1, pp. 81–96, 2002.

[9] Yang M-H, Ahuja N.: Gaussian mixture model for human skin color and its application in image and video databases, Proceedings of SPIE 99, 1999.

[10] Jayaram S., Schmugge S., Shin M.C., Tsap L.V.: Effect of Colorspace Transformation, the Illuminance Component, and Color Modeling on Skin Detection. IEEE Computer Society Conference on Computer Vision and Pattern Recognition CVPR'04, Vol. II, pp. 813–818, 2004.

[11] Wu H. Chen Q. Yachida M.: Face Detection From Color Images Using a Fuzzy Pattern Matching Method. IEEE Transactions on Pattern Analysis and Machine Intelligence, Vol. 21, No. 6, pp. 557–563, 1999.

[12] Vezhnevets V., Sazonov V., Andreeva A.: A survey on pixel-based skin color detection techniques, Proc. Graphicon-2003, pp. 85–92, 2003.

[13] Khan R., Hanbury A., Stöttinger J., Bais A.: Color based skin classification. Pattern Recognition Letters, Vol. 33, No. 2, pp. 157–163, 2012.

[14] Peer P., Kovac J., Solina F.: Human skin colour clustering for face detection. EUROCON 2003 – International Conference on Computer as a Tool, 2003.

[15] Cyganek B., Siebert J.P.: An Introduction to 3D Computer Vision Techniques and Algorithms, Wiley, 2009.

[16] Zadeh, L.A.: Fuzzy sets. Information and Control, 8, pp. 338–353, 1965.

[17] Kecman V.: Learning and Soft Computing: Support Vector Machines, Neural Networks, and Fuzzy Logic Models. MIT Press, 2001.

[18] Driankov D., Hellendoorn H., Reinfrank M.: An Introduction to Fuzzy Control. 2nd edition, Springer, 1996.

[19] http://opencv.org/

[20] Bradski G., Kaehler A.: Learning OpenCV. Computer Vision with the OpenCV Library. O'Reilly, 2008.

[21] Hsu R-L., Abdel-Mottaleb M., Jain A.K.: Face Detection in Color Image. IEEE Transactions on Pattern Analysis and Machine Intelligence, Vol. 24, No. 5, pp. 696–706, 2002.

[22] Cyganek B.: Soft System for Road Sign Detection. Analysis and Design of Intelligent Systems Using Soft Computing Techniques. Advances in Soft Computing 41, Springer, pp. 316–326, 2007.

[23] Polish Road Signs and Signalization. Directive of the Polish Ministry of Infrastructure (in Polish), Dz. U. Nr 170, poz. 1393, 2002.

[24] Chang C-C., Lin C-J.: LIBSVM, a library for support vector machines (2001). Software available at *http://www.csie.ntu.edu.tw/~cjlin/libsvm*

[25] Tax D.M.J.: One-class classification. PhD thesis, TU Delft University, 2001.

[26] Cyganek B.: One-Class Support Vector Ensembles for Image Segmentation and Classification. Journal of Mathematical Imaging & Vision, Vol. 42, No. 2-3, Springer, pp. 103–117, 2012.

[27] Berkeley Segmentation Database, *http://www.eecs.berkeley.edu/Research/Projects/CS/vision/grouping/segbench/*, 2010.

[28] http://www.wiley.com/go/cyganekobject

[29] Hough P.V.C.: Machine Analysis of Bubble Chamber Pictures, Proc. Int. Conf. High Energy Accelerators and Instrumentation, CERN, 1959.

[30] Duda, R.O. and Hart P.E.: Use of the Hough Transformation to Detect Lines and Curves in Pictures. Communications ACM, Vol. 15, pp. 11–15, 1972.

[31] Ballard, D.H.: Generalizing the Hough transform to detect arbitrary shapes. Pattern Recognition, Elsevier, Vol. 13, No. 2, pp. 111–122, 1981.

[32] Davies E.R.: Machine Vision. Theory, Algorithms, Practicalities. Morgan Kaufmann; 3rd edition, 2004.

[33] Jähne B.: Digital Image Processing. 6th edition, Springer-Verlag, 2005.

[34] McLaughlin R.A., Alder M.D.: The Hough Transform Versus the UpWrite. IEEE Transactions On Pattern Analysis And Machine Intelligence, Vol. 20, No. 4, pp. 396–400, 1998.

[35] Cyganek B.: Object Detection in Multi-Channel and Multi-Scale Images Based on the Structural Tensor. Lecture Notes in Computer Science, Vol. 3691, Springer, pp. 570–578, 2005.

[36] Cyganek B.: Combined Detector of Locally-Oriented Structures and Corners in Images Based, in Springer Lecture Notes in Computer Science, Vol. 2658, pp. 721–730, 2003.

[37] Aho A.V., Sethi R., Ullman J.D.: Compilers. Principles, Techniques, and Tools. Addison Wesley, 1988.

[38] Tadeusiewicz R, Ogiela M. Medical Image Understanding Technology: Artificial Intelligence and Soft-Computing for Image Understanding. Springer, 2007.

[39] Bigün, J., Granlund, G.H., Wiklund, J., Multidimensional Orientation Estimation with Applications to Texture Analysis and Optical Flow. IEEE Transactions on Pattern Recognition and Machine Intelligence, Vol. 13, No. 8, pp. 775–790, 1991.

[40] Jähne B.: Practical Handbook on Image Processing for Scientific Applications, CRC Press 1997.

[41] Press W.H., Teukolsky S.A., Vetterling W.T., Flannery B.P.: Numerical Recipes in C. The Art of Scientific Computing. Third Edition. Cambridge University Press, 2007.

[42] Nixon M., Aguado A.: Feature Extraction & Image Processing. Newnes, 2006.

[43] Huber R.J.: Robust Statistics. Wiley, 2004.

[44] Marona R.A., Martin R.D., Yohai V.J.: Robust Statistics. Theory and Methods. Wiley, 2006.

[45] Hartley R.I., Zisserman A.: Multiple View Geometry in Computer Vision. 2nd edition, Cambridge University Press, 2003.

[46] Faugeras O.D., Luong Q.-T.: The Geometry of Multiple Images. MIT Press, 2001.

[47] Lowe D.: Distinctive Image Features from Scale-Invariant Keypoints. International Journal of Computer Vision, Vol. 60, No. 2, pp. 91–110, 2004.

[48] Dalal N., Triggs B.: Histograms of Oriented Gradients for Human Detection. IEEE Computer Society Conference on Computer Vision and Pattern Recognition, Vol. 1, pp. 886–893, 2005.

[49] Tola E., Lepetit V., Fua P.: DAISY: An Efficient Dense Descriptor Applied to Wide-Baseline Stereo. IEEE Transactions on Pattern Analysis And Machine Intelligence, Vol. 32, No. 5, pp. 815–830, 2010.

[50] Bay H., Ess A., Tuytelaars T., Van Gool L.: SURF: Speeded Up Robust Features. Computer Vision and Image Understanding (CVIU), Vol. 110, No. 3, pp. 346–359, 2008.

[51] Ke Y., Sukthankar R.: PCA-SIFT: A More Distinctive Representation for Local Image Descriptors. Computer Vision and Pattern Recognition, Vol. 2, pp. 506–513, 2004.

[52] Sande van de K.E.A., Gevers T., Snoek C.G.M.: Evaluating Color Descriptors for Object and Scene Recognition. IEEE Transactions on Pattern Analysis and Machine Intelligence, Vol. 32, No. 9, pp. 1582–1596, 2010.

[53] Mikolajczyk K., Schmid C.: A Performance Evaluation Of Local Descriptors, Vol. 27, No. 10 pp. 1615–1630 2005.

[54] Winder S.A.J., Brown M.: Learning local image descriptors. IEEE Conference on Computer Vision and Pattern Recognition CVPR, pp. 1–8, 2007.

[55] http://pascallin.ecs.soton.ac.uk/challenges/VOC/

[56] Tuzel O., Porikli F., Meer P.: Pedestrian Detection via Classification on Riemannian Manifolds. IEEE Transactions on Pattern Analysis and Machine Intelligence, Vol. 30, No. 10, pp. 1713–1727, 2008.

[57] Cyganek B.: Real-Time Detection of the Triangular and Rectangular Shape Road Signs. Lecture Notes in Computer Science, Vol. 4678, Springer, pp. 744–755, 2007.

[58] Gonzalez R.C, Woods R.E.: Digital Image Processing, 2nd Edition, Prentice Hall, 2002.

[59] Cyganek B.: Circular Road Signs Recognition with Soft Classifiers. Integrated Computer-Aided Engineering, IOS Press, Vol. 14, No. 4, pp. 323–343, 2007.

[60] Cyganek B.: Road-Signs Recognition System for Intelligent Vehicles. Lecture Notes in Computer Science, Vol. 4931, Springer, pp. 219–233, 2008.

[61] Cyganek B.: Intelligent System for Traffic Signs Recognition in Moving Vehicles. Lecture Notes in Artificial Intelligence, Vol. 5027, Springer-Verlag Berlin Heidelberg, 2008, pp. 139–148.

[62] Hunt A., Thomas D.: The Pragmatic Programmer. Addison Wesley, 1999.

[63] McConnell S.: Code Complete. 2nd edition. Microsoft Press, 2004.

[64] Cyganek B.: Rotation Invariant Recognition of Road Signs with Ensemble of 1-NN Neural Classifiers. Lecture Notes in Computer Science, Vol. 4132, Springer, pp. 558–567, 2006.

[65] Forsyth D.A., Ponce J.: Computer Vision: A Modern Approach, Prentice Hall, 2003.

[66] Thrun S., Burgard W. Fox D.: Probabilistic Robotics. MIT Press, 2006.

[67] Cyganek B.: Neuro-Fuzzy System for Road Signs Recognition. Lecture Notes in Computer Science Lecture Notes in Computer Science, Vol. 5163, Springer, pp. 503–512, 2008.

[68] Ritter, G., X., Sussner, P., Diaz, J., L.: Morphological Associative Memories, IEEE Transactions on Neural Networks, Vol. 9, No. 2, pp. 281–293, 1998.

[69] Sussner P.: Observations on Morphological Associative Memories and the Kernel Method. Neurocomputing, Vol. 31, No. 1-4, Elsevier Science, pp. 167–183, 2000.

[70] Soille P.: Morphological Image Analysis. Principles and Applications. Springer, 2003.

[71] Myler H. R., Weeks A.: The Pocket Handbook of Image Processing Algorithms in C. Prentice-Hall 1993.

[72] Ritter, G.,X., Urcid, G., Iancu, L.: Reconstruction of Patterns from Noisy Inputs Using Morphological Associative Memories. Journal of Mathematical Imaging and Vision, Vol. 19, No. 2, pp. 95–111, 2003.

[73] Ohta Y-I., Kanade T., Sakai T.: Color Information for Region Segmentation. Computer Graphics and Image Processing, Vol. 13, No. 3, pp. 222–241, 1980.

[74] Pătrașcu V.: Fuzzy Image Segmentation Based on Triangular Function and Its n-dimensional Extension, in Soft Computing in Image Processing, Nachtegael M. Van der Weken D., Kerre E.E., Philips W. (editors), Studies in Fuzziness and Soft Computing, Springer, pp. 187–207, 2007.

[75] Poynton C.: Digital Video and HDTV. Algorithms and Interfaces. Morgan-Kaufmann, 2007.

[76] Lee H-C.: Introduction to Color Imaging Science. Cambridge University Press, 2005.

[77] Moon T.K., Stirling W.C.: Mathematical Methods and Algorithms for Signal Processing. Prentice-Hall 2000.

[78] Cyganek B.: Framework for Object Tracking with Support Vector Machines, Structural Tensor and the Mean Shift Method. Springer, Lecture Notes in Computer Science, Vol. 5863, pp. 399–408, 2009.

[79] Gavrila D.M., Munder S.: Multi-cue Pedestrian Detection and Tracking from a Moving Vehicle. International Journal of Computer Vision, Vol. 73, No. 1, 41–59, 2007.

[80] Wu B., Nevatia R.: Detection and Tracking of Multiple, Partially Occluded Humans by Bayesian Combination of Edgelet Based Part Detectors. International Journal of Computer Vision, Vol. 75, No. 2, pp. 247–266, 2007.

[81] Singh V., Wu B., Nevatia R.: Pedestrian Tracking by Associating Tracklets Using Detection Residuals. IEEE Workshop on Motion and Video Computing, pp. 1–8, 2008.

[82] Goubet E., Katz J., Porikli F.: Pedestrian Tracking Using Thermal Infrared Imaging. SPIE Conference on Infrared Technology and Applications, pp. 797–808, 2006.

[83] Li J., Gong W.: Real Time Pedestrian Tracking using Thermal Infrared Imagery. Journal of Computers, Vol. 5, No. 10, pp. 1606–1613, 2010.

[84] Assheton P., Hunter A.: A shape-based voting algorithm for pedestrian detection and tracking. Pattern Recognition, Vol. 44, No. 5, pp. 1106–1120, 2011.

[85] http://www.vision.caltech.edu/Image_Datasets/CaltechPedestrians/

[86] Dollár P., Wojek C., Schiele B., Perona P.: Pedestrian Detection: An Evaluation of the State of the Art. IEEE Transactions on Pattern Analysis and Machine Intelligence, Vol. 34, No. 4, pp. 743–761, 2012.

[87] Geronimo D., Lopez A.M., Sappa A.D., Graf T.: Survey on Pedestrian Detection for Advanced Driver Assistance Systems. IEEE Transactions on Pattern Analysis and Machine Intelligence, Vol. 32, No. 7, pp. 1239–1258, 2010.

[88] Nanda H., Davis L.: Probabilistic Template Based Pedestrian Detection in Infrared Videos. IEEE Conference on Intelligent Vehicle Symposium, Vol. 1, pp. 15–20, 2002.

[89] Broggi, A., Fascioli, A., Carletti, M., Graf, T., Meinecke, M.: A multi-resolution approach for infrared vision-based pedestrian detection. IEEE Intelligent Vehicle Symposium, pp. 7–12, 2004.

[90] Labayrade R., Aubert D., Tarel J.: Real Time Obstacle Detection in Stereovision on Non Flat Road Geometry through 'v- Disparity' Representation. IEEE Intelligent Vehicles Symposium, Vol. 2, pp. 17–21, 2002.

[91] Hu Z., Uchimura K.: U-V-Disparity: An Efficient Algorithm for Stereovision Based Scene Analysis. IEEE Intelligent Vehicles Symposium, pp. 48–54, 2005.

[92] Hu Z., Lamosa F., Uchimura K.: A Complete U-V-Disparity Study for Stereovision Based 3D Driving Environment Analysis. Fifth International Conference on 3-D Digital Imaging and Modeling, pp. 1550–6185, 2005.

[93] Sappa A.D., Dornaika F., Ponsa D., Gerónimo D., López A.: An Efficient Approach to Onboard Stereo Vision System Pose Estimation. IEEE Transactions on Intelligent Transportation Systems, Vol. 9, No. 3, pp. 476–490, 2008.

[94] Papageorgiou C., Poggio T.: A Trainable System for Object Detection. International Journal of Computer Vision, Vol. 38, No. 1, pp. 15–33, 2000.

[95] Sabzmeydani P., Mori G.: Detecting Pedestrians by Learning Shapelet Features. IEEE Conference on Computer Vision and Pattern Recognition, pp. 1–8, 2007.

[96] Walk S., Majer N., Schindler K., Schiele B.: New Features and Insights for Pedestrian Detection. IEEE Conference on Computer Vision and Pattern Recognition, pp. 1030–1037, 2010.

[97] Leibe B., Leonardis A., Schiele B.: Robust Object Detection with Interleaved Categorization and Segmentation. International Journal of Computer Vision, Vol. 77, No. 1-3, pp. 259–289, 2008.

[98] Ling B.: Crossing the Line. Vision Systems Design, pp. 27–30, 2009.

[99] Kalata P.R.: Alpha-Beta Target Tracking Systems: A survey. American Control Conference ACC/WM12, pp. 832-836, 1992.

[100] Enzweiler M., Gavrila D.M.: Monocular Pedestrian Detection: Survey and Experiments. IEEE Transactions on Pattern Analysis and Machine Intelligence, Vol. 31, No. 12, pp. 2179–2195, 2009.

[101] Wang J., Ma H.: MPL-Boosted Integrable Features Pool for Pedestrian Detection. IEEE International Conference on Image Processing, pp. 805–808, 2011.

[102] Maji S., Berg A., Malik J.: Classification Using Intersection Kernel SVMs Is Efficient. IEEE Conference on Computer Vision and Pattern Recognition, pp. 1–8, 2008.

[103] http://pascal.inrialpes.fr/data/human/

[104] http://www.vision.ee.ethz.ch/~aess/dataset/.

[105] http://www.mis.tu-darmstadt.de/tud-brussels

[106] http://www.science.uva.nl/research/isla/downloads/pedestrians/

[107] Dollár P., Belongie S., Perona P.: The Fastest Pedestrian Detector in the West. British Machine Vision Conference, 2010.

[108] Weimer, D., Kohler, S., Hellert, C., Doll, K., Brunsmann, U., Krzikalla, R.: GPU Architecture for Stationary Multisensor Pedestrian Detection at Smart Intersections. Intelligent Vehicles Symposium, pp. 89–94, 2011.

[109] Szeliski R.: Computer Vision. Algorithms and Applications. Springer, 2011.

[110] Koschan A., Abidi M.: Digital Color Image Processing. Wiley, 2008.

[111] Wyszecki G., Stiles W.S.: Corol Science. Wiley, 2000.

[112] Yoo T.S.: Insight into Images. Principles and Practice for Segmentation, Registration, and Image Analysis. A.K. Peters, 2004.

5

Object Recognition

5.1 Abstract

In this chapter various object recognition methods are discussed with examples and practical realizations, mostly oriented toward automotive systems. The methods discussed are strictly connected with the topics presented in the previous chapters. Many of them are based on the tensor processing discussed in Chapter 2, especially tensor filtering (Section 2.6), structural tensor (Section 2.7), eigendecomposition into tensor components (Section 2.9), as well as HOSVD decomposition of tensors with prototype patterns (Section 2.12.2). On the other hand, a background for object recognition is constituted in the classification methods discussed in Chapter 3. Many of the methods presented there are employed in object recognition, such as the means shift (Section 3.8), neural networks (Section 3.9), support vector machines (Section 3.12), as well as data clustering (Section 3.11), to name just a few. Special focus is placed upon kernel based methods and ensembles of classifiers which lead to new qualities in pattern recognition.

As alluded to previously, methods of object recognition in digital images are strictly related to object detection methods, discussed in Chapter 4. In many systems exemplars from these two groups are organized in a processing chain of cooperating classifiers. In others, object detection also means its recognition. In this book we assume, somewhat arbitrarily, that detection concerns answering the question of whether an object is present in an image, and sometimes we are interested in its position or in its movement if we are processing a sequence of images. Conversely, in recognition we are concerned with the specific type of an observed object, such as a particular type of a warning road sign, the concrete meaning of a text, specific persons or their specific behavior, and so forth. As mentioned, although somewhat arbitrary, the division can also be seen in ontological categories, as different levels of modeling based on available types of information. Related to object recognition is the even more challenging problem of scene comprehension. These are fascinating topics of computer vision.

5.2 Recognition from Tensor Phase Histograms and Morphological Scale Space

Place and orientation, or the phase, of an edge convey very distinctive information about the observed object. Therefore, it is no surprise that in many CV systems this kind of information has been proposed to be used for object recognition. For instance, computation of orientation histograms in edge points was already proposed by Freeman and Roth [1]. Strictly related histograms of local gradients then became building blocks for such sparse image descriptors as SIFT [2], HOG [3], and many others outlined in Section 4.4. In this section we discuss a variant of histograms of local orientation based on the structural tensor (Section 2.7) in connection with the nonlinear morphological scale space (MSS), details of which are provided in Appendix A.2. Properties, implementation, as well as applications of this approach are also discussed. These are based on publication [4].

We already know that ST provides distinctive information about local neighborhoods of pixels. This comes in the form of a dominating vector, its phase and magnitude, as well as a coherence measure which tells us what kind of a structure dominates in a given neighborhood (Section 2.7). These provide more information than a simple edge detector indicating positions of edge pixels. Hence, a proposed extension is to use the ST and to decompose it into two separate components, the so called the stick and plate components, discussed in Section 2.9. The former allows detection of local structures showing a strong linear response only, which in this application we are mainly interested in. Obviously computation of the phases of local structures makes sense exclusively in places with a strong linear response since in areas with a strong plate response there is no unique orientation of local neighborhood. Thus, the first idea of the presented method is to *compute histograms of phases (orientations) of local structures only in places with a high stick component* **S** *of the ST.*

The idea of object recognition from phase histograms relies on using the ST for both, finding areas with strong linear structure and then computing the phase of local orientation in these areas. With sufficiently precise gradient filters (Section 2.7.1), these parameters can be obtained with the desired accuracy. Subsequent classification is made by matching of the modulo shifted phase histograms thanks to which detection of the internal rotations of objects is possible. As a result the proposed method is resistant to shift, rotation, proportional change of scale, as well as to different lighting conditions and small occlusions of the test objects. Additionally, its computation is very fast and can be easily parallelized, as discussed in Section A.6. Hence, depending on the resolution of the input images and size of the detected object, its software implementation can easily reach real-time requirements. For instance in the system for road sign recognition, the basic version of the algorithm allows processing of 30 frames per second in the video stream of resolution 640×480 pixels.

Let us take a look at some of the internal properties of the objects which can be further utilized to improve histograms of gradients. There are many objects which have solid, complex, or fiber-like structures and for which description solely by their edges at one scale is not sufficient since they are not like wire-bounded. An example is presented in Figure 5.1.

It is obvious that one should take advantage of the additional knowledge on the type of an object, such as whether it is filled, as the one depicted in Figure 5.1(a), or only outlined (framed), as the one in Figure 5.1(b). However, if only the edges are taken into consideration then the two will have the same phase histograms and, in consequence, will be perceived as the same by a classifier.

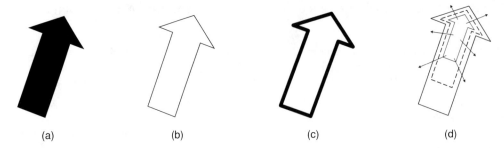

(a)   (b)   (c)   (d)

Figure 5.1 Histogram encoding of different versions of an arrow object. Perception of a black arrow object (a). A phase histogram with only the edges disregards the interior of the object (b). Edges can be computed at different scales (c). The object perceived at different scales (d) (from [4]).

When observing a rigid 2D object, like the one presented in Figure 5.1(a), a histogram of phases (orientations) of its edges conveys important information on the *mutual relation* and *support* of the phases. These properties will be the same irrespective of the object position, its rotation or a proportional scale change. Thus these invariant properties provide an important tag of a type of an object. The invariant is valid under any conformal transformation. However, support of different phases depends on the cumulative lengths of edges of the same phase. These edges correspond naturally to the border of an object. For instance, the two objects in Figure 5.2 will have histograms with the same mutual relations but with different support, since the two edges of the object in Figure 5.2(b) are shorter.

The last factor which should be mentioned here is that due to noise and distortions different objects can show very similar phase histograms which can lead to poor accuracy. Therefore, much stronger support can be gathered if edges are collected at different scales. Therefore a proposed improvement is to compute phase histograms *at different scales of the morphological scale space* (MSS) [4], which fundamentals are presented in Appendix A.2. This concept is depicted in Figure 5.1(c) for a coarse scale edge, and in Figure 5.1(d) for a series of images of the same object, viewed however at different scales.

To take advantage of a solid object, the two mechanisms are used:

1. Construction of the MSS and, at each of its levels, gathering support into one phase histogram describing an object. MSS can be obtained by consecutive application of the

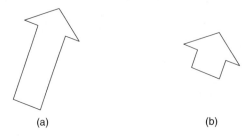

(a)   (b)

Figure 5.2 Explanation of the same mutual relation of phases with different support for the two objects (from [4]).

morphological dilation and erosion operators with certain structural element (usually set arbitrarily or guided by knowledge of shape of an object). This is discussed in Appendix A.2.

2. At each scale level, detection of the stick (elongation) component **S**, given in Equation (2.145), of the ST and the computation of the local phase only in these places.

In Section 2.7.3 it was shown that the ST by itself is also endowed with the scale-space properties. However, this is a linear scale which does not allow determination of the intrinsic structure of solid objects, for instance, in contrast to the nonlinear MSS. Therefore, all computations of the ST in the presented method are carried out at its one internal scale, i.e. ξ and ρ in Equation (2.108) do not change. Instead the input images are affected by the MSS.

As alluded to previously, local phases are computed at each scale level of the MSS. This leads to enhancement of the representation of the characteristic local features and boosting of their phases in the histogram. In effect greater accuracy of the method can be achieved compared to a version operating only at a single scale [5]. Summarizing, the two main stages of this method consists of building the phase histograms at different morphological scales and then matching them. Both are discussed in the next two sections.

5.2.1 Computation of the Tensor Phase Histograms in Morphological Scale

As alluded to previously, elongated areas, for which computation of their local phases is unique, correspond to the strong stick (linear only) component of the ST while keeping the other components at a much lower level. Their local phases are obtained from components of the ST. Thus, the proposed method can be used mainly for recognition of objects with strong linear structures (e.g. regular shapes like road signs [5], static views of human hands 1, 6], etc.).

Let us rewrite Equation (2.145) p. 71, for the 2D case in which $\lambda_3 = 0$, as follows

$$\mathbf{T} = \underbrace{(\lambda_1 - \lambda_2)\,\hat{\mathbf{u}}_1\hat{\mathbf{u}}_1^T}_{\mathbf{S}} + \underbrace{\lambda_2\left(\hat{\mathbf{u}}_1\hat{\mathbf{u}}_1^T + \hat{\mathbf{u}}_2\hat{\mathbf{u}}_2^T\right)}_{\mathbf{P}}. \tag{5.1}$$

In the above, **S** and **P** are the stick and a plate base tensors respectively, $\hat{\mathbf{u}}_1$ and $\hat{\mathbf{u}}_2$ are normalized eigenvectors, and $\lambda_{1,2}$ are eigenvalues of **T**, for which it holds always that $\lambda_1 \geq \lambda_2$. In 2D case, the eigenvalues are computed from Equation (2.92).

The first task is to find areas with strong linear component **S**. From the above formulas it is evident that these are characteristic of a high value of λ_1 with small λ_2 at the same time. Therefore our criteria for areas of interest can be formulated as

$$\frac{\lambda_2}{\lambda_1} \leq \kappa, \tag{5.2}$$

where κ is a threshold value, which in the experiments was less than 0.1. Inserting Equation (2.92) into Equation (5.2) yields

$$p \geq \frac{1 - \kappa}{1 + \kappa}, \tag{5.3}$$

where

$$p = \frac{\sqrt{(T_{11} - T_{22})^2 + 4T_{12}^2}}{T_{11} + T_{22}}, \text{ assuming } m = Tr(\mathbf{T}) = T_{11} + T_{22} \neq 0. \qquad (5.4)$$

In the above, T_{ij} are components of the ST. Algorithm 5.1 presents the steps of computation of the phase histogram at one scale.

```
1.  Set threshold κ (usually κ < 0.1).
    Initialize data structure for the histogram.

2. For each pixel do:

    3. Compute T from Equation (2.108).

    4. Compute trace m = T₁₁+T₂₂ (Equation (2.98)).

        5. If m > 0 (or above a threshold σ):

            6. Compute p from Equation (5.4).

            7. If Equation (5.3) holds:

                8. Compute φ from Equation (2.94) and add it to the histogram.
```

Algorithm 5.1   Computation of a phase histogram at one level of the scale-space.

Let us observe that large values of m in Equation (5.4) indicate the presence of an edge. However m, being a trace of the ST, denotes an averaged gradient in a neighborhood. Therefore it is more resistant to noise than other simple edge detection methods (such as the Sobel filter, etc.). In some applications, step 5 in Algorithm 5.1 can be more restrictive in order to suppress noise. In such a case we replace 0 with a positive threshold. This is also advised to avoid excessive numerical errors when dividing by m in Equation (5.4). Thus, a small threshold σ is always substituted for 0 in step 5 in Algorithm 5.1. A good hint here is to choose σ to be at least few orders of magnitude greater than a smallest value in the number representation (in our experiments it was 10e-6 since m was greater than 1 even for weak structures).

As already mentioned, phase histograms are proposed to be built at each level of the MSS. Each level is then added into one common histogram. Thus, the method does not suffer from excessive memory requirements. Additionally, at each processing level, nonlinear filtering of the histogram is proposed. It consists of suppressing to 0 of all bins which are below a threshold α.

The histogram building procedure in the MSS is presented in Algorithm 5.2. Its input consists of the reference patterns or test images, each denoted by **I**, and the filtering threshold α.

```
1.  Acquire an image I, setup filtering level α and initialize data structures.
2.  Compute orientation histogram H₀(I) from the image I based on Algorithm 5.1.
3.  Filter H₀: H₀=F(H₀, α).
4.  Create a copy J of the original image I: J = I.

5.  For each level i of the erosion scale do:

    6.  Compute morphological erosion of J: J = Eᵢ(J).

    7.  Compute orientation histogram Hᵢ(J) from Algorithm 5.1.

    8.  Filter the histogram Hᵢ: Hᵢ = F(Hᵢ, α).

    9.  Increment the histogram H₀: H₀ = H₀ + Hᵢ.

10. Copy original image to J: J = I.

11. Repeat steps 5 - 9 replacing erosion with dilation.
```

Algorithm 5.2 Histogram building in the morphological scale space. Input consists of an image I and the filtering level α. The phase histogram H_0 of an object is returned on output.

5.2.2 Matching of the Tensor Phase Histograms

Recognition with the tensor phase histograms belongs to the one-nearest-neighbor classification method, as discussed in Section 3.7.4.2. That is, the goal is to find the histogram of a prototype which resembles best the histogram of a test pattern. For this purpose the phase histograms need to be compared or matched, as discussed in Section 3.7.2.

To find the best prototype, a histogram **t** of an input image is checked against all S reference histograms **s** which were precomputed and stored in the database. Since rotations of the test objects are possible, the matches are computed for *the modulo shifted* versions of the histogram **t**. Thus, recognition with the phase histograms can be formulated as the minimization of the functional

$$E(\mathbf{s}, \mathbf{t}, r) = D_B(\mathbf{s}, \mathbf{t}\%r) + \Theta(\mathbf{s}, \mathbf{t}, r), \tag{5.5}$$

where **s** is a histogram of the *s-th* object in the database, D_B is a histogram match measure, $\mathbf{t}\%r$ denotes an r-modulo shifted histogram of the test object, r is in the range $\pm(0..r_{max})$, and finally $\Theta(r)$ is a rotation penalty factor. In some systems $\Theta(r)$ can be used to favor some specific rotations of the objects. For instance to favor the up-right position of the tested objects $\Theta(r)$ can be formulated as follows

$$\Theta(\mathbf{s}, \mathbf{t}, r) = \begin{cases} \gamma \dfrac{r}{r_{max}} D_B(\mathbf{s}, \mathbf{t}\%r), & for \quad r_{max} \neq 0 \\ \qquad 0, & for \quad r_{max} = 0 \end{cases}, \tag{5.6}$$

where γ denotes a constant (for upright objects this can be set from 0 to 10%). Obviously, setting γ to 0 rules out the rotation penalty factor from the optimization, i.e. all rotations are equally preferred.

The classifier answers a class s of the test object and its rotation r in respect to the best fit which is obtained from Equations (5.5) and (5.6), as follows

$$\underset{\substack{s \in S \\ -\Delta r \le r \le +\Delta r}}{\arg\min} \; E(s, t, r) = \underset{\substack{s \in S \\ -\Delta r \le r \le +\Delta r}}{\arg\min} \; \left[\left(1 + \gamma \frac{r}{r_{\max}} \right) D_B(s, t\%r) \right]. \qquad (5.7)$$

The above is simple to compute since histograms are one-dimensional structures.

```
1. Compute the database S of pattern histograms Hₛ.
   Compute histogram Hₜ of the test pattern t (Algorithm 5.2).
   Set up parameter γ of the rotation penalty and the separation parameter ε of the
      best match.

2. For each histogram s∈S do:

   3. For each rotation -Δr≤r≤Δr do:

      4. Compute E(s,t,r) from Equation (5.5);

      5. Store the two best values of E_{B1} and E_{B2} (E_{B1}>E_{B2}) and their parameters s
         and r.

6. If (E_{B1} - E_{B2}) > ε:

   7. Return class of s and rotation r found for E_{B1}.

   else

   8.  Respond: ''Unknown pattern''.
```

Algorithm 5.3 Recognition from the accumulated morphological phase histograms. Input consists of the database S of pattern histograms H_s and the histogram of the test object H_0. The system responds with a class of the pattern which fits best the object's histogram and its rotation, or an "Unknown pattern" message in case of insufficient separation of the best match values.

At the stage of finding the best matching histograms yet another modification is proposed. It allows control over the uniqueness of the match. For this purpose a method adopted from the stereo matching is used [7]. It consists of checking not only the value of the best match but also values of *the next two best matches*. If these two are sufficiently different from the former, then the response is accepted since the extreme is distinguishable. Otherwise, the classifier cannot unambiguously provide a class for the test pattern. Such a strategy decreases the rate of false positive responses, thus increasing the overall accuracy of the system.

Algorithm 5.3 presents the recognition process based on the accumulated morphological phase histograms. The input consists of the database S of pattern histograms H_s and the histogram of the test object H_0. Output of Algorithm 5.3 is a class of the pattern from the database that fits the test object. Additionally its rotation is also provided. However, if separation between the best matches is not sufficient then the method can finish with the

"*Unknown pattern*" answer. Nevertheless, results of the method depend on the chosen threshold parameters. Thus, their proper choice needs some experimentation.

For matching of a single pair of histograms a number of the histogram matching measures was tested, as presented in Section 3.5. However, the best results were obtained with the Bhattacharyya measure (3.113) and therefore this is used hereafter.

5.2.3   CASE STUDY – Object Recognition with Tensor Phase Histograms in Morphological Scale Space

The discussed method of object recognition based on their phase histograms can be used to recognize different road signs and static hand gestures, as will be shown.

Figure 5.3 and Figure 5.4 explain the behavior of the method for rotated patterns. As already mentioned, rotation causes a horizontal shift of its phase histogram. Let us note that the phase histograms are computed exclusively in the strong linear structures indicated by a large value of the stick component S of the ST. These are depicted in the middle column of Figure 5.3 and Figure 5.4. It is worth noticing that the joints of the edges are excluded from computation of the phases since these reveal many different orientations and therefore show a high plate response of the ST.

Each rotation of a prototype object, shown in the second columns of Figure 5.3 and Figure 5.4, directly causes a shift of their phase histograms computed in the areas of strong

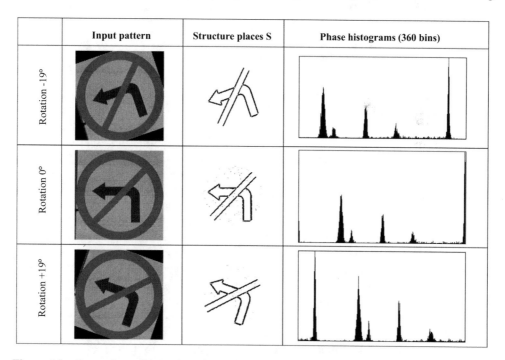

Figure 5.3   Comparison of rotated versions of a sign "No left turn" (B-21) shown in second column. Pictogram areas characteristic of a strong stick response S at one scale (third column), as well as their phase histograms (fourth column). Rotation of a sign denotes a horizontal modulo-shift of its phase histogram. Each histogram contains 360 bins (from [4, 5]).

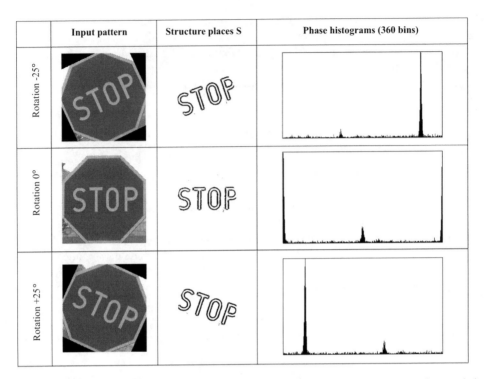

	Input pattern	Structure places S	Phase histograms (360 bins)
Rotation -25°			
Rotation 0°			
Rotation +25°			

Figure 5.4   Comparison of rotated versions of a "STOP" sign (B-20). Areas of pictograms characteristic of strong response **S** at one scale. Their phase histograms (fourth column) (from [4, 5]).

stick response **S**. This allows recognition of rotated patterns. Let us also note that changes of scale of the prototype patterns affect the heights of the bars in the histograms. However, if the histograms are normalized then the method is also resistant to a change of a scale. Thus, the main advantage of this method is its ability to recognize objects which are rotated and which are of different size. Finally, let us indicate that the requirements on the precision of detection of an object are also not too restrictive, since vertical and horizontal shifts do not affect the phase histograms as long as the object is still visible.

However, there are also some problems which should be mentioned. First let us note that when computing the phase histograms the spatial information about the edges is lost. This is an inherent feature of the histogram methods, discussed in Section 3.7.1. Thus many different objects can have very similar phase histograms. Moreover, an influence of the areas with strong linear characteristics (i.e. the strong **S** component) is limited since such points are relatively scarce. Thus, the overall signal to noise ratio in the histograms can be expected to be rather low. This is also related to the problem of recognizing occluded objects since in this case the histogram acquires different shape. Thus, if occlusions are significant then most probably the method will be less precise.

The aforementioned drawbacks can be alleviated somewhat by taking advantage of the MSS, as proposed in the previous section. This is exemplified in Figure 5.5 in which a "*No U turn*" road sign is shown in different erosive and dilative morphological scales. Each such representation adds to a common phase histogram thus amplifying the edge characteristics of the observed object.

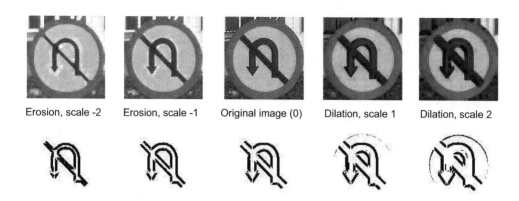

| Erosion, scale -2 | Erosion, scale -1 | Original image (0) | Dilation, scale 1 | Dilation, scale 2 |

Figure 5.5 An image of a road sign (top row) and its stick tensor components (bottom row) at different morphological scales (dilated versions to the left, eroded to the right). Scale 0 in the center (from [4]).

Figure 5.6a shows a real traffic scene with "*No U turn*" sign detected with the method discussed in Section 4.4.3. No registration is necessary, i.e. a detected place is cropped from the image and directly led to the presented phase classifier. However, in some cases registration can improve results, especially if the signs are not in the fronto-parallel position and exhibit strong projective deformation. The other interesting property is that the method can operate with *any dimensional signal*. This is due to the properties of the ST which can be computed in multidimensional signals, as discussed in Section 2.7.3. Thus, in our road sign example

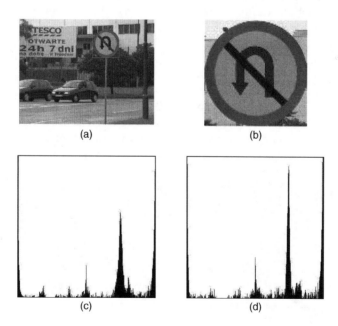

Figure 5.6 A real traffic scene with detected "No U turn" sign B-23 (a). Response of the system with a very similar sign from the database (b). Recognition was possible by matching of the phase histograms which for the signs in panes (a) and (b) are shown in (c) and (d) (from [4]).

the color or monochrome version of the cropped object can be used for classification. In this case, however, strong linear stick components are highly correlated and if present then they exist in all channels. Therefore there is no significant difference in accuracy between color and monochrome versions, whereas computations in the latter case are much faster. Nevertheless, the correlated channels allow removal of noise or distorted areas since frequently an unwanted signal, being independent, contaminates only one channel.

The system responds with a sign in Figure 5.6b from the database which is the most similar to the test pattern. Recognition was achieved in accordance with Algorithm 5.3. The cumulative phase histograms for the signs in panes Figure 5.6(a) and Figure 5.6(b) are shown in Figure 5.6(c) and Figure 5.6(d), respectively.

Nevertheless, experimental results show that the phase method is not as precise as the deformable models discussed in Section 5.5 with case examples presented in Sections 5.7 and 5.8. However, it requires much less memory and is faster (especially for rotation). As a side effect a value of rotation is computed in parallel to the classification process. At the same time some *a priori* knowledge on the allowable rotation of objects can be introduced into Equation (5.7).

For example, in the case of the road signs it is highly improbable that these will be rotated more than ~20°, since their up-right position is assumed. During the experiments it was observed that a fair trade-off between speed and accuracy is to use pictograms smaller than 100×100 pixels, since lower values cause insufficient numbers of stick pixels. The number of bins in the histogram can be in the range 180–360. As already mentioned, the method is fast and allows real-time operation in a serial C++ implementation. This depends also on the number of processed scales for each object.

The method was tested on another group of objects which are *static* gestures. Figure 5.7(a) depicts the used database with static hand gestures. An exemplary hand test pattern is shown in Figure 5.7(b).

Figure 5.8 depicts the process of building MSS from an image of a hand. When traversing the MSS some detail can be lost after reaching the so called an intrinsic scale level (Section A.2). Each image in Figure 5.7 is then used to add its phase values to the common histogram. Also, at each scale level the histogram is filtered out to remove noise.

As alluded to previously, application of the MSS at the stage of building phase histograms increases the discriminative abilities of the classifier. This can be observed in the plots of the match values of histograms (Figure 5.9(a)) which always increase after applying some levels of the scale-space in computation of the phase histograms. Value 0 of the scale in Figure 5.9

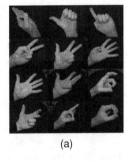
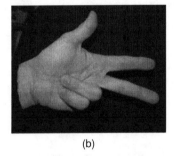

(a)   (b)

Figure 5.7   A data base with hand gestures (a). Exemplary test pattern (b) (from [4]).

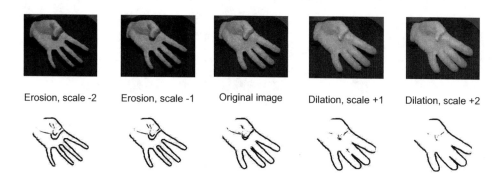

Erosion, scale -2   Erosion, scale -1   Original image   Dilation, scale +1   Dilation, scale +2

Figure 5.8 An image of a hand (top row) and its stick tensor components (bottom row) at different morphological scales (dilated versions to the left, eroded to the right). Input image in the center (from [4]).

indicates only one scale level, i.e. the scale space is not used. The question is however, how many levels of the scale should be employed? The plot in Figure 5.9(a) provides some hints. For some objects and after a certain value of the scale level there is no further improvement in classification. This is caused by reaching the intrinsic scale level of an object, which usually happens sooner for the erosion level since some details cease to exist after few erosions.

On the other hand, more scale levels increase the computation time. It was found in practice that two levels of erosions and up to three to four levels of dilations are sufficient. For example, in the case of "*No pass*" road sign, the match value fell after reaching two scale levels (a plot with circles in Figure 5.9(a)). This is caused by fine details which disappear at higher levels of the scale.

To lower the level of false positives the ratio of the best two matches is checked and if it is less than a certain threshold (in practice 5–10%), then the classifier responds "*Unknown pattern.*" The influence of the number of levels of the scale space on this separation ratio is different for different objects and does not always increase with higher scale level, as shown in the plot in

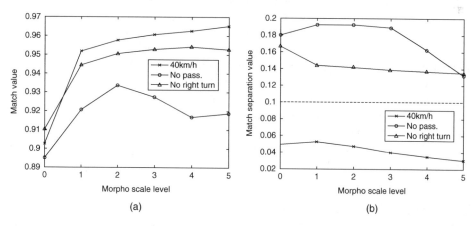

(a)   (b)

Figure 5.9 An influence of the level of the morphological scales on match value (a) and match separation value (b) for the winning patterns (road signs). The dashed line denotes a threshold value below which the classifier responds "Unknown pattern" to lower the false positive ratio (from) (from [4]).

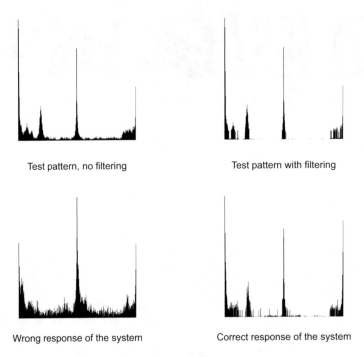

Test pattern, no filtering   Test pattern with filtering

Wrong response of the system   Correct response of the system

Figure 5.10   Effect of nonlinear histogram filtering (noise squeezing) on response of the system. Two test patterns (upper row) and two responses of the system (bottom row) are shown. Nonlinear filtering of the histograms (right column) allows correct response (bottom right) (from [5]).

Figure 5.9(b). For example for the sign "*40km/h speed limit*" this ratio is below the set threshold of 10%. This is caused by many signs which contain "0" in their pictograms, e.g. "30", "50", "70", and so on. Because of this their histograms are very similar due to their common parts.

Overall improvement in the recognition ratio, expressed in terms of the rate of true positives to all patterns, is in the order of 92–97% for the road signs daylight conditions and 75–82% for static hand gestures. This constitutes an improvement factor of about 5–10% over the method which does not utilize the MSS, i.e. the one operating exclusively in one scale.

In Algorithm 5.2 further improvement in classification was achieved thanks to the noise filtering of the histograms at each level of scale processing. Figure 5.10 depicts the effect of nonlinear histogram filtering (noise squeezing) on the response of the system. Two test patterns are shown (upper row) and two responses of the system (bottom row). Nonlinear filtering of the histograms (right) allows the correct response (bottom right) whereas unfiltered histograms result in a false positive.

5.3   Invariant Based Recognition

One of the main problems in object recognition is their variability in the observed scene. Therefore, methods which assume direct modeling frequently fail since a template for example does not match a real object very well. Therefore, one idea is to find such features which, on the one hand, are very characteristic of the sought object and, on the other hand, are as much *invariant* to object transformations as possible. Indeed, an analysis of geometric properties which remain invariant under different transformation groups constitute the foundation of modern mathematical geometry [8]. This famous postulate was formulated by Felix Klein in 1872. Thus, invariants are characteristic of different transformation groups. For instance the length of a line segment is an invariant of rotation, a cross ratio of four points is invariant to projective transformation, etc.

The other problem is application of invariants in CV for object recognition. In this respect there is still ongoing research which has resulted in development of such invariant features as the already mentioned SIFT descriptor [2] for sparse object representation or statistical moment invariants which are discussed in the following section.

5.3.1   CASE STUDY – Pictogram Recognition with Affine Moment Invariants

In this section a method for road sign recognition is presented which is based on statistical invariants. The first problem is that images of objects viewed by a camera are transformed projectively. Thus, recognition methods should be able to operate in the projective space. Fortunately, such stringent conditions can be slightly relaxed based on additional information on the specific object. One of the requirements imposed on the placement of the road signs is their good visibility to drivers. Thus, more often than not their planes are fronto-parallel to the sight direction of a driver. Thanks to this, the projective transformation can be approximated by the affine transformation [7, 9]. In the case of triangular and rectangular shapes the transformation can be simplified even further since the rotation can be compensated in the registration module based on the position of the characteristic corners, as discussed in Section 4.4. Unfortunately, this cannot be done in the case of circular signs. Therefore in this section *the affine moment invariants* (AMI) are analyzed for classification of this group of signs [10]. AMIs were proposed by Flusser and Suk based on the theory of algebraic moment invariants [11, 12]. They are more robust for larger group of transformations when compared with the well known Hu invariants [13]. However, similarly to the Hu invariants AMI are composed from the statistical moments of the image intensity signal, which were discussed in Section 2.8.

AMI are invariant to the general affine transformation which takes a 2D point $\mathbf{x}$ into the corresponding $\hat{\mathbf{x}}$ as follows

$$\mathbf{x}' = \mathbf{A}\mathbf{x} + \mathbf{B}, \text{ where } \mathbf{A} = \begin{bmatrix} a_{11} & a_{12} \\ a_{21} & a_{22} \end{bmatrix} \text{ and } \mathbf{B} = \begin{bmatrix} b_1 \\ b_2 \end{bmatrix}. \tag{5.8}$$

It was shown that the following first four AMI are sufficient for reliable pattern recognition in many vision tasks [12]

$$\pi_1 = c_{00}^{-4} \left(c_{20}c_{02} - c_{11}^2 \right),$$

$$\pi_2 = c_{00}^{-10} \left(c_{30}^2 c_{03}^2 - 6c_{30}c_{21}c_{12}c_{03} + 4c_{30}c_{12}^3 + 4c_{03}c_{21}^3 - 3c_{21}^2 c_{12}^2 \right),$$

$$\pi_3 = c_{00}^{-7} \left[c_{20} \left(c_{21}c_{03} - c_{12}^2 \right) - c_{11} \left(c_{30}c_{03} - c_{21}c_{12} \right) + c_{02} \left(c_{30}c_{12} - c_{21}^2 \right) \right],$$

$$\pi_4 = c_{00}^{-11} \left[c_{20}^3 c_{03}^2 - 6c_{20}^2 c_{11}c_{12}c_{03} - 6c_{20}^2 c_{02}c_{21}c_{03} + 9c_{20}^2 c_{02}c_{12}^2 + 12c_{20}c_{11}^2 c_{21}c_{03} + \right.$$

$$+ 6c_{20}c_{11}c_{02}c_{30}c_{03} - 18c_{20}c_{11}c_{02}c_{21}c_{12} - 8c_{11}^3 c_{30}c_{03} - 6c_{20}c_{02}^2 c_{30}c_{12} +$$

$$\left. + 9c_{20}c_{02}^2 c_{21}^2 + 12c_{11}^2 c_{02}c_{30}c_{12} - 6c_{11}c_{02}^2 c_{30}c_{21} + c_{02}^3 c_{30}^2 \right]. \tag{5.9}$$

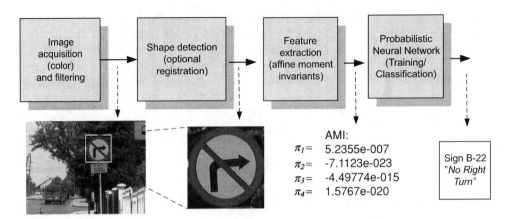

Figure 5.11 Processing path of the road sign classification system based on the affine moment invariants and the probabilistic neural network (from [10]). Color version of the image is available at www.wiley.com/go/cyganekobject.

In the above c_{ab} are central moments defined in Equation (2.121). The architecture of the classification system with AMI is depicted in Figure 5.11.

Signs are detected based on their colors and shape, as discussed in Section 4.4. Then their inner part is cropped and binarized to allow for immediate computation of the four AMI π_1-π_4. These are then fed to the PNN which was presented in Section 3.9.1.

A very important step in classification with the statistical moments is their proper normalization, since there are strong variations in exponents of particular invariants. In our system, each AMI in Equation (5.9) was multiplied by a common to its group exponent, so all of them had the same dynamic range. Values computed this way are in the same order of magnitude and in effect have a more balanced influence on the classification process. After the described normalization, π_1-π_4 constitute components of the feature vectors to the PNN depicted in Figure 3.18 (p. 267). The other possibility is to normalize them by their standard deviation.

Figure 5.12 presents a road sign from the prohibition group in different affine transformations and their corresponding first four AMI, computed in accordance with Equation (5.9). AMI computed in the three versions are very similar. The small differences are due to variations in the detection module, as well as being caused by noise. As experimentally checked, this can be improved by introducing the shape registration module in the processing path in Figure 5.11. This is at a cost of additional computations, however. Nevertheless in practice, such big deformations are unusual and for smaller deformations AMI are sufficient. Some errors are also introduced by the binarization module which operates directly on the intensity signal without any preconditioning. This can be improved by simple intensity histogram equalization technique, for instance.

To test behavior solely of the classifier, the first group of experiments relies on road signs that have been manually cropped from the real traffic scenes encountered on Polish roads (daylight conditions). Only circular signs were considered, i.e. the prohibitive "B" and obligation "C" groups of signs. To obtain even more examples the signs from the database were augmented by their deformed versions. The deformations were random affine with their maximal extent set as shown in Table 5.1. These patterns were used to train the PNN classifier in accordance with

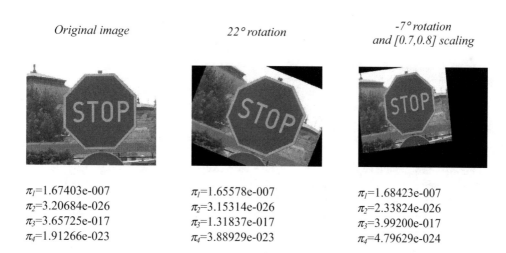

Original image   22° rotation   -7° rotation
                                                                      and [0.7,0.8] scaling

π_1=1.67403e-007 π_1=1.65578e-007 π_1=1.68423e-007
π_2=3.20684e-026 π_2=3.15314e-026 π_2=2.33824e-026
π_3=3.65725e-017 π_3=1.31837e-017 π_3=3.99200e-017
π_4=1.91266e-023 π_4=3.88929e-023 π_4=4.79629e-024

Figure 5.12   Affinely transformed views of the STOP (B-20) road sign and their four AMI.

the methodology described in Section 3.9. Table 5.1 presents experimental results obtained with the two groups of signs and in three levels of affine deformations.

Accuracy of recognition in Table 5.1 was measured simply as a ratio of the correct answers to the number of all answers (precision). The accuracy of this method is in general lower than other presented road sign recognition methods, such as for instance the tensor phase histograms (Section 5.2.3), ensemble of Hamming neural networks (Section 5.7), as well as the ensemble of tensors (Section 5.8). Closer scrutiny reveals that the problem is caused in the first degree by the outliers which are due to noise and imperfections of the shape detector and the binarization module. The second problem is variability of pictograms of the signs (i.e. *real* signs, even if of the same type such as "No left turn", differ slightly), as well as the variability of shapes of the pictograms represented on a relatively small discrete grid of an image. In other words the statistical moments, as well as their polynomials such as AMI, do not cope well with the discrete nature of digital images. This places an inherent limitation on the precision of the moment based techniques, such as the mentioned Hu invariants or even AMI.

The classifier always chooses the best class for an input pattern. However, this is not always correct since the input pattern can belong to false patterns (i.e. not a sign). To deal with such

Table 5.1   Accuracy of recognition of the system PNN-AMI for two groups of road signs "B" and "C" under different deformations (from [10]).

Type of RS Affine deformation group	Road signs group "B" Precision	Road signs group "C" Precision
Rotation up to ±5°, scaling ±5%	0.81	0.83
Rotation up to ±10°, scaling ±10%	0.70	0.77
Rotation up to ±15°, scaling ±15%	0.69	0.80

Table 5.2   Accuracy of recognition of the road signs from the two groups "B" and "C" (from [10]).

Road signs group "B"		Road signs group "C"	
Precision	Recall	Precision	Recall
0.71	0.85	0.70	0.91

situations a best match separation technique is employed again. It was found experimentally that if this separation of g_ω in Equation (3.170) is above 20% then the answer is correct with high confidence. The false positives usually have lower separation.

Table 5.2 shows the accuracy of the recognition system composed of the AMI-PNN classifier and the detector with the adaptive window growing technique, described in Section 4.4.3. Real road scenes for this experiment were randomly selected from a database that contains the signs from the "B" or "C" groups. Errors are mostly due to noise and occlusions in the input images, which in turn caused some improper sign locations by the detector. Such a situation almost always leads to wrong values of AMI and, in consequence, wrong classification results. Also the previously discussed limitations of the moment invariants influence the precision of this system.

Some classification errors in the "C" group are also introduced by signs which differ only by a rotation factor, e.g. these are the Polish signs C-1, C-3, C-5, C-9, and C-10, whose pictograms are defined in the regulation [15], as shown in Figure 5.44(c).

The affine moment invariants were also compared with the well known Hu invariants (the first four used) [13]. As expected, the obtained results of the Hu method were worse since these invariants cannot accommodate slanted objects well. This observation agrees with the results obtained by Flusser and Suk [12]. Classification times were of the order of milliseconds per single 320×240 image in the software implementation [10].

5.4   Template Based Recognition

Template matching belongs to the one of the most advanced and intensively developed areas of computer vision. The strategy of the methods belonging to this group is simple, at least in its very basic formulation: create an exemplary pattern and try to fit it to the signal. If there is a match which can be considered the best one, then one can assume the existence of a sought pattern in a signal. As seen in this explanation the method requires techniques of creating a template, and then means of its comparison with a signal. A template is a model of a sought pattern. In computer vision it is usually expressed in terms of intensity signals or feature vectors. A matching measure is used to point to a place in an input image which is the most similar to a model. However, in practice the method is not so simple. A first problem is the variability or vagueness of the objects one tries to find. An object can be seen from different viewpoints, at different scales, rotations, lighting conditions, in an image contaminated with noise, etc. Thus, in the first attempt one might think of generating all the possible variants of the prototype templates which are then used to match with a test pattern. However, even a short assessment reveals that dimensionality of template space grows enormously, restraining the usefulness of this approach.[1] The second problem is the choice of a matching criteria or, in other words, how to tell that the pattern is present in a signal, what is its most similar prototype, and finally how reliable such an answer can be. In this respect, the abilities of the biological systems observed in animals and humans seem formidable, especially when dealing with so many variations of patterns and providing such reliable answers!

The problem of template matching has been studied for years and in different scientific contexts. In computer vision template matching approaches were undertaken in almost all of its tasks like stereovision, camera calibration, recognition of objects such as faces, fonts, gestures, road signs, missiles, cars, animals, as well as for food inspection, behavior control, and many, many more. Considering all of the above, a statement that the template matching *is* computer vision itself is not that exaggerated, though we do not go that far and place template matching in the area of high level recognition. Literature on this subject is also ample and as a starting point some recent books can be recommended [7, 16, 17, 18].

More formally, template matching can be seen as a classification problem [18]. In mathematical terms template matching can also be formulated as hypothesis testing [16, 19]. Therefore the statistical methods of classification, discussed in Section (3.2), provide a first introduction to template matching.

As already mentioned the subject of template matching is very broad. In this section we discuss only a small section of it with majority of examples coming from automotive applications.

5.4.1   Template Matching for Road Signs Recognition

In the book we present vision systems oriented towards automotive applications, such as road sign recognition. Therefore a short and focused overview of this problem of template methods is presented in the rest of this section. In this respect Paclík *et al.* [20] propose a new method for road sign classification which is an extension of the well known cross-correlation template matching. A novelty, however, is matching of the selected areas of interest characteristic to the sought patterns, rather than the whole images. These areas are obtained from a classifier trained with real examples of road signs. However, only five classes of the prohibition signs were considered. The authors of this paper also tried to develop a matching measure which would be suitable for road sign recognition. Such a measure should be more resistant to the background pixels, as should also be faster in computation than the cross-correlation. Correlation of an image with a prototype is computed based on many partial correlations found in selected areas of the image. The set of these areas is updated during training to achieve maximal separation between prototype templates. The matching measure for two monochrome images I and J in an area R is defined as cross-correlation

$$S_r(I, J, R) = \frac{\langle \mathbf{R}(I), \mathbf{R}(J) \rangle}{\|\mathbf{R}(I)\| \cdot \|\mathbf{R}(J)\|}, \tag{5.10}$$

---

[1]These are sometimes called "brute force" methods. Usually these are not feasible in software sequential implementations. However, the advent of inexpensive programmable parallel devices (e.g. GPU, FPGA) caused an increased interest in such methods [21, 22].

where $\mathbf{R}(I)$, $\mathbf{R}(J)$ are vectors composed of pixels of I and J in R, $<,>$ denotes the scalar product. Then, to make Equation (5.10) resistant to lighting variations from each region its mean is subtracted. In this way the normalized cross-correlation measure is obtained

$$S_r (I, J, R) = \frac{\sum_{k \in R} \left(I_k - \bar{I}\right) \left(J_k - \bar{J}\right)}{\sqrt{\sum_{k \in R} \left(I_k - \bar{I}\right)^2 \cdot \sum_{k \in R} \left(J_k - \bar{J}\right)^2}}, \tag{5.11}$$

which is commonly used in stereovision [7]. In the above k iterates through all pixels in the region R. The value of S_r is in the range $[-1,+1]$, for which $+1$ denotes a perfect match, whereas -1 is a match with a negative of a pattern. However, a region R is further divided into subregions $r \in R$ and in each a correlation $s(I,J,r)$ is computed based on Equation (5.11). Having found $s(I,J,r)$, a resulting total match $S(I,J,R)$ can be determined as a function f of the component matches

$$S (I, J, R) = f [s (I, J, r_1), \ldots, s (I, J, r_N)]. \tag{5.12}$$

For f Paclík *et al.* propose one of the following functions [20]

$$S_{mean} (I, J, R) = \frac{1}{N} \sum_{k=1}^{N} s (I, J, r_k), \text{ or } S_{min} (I, J, R) = \min_{r_k} \{s (I, J, r_k)\}. \tag{5.13}$$

The set of subregions $r \in R$ can be adjusted to a specific prototype considering the class it belongs to. Given a training set Q of a class of certain signs, consisting of M prototypes, such a set $\{r\}$ is proposed which maximally separates the prototypes of Q from all other objects, such as the ones which are not road signs (e.g. background, etc.). Class separation is assessed with a well known Fisher measure [23]

$$C (Q, P, R) = \frac{(\mu_T - \mu_{NT})^2}{\sigma_T^2 + \sigma_{NT}^2}. \tag{5.14}$$

where P is a prototype image, R is a set of subregions, μ_T, μ_{NT}, and σ_T, σ_{NT} denote means and variance of the correlation measure $S(I,P,R)$ for images I of an object of interest (a sign) and other objects (not a sign), respectively. Paclík *et al.* in [20] propose a special training method to find partitioning R into subregions r for each class which at the same time maximizes Equation (5.14).

A separate problem is the choice of the right classifier for features that are correlation measures $s(I,J,r)$ for subregions r. One solution here is the one-nearest-neighbor method, discussed in Section 3.7.4.2. In this case, classification of a certain image I_x consists of computation of the correlation measure between I_x and a set of prototypes P_i, and then – if possible – finding the best match. This process can be stated as follows

$$k = \arg \max_{1 \leq i \leq N} \{S (I_x, P_i, R (P_i))\}, \tag{5.15}$$

where $R(P_i)$ denotes a region found for a prototype P_i. Nevertheless, the one-nearest-neighbor deteriorates in the case of distorted or noisy patterns, as well as for multimodal distributions or overlapping classes. A solution can be the application of the k-nearest-neighbors. However, a better proposition could be application of another type of classifier. Based on the works of Duin *et al.* [24] a new classifier was proposed which finds *dissimilarities* among patterns. Nevertheless, regardless of a chosen classifier gathering of a statistically valid set of prototypes acquired in different conditions becomes problematic.

The interesting part of the work by Paclík *et al.* [20] is the comparison of different classifiers. Apart from the already mentioned one-nearest-neighbor, the linear Fisher discriminator and the SIMCA (Soft Independent Modeling of Class Analogies) classifier were compared. The former projects the input feature space into the linear subspace, assuring maximal separation among classes. Then classification is performed by a linear classifier in the subspace. SIMCA also starts with PCA projections, however *separately* for each class, as also discussed in Section 3.3.2. Then a distance is computed between a test pattern and each of the subspaces. Finally, the pattern is attributed to the class for which its distance is the smallest. However, this distance is composed of two components. The first is the Euclidean distance to the center of the subspace. The second component is the Mahalanobis distance of the projected feature to a mean of that class. The two components are normalized in accordance with the distributions of the models.

Each of the tested classifiers in [20] was checked for its accuracy as well as its ability to reject objects not belonging to any of the classes. A final factor was speed of execution. The best results were reported by the similarity classifiers in contrast to the dissimilarity ones. The best values gave the SIMCA method working with adaptive subregions obtained in accordance with Equation (5.13).

Concluding our considerations on the system proposed by Paclík *et al.* in [20], their method seems to be problematic in the context of road sign recognition since it relies heavily on statistically numerous examples for each of the classes. These are hard to collect since some signs are quite rare compared to others. Also application of the 8-bit intensity values in this method seems to be its drawback for at least two reasons. The first is the relatively high dimensionality, whereas the signs can be classified in the binary fields, as shown in Section 4.4.5. The other problem with this representation is the strong influence of unfavorable factors such as lighting variations, noise, etc.

On the other hand, Gavrilla from the Daimler-Benz Research Group [25] proposed a system for road sign recognition based on the hierarchical correlation of templates in the space of distance transform (DT). DT transforms an image in which its pixels are labeled as belonging or not to an object of interest, into an image that's values of the pixels convey distance of that pixel to the closest pixel labeled as belonging to an object. Thus, DT requires prior detection of objects. Gavrilla proposes a simultaneous matching of N templates of the road signs. Thanks to this shifted signs can be detected. Additionally, to cope with different scales, matching is performed in the Gaussian scale-space, starting from the coarser and then going to the finer level. Speed acceleration is obtained by grouping N templates based on their mutual similarities. Thanks to this, similar templates can be matched together at the coarsest level of the scale-space, hence speeding up the computations.

However, correlation based on DT can be problematic. The first question is on the type of distance encoded by DT. Should it be the Euclidean or e.g. chamfer-2-3 metric? The other open problem is distance aggregation, i.e. the manner of computing a total distance of an object to a template pattern. Most of the known algorithms employ row aggregation independent of the placement of a feature. Other solutions prefer contours. In this case, a distance is aggregated starting from the first place in which a feature was detected. Weighted aggregation is also possible in which the most important feature has higher weights attributed.

5.4.2   Special Distances for Template Matching

Matching of a distance maps of a template S to an image I starts from feature detection. This results in two binary images B_S and B_I, respectively. Regarding the type of feature, it can be an edge, color, texture, etc. Then the map B_I is DT processed to obtain a distance map D_I. On the other hand, the template map B_S is firstly affinely transformed into a number of deformed maps B_{Si}. Each transformation is a combination of the horizontal and vertical shifts, rotations, and changes of scale. Then each such a map is anchored in all or selected points of the already prepared D_I. The last stage is computation of the aggregated distance for those pixels in D_I for which corresponding pixels (i.e. at the same index) of B_{Si} are set. Such pixels from D_I form a distribution of distances from the features of the template S to the closest features in a test image I. Naturally, the smaller the aggregated total distance, the more similar is S to I.

One of the aggregation schemes in the DT space is an averaged distance to the closest feature, called the chamfer distance. It is defined as follows [26]

$$M_{Ch}(S, I) = \frac{1}{\#S} \sum_{s \in S} m(I, s),$$ (5.16)

where $\#S$ denotes the number of features in a template S, $m(I,s)$ is a distance between a chosen feature $s \in S$ and the closest feature of this type in an image I.

The other is the Hausdorff measure defined for two sets A and B as follows [16]

$$M_H(A, B) = \max\left[h(A, B), h(B, A)\right],$$ (5.17)

where

$$h(A, B) = \sup_{a \in A} \inf_{b \in B} \|a - b\|,$$ (5.18)

$\|.\|$ is a certain norm defined in the space of images (usually the Euclidean one). Assuming that A and B are compact the above simplifies to

$$h(A, B) = \max_{a \in A} \min_{b \in B} \|a - b\|.$$ (5.19)

An advantage of M_H over M_{Ch} is its greater resistance to missing features, e.g. the ones which arise from occlusions, etc.

Gavrilla proposed a number of improvements to the basic formulation of his method [23]. One consists of correlation of more than one feature. For edges, their local orientations can be used as additional features, etc. The next improvement is to match more than one pattern at the same time. For this purpose similar templates are found and gathered together into one common template. Thus, the number of matches can be changed to one, hence speeding up computations. The reported accuracy of the method is 90% of positive recognitions, though speed of computations is not known.

5.4.3   Recognition with the Log-Polar and Scale-Spaces

In this section we discuss a template matching method which is more robust to the potential object rotations and changes of scale. This is achieved by dividing the recognition process into different scales of the Gaussian scale-space, as well as thanks to the log-polar transformation which allows template matching of the rotated or scaled patterns. The method was originally developed for the task of road sign recognition [27], then used for recognition of dental implants [28], as will be discussed. Thus the method is best fitted to recognizing well defined patterns, such as the circular prohibition and obligation road signs or dental implants, which in real observations can be rotated or scaled. Basically the method works with monochrome images, although color can help to narrow the search space and speed up computations, as discussed in Section 4.2.

The proposed method tries to overcome the abovementioned problems of pattern variability as well as the influence of such phenomena as partial occlusions, luminance variations or noise. The key techniques that help to solve this task are processing in the Gaussian scale space and search in the extended log-polar (LP) domain. Both spaces are inspired by biological vision systems [29]. Matching in the log-polar space has gained much attention in computer vision, mostly in object recognition and image registration. Interesting papers on these are works by Kara *et al.* [30] and Zokai *et al.* [31]. Computation of the Gaussian scale space and log-polar is presented in the book [7].

The best way to understand the method is to analyze the architecture of the system for road sign recognition shown in Figure 5.13. At first, from the monochrome version of the input image the Gaussian pyramid is created. An example for a real traffic scene is shown in

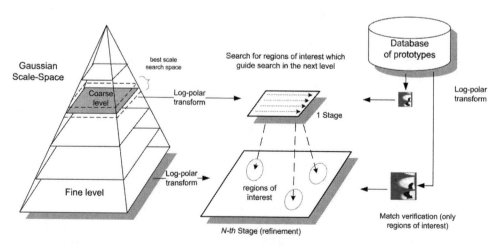

Figure 5.13   System for object recognition in the log-polar and Gaussian scale-space (based on [27]).

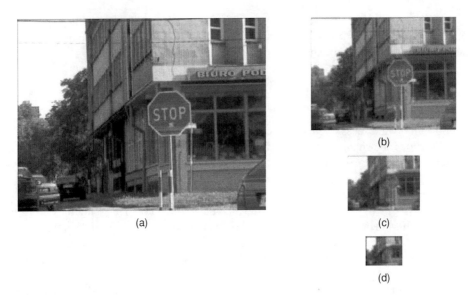

Figure 5.14   The Gaussian scale-space ($\sigma = 5$) of a real traffic scene: 320 × 240 (a), 160 × 120 (b),
80 × 60 (c), 40 × 30 (d).

Figure 5.14. Its coarsest level is then used for an initial search for templates from the database
DB containing the prototype templates, such as the prohibition road signs in our example.
However, rather than using original intensity values, the template object is LP transformed and
matched with each LP-transformed region in the coarse image (Stage 1 in Figure 5.13). In this
way the best matches found guide the search in the next, finer level. Additionally to the match
score the *scale* and *rotation* of the object is provided from the LP matching module, as will be
discussed. Taking this information the same type of search starts in the next finer level of the
pyramid but only in the places selected in the previous level. This time matching parameters
are just updated. In the system presented in [27] processing of two levels was sufficient for
reliable matching.

   As alluded to previously two transformations are involved in this method, both having
biological equivalents. The first is the scale-space constructed as the linear Gaussian pyramid.
It is built in accordance with the following iterative scheme [27, 32]

$$
\begin{aligned}
G^{(i=0)} &= I \\
G^{(i+1)} &= \downarrow_2 F\left(G^{(i)}\right)
\end{aligned}
\tag{5.20}
$$

where I is an input image, $G^{(i)}$ is an image at the i-th level of the pyramid, $\downarrow_2$ denotes operation
of signal down-sampling by a factor of two in this case, F is a smoothing mask i.e. the Gaussian
or binomial, as discussed in [7]. The finest, first level of the pyramid is given directly by the
original image; This corresponds to $i = 0$ in Equation (5.20).

   Choice of the proper scale for the LP and spatial matches is dependent on the scene-camera
setup. This requires some experimental data. Figure 5.14 depicts the Gaussian pyramid of one
of the real traffic scenes. In our acquisition modules images were always RGB color with

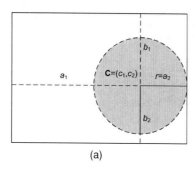 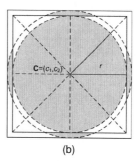

(a)   (b)

Figure 5.15 Role of the central point $\mathbf{C} = (c_1, c_2)$ and d_{max} for the log-polar transformation. Selection based on the smallest projected distance (a). For the geometrical center in a square search window, d_{max} is a radius of an inscribed circle (b) – the pictogram of a sign gets most attention, the outer ring is at coarse sampling, and the background corners (white) are excluded (from [27]).

initial resolutions of 640×480. Thus, for an average size of a circular road sign, the coarsest resolution which has enough data for initial recognition is about 160×120 (Figure 5.14c). This is a trade-off between object detail and computation time. Alternatively, *several scales around the initial one* can be searched and the best scale can be chosen. This can be determined from the optional detection module (Section 4.4.3) which can find the highest areas which contain the color of a sign.

The LP transformation is applied to the images of the pyramid and to the searched patterns. At a point $\mathbf{x} = (x_1, x_2)$ LP can be defined as follows [7]

$$ r = \log_B \left(\sqrt{(x_1 - c_1)^2 + (x_2 - c_2)^2} \right), \text{ and } \alpha = \text{atan2}\,(x_1 - c_1, x_2 - c_2). \quad (5.21) $$

In the above $\mathbf{C} = (c_1, c_2)$ is a center of transformation, B is a base of a logarithm which can be any value > 1.0. The function *atan2* has been already used in formula (2.94), on p. 48, and explained therein. In practice B is chosen to fit the value of r_{max} which is the maximal distance from the center $\mathbf{C}$ and points in the area of interest around $\mathbf{C}$, as follows

$$ B = \exp \left[\frac{\ln\,(d_{max})}{r_{max}} \right], d_{max} > 1, r_{max} > 1, \quad (5.22) $$

where $d_{max} = min\{d_{1max}, d_{2max}\}$ denotes a minimal distance from a chosen center $\mathbf{C}$ and surrounding window, as depicted in Figure 5.15a. r_{max} is the maximal number of columns in the LP image. Thus, LP results in a nonlinear and nonuniform sampling of the spatial intensity signal. Nonlinearity is caused by the polar mapping, whereas log operation leads to the nonuniform sampling (5.21).

The LP space is used at the coarsest level of the image pyramid to search for "promising" sign locations. Since in this case the sought objects are circular road signs, the search window was set to be rectangular and the center $\mathbf{C}$ of the log-polar is set to be a geometrical center of each search window (Figure 5.15(b)). Thus, the most important part of a sign, i.e. its center with a pictogram, will be given most attention.

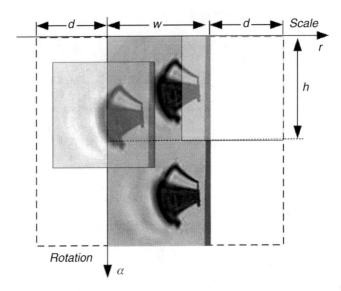

Figure 5.16   Four dimensional search space in the log-polar domain. The test pattern $w \times h$ is wrapped around to allow proper match of the rotation and scale. Scale search is extended by d (from [27]).

Search for the best matches in the LP space is carried out in the four dimensions as depicted in Figure 5.16. At first a template is transformed into the LP representation. Then, for each position (x_{1b}, x_{2b}) in the background image, a region is selected of the same size as the template [25]. This region is also LP transformed. Next, the two LP images are matched. This match, however, is done in *an extended space* [31], depicted in Figure 5.16. The background pattern is *wrapped around to 2h* to allow proper search of the rotation parameter. The scale search span is also *extended by a distance d*, which can be up to the template width w (in practice it was set to $0.8w$). Further discussion on the search method in the LP space, as well as implementation details can be found in [7, 31].

The best match position in this extended LP space corresponds to the scale and rotation (r, α) between background and the template. For the best match(es) the four parameters $(x_{1b}, x_{2b}, r, \alpha)$ are saved. Two methods for computing the best matches were used. The first one is to set a fixed threshold and accept only values with correlation measures above this threshold. In practice this threshold is in the range [0.6–0.8]. The second technique relies on a priority queue of a fixed length L (5 in our experiments), which stores N positions of the best matches. In the latter case, however, we can allow areas which are actually false positives. This is not a problem since the second classifier can resolve this difficulty.

As already pointed out, the next matching stage is performed also in the LP domain, however at the finer scale. This time, *only* the interest areas, already selected by the previous matching process, are checked. Thus, the purpose of this run is to verify the already found objects. Nevertheless, this matching has to be preceded by registration of the DB stored template to the finest scale. This is done by an affine warping based on the already found scale s from the tuple $(x_{1b}, x_{2b}, r, \alpha)$ for this area. Additionally we have to adjust for the right scale in the pyramid. The match value is checked and if it is above a threshold (in practice set to 0.9–0.95) the match is accepted, and thus a sign is recognized.

(a)   (b)

Figure 5.17   Recognition results of the road signs. The red segmented image used to guide the first stage of LP matching (a). The found STOP sign (b).

The choice of the matching measure used in the aforementioned searches is very important. Because of the luminance variations, the only possible solutions here are the correlation measures which can compensate for this effect. One of them, also suggested by Zokai *et al.* [31], is the cross-correlation (5.11). However, in our experiments the Census transformation was used since it is less computationally demanding and can cope with intensity variations, as discussed e.g. in [7]. Nevertheless and regardless of the used matching measure, the real computational burden is the search in the four-dimensional space. Therefore red color information from the segmentation module was additionally used to improve computations. That is, matching was done only in the areas which contain at least 10% of red pixels, as shown in Figure 5.17(a). However, since these are relatively sparse, great saving in computations is usually obtained. Figure 5.17(b) shows the found STOP sign. The method allows detection of the internal rotation of the sign, which in this case is $-5°$.

An inevitable problem of object recognition is the decision on how reliable an answer can be. In the case of template matching the only measure of this factor is the value of the match. Even if it is high, i.e. close to its maximal value, there are no guarantees that the right object was detected. This can result in the wrong position of an object, wrong class, or false positives. In other words, an answer on the class and position of an object is based exclusively on match value. Therefore usually a threshold is set which allows the rejection of answers which are below such a threshold. However, in practice determination of this value is always problematic. If its value is too restrictive, then some valid objects can be rejected thus lowering the recall factor. On the other hand, too low a value will increase the ratio of false positives thus lowering precision. In our experiments the threshold was set experimentally to a value ~0.9, whereas 1.0 is a maximum score. Table 5.3 presents experimental measurements of the accuracy of the method.

The method was tested with a database containing images in daylight conditions. Additional tests were made for the same images but with changed scales and rotation. Obtained accuracy results expressed by the *F*-measure (6.25) are presented in Table 5.3. The methods show high accuracy, for distorted patterns too. However, its computation time is rather prohibitive and does not allow real-time processing in serial software implementation. Nevertheless, a speed

Table 5.3 Accuracy of the log-polar scale space method applied to recognition of the road signs.

Conditions of experiments	F-measure
A database with real traffic scenes in daylight conditions	0.97
As above with additional rotation ±5° and scale ±5%	0.95
As above with additional rotation ±15° and scale ±15%	0.94

up of at least an order of magnitude can be expected if this algorithm – especially the modules of log-polar transformation and search in the four-dimensional space – is ported to the parallel platform (e.g. GPU or FPGA), since its operations are highly concurrent.

As already mentioned, the recall parameter depends on the set matching threshold. In our case it was 0.9 which was found to be a fair compromise between recall and precision. Matching with the Census outperforms other measures since it can cope with signal variations and performs the fastest [7].

Based on the same methodology a system was built for automatic examination of implant placements from the maxillary radiograph images [28]. Again, to find rotated and scale changed implants the system carries out template matching in the extended log-polar space. However, in this case matching is done in the anisotropic, rather than in the linear Gaussian, scale-space. As discussed in Section 2.6.2, anisotropic diffusion allows selective filtering which preserves object details, such as its edges. The precise location of an implant is then refined based on the fine level of this space. The two processes are additionally controlled by the contour images which delineate exact positions of implants and other dental objects.

Figure 5.18 shows a radiograph image of a patient (a) and a dental implant. Radiograph images are then used to build the anisotropic scale space which will be used for object matching. Figure 5.19 depicts a morphological gradient of the radiograph image from Figure 5.18(a). Its binary contour map from the morphological gradient is shown in Figure 5.19(b). As already mentioned, LP matching is carried out only around these areas.

Algorithm 5.4 presents detailed steps of the method for detecting tissues around implants in the maxillary images.

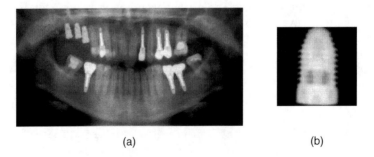

(a)   (b)

Figure 5.18   Radiograph image of a patient (a) and radiograph image of an implant (b) (from [28]).

(a)   (b)

Figure 5.19   A morphological gradient of the radiograph image in Figure 5.18(a). Binary contour map from the morphological gradient which guides the LP matching (b) (from [28]).

1. Construct the anisotropic scale space for both dental and implant images.

2. **For each** coarse level **do**:

   3. Find contours of the high contrast objects.

   4. Create a binary map by thresholding contour areas.

   5. Match implant patterns and the maxillary image in the extended log-polar space only around the areas of contours found in the previous step.

   6. Select and store pixel positions of the found best matches.

7. **For each** fine level **do**:

   8. Scale positions of the best matches from step 6 to fit to the fine level coordinates.

   9. Match implant patterns and the maxillary image in the extended log-polar space, exclusively in the found best match positions from the coarse level.

   10. Save the best matches (above a preset threshold).

11. In each of the matched areas select darker regions of low contrast outside the edge contours which correspond to radiograph of the tissue for examination

Algorithm 5.4   Steps of the method of detecting tissues around implants in the maxillary images.

Figure 5.20 shows the results of implant detection with the presented method. All the places found are further scrutinized since what needs to be examined is the state of a tissue around the implants. The obtained detection accuracy is 97% which is similar to the accuracy obtained for road signs recognition. Further details of this method are described in the paper [28].

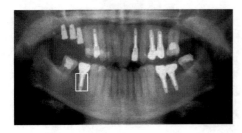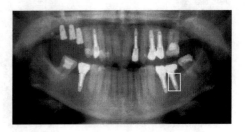

Figure 5.20   Results of dental implant detection in the maxillary images (from [28]).

A similar technique based on LP matching was used for the matching of stereo images, as described in [7,33]. At first corners are found with the structural tensor. Then matching is carried out in the log-polar space around these corners. The internal geometry of a scene is then obtained with the help of the multi-focal tensors, as discussed in Section 2.11. The method also allows detection of differences in internal rotations and scale around the corresponding corners which, in this case, indicates a highly probable false match.

5.5   Recognition from Deformable Models

As many recognition methods, recognition with deformable models is also based on mathematical statistics [17]. It addresses the problem of the variability of parameters in recognition which is due either to imprecise initial models or variations of the measurements (observations), which are additionally affected by such factors as distortions, occlusions, and noise. As pointed out by Amit, the term deformable models, or deformable-template model, can refer to deformable contours, deformable curves, or deformable images [34]. Classifiers built based on deformed versions of the prototype patterns are discussed in the following Sections 5.7 and 5.8.

Finding objects in images can be explained as fitting a curve, or a surface in the 3D case, to a set of characteristic areas in an image. This process is governed by models of edges, texture, 3D vertices, etc., expressed in terms of the energy minimization or dynamic force [35]. The latter was proposed by Terzopoulos as a method of surface evolution [36]. This method is known also as *the active contour* or *snakes* [37]. The appropriate energy functional can be approximated as follows

$$E = \int \left[E_{internal}(\mathbf{d}) + E_{image}(\mathbf{p} + \mathbf{d}) + E_{other}(\mathbf{p} + \mathbf{d}) \right] ds, \qquad (5.23)$$

where $\mathbf{p}(s)$ is the initial point from the set of points which define an active contour, $\mathbf{d}(s)$ denotes a displacement vector along the contour, and s is the arc length. The first component $E_{internal}$ assures continuity and regularity of the contour or surface in the 3D case. It controls stretching and bending of the contour. The second one, E_{image} is responsible for attracting the contour points close to the large gradient variations in an image which correspond to edges. Finally, E_{other} can be used to include other constraints, usually to push the model from the local minima [35].

The idea of classification with deformable images relies on the assumption of a geometrical and topological similarity between a real object and corresponding different variants of its model. Such an approach has an advantage over other methods, like deformable contours or curves, since it takes into consideration topology of the whole image, not just its silhouette. We explore this property in many of the discussed methods.

Let us now consider a certain continuous image space Ω which is usually a unit square or, sometimes, a unit circle. Let a certain prototype image be a function $I_P(\mathbf{x})$, for which $\mathbf{x} \in \Omega$. Let θ denote a smooth deformation Ω into Ω [34]. An image $I_P{}'(\mathbf{x}) = I_P(\theta(\mathbf{x}))$, which is obtained from the prototype after warping driven by θ, has a number of specific properties. If a deformation θ is smooth then the extreme points are preserved, i.e. maxima are mapped onto maxima, minima onto minima, saddle points onto saddles, etc. This way, despite shape changes of some structures of an image, its topological properties are preserved. In the registration task the problem is to find such θ, or its inverse, which when applied to the two-dimensional spatial domain of a prototype image leads to an image which is in some sense the closest to a real image $I(\mathbf{x})$ [38]. This is an optimization problem which can be written as follows

$$\Psi(\theta) = \min_{\theta} \int [I_P(\theta(x)) - I(x)]^2 \, dx, \qquad (5.24)$$

where $\theta(\mathbf{x}) = \mathbf{x} + U(\mathbf{x})$ and U denotes the so called *a pixel displacement field*. However, the solution to the above is not trivial. Even if a transformation is found which best approximates the template model of an object the question remains whether such an object is really observed. This can be reformulated by defining a set of allowable transformations θ for the objects. For example, a human face or hands can deform, but exclusively and in a specific way, and so on. Such knowledge is usually expressed by adding a second term to (5.24) which places some constraints on θ.

In many applications, such as recognition of road signs, θ is constrained to a group of affine transformations $\mathbf{A}$. In the road sign recognition system discussed in the next sections, the input patterns are already registered thanks to the found affine transformation based on the characteristic points. Therefore it is not necessary to directly compute a solution to Equation (5.24). However, for real cases it appears that the found transformation $\mathbf{A}$ is not perfect and some *small* deviations $\Delta \mathbf{A}$ are still possible due to the imperfection of the detection module, discussed in Section 4.4.5. These can be horizontal or vertical shifts up to few pixels in both directions, as well as unknown rotations in the case of the circular shapes. Therefore the simple idea is to generate a small subset of the most probably expected deformations $\Delta \mathbf{A}$ of the prototype templates and then to perform classification independently in each of the groups of deformations. This technique has been realized with the help of the neural networks organized in a group of cooperating expert classifiers, as well as tensor space, as discussed in the next sections.

The other method which was reported to have very good recognition properties is the method of invariant metric and the *tangent distance* proposed by Simard *et al.* [39], which is invariant under the group of affine transformations [40]. The method was tested in the task of handwritten digit recognition. Their approach assumes representation of images as points in the high dimensional space, as discussed in Section 2.3. For example each prototype image of a digit is represented as an $16 \times 16 = 256$ icon. Thus, each image is a point in the $\Re^{256}$ space. Moreover, all of the rotated versions of a prototype pattern are points in this space close to the original one. All of them create a curve in this space, called *the invariance manifold*. Now, the test pattern is classified by measuring the shortest distance between the curves of its rotated version and rotated prototypes. This distance is coined *an invariant metric* [40].

Finally it is worth noticing that the deformable models are sometimes joined to the sparse models. The latter are used to obtain the initial set of points for the former. The recognition tasks are then performed with the help of deformable models [34].

5.6   Ensembles of Classifiers

Frequently answers provided by many persons, or experts, are more reliable than those given by just one person. The answer need not even be provided by a group of experts. If people gathered in a group of supporters have a common point on a certain matter, this has an impact. Such situations are common in daily life. For instance in severe health conditions it is common practice to consult more than one doctor to choose the best treatment procedure. Another illustrative example is the popular TV show in which the players have to answer questions of differing complexity from different domains. Say, at each of the sixteen levels there are four answers from which only one is correct. The player who wants to win the top prize has to give a correct answer to all of them. However, there are some "lifelines" which can be used if an answer is not known. These are the possibility of removing two incorrect answers leaving a 50/50 choice, calling a knowledgeable friend, or asking the audience to poll an answer. These, despite their sociological and medial aspect, have interesting statistical properties as discussed by Policar [41]. Interestingly enough it can be shown that under certain conditions the last "lifeline," i.e. asking the audience for a poll, has the greatest chance of success even if the audience is composed of "ordinary" people, each with potentially the same abilities as the player. Regardless, their cooperative response has different statistical properties, greatly increasing the chances of success. The same idea can be used when combining even weak classifiers to cooperate in an ensemble toward solving a recognition task. This can be shown by considering *majority voting* in which an answer is attributed to the largest group of unanimous answers among experts. Thus, assuming that each classifier has to choose one correct response *true* or *false*, success of ensemble of N classifiers is obtained if half of them plus one give the correct answer, i.e. the integer division $\lfloor N/2 \rfloor + 1$. Assuming that a probability of a correct answer for a single classifier is p and that the votes are given independently, the Bernoulli probability scheme for this problem can be written as

$$P_{ensemble\_success} = \sum_{k=\lfloor \frac{N}{2} \rfloor + 1}^{N} \binom{N}{k} p^k (1-p)^{N-k}. \qquad (5.25)$$

From the above it can be shown that in the extreme case if N tends to infinity, then if only $p > 0.5$, probability $P_{ensemble\_success}$ tends to 1 [42]. This is why audience polling works. In practice it is sufficient to have N sufficiently large and p larger than 0.5 to get the correct answer from the ensemble. These are rather mild requirements since the probability of 0.5 corresponds to blind coin tossing, when in practice we can be more reliable with the classifiers. However, let us remember that this result is for the binary problem. Further analysis of other cases can be found in the book by Kuncheva [42].

Despite the simplest majority voting there are other schemes for combining the classifiers into an ensemble and their responses. The question of how to treat the answers of the expert classifiers and how to draw a single statement from multiple responses is not a trivial one. The problem has been analyzed by many researchers. For instance Kittler *et al.* [43] developed a common theoretical framework for analyzing the operation of combined classifiers. Their results show that the sum rule, which comes with the most restrictive assumptions, outperforms other schemes. On the other hand, Alexandre *et al.* [44] present a comparative study of the arithmetic and geometric means when applied to a combination of classifiers. They showed that the two schemes are equivalent only in the case of a binary classification problem solved with two classifiers which give estimates of the *a posteriori* probabilities that sum to 1, such as the k-nearest-neighbors (Section 3.7.4.2). There are many other schemes discussed e.g. in [42]. However, as shown by many authors there is not a best solution to this problem. This comes as no surprise since it stays in perfect accordance with the *no-free-lunch* theorem [23].

Thus the classifiers can be organized into different configurations, starting from the serial "production line," to a completely independent parallel operations. An example of the former is the system for face detection based on a cascade of weak classifiers, discussed in Section 4.2.2. On the other hand, an example of the latter approach is the presented road sign recognition system, as reported in [45, 46, 47, 48]. One of them is an ensemble consisting of a mixture of experts which are Hamming neural networks (Section 3.9.2), each specializing in object recognition solely in one group of deformable template models (Section 5.4). The second discussed ensemble for road sign recognition explores properties of tensor decomposition (Section 5.8.2).

The second question is how to choose the classifiers constituting an ensemble and how to train them given the, frequently very scarce, set of training patterns. Choice of the individual classifiers is usually dictated by the domain and available solutions. More often than not this is a heuristic and/or experimental process based on human expert knowledge and available solutions for a given domain (e.g. taking into account dimensionality of the classified data, required accuracy, execution speed, etc.). There is a question of how to measure the competence of such a group. An interesting idea is to compare the response of a classifier with random guessing, proposed by Wołoszyński and Kurzyński [49]. As was shown, classification accuracy always increases when a measure of competence is taken into consideration which allows rejection of the weakest classifiers, leaving only competent ones. This was tested on a few different sets of training data and with such response schemes as sum, product, and majority voting.

Regarding the training procedure it should be organized in such a way as to increase the *diversity* of the classifiers given a training set, since diversity increases generalization properties and accuracy of the whole ensemble. In other words, considering real data which are frequently contaminated with noise and outliers, the idea is to make each classifier react as much as possible unanimously when observing known data and at the same time to make them as different as possible when making errors. This certainly takes advantage of the statistical correlation between the true patterns whereas occurrence of outliers and noise usually do not show such correlation (i.e. errors are "more" unpredictable). To achieve the required diversity, each classifier, even if they are of the same type, should be trained with a different set of training examples. For this task different resampling methods have been devised. One of them is *bootstrapping* or *bagging* when the data sets are drawn randomly but with replacements from the original training set. If this drawing has no replacements then the method is called *jackknife* of *k-fold data split*, in which a different subset of the training set is used for each individual classifier [23, 41, 50]. Such an approach was undertaken in the system of handwritten digit recognition with an ensemble of tensor based classifiers, which is discussed in Section 5.8.3.

In many cases another method, called *boosting*, was reported to provide even better results. It consists of training consecutive classifiers with data sets which have caused problems (misclassifications) for the previously trained classifiers, and so on. A generalization of the boosting method was proposed by Freund and Schapire and called the *AdaBoost* method [51].

The methods combine a set of generated hypothesis based on the weighted majority voting. The novelty comes from the iterative update of the training data during generation of the hypothesis. As already mentioned, the main idea when training a series of consecutive classifiers is to provide instances of data which caused misclassification problems for the previously trained classifiers. However the real novelty in the *AdaBoost* is formation of a weight distribution on the training data from which new training subsets are chosen. Obviously, the patterns which cause more misclassifications effectively receive more attention during the training stage, i.e. they are assigned higher weights. By this token, *AdaBoost* iteratively focuses on the more troublesome patterns. In result *AdaBoost* follows the weighted majority voting scheme in which classifiers which exhibit good performance during training are attributed higher voting weights (this is so called AdaBoost.M1 algorithm). It was observed that such a training strategy is very resistant to the overfitting problem, which manifests with a relatively poor performance on test data, whereas training errors remain very small [41].

Finally, Jacobs *et al.* proposed a combination of expert classifiers orchestrated by an external arbitration system controlling weights attributed to each of the experts [52]. This system, called *a mixture of experts*, usually employs *the gating network* which is responsible for determining the distribution of weights assigned to single classifiers of the system. More often than not the gating module is realized as a neural network trained with the same training set by means of the backpropagation (gradient descent) method [53]. As a result the method can be seen as the dynamic selection method of the best classifier(s).

5.7   CASE STUDY – Ensemble of Classifiers for Road Sign Recognition from Deformed Prototypes

A module for automatic road sign recognition can become a part of a larger Driving Assisting System (DAS), discussed in many sections of this book. Its role is to recognize the road signs ahead and inform the driver of a particular sign or, after an analysis of the current situation on the road, information of incoming threads. For instance, if a "Speed limit" sign is recognized by the system of a car moving with a higher speed than allowed, a driver can be alerted to the situation or speed can be adjusted accordingly if a speed control system is active.

The problem for road sign recognition lies mainly in the conditions for image acquisition, such as vibrations, poor visibility, occlusions, weather, dust, etc. Also, the large number of different signs does not facilitate the task. There are the four main and four additional groups of signs,[2] as well as diversity in their real appearance. An example is presented in Figure 5.21, for a single "Road works" warning sign (A-14). After closer observation it is easy to observe different parameters of the signs. Thus, it is not uncommon to have different paint used even in the same group of signs. Some of them are also covered with fluorescent foil which changes their reflectance properties. Last but not least, time and weather conditions change their appearance. For example signs become paler over time.

Although efforts are made to place the signs so they are well visible, frequently they occupy available positions near the road together with commercial banners, for instance. Also, signs are observed in motion, usually with cameras assembled behind the windshield, at different viewpoints, in different lighting and reflection conditions, at day and night time, in different

---

[2]Concerns Polish regulations [15].

Figure 5.21 Variability of the "Road works" warning sign. Color versions of this and subsequent images are available at www.wiley.com/go/cyganekobject.

seasons of a year and in different weather conditions. It is also very difficult to gather a significantly meaningful set of real exemplary prototypes for each sign, which number more than two hundred. Therefore in many of the presented road sign recognition systems classifiers we have made an assumption of using just one dominant prototype for each sign. This is in agreement with our daily experience – when a person is presented with a new symbol, for instance, even in a single handdrawing, he/she has no problem with its recognition in real conditions, even if that sign is greatly deformed. This astonishing ability of the human visual system has been analyzed by many research groups and has resulted in some theories, such as the gestalt [54]. The prototypes for experiments can come e.g. from the formal specification [15], or from images of well visible real exemplars. However, many classification methods cannot cope with this variability to any comparable degree compared to the human visual system. Our aim is to design such systems which will be highly tolerant to the aforementioned variability.

The time available for reaction in automotive applications is very limited. This concerns the signal processing path, from image acquisition, its processing by the computer, up to the time period for the driver's reaction [55]. Figure 5.22 shows a possible scenario of DAS operation in a moving vehicle.

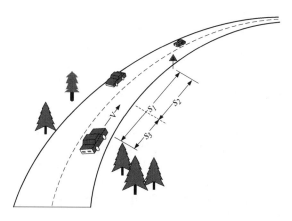

Figure 5.22 Response timings analysis of the object recognition system for Driving Assisting Systems (from [55]).

Table 5.4   Evaluation of the maximum computation time with respect to the distance of an obstacle s_1 and time left for the reaction of the driver which is t_{2min} 2 s or more realistic 4 s (in parenthesis). Based on [57].

s_1 \ v	60 km/h	100 km/h	120 km/h
60 m	1.6 (-)	0.2 (-)	(-)
80 m	2.8 (0.8)	0.9 (-)	0.4 (-)
100 m	4 (2)	**1.6** (-)	1 (-)
120 m	5.2 (3.2)	**2.3** (0.3)	**1.6** (-)

A simple timing analysis allows assessment of requirements on the time response of the system. Using the symbols shown in Figure 5.22 we easily obtain

$$t_3 = (s_1 - s_2)/v, \qquad s_2 = t_{2min} v, \tag{5.26}$$

where v is an average speed of a vehicle, s_1 denotes a distance of a car from an object (a sign, pedestrian, etc.), and s_2 is a distance passed by a vehicle during operation of DAS. The speed is assumed not to change very rapidly during the computation time.

However, the most important factor is the time left for a driver for reaction. This is defined as the minimal allowable period of time from the moment of the system response to the moment of reaching an object, denoted by t_{2min}. From the above we see that the DAS computation time has to fulfill the following condition

$$t_3 = s_1/v - t_{2min}. \tag{5.27}$$

Table 5.4 presents selected values of a time t_3 for system total response with respect to the car speed in order to allow driver's response t_{2min} of 2s and 4s, respectively.

Values in Table 5.4 indicate that with the speed of a car being 100 km/h, even assuming a very precise acquisition system which would be able to detect obstacles from a distance of 120 m or further, the time left for reaction of the system is about 1.6 s, which leaves only 2 s for a driver for reaction. However, in practice the working distance of the acquisition system is much shorter, depending heavily on weather conditions. Thus, the run-time requirements placed on the system are very stringent. This greatly affects design of the vision processing path.

5.7.1   Architecture of the Road Signs Recognition System

Figure 5.23 depicts the architecture of the system which follows the chain of image acquisition, object detection and tracking, up to object classification. These modules can be realized with the different methods discussed in sections of this book.

Figure 5.24 depicts an activity diagram of the system. After image acquisition there are seven paths of detection and classification. In some cases, detection of a particular shape also means recognition of a sign. An example here is the "Yield" sign. However, for other groups, characteristic shapes are first detected, then their pictograms are classified. Detection is underpinned by tracking. The groups of warning, information, obligation, and prohibition signs, after detection of their characteristic shapes, are fed into the classification module. To

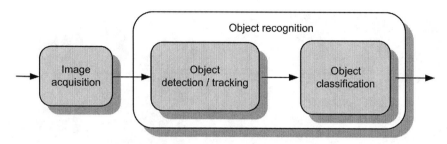

Figure 5.23   Architecture of the system.

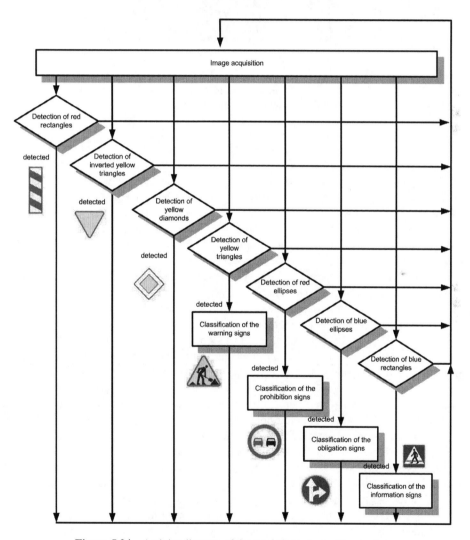

Figure 5.24   Activity diagram of the road signs recognition system.

 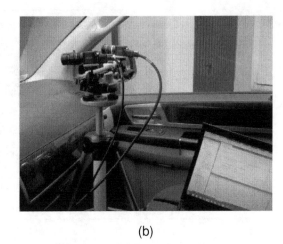

(a)   (b)

Figure 5.25   A laboratory system for road sign recognition. The acquisition module mounted in a car (a). The software processing platform running on a laptop computer (b).

some degree detectors can facilitate the process of sign registration. However, this is more difficult in the case of circular signs since there are no salient points which allow correction of their internal rotation. Thus, the classification module has to be able to correctly recognize pictograms of the signs which are geometrically deformed, as will be discussed. In this version of the system detection modules were employed which were discussed in Sections 4.4.1 to 4.4.4. A complete description of the road sign detection module is contained in Section 4.4.5. Tracking modules were discussed in Section 4.5.

Figure 5.25(a–b) depict the acquisition system assembled in a car. It consists of two color cameras mounted on a tripod. The images are acquired and then processed in the software framework run on the laptop computer visible in Figure 5.25(b). Such a configuration of cameras also allows acquisition of stereo-images.

In the following part of this section we focus on the second module in Figure 5.23, i.e. classification module organized as an ensemble of cooperating classifiers (mixture of experts).

Figure 5.26 depicts the architecture of the presented system for road sign recognition. After acquisition and detection modules which were discussed in Section 4.4.5, there are two classification blocks. One operates in the spatial, the other in the log-polar, domains. As discussed in Section 5.4.3, the log-polar transformation allows recognition of rotated and scale transformed patterns since in the log-polar space both transform into simple horizontal and vertical shifts. Let us now focus upon the structure of the ensemble of classifiers, that's architecture is shown in Figure 5.27. It is composed of a number of the Hamming neural networks, presented in Section 3.9.3. Each of them carries out classification within a single group of signs and their deformable template models, as discussed in Section 5.5. The set of classifiers is controlled by the arbitration unit operating in the winner-takes-all fashion. It is augmented with the boosting mechanism based on a group of unanimous classifiers, discussed in Section 5.7.3. Let us note that we have two such ensemble blocks, as shown in Figure 5.26.

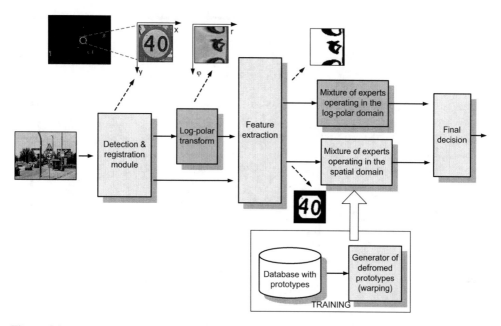

Figure 5.26 Block diagram of a compound system of road signs recognition with two ensembles of expert-classifiers. The first one operates in the log-polar space. The second in the spatial domain (from [48]).

Training of each of the HNN classifiers is based on the formal specification of Polish road signs [15], and follows the method described in Section 3.9.3. This is different compared to other systems which rely heavily on the statistically dominating training sets, such as the probabilistic neural networks (Section 3.9.1). As already mentioned, each of the prototype patterns is slightly deformed to accommodate possible deformations of the observed patterns. In consequence each of the composing HNN classifiers specializes in recognition of a single deformation.

During recognition a binary feature vector is passed from the detection module to each of the HNN member of the ensemble, as discussed in Section 4.4.5. The vector encodes a pictogram of the sign. Each HNN classifier in the ensemble responds with the winning pattern together with its score s_i, obtained in the iterative winner-takes-all process described in Equation (3.178). However, if a real pattern is observed, then many of these classifiers will tend to have the same response. This property is accounted for in the arbitration unit which promotes a group of similar answers among experts. The arbitration unit, discussed in Section 5.7.3, responds with a single answer that represents a sign recognized by the system. However, it is also possible that the reliable selection of a single winner is not possible. Thanks to this, the false positive responses are reduced. Further details of the system are discussed in the following sections. This presentation is based on the publications [45, 46, 47, 48].

As already mentioned, exactly the same structures of ensembles of HNN classifiers are used in the spatial and log-polar domains. Both are trained in such a way as to accommodate small horizontal and vertical shifts of the recognized patterns. However, in the case of log-polar

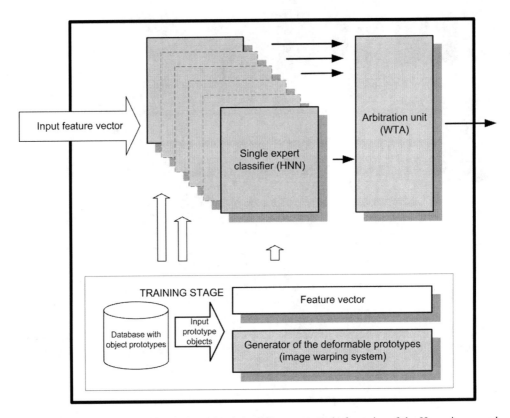

Figure 5.27 Architecture of the ensemble of classifiers composed of a series of the Hamming neural networks. Each of them classifies pictograms within a single group of deformable prototypes. The set of classifiers is controlled by the arbitration unit operating in the winner-takes-all process. In the system shown in Figure 5.26 there are two such ensembles. One operates in spatial, the other in the log-polar, domains.

domain, horizontal and vertical shifts of patterns represent their rotation and change of scale in the spatial domain. The two ensembles of classifiers support each other. That is, two are used to classify the patterns and the role of the arbitration unit is to select a uniform answer.

The real advantage of HNN is their speed of execution despite the iterative process of selecting a winning output neuron. They are also well suited for processing binary patterns which are pictograms in this case. In effect HNN offers a good balance of accuracy and speed – the two paramount factors of automotive systems.

5.7.2 Module for Recognition of Warning Signs

As shown in Figure 5.26, pictograms for classification are supplied from the detection and registration module. After detection, a region with the potential sign is cropped from the image and its dimensions are adjusted by a warping procedure driven by the affine homography, since

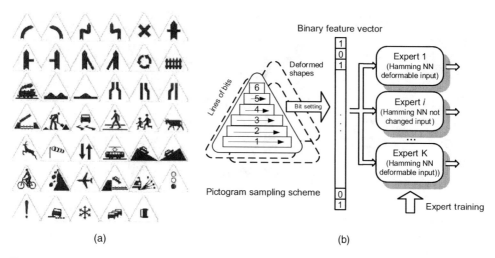

(a)

(b)

Figure 5.28 Pictograms of the Polish warning signs (group "A") from the formal specification (a). Preparation of the feature vector for the group of HNN classifiers (b) (from [45])

signs are planar rigid objects. Practical details of this procedure are described in [7]. Once again, registration is straightforward in the case of triangular and rectangular shapes, since positions of their characteristic points are well known from the detector. This is complicated slightly in the case of circular objects, since only their circumscribed rectangle can be used to adjust the size. Thus, their internal rotation needs to be accounted for in the classifier. Furthermore, due to geometric deformations, noise, weather condition, etc., pictograms of even triangular or rectangular signs are far from being exactly as specified by the formal prototypes. Therefore each expert HNN is responsible for classification within a single group of slightly deformable prototype pictograms. During operation, each expert responds with its proposition of a winner, providing also a value of its response $s_i[t]$ given in Equation (3.178).

Figure 5.28(a) shows pictograms of the Polish warning signs from the formal specification (group "A" in [15]). Preparation of the feature vector for the classifiers presents Figure 5.28(b). A pictogram is first binarized and then sampled depending on the shape of a sign. Figure 5.29 contains three sets of prototype pictograms of the warning signs which were geometrically deformed by horizontal and vertical shifts. Each is then used to train a *single* expert HNN which responds with a best fitting prototype. However, a single HNN expert is trained with all types of pictograms. Thus, an expert specializes in all signs with specific one deformation. All responses are finally resolved by an arbitration module, discussed in Section 5.7.3.

A complete recognition process for a real traffic scene is presented in Figure 5.30. An image containing a warning sign "Right turn" (A-1) and its detected shape are depicted in Figure 5.30(a) and (b), respectively. Detection of the regular-shaped objects follows the method described in Section 4.4. A sign is then cropped from the rest of the image and registered which results with an image in Figure 5.30(c). Finally, the system returns a prototype pattern, visible in Figure 5.30(d), which is the nearest prototype pattern in the sense of the Hamming distance.

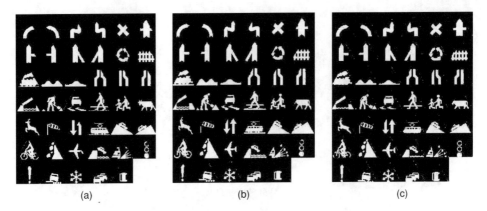

(a)   (b)   (c)

Figure 5.29 Three examples of geometrically deformed pictograms (horizontal and vertical shifts) of the warning signs from the formal specification. Each is used to train a separate expert Hamming neural network (from [45]).

Figure 5.31(a) shows another traffic scene which this time contains information signs belonging group of rectangular shapes. However, their pictograms need to be sampled differently to the triangular shapes shown in Figure 5.28(b). This process is presented in Figure 5.31(b).

The formal specification of the information signs (group "D") is depicted in Figure 5.32(a) [15]. The database created from the specification is visible in Figure 5.32(b). Their pictograms

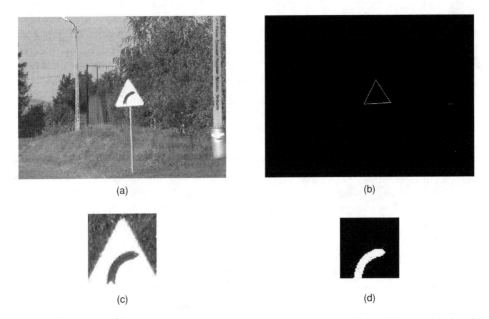

(a)   (b)

(c)   (d)

Figure 5.30 Steps of sign recognition. An image of a real road scene with the "Right turn" sign (a). Detected triangular shape of a sign (b). Cropped and registered region (c). The strongest response from the database (d) (from [45]).

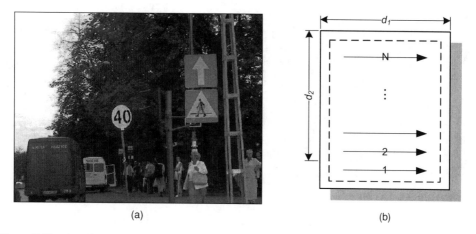

(a)   (b)

Figure 5.31 A real road scene with information signs (a). Sampling scheme for rectangular objects (b) (from [45]).

after binarization and sampling are shown in Figure 5.32(c). From experiments on real scenes we noted that binarization around the mean intensity value gives acceptable results for most of the scenes under different lighting conditions and such a method was used in Figure 5.32(c) [46, 45, 48].

Figure 5.33 and Figure 5.34 depict steps in sign detection for the scene in Figure 5.31(a). Areas that have been found are shown in Figure 5.33(a), as well as in Figure 5.33(b), this time shown together with the original image. It is clear that the detection process is not perfect and slightly larger areas were selected. Apart from this the signs are somewhat left rotated.

Figure 5.34 shows the results of classification of the information signs in Figure 5.31(a). Circular signs are recognized by a different process (Figure 5.24), as will shortly be discussed. First, the test prototypes are selected from the original image (left column in Figure 5.34).

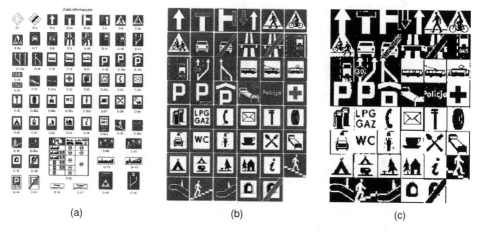

(a)   (b)   (c)

Figure 5.32 The formal specification of the "D" group of the Polish road signs (a). The database created from the specification (b). The pictograms after binarization and sampling (from [45]).

(a)   (b)

Figure 5.33   Detection of the two information signs in a scene from Figure 5.31(a) (from [45]).

Preprocessed pictograms which are fed into the set of HNN are shown in the middle column of Figure 5.34. Finally, responses of the system – in the form of the best fitting prototypes – are shown in the last column of Figure 5.34.

As shown in Figure 5.24, the recognition process follows in different threads in order to detect all possible signs in a scene. Figure 5.35 shows the process of detection of the "Speed limit 40km/h" sign (B-33) in the scene from Figure 5.31(a). Details of the detection module for circular signs were discussed in Section 4.4.3. Objects of interest are searched based on color segmentation (Section 4.2). Places with the found potential locations of objects of interest are depicted in Figure 5.35(a). However, after figure verification, that's intention is to eliminate all objects which do not comply with a model of a circular sign in this case – see Section 4.4.4 – only one verified place is left for further classification (Figure 5.35(b–c)).

Figure 5.36(a) presents a database of prototype patterns, created again from the formal Polish specification of prohibition signs [15]. The system responds with a found pattern from this database – shown in Figure 5.36(b).

Figure 5.34   Response of the system for the two signs visible in the scene in Figure 5.33(b).

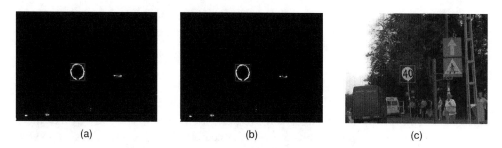

(a)   (b)   (c)

Figure 5.35 Detection of the B-33 sign in the scene from Figure 5.31(a). Places with a potential location of the circular sign (a). Only one place left after figure verification (b). The found sign in the scene (c) (from [45]).

The other parameters of the recognition process, such as match score and found internal rotation of a sign, are also provided since these are used for further validation of the response.

The main problem with classification based on deformable models is the large number of prototype patterns. In consequence, this entails higher memory and computational requirements. For example, considering 50 signs in specification, and allowing horizontal and vertical deformations of ± 8 pixels in steps of 2, results in $9 \cdot 9 \cdot 50 = 4050$ patterns which need to be stored in the ensemble of HNN. However, the presented system with expert HNN, despite a significant number of prototype patterns, exhibits very short training and response times even in the software implementation, thus allowing for real-time response. Depending on the image contents, this is 10–15 classified frames of 640×480 pixels per second in a serial C++ implementation. Thus, there is room for optimization with OpenMP or GPU which would allow a much faster response. The accuracy of the system is also satisfactory although HNN

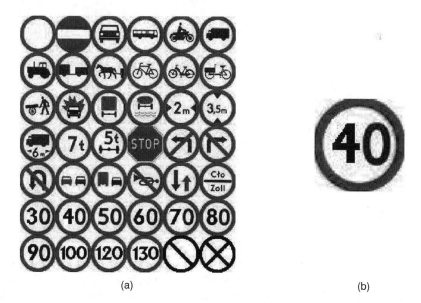

(a)   (b)

Figure 5.36 Formal specification of the prototype patterns (a). A correct response of the system for the scene in Figure 5.31(a) (b) (from [45]).

follows the simple one-nearest-neighbor classification rule, as discussed in Section 3.7.4.2. In our realization this factor is in the order of 91–93% of correct answers for the test signs in daily conditions.

5.7.3   The Arbitration Unit

As alluded to previously in Section 3.9.3, the response of the HNN is obtained by the winner-takes-all (WTA) mechanism selecting the winning neuron. However, a similar situation happens at the output of the ensemble of HNN, as shown in Figure 5.27. Thus, the ensemble requires its own arbitration unit which selects a winner for the whole system. Not surprisingly, for this purpose the same arbitration mechanism WTA was chosen, as shown in Figure 5.27.

For a total number of E experts, inputs to this layer come from the experts $e \in \{1, \dots, E\}$ after reaching their final states with selected winner class w_e, and the final value "strength" of the winning neuron, denoted as a_w. Weights of the arbitration unit for the ensemble are initialized in accordance with the same Equation (3.172) as for a single HNN, setting $C = E$. This randomized scheme appears to be the most effective for initialization, especially for similar output values a_w coming from the experts. Finally, an answer is accepted if and only if a winning neuron has a score greater than a predefined threshold t_H. Choice of its value depends on many factors, such as the size of the pictograms in the database, number of experts, assumed level of acceptance/rejection ratio, etc.

However, to improve the quality of the aforementioned voting mechanism a special strategy was applied to cope with situations when many experts select the same winner. This is more probable for a limited number of experts. Such common voting can be a clue to selecting a dominating class, i.e. a scheme which mimics human experts or a democratic voting system. For this purpose a promotion mechanism is proposed which favors a group of unanimous classifiers – members of the ensemble. At first such a group G of unanimous experts is located (these are experts voting for the same class) which then has the privilege of relaxing their inhibition weights to the neurons belonging to their group. Thanks to this a cooperating group of support emerges. If such a dominating group does not exist then this mechanism is not applied. To use this rule the weights (3.172) of neurons belonging to the group of support are modified as follows

$$m_{kl} = \begin{cases} m_{kl} + z_G & if \ \ m_{kl} < -z_G \\ m_{kl} & otherwise \end{cases}, \quad k \neq l, \{k, l\} \in G, \tag{5.28}$$

where z_G is a positive constant value related to the mean of the cooperating group G, as follows

$$z_G = \frac{q}{N} \sum_{\{k,l\} \in G, k \neq l} |m_{kl}|, \tag{5.29}$$

in which N is a number of pairs $\{k,l\}$ of indices such that $\{k,l\} \in G$, $k \neq l$, q is a constant (0.05–0.3). In other words, since all m_{kl} connected to other neurons are negative, the procedure described by Equation (5.28) makes them less negative, but only with respect to their group of support. In effect, the neurons in this group cooperate, while strongly extinguishing responses of neurons not belonging to that group.

5.8   Recognition Based on Tensor Decompositions

In this section we discuss recognition methods which rely on subspaces built from decomposition into orthogonal bases of the tensors containing prototype patterns. We start with a short theoretical introduction, then follow with its applications and implementation.

5.8.1   Pattern Recognition in SubSpaces Spanned by the HOSVD Decomposition of Pattern Tensors

In Section 2.12.4 we discussed HOSVD decomposition of tensors and showed that HOSVD leads to an orthogonal space spanned by tensor basis, given by Equations (2.217) and (2.218). The obtained tensors $\mathcal{T}_n$ are orthogonal and thanks to this they constitute a basis. Thus, pattern recognition with HOSVD can follow the idea of building an orthogonal subspace representing each prototype pattern and then testing a distance of projections of an unknown object into each of the subspaces. More details on subspace pattern classification is discussed in Section 3.3.2. A road sign recognition system exploiting this idea is discussed in Section 5.8.2.

As alluded to previously, in the case of C different prototypes, each of the corresponding tensors is decomposed into a set of orthogonal tensor bases (Section 2.12.4). During pattern classification a given test pattern $\mathcal{X}$ is projected into each of the spaces spanned by the set of the bases in Equation (2.217). Then, a closest subspace c is searched. This can be stated as solving the following optimization problem [56]

$$\underset{i,w_n^c}{\arg\min} \underbrace{\left\| \mathcal{X} - \sum_{n=1}^{N} w_n^c \mathcal{T}_n^c \right\|^2}_{Q_c}, \tag{5.30}$$

where w_n^c are the coordinates (as yet unknown) of $\mathcal{X}$ in the space spanned by $\mathcal{T}_n^c$, $N \leq N_P$ denotes a number of chosen dominating components; C different classes are assumed, each represented by a tensor base $\mathcal{T}_n^c$, where $1 \leq c \leq C$.

To solve Equation (5.30) let us analyze the squared norm Q_i in Equation (5.30) for a chosen c, as follows

$$Q = \left\| \mathcal{X} - \sum_{n=1}^{N} w_n \mathcal{T}_n \right\|^2 = \left\langle \mathcal{X} - \sum_{n=1}^{N} w_n \mathcal{T}_n, \mathcal{X} - \sum_{n=1}^{N} w_n \mathcal{T}_n \right\rangle \tag{5.31}$$

In order to find a minimum of Equation (5.30) for each w_n, the set of derivatives with respect to each w_n is computed

$$\frac{\partial Q}{\partial w_n} = \frac{\partial}{\partial w_n} \left[\langle \mathcal{X}, \mathcal{X} \rangle - 2 \sum_{n=1}^{N} w_n \langle \mathcal{T}_n, \mathcal{X} \rangle + \sum_{n=1}^{N} w_n^2 \langle \mathcal{T}_n, \mathcal{T}_n \rangle \right]$$
$$= -2 \langle \mathcal{T}_n, \mathcal{X} \rangle + 2 w_n \langle \mathcal{T}_n, \mathcal{T}_n \rangle \tag{5.32}$$

In the equation above we relied on the following properties of the inner product of tensors

$$\langle \mathcal{T}_n, \mathcal{X} \rangle = \langle \mathcal{X}, \mathcal{T}_n \rangle,$$
$$\langle \mathcal{T}_g, \mathcal{T}_n \rangle = 0 \qquad for \ g \neq n.$$
(5.33)

Now equating (5.32) to zero for each n we obtain

$$w_n = \frac{\langle \mathcal{T}_n, \mathcal{X} \rangle}{\langle \mathcal{T}_n, \mathcal{T}_n \rangle},$$
(5.34)

which, due to the fact that the second derivative of Q

$$\frac{\partial^2 Q}{\partial w_n^2} = 2 \langle \mathcal{T}_n, \mathcal{T}_n \rangle = 2 \, \| \mathcal{T}_n \|^2$$

is positive, gives sufficient conditions on a minimum of Equation (5.30). Finally, taking Equation (5.34) into Equation (5.30), for each pattern c, the following residual value is obtained

$$\rho(c) = \left\| \mathcal{X} - \sum_{n=1}^{N} \frac{\langle \mathcal{T}_n^c, \mathcal{X} \rangle}{\langle \mathcal{T}_n^c, \mathcal{T}_n^c \rangle} \mathcal{T}_n^c \right\|^2 = \left\| \mathcal{X} - \sum_{n=1}^{N} \frac{\langle \mathcal{T}_n^c, \mathcal{X} \rangle}{\| \mathcal{T}_n^c \|^2} \mathcal{T}_n^c \right\|^2$$

$$= \left\| \mathcal{X} - \sum_{n=1}^{N} \langle \hat{\mathcal{T}}_n^c, \mathcal{X} \rangle \hat{\mathcal{T}}_n^c \right\|^2,$$
(5.35)

where the *hat* notation relates to the normalized tensors. To make $\rho(c)$ fair to all kinds of input patterns it is next divided by a square norm of $\mathcal{X}$, as in the formula (3.48). The following is obtained

$$\hat{\rho}(c) = \frac{\rho(c)}{\|\mathcal{X}\|^2} = \left\| \hat{\mathcal{X}} - \sum_{n=1}^{N} \langle \hat{\mathcal{T}}_n^c, \hat{\mathcal{X}} \rangle \hat{\mathcal{T}}_n^c \right\|^2$$

$$= \langle \hat{\mathcal{X}}, \hat{\mathcal{X}} \rangle - 2\hat{\mathcal{X}} \sum_{n=1}^{N} \langle \hat{\mathcal{T}}_n^c, \hat{\mathcal{X}} \rangle \hat{\mathcal{T}}_n^c + \left\langle \sum_{n=1}^{N} \langle \hat{\mathcal{T}}_n^c, \hat{\mathcal{X}} \rangle \hat{\mathcal{T}}_n^c, \sum_{n=1}^{N} \langle \hat{\mathcal{T}}_n^c, \hat{\mathcal{X}} \rangle \hat{\mathcal{T}}_n^c \right\rangle \quad (5.36)$$

$$= 1 - 2 \sum_{n=1}^{N} \langle \hat{\mathcal{T}}_n^c, \hat{\mathcal{X}} \rangle^2 + \sum_{n=1}^{N} \langle \hat{\mathcal{T}}_n^c, \hat{\mathcal{X}} \rangle^2 = 1 - \sum_{n=1}^{N} \langle \hat{\mathcal{T}}_n^c, \hat{\mathcal{X}} \rangle^2.$$

Thus, to minimize Equation (5.30) we need to maximize the following value

$$\arg \max_{c} [\hat{\rho}(c)] = \sum_{n=1}^{N} \langle \hat{\mathcal{T}}_n^c, \hat{\mathcal{X}} \rangle^2.$$
(5.37)

In other words, the HOSVD based classifier returns a class c for which the corresponding $\hat{\rho}(c)$ from Equation (5.37) is the largest.

The number N of components in Equation (5.37) depends on the task. It might seem that higher values of N allow better accuracy. However, what was observed in practice is that for different recognition tasks there is a certain optimal value of N. This can be connected with the choice of an optimal number of principal components in the principal component analysis, discussed in Section 3.3.1.

Finally, to also allow reasonable behavior for unknown patterns, for example the ones for which the system was not trained for, a threshold can be set on the minimal allowable value of a winning $\hat{\rho}(c)$ in Equation (5.37), below which an unknown class can be indicated. This is a broader problem in pattern recognition of how to prepare a classifier on unknown patterns and how to set its confidence region.

The classifier for handwritten character recognition based on the described tensor subspace building was proposed by Savas and Eldén [56]. This idea applied to the space of deformable patterns was then used to build a classification stage of the system for road sign recognition [57]. A detailed description of its modules follows in further sections of this book.

5.8.2 CASE STUDY – Road Sign Recognition System Based on Decomposition of Tensors with Deformable Pattern Prototypes

Recognition of objects in images requires their prior definition. This can be done in a number of ways, providing their models in the form of templates, sets, or characteristic points, or a number of their prototypes. The problem is that what we see in images can differ from the object definition expressed with its model. One of the reasons is different observation conditions than those foreseen when defining a model. Therefore, an idea is to foresee how different observations can be in real situations and then artificially generate additional views from the object models. This way we obtain a version of the method of deformable models, as discussed in Section 5.5. What is now interesting to observe is that for images each of the deformable models is a separate view of the same object. Thus, if used together they form a tensor which is 3D for monochrome and 4D for color images, respectively.

The idea of deformable prototypes is exemplified by a pictogram of an obligation road sign. Its slightly rotated versions are shown in Figure 5.37. These are artificially generated views which are expected in a real situation. All of them, when stacked together, form a 3D tensor representing the same pattern, i.e. a road sign. Below we analyze how this tensor can be used in the object recognition process.

The method of building orthogonal tensor spaces and object recognition with pattern projections into these spaces, described in the previous sections, was used for pictogram classification. It is part of the road sign recognition system that's architecture is depicted in Figure 5.38.

Figure 5.37 Deformed versions of an obligation road sign obtained with small rotations and additions of Gaussian noise. These form a 3D tensor which after decomposition spans a space relevant to this sign.

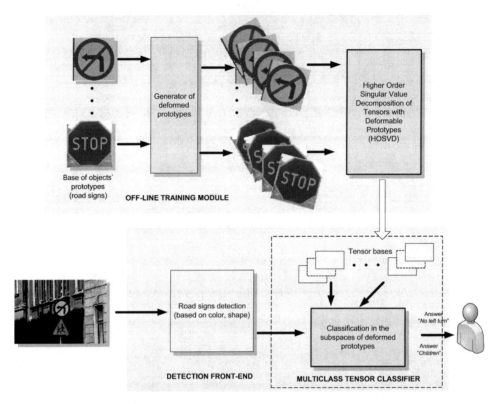

Figure 5.38   Architecture of the system for road sign recognition operating with deformable pattern prototypes. Color version of this image is available at www.wiley.com/go/cyganekobject.

The system consists of two main blocks. The first one (top path in Figure 5.38) is the off-line training module which follows the procedure described in Section 5.8.1. The second block is responsible for object recognition in a real traffic scene. This, in turn is composed of the detection and classification modules (lower path in Figure 5.38). The detection module accepts as input color images and returns rectangular outlines of the compact red objects, as described in Section 4.4.3. Such rectangles are then cropped from the image and their color signal is converted into a monochrome version, simply taking just the blue out of the RGB channels. For red rim signs (the prohibitive and warning signs) this method provides better contrast than channel averaging. Then, the detected rectangle is registered to the size expected by the classification module, with the geometrical warping with bilinear interpolation [7].

Each tensor of deformable prototypes is built from only one prototype extracted from a real traffic scene. It also works fine for patterns taken from the original law regulation containing exemplary road signs [15]. In the same fashion other objects can be provided which will then be classified. Experiments showed that with this method recognition of static hand gestures is also possible [58].

The patterns from the chosen set are then affinely transformed by the prototype generator module (Figure 5.38). In this system because the images are already registered to some

(a)   (b)

(c)   (d)

Figure 5.39 Steps of recognition of the STOP road sign. Real traffic scene (a). A region of a sign detected in the red color segmented map (b). Blue channel used in the recognition stage (c). Correctly recognized STOP sign outlined in the scene (d). (For a color version of this figure, please see the color plate section.)

common size by the detector, it is sufficient to constrain the geometric transformations to small rotations. Other transformations are not necessary since the method has been shown to work fine with some small variations in horizontal/vertical positioning (by few pixels).

Figure 5.39 depicts recognition stages of the signs visible in a real traffic scene, shown in Figure 5.39(a). First, red oval regions are detected based on pixel based color segmentation of the images, as discussed in Section 4.2.3. These are shown in Figure 5.39(b). Such a strategy allows fast operation. However, additional shape detection would improve detection accuracy. Then, potential sign regions are cropped from the blue channel of the color image, as shown in Figure 5.39(c). The inner region of the cropped rectangle is fed to the tensor based classifiers, as presented in Figure 5.38. A correctly recognized STOP sign is outlined, as shown in Figure 5.39(d), and a message is passed from the system.

The upper row of Figure 5.40 shows the first five base tensors $\mathcal{T}_n$ of the STOP road sign obtained after the HOSVD decomposition of the 3D tensor of geometrically deformed versions of this sign. Their corresponding slices of the core tensors $\mathcal{Z}_n$ are visualized in the lower row in Figure 5.40. It is worth noting, that in the slices of the core tensor signal energy tends to concentrate in the upper-left corner, as discussed in Section 2.12.2.

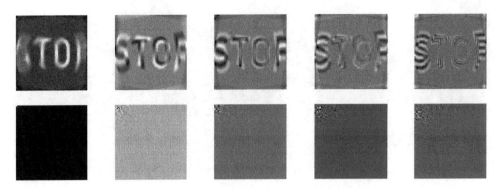

Figure 5.40   First five base tensors T_n of the STOP road sign (upper row) and the corresponding five core tensors Z_n (lower row) obtained after the HOSVD decomposition of the 3D tensor of geometrically deformed versions of the sign pictogram. In the slices of core tensor signal energy tends to concentrate in the upper-left corner.

Figure 5.41 depicts recognition of two prohibition signs from a real traffic scene. In this case, each of the detected oval regions is fed independently to the tensor classifier. Results are outlined as shown in Figure 5.41(d). From this scene it is evident that the detection module is able to outline compact red regions, which after verification of its size, fill aspect, and proportions, is fed to the tensor classifier (Section 4.4.5). However, the pictogram to be recognized can be distorted, as the "No Left Turn" in Figure 5.41.

Stages of recognition in a group of obligation signs is visualized in Figure 5.42. Compared to the red prohibition signs shown in Figure 5.39 and Figure 5.41, in this case the only change is detection in the blue color segmentation map. The methods of segmentation are discussed in Section 4.2.3. The recognition process is carried out with the pictogram cropped from the red channel of the color image. Selection of only this channel for obligation signs provides higher contrast compared to the monochrome version obtained with channel averaging. Correctly recognized "Turn Right" obligation sign is outlined in Figure 5.42(d). Other groups of road signs are recognized with the identical training procedure with only the database of the relevant pictograms changed. However, the main difference lies in the detection which is different for the triangular and rectangular objects. In this case also, a corner based registration allows almost perfect alignment to the reference position which was not possible for circular signs. Therefore, for the triangular and rectangular signs their tensors of deformable versions can contain only few exemplars. It is also possible to use a simpler classifier, such as the Hamming neural network, presented in Section 3.9.3.

Figure 5.43 in the upper row visualizes the base tensors T_n of the "Turn Right" obligation sign, which steps of the recognition process is shown in Figure 5.42. Their corresponding core tensors Z_n obtained after the HOSVD decomposition of the 3D tensor of geometrically deformed versions of the sign pictogram are shown in the lower row of Figure 5.43.

The base patterns, i.e. road sign pictograms in this case, came from the databases of prohibition signs shown in Figure 5.44(a) as well as the complete formal specification already presented Figure 5.36(a). Used databases for obligation signs are depicted in Figure 5.44(b–c). In all of the presented experiments geometric deformations consisted in pattern rotation from $-12°$ up to $+12°$ in steps of $2°$ of the real exemplars. This makes a total of 13 deformed versions for each prototype pictogram, each of size 64×64 pixels. Thus, the prototype tensors

Figure 5.41 Steps in recognition of the two prohibition signs. A traffic scene (a). Regions of separate red circles detected in the color segmented map (b). Blue channel used in the recognition stage (c). Correctly recognized and outlined signs "No Parking" and "No Left Turn" (d). (For a color version of this figure, please see the color plate section.)

are of dimensions $64 \times 64 \times 13$ in this case. Higher values of pictogram resolution usually lead to slightly better accuracy, however at the cost of longer computations. In the experiments this resolution was found as an acceptable trade-off. An interesting conditioning technique was applied in the system by Savas and Eldén which consists of adding some Gaussian noise to the patterns in the tensor [56]. In some cases, this leads to the improvement of the recognition rates. In the presented experiments the Gaussian noise was added at the level of 5–8%. The method of generating such types of noise is discussed in [7]. Finally, the optimal number N of components in Equation (5.37) was found to be in the range of 8–10.

Qualitative measurements were made on a database of a few hundred real traffic scenes taken in good weather. Other conditions, such as rain or night, greatly affect detection and require different front-end processing methods. Figure 5.45 shows a plot of an average accuracy A, as defined in Equation (A.26) p. 493, with respect to the number of components N in Equation (5.37) of the tensor based recognition module only. In other words, accuracy of the detection module is not considered here. It is clear that recognition accuracy reaches 95–96% for the two groups of the circular shape signs. The usual errors occurred for partially visible objects or in views greatly affected by the perspective distortions. Some of the patterns differ only by

(a)   (b)

(c)   (d)

Figure 5.42   Recognition of the obligation sign from a road scene (a). A blue object detected from the color segmented map (b). Red channel used in recognition stage (c). Correctly recognized "Turn Right" obligation sign (d). (For a color version of this figure, please see the color plate section.)

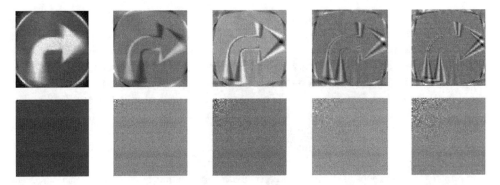

Figure 5.43   First five base tensors $\mathcal{T}_n$ of the "Turn Right" obligation sign (upper row), and its corresponding five core tensors $\mathcal{Z}_n$ (lower row) obtained after the HOSVD decomposition of the 3D tensor of geometrically deformed versions of the sign pictogram.

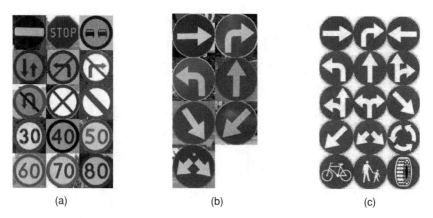

(a)   (b)   (c)

Figure 5.44 The road sign databases used in the presented experiments. A database of the most frequent prohibition signs created from real exemplars (a). A database of most frequent obligation signs from real images (b). A database of obligation signs from the formal specification (c).

a percentage of their area, such as "30," "50," and "80" speed limit signs, for example (see Figure 5.44).

Operation of the presented tensor subspace classifier is very fast since only N inner products in Equation (5.37) need to be computed which are then squared and summed up. Training of the system takes around 8–45 s in the serial C++ software implementation, depending on the size of the database, as well as on the number of chosen deformations. Classification time is in the order of 25 frames of resolution 640×480 per second. The operation can be easily parallelized since the distance parameter $\hat{\rho}(c)$ in Equation (5.37) can be computed independently for each of the classes c. Also, the training of each of the classifiers can be carried out independently. Implementation of the HOSVD based classifiers is discussed in Section 5.8.4. Software is available at the book web page [14].

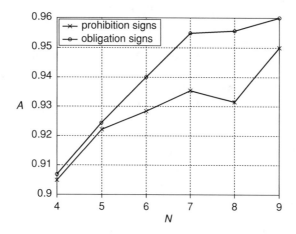

Figure 5.45 Accuracy A in respect to the number N of the components in Equation (5.37) in recognition of the prohibition "x" and obligation "o" signs.

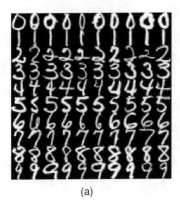 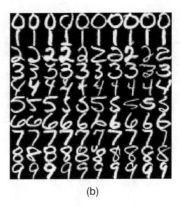

          (a)   (b)

Figure 5.46 Visualization of the two data sets from the ZIP database (from data). Ten exemplars of each digit from the training set (a), and from the testing set (b).

5.8.3 CASE STUDY – Handwritten Digit Recognition with Tensor Decomposition Method

Recognition of handwritten characters belongs to the one of the classical problems of CV. Figure 5.46(a) and Figure 5.46(b) depict respectively ten exemplars of ten digits from the training and test handwritten characters, contained in the "ZIP code" data set [59, 60]. As reported in the note on the web page[3] [59], these were originally collected and preprocessed (deslanted and size normalized) by LeCun in the project launched by the US Postal Service (USPS) on character recognition from handwritten envelopes. Each character is represented as a 16 × 16 grayscale image. There are 7291 training and 2007 test observations in this data set. The images in Figure 5.46 were created from the "ZIP code" data sets. It is worth noticing that this data set is very difficult to classify without error even for humans. In this case the reported human error rate is 2.5% [61].

Another big database of handwritten digits is MNIST, created by LeCun *et al.* [62]. It contains a training set consisting of 60 thousand examples, and a test set with 10 thousand examples. MNIST is a subset of a larger database from NIST. The original scanned digits had been first warped to the 20×20 fixed size, preserving their aspect ratio. Then they have been centered around their center of mass (see Section 2.8), and finally centrally framed into fixed size images of 28 × 28 pixels. Although the digits could be represented by only two black and white values (or 0/1 bits), the images contain gray levels as a result of the anti-aliasing filtering of the warping process.

The paper by LeCun *et al.* presents a discussion of different gradient-based learning techniques applied to the problem of handwritten digit recognition [63]. The authors devised the convolutional neural network [64] and compared its performance with other known methods, such as *k*-nearest-neighbors, RBF neural networks, PCA decomposition (Section 3.3.1), Tangent Distance [39], and SVM with different kernels (Section 3.10), etc. The paper also presents a document recognition system composed of modules for field extraction, segmentation, recognition, and finally language modeling. For a global training of such a compound system the authors of [63] propose the graph transformer network which also relies on a gradient-based

---

[3]This is the web page accompanying the book by Hastie *et al.* [41].

learning aimed at minimizing overall system error. The described system was employed to a real task for an analysis of bank checks.

Maji and Malik present their classifier for handwritten digit recognition using blockwise histograms of locally oriented gradients and a relatively simple SVM classifier with the intersection kernel [65, 66]. Thanks to their choice of more robust features than simple intensity values, as well as to the hierarchical collection of features, their system achieves high accuracy with comparatively simple classifiers. Their system is based on spatial pyramids constructed over various information sources gathered in the input images. The simplest source of information is pure intensity signal. However, as suggested by Maji and Malik, the best results were obtained with the signals from orientation filtering of the images. The original idea was to successfully add responses of increasing blocks of images. For each pixel the orientation and magnitude responses are computed, from which the histograms are created but for overlapping regions of different size. In this respect Maji and Malik use regions 4×4, 7×7, as well as 14×14, overlapping by 50%. The histograms are then weighted appropriately to account for their different support, and finally concatenated to a single histogram. Then the SVM classifier with the intersection kernel, already discussed in Section 3.10.1, allows handwritten character recognition with one of the highest accuracies among the methods reported in literature. This becomes evident especially when compared with the state-of-the art solutions which employ polynomial kernels which require more computations. The reported accuracy of the system by Maji and Malik reaches 99.2% and 96.6% for the MNIST [62] and USPS [59] data sets, respectively.

As alluded to in Section 5.8.1, Savas and Eldén proposed a HOSVD based method for handwritten digit recognition [56]. For each digit from 0 to 9, their training handwritten exemplars are stacked into a 3D tensor which is then decomposed with the HOSVD method and based on the obtained core tensor the tensor subspace is built. In recognition, a test digit is projected onto each of the tensor subspaces. Their method was an inspiration for the road sign recognition module presented in Section 5.8.2.

However, for large data sets the associated pattern tensor is very large. Hence, after an analysis of the HOSVD algorithm performed in Section 2.12.3 we can easily see that this can pose memory problems, as well as a significant computation time, especially during training. This was the reason behind the development of the ensemble of multiclass tensor classifiers that's architecture is shown in Figure 5.47. The idea is to split the initial large data set into a number E of smaller, but computationally tractable, data sets called data partitions. Then, each of the partitions is used to train a single multiclass tensor classifier, of an exactly the same structure as the one used for classification of multiple road signs, as discussed in Section 5.8.2. However, for each tensor classifier a different data partition is used which naturally leads to diversification inside the system [42]. When an unknown pattern is classified, each of the classifiers responds with its own candidate class. In the simplest case majority voting is used to select a single answer from the whole system.

There are at least a few questions associated with the presented method. The first is the choice of a number of classifiers E. The second is how to split the data. The next one concerns the best way to join their responses in the combiner module, as shown in Figure 5.47. For a particular data set the question is also how to preprocess data to achieve the best results. A number of tests were run to answer these questions, as described in the paper [67].

Partitioning of the input data is performed in the random data splitter, shown in Figure 5.47. For this purpose the *bagging* method can be used which consists in creating a number of variants of the training set by a uniform data sampling with replacement (a bootstrap

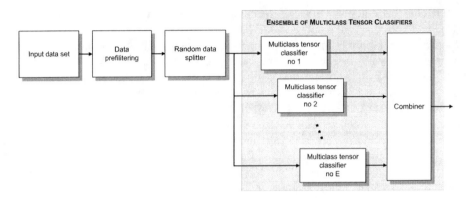

Figure 5.47   Block diagram of an ensemble of tensor based classifiers for recognition of multiclass patterns.

aggregation method) [68]. This strategy helps in reducing variance of a classifier and improves its generalization properties [69]. As alluded to previously, each data partition is used to train a separate member of the ensemble. A number of data in each partition is a parameter that needs to be chosen.

Accuracy can be affected by data preprocessing, which in Figure 5.47 is done in the prefiltering block. In [67] a number of preprocessing filters was tested, such as nonparametric Census transformation, edge detection, as well as image enlargement. In the presented experiments only the latter showed enhanced accuracy, however at the cost of longer computation. Also, the latest experiment shows that application of the IMED distance, discussed in Section 2.5.2, improves accuracy in the order of 1% depending on a data type.

Figure 5.48(a) shows a plot of the accuracy A of the ensemble with respect to the number of classifiers E in the ensemble, as well as for different data partitions obtained in the bagging of the original ZIP database. The highest recognition was achieved for the largest number of data in the partitions. However, what is really important is the steepest ascent of the accuracy curve in response to the increased size of members of the ensemble, starting from a single classifier, i.e. $E = 1$. This tendency indicates that application of even 4 or 5 classifiers of the same structure but with slightly different data sets leads to an over 1% increase of accuracy. The second plot, depicted in Figure 5.48(b), shows dependency of the accuracy A in respect to the number N of components chosen in the tensor classifiers (Equation (5.37)). It can be observed that initially an increase of N results in a rapid increase in accuracy. Then a plateau is achieved which indicates a kind of internal dimensionality of data. A conclusion is that depending on the problem, all the parameters need to be chosen experimentally. However, for a given problem usually an optimal set of these can be found. Also, for each class c, its own best number N_c of components can lead to even better results.

The majority voting strategy implemented in the combiner block in Figure 5.47 outputs simply a label of a class with the highest number of votes in the ensemble. However, a more advanced strategy, such as the weighted majority voting can lead to better results [42].

As the presented results show a great advantage of using an ensemble of tensor based classifiers is that each of them is trained with a less numerous data partition than the whole data set and still the overall accuracy of the ensemble is usually higher than for a single

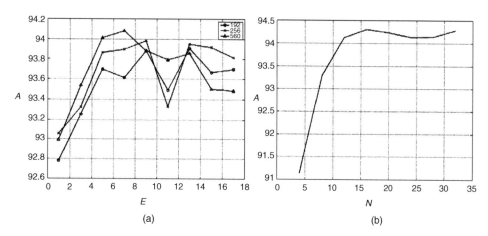

(a)   (b)

Figure 5.48   Plot (a) – accuracy A in respect to the number of classifiers E in the ensemble for 3 bagging partitions (192, 256, 560). Input patterns of size 16×16. Number of components N in Equation (5.37) set to 16. Plot (b) – A in respect to the number of components N. Input patterns warped to size 32×32, number of classifiers $E = 15$, bagging partitions of 192 patterns (after [67]).

classifier. As a side effect, training of each of the member classifiers is simpler and shorter due to the smaller data set. It is also possible to extend the ensemble with new members if new training data become available at a later time.

5.8.4   IMPLEMENTATION of the Tensor Subspace Classifiers

Figure 2.42 shows the class diagram for the *T_HOSVD_SingleClass_Classifier* and *T_HOSVD_MultiClass_Classifier* classes which implement the tensor classifier described in Section 5.8.1. The former is capable of creating tensor bases for a single pattern class which

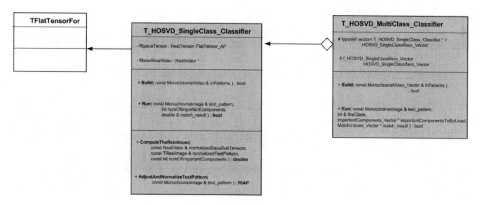

Figure 5.49   Relations and the most important members of the *T_HOSVD_SingleClass_Classifier* and *T_HOSVD_MultiClass_Classifier* classes implementing tensor based classifier for single and multiple pattern classes.

constitutes the tensor subspace. The latter builds tensor subspaces for multiple tensors, each containing exemplars of a single class.

The most important members of the two aforementioned classes are *Build* and *Run*. The first one is used to build a classifier, whereas the second to classify an unknown pattern passed as an argument. It is no surprise that each pattern is represented by the monochrome image object (the *MonochromeImage* class). Therefore the *Build* member in the *T_HOSVD_SingleClass_Classifier* class takes a video of monochrome image objects (*MonochromeVideo*), in which each frame contains consecutive patterns of a single class. These can be any set of patterns that represent a single class, as discussed in Section 5.8.3, or these can be deformed versions of a single pattern (Section 5.8.2). In the *Build* member of the *T_HOSVD_MultiClass_Classifier* class a number of such videos with training patterns is necessary. Therefore in this case *Build* takes as its argument a vector of monochrome images (*MonochromeVideo_Vector*). Although a multiclass object can be seen as an extension of the single-class object it is interesting to observe that the relation between the *T_HOSVD_MultiClass_Classifier* and *T_HOSVD_SingleClass_Classifier* classes is of the "has a" type. Thanks to this all numbers of classes in the multiclass object are treated in the same way (for the "is as" one class would be treated differently from the case of two or more classes).

```
///////////////////////////////////
// HOSVD classifier for a single
// pattern class
///////////////////////////////////
class T_HOSVD_SingleClass_Classifier
{
    private:
            // this is a 3D tensor containing the prototypes,
                each set as a separate slice
            RealTensor::FlatTensor_AP fSpaceTensor;

            RealVideo * fBaseSliceVideo; // contains base slices
                                            for classification
    public:
            ///////////////////////////////////////////////////////////////
            // This function computes the best approximation of the
            // test pattern (which needs to be normalized to sum up
            // to 1.0) in the coordinate space provided by the
            // normalized base subtensors.
            ///////////////////////////////////////////////////////////////
            //
            // INPUT:
            //          normalizedBaseSubTensors - a series of the normalized
            //              bases
            //          normalizedTestPattern - a normalized test pattern
            //          numOfImportantComponents - number of components
            //              taken into consideration (should be less or equal
            //              to the number of bases, i.e. size of
            //              the normalizedBaseSubTensors)
            //
            // OUTPUT:
            //          the residuum (the lower, the closest the test pattern
```

```
//              to the bases)
//

// REMARKS:
//
//
virtual double ComputeTheResiduum(
            const RealVideo & normalizedBaseSubTensors,
            const TRealImage & normalizedTestPattern,
            const int numOfImportantComponents ) const;

/////////////////////////////////////////////////////////////
// This function adjust size of the input pattern to the
// size of the prototypes used in training.
/////////////////////////////////////////////////////////////
//
// INPUT:
//              test_pattern - monochrome test pattern
//
// OUTPUT:
//              adjusted & normalized real test pattern
//
// REMARKS:
//
//
virtual RIAP AdjustAndNormalizeTestPattern(
            const MonochromeImage & test_pattern ) const;
public:
/////////////////////////////////////////////////////////////
// This function builds the classifier from the input
// patterns which are assumed to belong to a SINGLE class.
/////////////////////////////////////////////////////////////
//
// INPUT:
//              inPatterns - a video in which each frame
//                      contains an exemplar of the pattern
//                      of (assuming) the ** same ** class
//
// OUTPUT:
//              true if ok, false otherwise
//
// REMARKS:
//
//
virtual bool Build( const MonochromeVideo & inPatterns );

/////////////////////////////////////////////////////////////
// This function checks a given test pattern with the
// already built classifier, returning the fit measure.
/////////////////////////////////////////////////////////////
//
// INPUT:
//              test_pattern - the monochrome test pattern
//                      which can be in any size
```

```
           //            numOfImportantComponents - number of important
           //                components that will be used to compute
           //                the residuum (the bigger, the bigger accuracy
           //                at a cost of computation time)
           //            match_result - on success contains the match
           //                value
           // OUTPUT:
           //            true if ok, false otherwise
           //
           // REMARKS:
           //
           //
           virtual bool Run( const MonochromeImage & test_pattern,
           int numOfImportantComponents, double & match_result ) const;
};
```

Algorithm 5.5 Definition of the *T_HOSVD_SingleClass_Classifier* (most important members shown).

Algorithm 5.5 presents definitions of the most important members of the *T_HOSVD_SingleClass_Classifier* class. The auxiliary function *ComputeTheResiduum* is responsible for computation of the distance between a pattern and a subspace, given in Equation (5.37). Its parameter *numOfImportantComponents* refers to *N* in (5.37). However, it is called by the *Run* function and rarely outside it. The second auxiliary function *AdjustAndNormalizeTestPattern* adjusts the size of the input pattern to the size of the prototypes in the bases. Such registration is required by this type of classifiers.

Algorithm 5.6 presents important members of the *T_HOSVD_MultiClass_Classifier* class used for training multiple patterns. For this purpose this class contains a number of *T_HOSVD_SingleClass_Classifier* objects, each used to build and then classify its specific pattern class.

```
/////////////////////////////////////
// The HOSVD classifier for
// multiclass patterns
/////////////////////////////////////
class T_HOSVD_MultiClass_Classifier
{
   protected:
        typedef
            vector< T_HOSVD_SingleClass_Classifier* >
                              HOSVD_SingleClassifiers_Vector;

        HOSVD_SingleClassifiers_Vector    f_HOSVD_SingleClassifiers_Vector;

   public:
        typedef vector< MonochromeVideo * >   MonochromeVideo_Vector;
        typedef vector< int >                 ImportantComponents_Vector;
        typedef vector< double >              MatchValues_Vector;
```

```
        public:
                /////////////////////////////////////////////////////////////
                // This function builds the classifier from the input
                // patterns which are assumed to belong to MULTIPLE classes.
                /////////////////////////////////////////////////////////////
                //
                // INPUT:
                //
                //
                // OUTPUT:
                //
                //
                // REMARKS:
                //
                //
                virtual bool Build( const MonochromeVideo_Vector & inPatterns );

                // This adds a single classifier to the collection
                virtual bool IncrementalBuild( const MonochromeVideo & inPatterns );

                /////////////////////////////////////////////////////////////
                // This function checks a given test pattern with the
                // already built classifier, returning the fit measure.
                /////////////////////////////////////////////////////////////
                //
                // INPUT:
                //
                //
                // OUTPUT:
                //
                //
                // REMARKS:
                //
                //
                virtual bool Run(
                        const MonochromeImage & test_pattern, int & theClass,
                        ImportantComponents_Vector * importantComponentsToBeUsed,
                        MatchValues_Vector * match_result ) const;
};
```

Algorithm 5.6 Definition of the *T_HOSVD_MultiClass_Classifier* (shown most important members).

The *IncrementalBuild* member in the *T_HOSVD_MultiClass_Classifier* class allows the addition of a new pattern to the classifier already trained with other pattern classes.

The presented classes can be directly used to build and run tensor subspace classifiers for any pattern, especially those which are two-dimensional (i.e. all types of images). For some applications requiring image recognition, additional preprocessing or transformation of the intensity signal into other domains (e.g. wavelets or log-polar) can be beneficial, as shown in the case of the classifier for handwritten digit recognition (Section 5.8.3).

5.9   Eye Recognition for Driver's State Monitoring

In this section a system for eye and driver's state recognition in a moving vehicle is discussed. The main idea behind this module is to recognize the state of a driver based on observation of his/her eyes. First, eye regions are detected, then a recognition process takes place to tell if they are open and for how long. All this information can be used to monitor a driver's state to predict his/her tiredness or to alert if a driver is falling asleep, since all these situations can lead to a serious accident. This system could become a part of DAS comprising other subsystems too such as the discussed road sign recognition module [70].

Research on driver monitoring systems is active and many similar systems are reported. For instance, Lee *et al.* discuss a forward collision warning system with a real-time gaze estimator based on the driver's head orientation [71]. A system operating in the infra-red spectrum was proposed by Bergasa *et al.* [72]. In this case eye detection relies on a well known effect of white spots in images caused by the IR light reflected by the retina. However, for this purpose the system needs to be equipped with two rings of light emitting diodes (LED). This, in turn, requires central placement of the camera and the rings. Pupils in their system are detected by searching the entire image and locating two bright blobs. Then, the best fitted pair is processed by the Kalman tracker (Section 3.6). The driver's behavior is then analyzed with the state machine.

A system with cooperating Kalman and mean-shift trackers is proposed by Zhu and Ji [73]. Their method joins appearance-based object recognition and tracking with active IR illumination, using the abovementioned effect of high reflectance of pupil. Eye classification is done with the SVM. The method allows robust real-time eye detection and tracking under variable lighting conditions and various orientations of faces. However, the system was not designed to operate in a moving car.

A three stage system is described in the paper by García *et al.* [74]. At first face and eye detection is performed with method by Viola and Jones [75]. Detected objects are tracked with the Kalman filter. Positions of the pupils are then detected and combined with an adaptive light filtering. Finally, the percentage of eye closure parameter (PERCLOSE) is computed which can indicate the driver's state.

D'Orazio *et al.* report a system for driver inattention based on eye recognition from visible spectrum images [76]. Iris regions are found with the Hough transform operating on edge pixels. Then eye candidates are detected through analyzing symmetrical regions. These are next validated with the back-propagation neural network trained with the wavelet signals of the off-line collected eye and non-eye prototype patterns. The driver's conditions is then inferred by the probabilistic model, built upon the multivariate Gaussian mixture, that characterizes the normal behavior of a driver. For this purpose the percentage of time the eyes are closed and eye closure frequency are used. Parameters of the Gaussian mixture were found with the expectation minimization method. As reported, their system attained 95% detection rate and operates in near real-time on a 320×240 video from the webcamera mounted in a car.

The mentioned PERCLOSE parameter is frequently used to assess driver fatigue. Senaratne *et al.* compare two methodologies for its measurement [77]. The first is based on the classification approach. In this approach eyes are detected and tracked with a trained classifier indicating if the eyes are closed or open. In the second approach an optical flow is employed which allows upper eyelid detection from the proposed eye model. The experiments show that the classification method is more accurate but requires a number of training patterns which are

 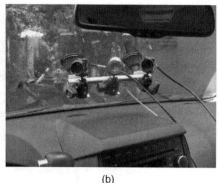

(a)   (b)

Figure 5.50   A car equipped with the visual system for monitoring of a driver (a). View of the cameras and illumination setup connected to the on-board computer. From left – NIR camera, NIR LED illumination module, color camera (b). Color versions of these and subsequent images are available at www.wiley.com/go/cyganekobject.

not always available or representative to a large number of drivers. In such cases, the optical flow method is more practical. The paper by Senaratne *et al.* contains a number of useful links to other systems for driver monitoring.

An eye movement monitoring system was proposed by Lenskiy *et al.* [78]. Eye detection is performed in the skin segmented images obtained with a neural network and segmentation algorithm which uses SURF facial features. Detected features are then used to estimate the probability density function for eyes, eyebrows, lips, and nose classification. Iris candidates are then found with the circular Hough transform. Their final position is validated from correlation with a predefined mask. Then the Kalman filter is used for tracking, which allows estimation of the duration of the phases of eye closure as well as blinking frequency. These factors are used to assess the driver's state. The accuracy of the system is reported to be 96%, as well as its near real-time performance in MatLab® implementation.

Figure 5.50 shows our system for driver monitoring mounted in a car. Requirements placed upon the system assume its real-time operation in all lighting conditions expected in a car. For this purpose a special camera setup has been designed which contains two cameras, one operating in visible, and the other in the near infra-red (NIR) spectra. The camera setup is augmented by the NIR lighting module composed of only one LED diode that complies with the allowable power level [79].

Because of the assumed conditions the main operating camera is NIR, whereas the color camera can carry out auxiliary processing in daylight conditions, e.g. to avoid NIR exposure to a driver. Figure 5.51 shows a block diagram of the NIR processing path for driver's eye recognition. Detection is based on the human eye model shown in Figure 5.52. However, we do not rely on the aforementioned effect of a bright spot of the reflected NIR rays from the retina. Instead, based on the histogram analysis, the camera is set to operate in a little overexposure. Thanks to this skin areas are bright, whereas the pupil area remains relatively dark. This greatly simplifies detection.

As shown in the diagram in Figure 5.51, NIR frames are morphologically eroded to remove noise. Then the integral image is constructed which allows fast computation of a sum of pixels in any rectangular region of the image. Implementation of this technique is described e.g. in [7].

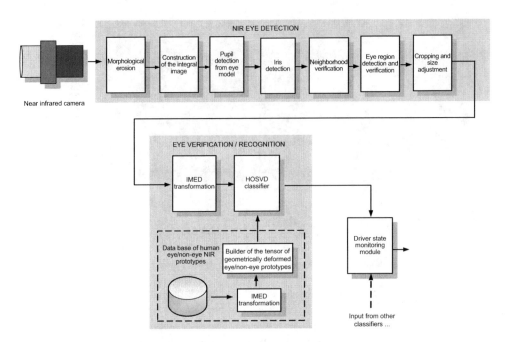

Figure 5.51 Block diagram of the system for eye recognition from the NIR images. The system consists of the detection and recognition modules. Detection is based on the human eye model. Detected areas are fed to the tensor based classifier responsible for the final decision. Results are passed to the driver state monitoring block which carries out analysis of the time eyes are closed and open.

The aforementioned assumptions of eye detection in the NIR images were put into a detection method that's steps are shown in Algorithm 5.7. The method can be characterized as a set of filters which sift out all areas which do not comply with the rules defining the human eye model. At first each rectangular region is checked to see if its average intensity is below a preset threshold, as described in Equation (5.38). Then only those regions that fulfill Equation (5.38) are checked to fit into the simplified iris model shown in Figure 5.52(b). This is achieved by consecutive checking of the conditions (5.39) and (5.40). Only if all these

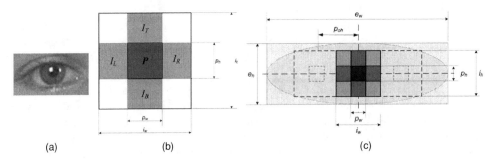

Figure 5.52 Human eye (a). Iris (b), and eye model (c) for eye detection in the NIR spectrum (based on [80]).

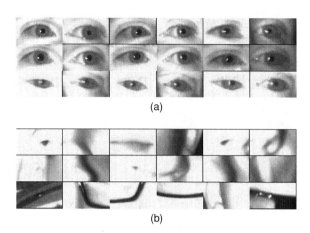

(a)

(b)

Figure 5.53 A training database containing eye (a) and not-eye (b) patterns in the NIR spectrum. These are used to train the second classifier (HOSVD) in the chain shown in Figure 5.51.

rules are fulfilled for a given region is it put into the queue of length L, ordered by values of the average pupil intensity s_P, computed in Equation (5.38). The reason for using this data structure is to select up to L rectangles which fit best to the eye model in Figure 5.52(c). In the next step the overlapping regions are removed, leaving only the best fitted ones.

Areas detected in the previous step are then cropped, adjusted to the predefined dimension by warping, and fed to the tensor based classifier responsible for final decision, as shown in Figure 5.51. This is the same version of the HOSVD based ensemble of classifiers which is also used for road sign detection. HOSVD decomposition is discussed in Section 2.12.2, whereas a system of classifiers based on HOSVD is described in Section 5.8. HOSVD based classifiers are trained with eye and non-eye prototype patterns shown in Figure 5.53. Additionally, to improve accuracy, the patterns are processed with the IMED transformation, as shown in Figure 5.51. IMED is discussed in Section 2.5.2.

Parameters listed in Table 5.5 define the basic relations of dimensions in the eye model. However, the input parameters are pupil horizontal and vertical dimensions p_w and p_h, respectively, whereas all others are related to these two.

Table 5.6 contains definitions of the parameters that are used in Algorithm 5.7. These are mostly thresholds and ratios of dimensions and intensity values. These were determined

Table 5.5 Parameters that define eye dimensions in the model shown in Figure 5.52(c) (from [80]).

Parameter	Description	Values
p_w	Pupil width	11–13 pixels
p_h	Pupil height	11–13 pixels
i_w	Iris width	$[0.85–1.1] * p_w$
i_h	Iris height	$[0.85–1.1] * p_h$
e_w	Eye width	$[6.9–9.7] * p_w$
e_h	Eye height	$[79.1–4.7] * p_h$
p_{oh}	Horizontal pupil distance	$2 * p_w$

Table 5.6 Eye detection parameters (from [80]).

Parameter	Description	Values
t_p	Threshold on a maximal average intensity in a pupil area	128
R_{ip}	Ratios of average intensity in iris area to average intensity in pupil area	[1.5, 2.75]
$r_{wip} = i_w / p_w$	Ratio of iris to pupil width	1.9 ($\pm$10%)
$r_{hip} = i_h / p_h$	Ratio of iris to pupil height	1.8 ($\pm$15%)
r_{LR}	Ratio of average intensity in iris region I_L to I_R	0.2
r_{BT}	Ratio of average intensity in iris region I_T to I_B	0.3
r_{TL}	Ratio of average intensity in iris region I_L to I_T	0.25
r_{BR}	Ratio of average intensity in iris region I_R to I_B	0.25

experimentally based on the NIR test sequences. The parameters should be adjusted accordingly if changing the settings of the image acquisition module.

---

1. Erode NIR frame with the structural element of size $m_w \times m_h$.

2. Compute the integral image.

3. **For** each pixel **do**:

    4. **If** average intensity in P below t_p, i.e., $s_P = S_{av}(P) < t_p$:   (5.38)

      5. **If** relation of average intensities is in the range, i.e. $\dfrac{S_{av}(I)}{S_{av}(P)} \in R_{ip}$:   (5.39)

        6. **If** the mutual ratios of average intensities in

        the iris regions I_T, I_B, I_L, I_R, i.e. $\dfrac{|S_{av}(I_a) - S_{av}(I_b)|}{|S_{av}(I_a) + S_{av}(I_b)|} < r_{av}$,   (5.40)

        are below the thresholds $r_{LR}, r_{BT}, r_{TL}, r_{BR}$ for all a, b $\in \{T, B, L, R\}$:

          7. Insert the region to the queue of length L
            ordered by the values of s_P in (5.38).

8. Remove overlapping eye regions leaving the ones with lower value of s_P.

9. Compute eye regions extending the found pupil-iris rectangle to dimensions e_w, e_h.

Algorithm 5.7 Steps of the eye detection method from the near infra-red images (based on [80]). $S_{av}(P)$ denotes average intensity in a window P.

The last group, shown in Table 5.7, consists of other control parameters, such as a chosen length of the queue of the eye region candidates, as well as parameters of the input filter.

Figure 5.54 shows examples of eye detection in a test sequence. Figure 5.54(a) depicts all pupil candidates computed in steps 1–7 of Algorithm 5.7. The detected places after removing

Table 5.7 Parameters that control eye detection (from [80]).

Parameter	Description	Values
L	Max number of pupil candidates (length of the ordered queue)	150
m_w	Width of the structural element of the erosion filter	3
m_h	Height of the structural element of the erosion filter	3

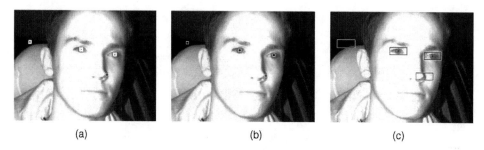

(a)   (b)   (c)

Figure 5.54 Examples of eye detection in a test sequence. Image (a) contains all pupil candidates computed in steps 1–7 of the detection algorithm. Detected places after removing overlapped regions (b). Candidate eye regions which are fed to the HOSVD eye recognition module (c) (from [80]).

overlapped regions are shown in Figure 5.54(b). Finally, candidate eye regions which are fed to the HOSVD eye recognition module are shown in Figure 5.54(c). Some of them are false positives. These usually are rejected by the HOSVD classifier.

Figure 5.55 shows the results of eye recognition of the test sequence in Figure 5.54. Eye regions that passed the HOSVD classifier are shown with black rectangles.

The achieved average true positive and false positive ratios are TP = 97% and FP = 0.3%, respectively (Table A.1). Let us observe that in this case keeping the FP ratio at the lowest level is obligatory, since false recognitions can make the system fail to react when a driver falls asleep, for instance. Eye recognition results are passed to the driver state monitoring block

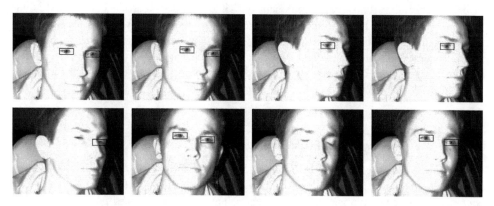

Figure 5.55 Results of eye recognition in a test sequence. Eye regions that passed the HOSVD classifier are shown in black (from [80]).

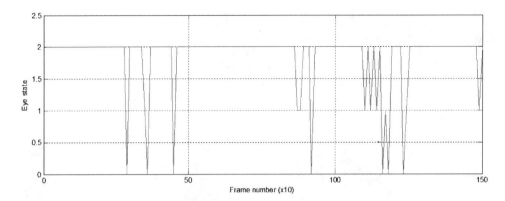

Figure 5.56   PERCLOSE plot of a test sequence. Periods of open eyes encoded with level 2. Partially open eyes or conditions with only single eye visible are marked at 1. Closed eyes periods are at level 0 (from [80]).

which carries out analysis of the time periods in which eyes are closed and open. These, in the form of the PERCLOSE plot are shown in Figure 5.56. Periods of open eyes are encoded as level 2. Partially open eyes or conditions with only single eye visible are marked at 1. Finally, periods with eyes closed are encoded with level 0.

As alluded to previously, the described eye recognition module from NIR images can be augmented with the analogous system operating in the visible spectrum. However, it requires slightly different formulation of the eye model since in this case detection is based on skin segmentation, as described in [80].

5.10   Object Category Recognition

In this section we briefly discuss recent approaches to the general problem of object categorization for large databases of images, as well as the related problems of image categorization, visual search engines, and scene comprehension. These are very broad subjects that have found reflection in dozens of recent conference and journal papers. Therefore an interested reader is referred to the publications indicated here, as well as those cited in their further references.

The two most distinctive methods are part-based and bag-of-words, which have already been mentioned in previous sections, such as the one on pedestrian detection (Section 4.7). These are addressed in the following sections.

5.10.1   Part-Based Object Recognition

Object detection and recognition by finding their specific parts or characteristic areas belongs to one of the fundamental methods of computer vision and pattern recognition. This is strictly connected with template matching, object segmentation, as well as key point detectors, etc. The main idea behind this group of methods is detection of characteristic parts of objects, as well as their mutual relations. Therefore the methods offer good models of real objects, such as human silhouettes, moving cars, etc.

Object detection using the statistics of parts was proposed by Schneiderman and Kanade [81]. As well those known, objects in images can be of any size, location, pose, as well as with distortions, noise, etc. To cope with these variations the method by Schneiderman and Kanade is based on a number of classifiers, each operating at different range of local orientation. Each of the classifiers scans the whole image and answers if an object of interest is present at a given image region. However, the interesting part of this method is that each classifier is based on the statistics of localized parts of objects. Each part, in turn, is obtained from the wavelet coefficients of the image. Thus, the classifiers are trained using the AdaBoost approach with wavelet features coming from objects of interest, as well as non-object images. During detection the classifiers look for the parts and, if found, assess their conditional probabilities. Then, the likelihood ratio is computed in steps. At each step, a classifier decides whether to continue the process or to break if the checked part belongs to the nonobjects. Additionally, the whole process is performed in the scale-space of the images, starting from the coarse level. The method was tested for detection of faces and cars of any size, location, and pose.

Fergus *et al.* propose a similar method of learning object categories with a series of weak classifiers [82]. Models of each category are compositions of specific parts whose parameters are obtained by maximizing the probability of the training data. It is interesting to note that the proposed models comprise not only the appearance of the parts but also their mutual positions and relative scale. During classification, at first areas in a test image are sifted out to leave only the promising locations. Then, in the next stage, the category model is attempted to be fitted to the input data. The method is reported to provide good results in six diverse categories of objects with rigid but also flexible objects, such as animals.

A very interesting semi-supervised approach is proposed by Nguyen *et al.* which presents a feedback-like method of learning weak classifiers for supervised discriminative localization and classification [83]. The method tries to circumvent the problem of learning classifiers from large data sets of human annotated objects, due to difficulties in their creation. The method by Nguyen *et al.* trains the classifiers from patterns annotated with just binary labels, i.e. object vs. nonobject, not requiring their location, however. During training the method simultaneously localizes the regions of interest in the training patterns, as well as learns a region-based SVM classifier. In effect robustness to background and uninformative signal is achieved. The method was evaluated on recognition of time series patterns, which is carried out by finding the most discriminative set of temporal segments in the input signal. Experimental results show high accuracy despite the mild requirements imposed on annotations of the training data set.

A part-based approach is undertaken in many of the methods for pedestrian detection, as discussed in Section 4.7. For instance, an implicit shape model approach was proposed by Leibe *et al.* [84]. The key points and their local neighborhoods are computed. Then, they are clustered to construct a codebook. From this codebook an implicit shape model is built, which is then used for object recognition.

5.10.2   Recognition with Bag-of-Visual-Words

A method for recognition of categories of objects called bag-of-keypoints was originally proposed by Csurka *et al.* [85]. The main idea consists of finding keypoints – visual words –

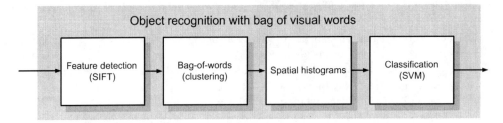

Figure 5.57   Common stages of object categorization with bags of visual words.

which are cluster centers of the affine invariant descriptors of image patches, such as SIFT, etc. For classification the SVM proves to provide the best results. The idea is depicted in Figure 5.57.

Distribution of visual words in a query image is compared to distributions of words found in the training images. This idea was then extended by many authors. See the paper by Zhang *et al.* for an in-depth discussion of this technique [86].

Lazebnik *et al.* propose a method for scene categorization with the help of the global geometric correspondence [87]. At first an image is clustered into increasingly fine subregions. Then, in each of the regions, a histogram of local features is computed. In effect, a spatial pyramid is obtained in which each region of an image is associated with its specific histogram of local features. The method by Lazebnik *et al.* extends the common bag of features strategy, as was shown experimentally on the Caltech-101 benchmark database. The extended version of this database, Caltech-256, was analyzed by Griffin *et al.* in their method for object categorization [88]. This extends the Caltech-101 data set, for each category there are at least 80 images, images have internal rotations removed, and there is a new clutter category for testing background rejection abilities of the methods. Griffin *et al.* suggest a number of testing paradigms to measure classification performance. The data set is benchmarked with two metrics as well as a state-of-the-art spatial pyramid matching method .

Table 5.8 presents information and references to the common data sets used for evaluation of various methods of object recognition and categorization, as well as scene analysis and understanding.

Everingham *et al.* introduced the PASCAL database as a benchmark for visual object detection and recognition tasks [89]. In their paper, apart from a description of the evaluation

Table 5.8   Common data sets for object categorization.

1	Caltech-101 [90]	Collected examples from Google Images, manually selected to the appropriate categories
2	Caltech-256 [88, 91]	Extended version of Caltech-101, contains at least 80 images in each category, removed internal rotations, introduced clutter category for background rejection
3	PASCAL [89, 92]	A benchmark database for visual object detection and recognition tasks
4	LabelMe [93, 94]	A large and still growing database for image annotation by volunteers
5	Flickr [95]	Huge image and video hosting website for online and mobile community

procedure, Everingham *et al.* review other existing databases, as well as state-of-the-art methods (as for the year 2009) with their statistical properties. The paper is highly recommended reading on the subject as well as a valuable source of hints on future directions which briefly are as follows:

- More object classes – the variability of images and objects puts ever growing requirements on statistically meaningful collections of objects, which can be easily done with Internet search tools, but most of their annotations have to be done with the help of human operators;
- Object parts – annotation of object parts allows the testing of more precise or specialized methods, as well as supporting methods which rely on interaction with objects (such as robots, etc.);
- Beyond nouns – available data sets focus on the annotation of specific objects, i.e. nouns; however, more advanced scene analysis leads to action discoveries, which would be useful in many vision systems such as surveillance, etc.
- Scene dynamics – suggest using video instead of still images; these provide new types of information which can be used to develop and test diverse methods for action recognition, learning rich 3D, multiaspect object models, etc.
- Alternative annotation methods – this is an important problem since database annotation requires a great deal of work of many people; however, tedious work often leads to inaccuracies, errors, etc. With help come tools for ground-truth data collection, see Section 4.2.1, as well as the Internet community such as in the case of the discussed LabelMe project.
- The future is bright – indeed, agreeing with Everingham *et al.*, we live in times of incredible technological revolution which frequently make possible applications which were unthinkable a year ago; this is what makes computer vision so fascinating but also important.

Russel *et al.* present their system LabelMe which is a large and still growing database and web-based tool for image annotation by volunteers [93]. Such a huge and diversified object categories database allows new types of experiments on supervised learning and quantitative evaluation of the new algorithms. Russel *et al.* propose an extension to that data set, called WordNet, which allows automatic enhancement of object labels, discovery of object parts, recovery of depth ordering of objects in a scene, with minimal need for human intervention. The tool is available at the Internet address after user registration [94]. Interestingly, such huge databases can be very useful for development of other methods, such as human skin detection, etc.

Conversely, Flickr is a website with a huge repository of pictures and videos uploaded by millions of members. The database is frequently used to test methods of object detection, recognition, as well as scene categorization and comprehension, taking advantage of the statistical diversity of the contained there images [95].

As we saw, most methods for object categorization require training on large sets of training images. However, Fei-Fei *et al.* investigated a way of learning significant information on object categories from very limited, possibly just one, exemplar prototype [90]. Rather than learning from scratch, the key idea is to get usage of information coming from previously learned categories. In other words, the categories already learned by the system provide information that helps in learning new categories with fewer training examples. Fei-Fei *et al.* use this concept to build a Bayesian framework in which object categories are represented with probabilistic models. Prior knowledge is represented as a PDF on the parameters of these models. On the other hand, the posterior probability for an object category is computed by updating the prior based on incoming observations. As shown, the proposed Bayesian framework leads to informative object models for databases with a very limited number of exemplars and outperforms other approaches which build models with the help of the ML and MAP schemes (Section 3.4). In their (optimistic) conclusion the authors indicate the important role of best representation of prior knowledge. This aspect can be analyzed from different points of view, such as the statistical, or as a design methodology exploiting expert knowledge. The latter was undertaken in many road sign recognition methods presented in this book when only single prototype exemplars are assumed following the idea of human generalization ability.

Some further details of the method by Fei-Fei *et al.* can be of interest here. In their work features are found in the gray images in their areas characteristic of high saliency of location and scale. A number of features is then reduced as a result of clustering of the most salient regions over location and scale simultaneously. Then, this way found regions are cropped from the image and warped to the patches, each of dimensions 11×11 pixels. These 121-dimensional features are further reduced with the PCA with basis previously calculated from the background data sets. In the experimental setup Fei-Fei *et al.* start from the four object categories which are human faces, motorbikes, airplanes, as well as cats. Then, the initial set is combined with other categories which names were generated with help of the Webster Collegiate Dictionary and which image examples were found with the Google image search. In result, each category of the foreground objects contains from 45 up to 400 images. However, there are also objects of the background clutter category, which were obtained entering "things" keyword to the Google search. The experiments conducted by Fei-Fei *et al.* on images from 101 categories show that only few training examples can be sufficient to build models that are able to achieve high detection rate in order of 70–95%.

As already discussed in Section 2.12, Liu *et al.* present a method for nonparametric scene parsing via label transfer [96]. For each of the input images, their nearest neighbors from the annotated databases are first located. Then the dense correspondences are computed with the SIFT flow. This aligns the images on their local structures. Finally, the test image is segmented with the Markov random field method using annotations of the nearest found images from the databases.

Shotton *et al.* present a novel approach to efficient learning of a discriminative model for classification and understanding of real photographs [97]. Their method incorporates texture, spatial, as well as context information. Classification and feature selection rely on a shared boosting method trained on a large database of images.

As alluded to previously, the subject is very broad and dynamically changing. A systematic introduction can be found in the book by Szeliski [98], as well as recent conference and journal papers.

5.11 Closure

5.11.1 Chapter Summary

This chapter was devoted to a group of object recognition methods. These are strictly connected with detection methods discussed in previous chapter, since in some cases detection also denotes recognition. This somewhat arbitrary division serves better understanding of the main ideas behind each method, as well as their fields of applications. Also a suitable background is provided by the tensor and classification methods described in Chapters 2 and 3, respectively.

We started with the method of histograms of oriented structures coming from the structural tensor. Its connection with the morphological scale-space makes it more robust. The method stems from a common ground of methods based on local gradients, such as the well known image descriptors SIFT, HOG, etc. However, the main difference is that objects are encoded globally rather than in a sparse fashion.

Invariant based recognition methods rely on specific object definition which allows their recognition despite variations of appearance in images. Affine moment invariants are here proposed to cooperate with the probabilistic neural network applied to road sign recognition.

Template based methods were discussed next. In this group the log-polar transformed objects are matched in the Gaussian scale-space. Thanks to the properties of this transformation, scale changed and rotated objects can be reliably detected. However, computation requirements are significant since the 4D template search is required. Thus, the method is improved by a guided search for areas in an image which are likely to contain the object of interest. The method was tested in road sign recognition, as well as in finding dental images in radiograph maxillary images.

Consecutive methods exploit the domain of deformable models, as well as ensembles of classifiers. These show superior results compared to traditional realizations. In this respect two systems for road sign recognition were discussed. The first operates with an ensemble of the Hamming neural networks. The second relies on tensor representation and decomposition of tensors composed of deformed prototypes. The latter approach is tested in the context of road sign recognition, as well as in recognition of handwritten digits. The method shows its superior accuracy and response time. C++ implementation of the tensor based method is also provided. The method is also very flexible and can be used for different groups of objects of interest. In many cases it is sufficient only to change the definition and re-train the classifier.

Eye recognition for driver fatigue monitoring was then discussed. This method relies on NIR images and follows the cascade architecture of classifiers. The last in the chain is the aforementioned tensor based classifier.

The chapter concluded with discussion on groups of methods for general object categorization, as well as scene analysis. The first one – called part based recognition – relies on finding specific parts, and frequently their mutual locations, in objects of interest which are then used to train the classifiers. The second group of methods perform object recognition with a bag of visual words. These are simply clustered sets of distinctive sparse features detected in the training data set. The sets of annotated data sets for evaluation of object categorization methods are also provided.

5.11.2 Further Reading

Recent advances in vision based object recognition can be traced in prominent conference and journal publications. In this respect we can mention the highly recommended *IEEE Transactions on Pattern Analysis, Image* Processing, and Neural Networks, *International Journal of Computer Vision, Pattern Recognition*, to name just a few. They are just click away on the Internet.

From the bookshelf category an overview of the majority of the main CV tasks, including object recognition, is presented in the book by Szeliski [98]. However, in respect of explanations of the most important pattern classification methods, recommended reading are the books by Duda *et al.* [23], as well as the one by Theodoridis and Koutroumbas [68] and Hastie *et al.* [40].

On the other hand, the book by Müller *et al.* [99] provides information on image indexing, information retrieval, and object detection. A graphical search method for images by picture-finder is presented in the paper by Hermes *et al.* [100].

One of the main issues of CV and neurophysiology is to push our knowledge further and find a strategy for understanding how our brains accomplish the task of object recognition. Research achievements in this area can inspire development of better computer algorithms. However, in general computers operate under different principles than our brains which can suggest different ways of development. For instance, the statistical approach of learning parts of local patterns of objects which are then used to train a cascade of classifiers with AdaBoost allows high accuracy and speed in recognition of specific objects. An example here is the system developed by Viola and Jones for human face detection [75]. Another issue is relating what it means to see as a human and as a computer. A broad spectrum of these and other issues related to object recognition by computers and by the human brain are addressed in the very unique book edited by Dickinson *et al.* [101].

Computer vision systems (CVS) are addressed in many publications. Here, discussion can be split into hardware and algorithmic parts. The former concerns construction of the acquisition system, type of cameras, connectors, then processing units. Conversely, the algorithmic part is mostly what we are dealing with in this book. In industry practice both parts need to be designed together to fit requirements. For instance different aspect of CVS for inspection in food quality control are discussed in the book edited by Sun [102]. Chellappa *et al.* discuss CVS for recognition of humans and their activities [103]. The classical book on machine vision is the one by Davies [104]. It provides an explanation of basic methods with some algorithms. Some techniques of robot vision systems can be accessed in the book by Florczyk [105]. Kisačanin *et al.* discuss many aspects of embedded computer vision [106].

However the most challenging part is the attempt to build a form of artificial intelligence into computer vision systems. The goal is to make the machine understand what it sees, infer new information based on built in inferring rules and acquired information, or even make it able to learn new things or new inference rules which have not been directly provided to them. Such cognitive systems could find broad applications, such as in the mission strategic systems, Mars voyager, etc. or serve as intelligent support for experts, medical image processing, and so forth [99, 107].

Problems and Exercises

1. Apply a histogram intersection method to compare histograms discussed in Section 5.2.2.
2. Check the method of affine moment invariant for recognition of selected icons. How discriminative are the moments?
3. Implement a method for finding keypoints in images. Then use the Hausdorff measure (5.19) to compare the images. Measure performance of the method.
4. Check the Hausdorff measure for comparison of partially occluded images.
5. Analyze the possible ways of speeding up computation of the log-polar transform discussed in Section 5.4.3.

References

[1] Freeman W.T., Roth M.: Orientation Histograms for Hand Gesture Recognition. Mitsubishi Electric Research Laboratories, TR-94–03a, 1994.

[2] Lowe D.: Distinctive Image Features from Scale-Invariant Keypoints. International Journal of Computer Vision, Vol. 60, No. 2, pp. 91–110, 2004.

[3] Dalal N., Triggs B.: Histograms of Oriented Gradients for Human Detection. IEEE Computer Society Conference on Computer Vision and Pattern Recognition, Vol. 1, pp. 886–893, 2005.

[4] Cyganek B.: Recognition of Solid Objects in Images Invariant to Conformal Transformations. International Conference on Computer Recognition Systems CORES'09, Advances in Soft Computing, Vol. 57, Springer, pp. 247–255, 2009.

[5] Cyganek B.: A Real-Time Vision System for Traffic Signs Recognition Invariant to Translation, Rotation and Scale. ACIVS 2008, France, Lecture Notes in Computer Science, Vol. 5259, Springer, pp. 278–289, 2008.

[6] Freeman W.T., Tanaka K., Ohta J., Kyuma K.: Computer vision for computer games. IEEE 2nd Int. Conf. on Automatic Face and Gesture Recognition, 1996.

[7] Cyganek B., Siebert J.P.: An Introduction to 3D Computer Vision Techniques and Algorithms, Wiley, 2009.

[8] Mundy J.L., Zissermann A.: Geometric Invariance in Computer Vision. MIT Press, 1992.

[9] Hartley R.I., Zisserman A.: Multiple View Geometry in Computer Vision. 2nd edition, Cambridge University Press, 2003.

[10] Cyganek B.: Circular Road Signs Recognition with Affine Moment Invariants and the Probabilistic Neural Classifier. Lecture Notes in Computer Science, Vol. 4432, Springer, pp. 508–516, 2007.

[11] Flusser J., Suk T.: Pattern recognition by affine moments invariants. Pattern Recognition, Vol. 26, No. 1, pp. 167–174, 1993.

[12] Flusser J., Suk T.: Affine moments invariants: A new tool for character recognition. Pattern Recognition Letters, Vol. 15, No. 4, pp. 433–436, 1994.

[13] Hu, M.K.: Visual pattern recognition by moment invariants. IRE Transactions on Information Theory, No. 8, pp. 179–187, 1962.

[14] http://www.wiley.com/go/cyganekobject

[15] Polish Road Signs and Signalization. Directive of the Polish Ministry of Infrastructure (in Polish), Dz. U. Nr 170, poz. 1393, 2002.

[16] Brunelli R.: Template Matching Techniques in Computer Vision. Theory and Practice. Wiley, 2009.

[17] Grenander U., Miller M.: Pattern Theory. From Representation to Inference. Oxford University Press, 2007.

[18] Forsyth D.A., Ponce J.: Computer Vision: A Modern Approach, Prentice Hall, 2003.

[19] Papoulis A.: Probability, Random Variables, and Stochastic Processes. Third Edition, McGraw-Hill, 1991.

[20] Paclík P., Novovičová J., Duin R.P.W.: Building road sign classifiers using a trainable similarity measure. IEEE Transactions on Intelligent Transportation Systems, Vol. 7, No. 3, pp. 309–321, 2006.

[21] https://developer.nvidia.com/

[22] Plaza A., Plaza J., Paz A., Sánchez S.: Parrallel Hyperspectral Image and Signal Processing. IEEE Signal Processing Magazine, Vol. 28, No. 3, pp. 119–126, 2011.

[23] Duda R.O., Hart P.E., Stork D.G.: Pattern Classification. Wiley, 2001.

[24] Duin R.P.W., de Ridder D., Tax D.M.J.: Experiments with object based discriminant functions: a featureless approach to pattern recognition. Pattern Recognition Letters, Vol. 18 No. 11–13, pp. 1159–1166, 1997.

[25] Gavrila D.M.: Multi-feature Hierarchical Template Matching Using Distance Transforms. Proc. of the Int. Conf. on Pattern Recognition, Brisbane, pp. 439–444, 1998.

[26] Marchand-Maillet S., Sharaiha Y.M.: Binary Digital Image Processing. A Discrete Approach. Academic Press, 2000.

[27] Cyganek B.: Road Signs Recognition by the Scale-Space Template Matching in the Log-Polar Domain. Lecture Notes in Computer Science, Vol. 4477, Springer, pp. 330–337, 2007.

[28] Cyganek B., Malisz P.: Dental Implant Examination Based on the Log-Polar Matching of the Maxillary Radiograph Images in the Anisotropic Scale Space. IEEE Engineering in Medicine and Biology Conference, EMBC 2010, Buenos Aires, Argentina, pp. 3093–3096, 2010.

[29] Wandell B.A.: Foundations of Vision. Sinauer Associates Publishers Inc. 1995.

[30] Kara L.B., and Stahovich T.F.: An image-based, trainable symbol recognizer for hand-drawn sketches. Computers & Graphics, Vol. 29, No. 4, pp. 501–517, 2005.

[31] Zokai S., Wolberg G.: Image Registration Using Log-Polar Mappings for Recovery of Large-Scale Similarity and Projective Transformations. IEEE Transactions on Image Processing, Vol. 14, No. 10, pp. 1422-1434, October 2005.

[32] Jähne B.: Digital Image Processing. 6th edition, Springer-Verlag, 2005.

[33] Cyganek B.: An Algorithm for Computation of the Scene Geometry by the Log-Polar Area Matching Around Salient Points. Lecture Notes in Computer Science, Vol. 4910, Springer, pp. 222–233, 2008.

[34] Amit, Y.: 2D Object Detection and Recognition, MIT Press, 2002.

[35] Metaxas D.N., Chen T.: Deformable Models, in Insight into Images [109], pp. 219–235, 2004.

[36] Terzopoulos D.: Regularization of Inverse Visual Problems Involving Discontinuities. IEEE Transactions on Pattern Analysis and Machine Intelligence, Vol. 8, No. 4, pp. 413–424, 1986.

[37] Kass M., Witkin A., Terzopoulos D.: Snakes: Active Contour Models. International. Journal of Computer Vision, Vol.1, No. 4, pp. 321–331, 1987.

[38] Goshtasby A.A.: 2-D and 3-D image registration. Wiley Interscience, 2005.

[39] Simard D., LeCun Y, Denker J.: Efficient pattern recognition using a new transformation distance. Advances in Neural Information Processing Systems, Morgan Kaufman, pp. 50–58, 1993.

[40] Hastie T., Tibshirani R., Friedman J.: The Elements of Statistical Learning. Springer, 2nd edition, 2009.

[41] Polikar R.: Ensemble Based Systems in Decision Making. IEEE Circuits and Systems Magazine, Vol. 6, No. 3, pp. 21–45, 2006.

[42] Kuncheva L.I.: Combining Pattern Classifiers. Methods and Algorithms. Wiley Interscience, 2005.

[43] Kittler J., Hatef M., Duin R.P.W, Matas J.: On Combining Classifiers. IEEE Transactions on Pattern Analysis and Machine Intelligence, Vol. 20, No. 3, pp. 226–239, 1998.

[44] Alexandre L.A., Campilho A.C., Kamel M.: On combining classifiers using sum and product rules. Pattern Recognition Letters, Vol. 22, No. 12, pp. 1283–1289, 2001.

[45] Cyganek B.: Recognition of Road Signs with Mixture of Neural Networks and Arbitration Modules. Lecture Notes in Computer Science, Vol. 3973, Springer, pp. 52–57, 2006.

[46] Cyganek B.: Committee Machine for Road-Signs Classification. Lecture Notes in Artificial Intelligence, Vol. 4029, Springer, pp. 583–592, 2006.

[47] Cyganek B.: Rotation Invariant Recognition of Road Signs with Ensemble of 1-NN Neural Classifiers. Lecture Notes in Computer Science, Vol. 4132, Springer, pp. 558–567, 2006.

[48] Cyganek B.: Circular Road Signs Recognition with Soft Classifiers. Integrated Computer-Aided Engineering, IOS Press, Vol. 14, No. 4, pp. 323–343, 2007.

[49] Wołoszyński T., Kurzyński M.: On a New Measure of Classifier Competence in the Feature Space, in Computer Recognition Systems 3, Springer, pp. 285–292, 2009.

[50] Bishop C.M.: Pattern Recognition and Machine Learning. Springer 2006.

[51] Freund Y., Schapire R.E.: Decision-theoretic generalization of on-line learning and an application to boosting. Journal of Computer and System Sciences, Vol. 55, No. 1, pp. 119–139, 1997.

[52] Jacobs R.A., Jordan M.I., Nowlan S.J., Hinton G.E.: Adaptive mixtures of local experts. Neural Computation, Vol. 3, No. 1, pp. 79–87, 1991.

[53] Haykin S.: Neural Networks and Learning Machines. 3rd Edition, Pearson Education, 2009.

[54] Desolneux A., Moisan L., Morel J-M.: From Gestalt Theory to Image Analysis. Springer, 2008.

[55] Cyganek B.: Road-Signs Recognition System for Intelligent Vehicles. Lecture Notes in Computer Science, Vol. 4931, Springer, pp. 219–233, 2008.

[56] Savas B., Eldén L.: Handwritten digit classification using higher order singular value decomposition. Pattern Recognition, Vol. 40, No. 3, pp. 993–1003, 2007.

[57] Cyganek B.: An Analysis of the Road Signs Classification Based on the Higher-Order Singular Value Decomposition of the Deformable Pattern Tensors. Springer, Lecture Notes in Computer Science, Vol. 6475, pp. 191–202, 2010.

[58] Cyganek B.: Software Framework for Efficient Tensor Representation and Decompositions for Pattern Recognition in Computer Vision. Advances in Soft Computing Series "Image Processing & Communications Challenges II", Springer-Verlag, pp. 185–192, 2010.

[59] http://www-stat.stanford.edu/~tibs/ElemStatLearn

[60] Hull J.: A database for handwritten text recognition research. IEEE Transactions on Pattern Analysis and Machine Intelligence, Vol. 16, No. 5, pp. 550–554, 1994.

[61] Bromley J., Sackinger E.: Neural-network and k-nearest-neighbor-classifiers. Technical Report 11359-910819-16TM, ATT, 1991.

[62] http://yann.lecun.com/exdb/mnist/

[63] LeCun Y., Bottou L., Bengio Y., Haffner P.: Gradient-Based Learning Applied to Document Recognition. Proc. IEEE on Speech & Image Processing. Vol. 86, No. 11, pp. 2278–2324, 1998.

[64] LeCun Y, Bengio Y. Convolutional Networks for Images, Speech, and Time-Series in The Handbook of Brain Theory and Neural Networks. MIT Press, 1995.

[65] Maji S., Malik J.: Fast and Accurate Digit Classification. Technical Report No. UCB/EECS-2009-159, University of California at Berkeley, Electrical Engineering and Computer Sciences (http://www.eecs.berkeley.edu/Pubs/TechRpts/2009/EECS-2009-159.html). November 25, 2009.

[66] Maji S., Berg A., Malik J.: Classification Using Intersection Kernel SVMs Is Efficient. IEEE Conference on Computer Vision and Pattern Recognition, pp. 1-8, 2008.

[67] Cyganek B.: Ensemble of Tensor Classifiers Based on the Higher-Order Singular Value Decomposition. Springer, Lecturer Notes in Computer Science, Vol. 7209, pp. 578–589, 2012.

[68] Theodoridis S., Koutroumbas K.: Pattern Recognition. Fourth Edition. Academic Press, 2009.

[69] Grandvalet Y.: Bagging equalizes influence. Machine Learning, Vol. 55, No. 3, pp. 251–270, 2004.

[70] Dong Y., Hu Z., Uchimura K., Murayama N.: Driver Inattention Monitoring System for Intelligent Vehicles: A Review. IEEE Transactions on Intelligent Transportation Systems, Vol. 12, No. 2, pp. 596–614, 2011.

[71] Lee S.J., Jo J., Jung H.G., Park K.R., Kim J.: Real-Time Gaze Estimator Based on Driver's Head Orientation for Forward Collision Warning System. IEEE Transactions on Intelligent Transportation Systems, Vol. 12, No. 1, pp. 254–267, 2011.

[72] Bergasa L. M., Nuevo J., Sotelo M. A., Barea R., Lopez E.: Visual Monitoring of Driver Inattention, In D. Prokhorov (Ed.): Computational Intelligence in Automotive Applications, SCI 132, pp. 25–51, 2008.

[73] Zhu, Z., Ji, Q.: Robust real-time eye detection and tracking under variable lighting conditions and various face orientations. Computer Vision and Image Understanding, Vol. 98, No. 1, pp. 124–154, 2005.

[74] García I., Bronte S., Bergasa L. M., Almazán J., Yebes J.: Vision-based Drowsiness Detector for Real Driving Conditions. 2012 Intelligent Vehicles Symposium. Alcalá de Henares, Spain, 2012.

[75] Viola P., Jones M.: Robust real-time face detection, International Journal of Computer Vision, Vol. 57, No. 2, pp. 137–154, 2004.

[76] Orazio De, T., Leo, M., Guaragnella, C., Distante, A.: A visual approach for driver inattention detection. Pattern Recognition, Vol. 40, No. 8, pp. 2341–2355, 2007.

[77] Senaratne R., Jap B., Lal S., Hsu A., Halgamuge S., Fischer P.: Comparing two video-based techniques for driver fatigue detection: classification versus optical flow approach. Machine Vision and Applications, Vol. 22, No. 4, pp. 597–618, 2011.

[78] Lenskiy A.A., Lee J-S.: Driver's Eye Blinking Detection Using Novel Color and Texture Segmentation Algorithms. International Journal of Control, Automation, and Systems, Vol. 10, No. 2, pp. 317–327, 2012.

[79] Agilent Technologies Inc.: Compliance of Infrared Communication Products to IEC 825-1 and CENELEC EN 60825-1, Application Note No. 1118, 1999.

[80] Cyganek B., Gruszczyński S.: Hybrid Computer Vision System for Drivers' Eye Recognition and Fatigue Monitoring. Accepted to Neurocomputing, 2012.

[81] Schneiderman H., Kanade T.: Object Detection Using the Statistics of Parts. International Journal of Computer Vision, Vol. 56, No. 3, pp. 151–177, 2004.

[82] Fergus R., Perona P, Zisserman A.: Weakly supervised scale-invariant learning of models for visual recognition. International Journal of Computer Vision, Vol. 71, No. 3, pp. 273–303, 2007.

[83] Nguyen M.H., Torresani L., de la Torre L., Rother C.: Weakly supervised discriminative localization and classification: a joint learning process. IEEE International Conference on Computer Vision ICCV 2009: pp. 1925–1932, 2009.

[84] Leibe B., Leonardis A., Schiele B.: Robust Object Detection with Interleaved Categorization and Segmentation. International Journal of Computer Vision, Vol. 77, No. 1-3, pp. 259–289, 2008.

[85] Csurka G., Dance C.R., Fan L., Willamowski J., Bray C.: Visual categorization with bags of keypoints. Workshop on Statistical Learning in Computer Vision, European Conference on Computer Vision, pp. 1–22, 2004.

[86] Zhang J., Marszałek M., Lazebnik S., Schmid C.: Local Features and Kernels for Classification of Texture and Object Categories: A Comprehensive Study. Vol. 73, No. 2, pp. 213–238, 2007.

[87] Lazebnik S., Schmid C., Ponce J.: Beyond bag of features: Spatial pyramid matching for recognizing natural scene categories. IEEE Conference on Computer Vision and Pattern Recognition, pp. 2169–2176, 2006.

[88] Griffin G., Holub A., Perona P.: Caltech-256 Object Category Data Set. Technical Report 7694, California Inst. of Technology, 2007.

[89] Everingham M., Van Gool L., Williams C.K.I., Winn J., Zisserman A.: The PASCAL Visual Object Classes (VOC) Challenge. International Journal of Computer Vision, Vol. 88, No. 2, pp. 303–338, 2010.

[90] Fei-Fei L., Fergus R., Perona P.: One-Shot Learning of Object Categories. IEEE Transactions on Pattern Analysis and Machine Intelligence, Vol. 28, No. 4, pp. 594–611, 2006.

[91] http://www.vision.caltech.edu/Image_Datasets/Caltech256/

[92] http://pascallin.ecs.soton.ac.uk/challenges/VOC/

[93] Russel B.C., Torralba A., Murphy K.P., Freeman W.T.: LabelMe: A database and web-based tool for image annotation. International Journal of Computer Vision, Vol. 77, No. 1-3, pp. 157–173, 2008.

[94] http://new-labelme.csail.mit.edu/Release3.0/index.php?message=1

[95] . Flickr, 2011. http://www.flickr.com

[96] Liu C., Yuen J., Torralba A.: Nonparametric Scene Parsing via Label Transfer. IEEE Transactions on Pattern Analysis and Machine Intelligence, Vol. 33, No. 12, pp. 2368–2382, 2011.

[97] Shotton J., Winn J., Rother C., Criminisi A.: Textonboost for image understanding: Multi-class object recognition and segmentation by jointly modeling appearance, shape and context. International Journal of Computer Vision, Vol. 81, No. 1, pp. 2–23, 2009.

[98] Szeliski R.: Computer Vision. Algorithms and Applications. Springer, 2011.

[99] Müller, H., Clough, P., Deselaers, T., Caputo, B.: ImageCLEF. Experimental Evaluation in Visual Information Retrieval. Springer, 2010.

[100] Hermes, T., Miene, A., Herzog, O.: Graphical Search for Images by Picture-Finder. Multimedia Tools and Applications. Special Issue on Multimedia Retrieval Algorithmics, Vol. 27, No. 2, pp. 229–250, 2005.

[101] Dickinson S.J., Leonardis A., Schiele B., J. Tarr M.J. (Editors): Object Categorization. Computer and Human Vision Perspectives, Cambridge University Press, 2009.

[102] Sun D-W.: Computer Vision Technology for Food Quality Evaluation. Academic Press, 2008.

[103] Chellappa R., Roy-Chowdhury A.K., Zhou S.K.: Recognition of Humans and their Activities Using Video. Morgan & Claypool Publishers, 2005.

[104] Davies E.R.: Machine Vision. Theory, Algorithms, Practicalities. Morgan Kaufmann; 3rd edition, 2004.

[105] Florczyk S.: Robot Vision. Video-based Indoor Exploration with Autonomous and Mobile Robots. Wiley, 2005.

[106] Kisačanin B., Bhattacharyya S.S., Chai S.: Embedded Computer Vision, Springer, 2009.

[107] Tadeusiewicz R, Ogiela M. Medical Image Understanding Technology: Artificial Intelligence and Soft-Computing for Image Understanding. Springer, 2004.

[108] Yoo T.S.: Insight into Images: Principles and Practice for Segmentation, Registration, and Image Analysis. A.K. Peters, 2004.

A

Appendix

A.1 Abstract

This chapter contains short discussions on selected problems which can be useful for understanding or applying the methods and techniques discussed in previous chapters of this book.
We start with two sections devoted to morphological methods. The first one presents the
morphological scale-space which shows some nice properties and in some applications can
be more helpful than the well known Gaussian or Laplacian scale-spaces. Then we make a
brief introduction to the domain of morphological tensor operators. The main problem with
morphological processing is definition of the order relation, as well as definition of maximum
and minimum elements. Many works were devoted to this problem in the context of nonscalar
objects. Next, geometry of quadratic forms is outlined which was used for instance in the
CamShift method. Then the problem of testing classifiers is discussed. The main goal is to
gather different approaches in this respect, as well as to provide a concise list of the most
popular classifier performance parameters and measures.

The rest of this chapter contains a short introduction to the OpenMP library which facilitates
transformation of the serial code into a concurrent one. With this technology code accelerations
of 2–3 times are easily achievable. In the next section some useful MATLAB® functions for
matrix and tensor processing are summarized. Finally, a short guide to the attached software
is provided.

A.2 Morphological Scale-Space

Image processing at different scales deals with such a representation of an image which allows
detection of features with certain level of details. Relatively well known are the linear Gaussian
or Laplacian scale-spaces. In these a series of linearly filtered images is created which changes
the spectral properties of an image. Each image can then be subsampled at a rate which
depends on the size of the detected features. In effect a pyramid of images is obtained starting
from the largest image with fine features up to the smallest image with coarse features. One
of the most important assumptions in this process is causality. This means that each feature
in a coarse scale has a corresponding feature at a fine level. In other words features are not

---

generated by the scale-space algorithm. In Section (5.2) the technique of phase histograms is proposed which is augmented with the morphological scale-space that complies with the mentioned assumptions of causality and is nonlinear at the same time. In this section we give a brief overview of the morphological scale-space theory based on the paper by Jackway and Deriche [1].

Assuming that $f(\mathbf{x})$ is a continuous function representing an image and $s(\mathbf{y})$ is a function called a structuring element then the following equations define respectively morphological dilation and erosion, as follows

$$d(f, s) = \sup_{\mathbf{y} \in S} [f(\mathbf{x} + \mathbf{y}) + s(\mathbf{y})], \qquad (A.1)$$

$$e(f, s) = \inf_{\mathbf{y} \in S^*} [f(\mathbf{x} + \mathbf{y}) - s^*(\mathbf{y})], \qquad (A.2)$$

where S and S^* denote the support for the structural elements $s(\mathbf{y})$ and $s^*(\mathbf{y})$, respectively. By definition, $s^*(\mathbf{y}) = s(-\mathbf{y})$ for all $\mathbf{y} \in S^*$ [2].

As pointed out by Jackway and Deriche the structuring element has to also comply with additional conditions to avoid unwanted shifts in the features in the processed image. These are fulfilled by the following zero-shift structural element

$$\sup_{\mathbf{y} \in S} [s(\mathbf{y})] = 0 \ and \ s(0) = 0 \qquad (A.3)$$

Usually the structural element is taken to be a sphere of radius σ

$$s(\mathbf{x}, \sigma) = |\sigma| \left(\sqrt{1 - \left\| \tfrac{\mathbf{x}}{\sigma} \right\|^2} - 1 \right), \ and \ \|\mathbf{x}\| \le \sigma. \qquad (A.4)$$

Now the multiscale dilation-erosion operator is defined as follows

$$m(f, s_\sigma) = \begin{cases} d(f(\mathbf{x}), s_\sigma) & if & \sigma > 0 \\ f(\mathbf{x}) & if & \sigma = 0 \\ e(f(\mathbf{x}), s_\sigma) & if & \sigma < 0 \end{cases} \qquad (A.5)$$

Let us observe that in the above we have a notion of the positive scale ($\sigma > 0$), which corresponds to the morphological dilation, whereas for negative scales ($\sigma < 0$) this boils down to erosion. As a consequence, positive scales pertain to the local maxima of an image, while the negative pertain to the local minima. Obviously, for larger modules of σ the filtered image contains less detail.

The interesting and very important property of the just defined erosion-dilation operator m with the structural element that complies with Equation (A.3) is that it preserves the local extremes of the original signal $f(x)$. More specifically we define the extremes E_{max} and E_{min} of $f(\mathbf{x})$ as

$$E_{max}(f) = \{\mathbf{x}: f(\mathbf{x}) \text{ is a local maximum}\}, \ E_{min}(f) = \{\mathbf{x}: f(\mathbf{x}) \text{ is a local minimum}\}. \quad (A.6)$$

From the postulated causality property the following theorem holds [1]:

Theorem A.1   Dilation-erosion scale-space monotonic property.

For the dilation-erosion operator (A.5) and the structural element which complies with Equation (A.3), if an increasing sequence of scales $\sigma_1 < \sigma_2 < 0 < \sigma_3 < \sigma_4$ is given, then

$$E_{\min}\left(m\left(f, s_{\sigma_1}\right)\right) \subseteq E_{\min}\left(m\left(f, s_{\sigma_2}\right)\right) \subseteq E_{\min}\left(f\right), \tag{A.7}$$

and

$$E_{\max}\left(m\left(f, s_{\sigma_4}\right)\right) \subseteq E_{\max}\left(m\left(f, s_{\sigma_3}\right)\right) \subseteq E_{\max}\left(f\right). \quad \square$$

Proof of this theorem is provided in [1]. It follows from the above that the morphological scale-space operator $m(f, s_\sigma)$ preserves the number and positions of each of the image extremes up to a certain scale which is called *an intrinsic scale*. This paves the way towards an application of the morphological scale-space pyramids in an analogous way as the linear scale-spaces (such as a Gaussian). This means also that the morphological scale-space is suitable for coarse-to-fine tracking of features. That is, if a signal feature – which usually corresponds to some form of extreme in an image – appears at a scale $\sigma_i < 0$ then it also appear at a zero scale and all scales up to a certain intrinsic scale, which depends on the size of a given feature.

Figure A.1 depicts the smoothing effect of the multiscale erosion and dilation when applied to an edge of an irregular shape. In this example all three structural elements are circles but can be any other shape as well. It is sometimes assumed that the structural circle operating outside a shape has a positive radius, whereas its inner erosive counterpart has a negative one. Thus, dilation corresponds to positive, and erosion to negative, scales, respectively. Smoothing of a curve is achieved by tracing the *position of the center* of a structural circle rolling alongside the edge of a figure.

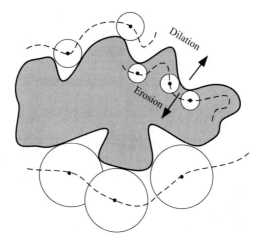

Figure A.1   Curve evolution by dilation and erosion in the multiscale morphological space.

Positions of the signal features can be assessed by means of the scale-space fingerprints which are plots of the scale dependent point sets versus scale. However, since these require time consuming computations, reduced fingerprints were proposed in [1] instead, which are defined as

$$F(f, s_\sigma) = \begin{cases} E_{\max}\left[m\left(f, s_\sigma\right)\right] & if \quad \sigma > 0 \\ E_{\min}\left(f\right) \cup E_{\max}\left(f\right) & if \quad \sigma = 0 \\ E_{\min}\left[m\left(f, s_\sigma\right)\right] & if \quad \sigma < 0 \end{cases} . \tag{A.8}$$

The property of a continuity of features over scale is preserved also for the reduced fingerprints which makes them useful for analysis of the morphological scale-spaces. For a 2D signal, such as a digital image, the reduced fingerprints can be represented as a stream of triples (x_k, y_k, σ_k), where x_k, y_k are spatial coordinates of a feature with index k and under scale σ_k. Such a set of triples does not uniquely represent an image but can be used for efficient feature tracking or matching [1].

A.3   Morphological Tensor Operators

Morphological operators have been found to be of great interest in image filtering and analysis due to many desirable properties which cannot usually be achieved with linear methods [3, 4]. However, these methods have a severe limitation when applied to non-scalar data such as color images, as they do not mention higher dimensional values, such as the tensor fields. Once again the limitation results from the lack of a well defined order relation for nonscalars, and, related to this, definitions of a maximum or a minimum element. Let us recall that the median filter was defined indirectly after choosing a distance (a metric) among nonscalar data (Section 2.6.1). However, such a procedure is burdened with ambiguity in the choice of the mentioned metric.

In the quest to find "good" relation properties for nonscalar data, Burgeth *et al.* [5] applied so called Loewner ordering to define morphological operators for positive semidefinite tensor data. Based on this it became possible to define the concept of the maximal and minimal elements which are rotationally invariant, preserve semidefinitness of data, as well as exhibit a continuous dependence on the input data – conditions required from the morphological operators.

The Loewner ordering for the symmetric $n \times n$ real matrices $\mathbf{R}, \mathbf{S} \in Sym(n)$ is defined as [5]

$$\mathbf{R} \geq \mathbf{S} \Leftrightarrow \mathbf{R} - \mathbf{S} \in Sym^+(n). \tag{A.9}$$

The $Sym(n)$ is a cone of positive semidefinite matrices of size $n \times n$. A cone is a subset of a vector space which is stable under addition and multiplication with a positive scalar. However, the above ordering does not provide notions of the unique supremum and infimum which are necessary for morphological operators. These, in turn, can be introduced with the help of the notion of *penumbra* $P(\mathbf{R})$ of a matrix $\mathbf{R} \in Sym(n)$. This is defined as a set of all matrices $\mathbf{S}$ which are smaller from $\mathbf{R}$ in the light of the Loewner ordering (A.9), that is

$$P(\mathbf{R}) = \{\mathbf{R}, \mathbf{S} \in Sym(n), \mathbf{R} \leq \mathbf{S}\} = \mathbf{S} - Sym^+(n). \tag{A.10}$$

Thus, the problem of determining the maximum of a set $\{\mathbf{Q}_i\}$ of matrices can be stated as finding the minimal penumbra covering penumbras $P(\mathbf{Q}_i)$ of all the matrices from the set. A vertex of the found minimal penumbra corresponds to the sought maximal matrix $\mathbf{Q}$ that dominates all $\mathbf{Q}_i$. However, since determination of the cones is difficult, Burgeth *et al.* propose searching for the smallest sphere enclosing the spheres associated with the set $\{\mathbf{Q}_i\}$. Details of this procedure can be found in [5]. In this way the defined morphological operators were successively applied to the filtering of the DT-MRI images.

A.4   Geometry of Quadratic Forms

A quadratic form Θ of a Hermitian (i.e. self-adjoint, self-conjugate) matrix $\mathbf{A}$ is defined as follows:

$$\Theta(\mathbf{A}, \mathbf{x}) = \mathbf{x}^T \mathbf{A} \mathbf{x}, \tag{A.11}$$

where $\mathbf{x}$ is an argument of the form Θ. This can be also expressed in the index notation

$$\Theta(\mathbf{A}, \mathbf{x}) = a_{kl} x^k x^l. \tag{A.12}$$

Since $\mathbf{A}$ is symmetric, it can be diagonalized as follows [6, 7]

$$\mathbf{A} = \mathbf{S} \mathbf{D} \mathbf{S}^T, \tag{A.13}$$

where $\mathbf{D}$ is a diagonal matrix. Substituting Equation (A.13) into Equation (A.11) yields

$$\Theta(\mathbf{A}, \mathbf{x}) = \mathbf{x}^T \mathbf{S} \mathbf{D} \mathbf{S}^T \mathbf{x} = \left(\mathbf{S}^T \mathbf{x}\right)^T \mathbf{D} \left(\mathbf{S}^T \mathbf{x}\right). \tag{A.14}$$

Thus a new quadratic form is obtained, as follows

$$\Theta(\mathbf{A}, \mathbf{y}) = \mathbf{y}^T \mathbf{D} \mathbf{y}. \tag{A.15}$$

where $\mathbf{y} = \mathbf{S}^T \mathbf{x}$ is a *new* variable. The above equation can be written as

$$\Theta(\mathbf{A}, \mathbf{y}) = \mathbf{y}^T \mathbf{D} \mathbf{y} = \sum_{i=1}^{N} \lambda_i y_i^2. \tag{A.16}$$

The level sets of a quadratic form Θ are those points which fulfill the condition

$$\Theta(\mathbf{A}, \mathbf{y}) = C, \tag{A.17}$$

where C is a constant value. From the above the following is obtained

$$\sum_{i=1}^{N} \lambda_i y_i^2 = C. \tag{A.18}$$

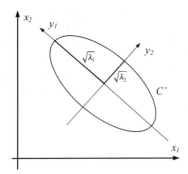

Figure A.2   Level sets of the 2D quadratic form.

Specifically, for $N = 2$ we have

$$\lambda_1 y_1^2 + \lambda_2 y_2^2 = C \tag{A.19}$$

which after expanding yields

$$\frac{y_1^2}{\lambda_1} + \frac{y_2^2}{\lambda_2} = C'. \tag{A.20}$$

When $C' = 1$ then the above equation denotes an ellipse with major and minor axes given by $\sqrt{\lambda_1}$ and $\sqrt{\lambda_2}$, respectively, as depicted in Figure A.2.

Finally, it is a well known fact from statistics that the exponent of the multimodal Gaussian distribution is a quadratic form (A.11) in which $\mathbf{A}$ is an inverse of a covariance matrix [8, 9, 10].

A.5   Testing Classifiers

A classifier trained with a partition of data with known classes, called *a training set*, can be tested with another partition of data with known classes, called *a testing set*. Then performance of a classifier can be assessed based on the numbers of good and bad responses. In the simplest case there are two tags (classes) which can be assigned to an object, i.e. whether it does or does not belong to a given class. This corresponds to the binary classification problem, discussed in Section 3.4.5.

In the group of classifiers which are trained from the set of training patterns the general problem is a trade-off between the proper response of a trained classifier on the patterns drawn from this training set and the generalization properties of this classifier. If the training parameters of a classifier, which obviously depend on its type, are chosen in such a way that makes it very tightly encompassing of training data, then the classifier tends to have a perfect response only in this training set. However, other objects from the same class but from outside the training set are usually misclassified. Such a classifier is said to be *overtrained*, thus losing its generalization capabilities. On the other side there are classifiers which allow too many data points, which usually results in high rate of false positive responses. To measure

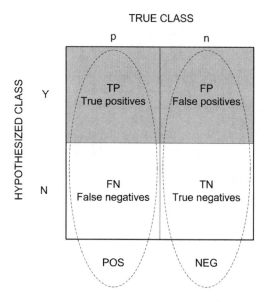

Figure A.3   The confusion matrix.[1]

such properties and to compare performance of different types of classifiers, various test measures have been proposed. From these can be found two broad applications in computer vision.

The first method relies on measuring precision and recall parameters of the system. Precision is a fraction of the true positive responses over all positive responses of a classifier, which can also contain some false positives. It gives the probability that a system's response is valid. On the other hand, recall gives a fraction of the true positives answers of a classifier as compared to all positive exemplars in the training set. It provides the probability that the ground-truth data was detected by the system.

The other useful parameter for assessing a classifier is the F-measure (A.25). This can be used to show a trade-off between the success rate of a classifier expressed by the recall parameter R in contrast to the amount of errors that can be tolerated, as expressed by precision parameter P. The ratio of this compromise is expressed by the parameter α, which is usually set to $\frac{1}{2}$. A location of the maximum value of F allows determination of the optimal parameters of a classifier [11].

Basic definitions and explanations of the most common parameters and measures used when testing classifiers are presented in Figure A.3 and Table A.1.

On the other hand the plots of the true positive rate TPR (A.22) versus the false positive rate FPR (A.23) can be used.[2] These are called *the receiver operating characteristics* (ROC) graphs, which depict the aforementioned trade-off between benefits and costs in classification.

A ROC plane is depicted in Figure A.4. A point with $TPR = 1$ and $FPR = 0$ is a perfect point. A diagonal line corresponds to the random guessing classifier, i.e. the one which correctly

---

[1] In some papers a confusion matrix is called a contingency table.
[2] Some authors prefer to use the true negative rate TNR instead, which is just a complement of the FPR [12].

Table A.1   Definitions of the most common performance parameters and measures of classifiers.

Number of positive *POS* and negative *NEG* exemplars (see Figure A.3):

$$POS = TP + FN, \quad NEG = FP + TN \tag{A.21}$$

True positive rate *TPR* (hit rate, recall *R*, sensitivity):

$$R = TPR = \frac{TP}{POS} = \frac{TP}{TP + FN}, \tag{A.22}$$

False positive rate *FPR* (A fallout, a complement of the true negative rate *TNR*, i.e. specificity):

$$FPR = \frac{FP}{NEG} = \frac{FP}{FP + TN}, \quad \left(FPR = \frac{FP}{NEG} = \frac{NEG - TN}{NEG} = 1 - \frac{TN}{NEG} = 1 - TNR \right) \tag{A.23}$$

Precision *P*:

$$P = \frac{TP}{TP + FP}, \tag{A.24}$$

F-measure:

$$F\text{-}measure = \frac{PR}{\alpha R + (1 - \alpha) P}, \tag{A.25}$$

where α denotes a relative cost of a success, in terms of recall *R*, versus errors that can be tolerated, as described by precision *P*. Usually $\alpha = 0.5$.

Accuracy (a ratio of good answers to all answers, Percentage of Correct Classifications – *PCC*):

$$A = \frac{TP + TN}{TP + TN + FP + FN}, \tag{A.26}$$

Geometric mean of *TPR* and *TNR*:

$$G = \sqrt{TPR \cdot TNR}, \tag{A.27}$$

Jaccard coefficient:

$$JC = \frac{TP}{TP + FP + FN}, \tag{A.28}$$

classifies objects in about 50% of answers. This corresponds to the coin tossing classifier, for instance. Thus, to get better results, a classifier needs to have some "knowledge" of the classified objects.

A classifier trained with a training set answers a point on the ROC graph when tested with a testing set. Different classifiers, or the same classifier but trained with different settings,

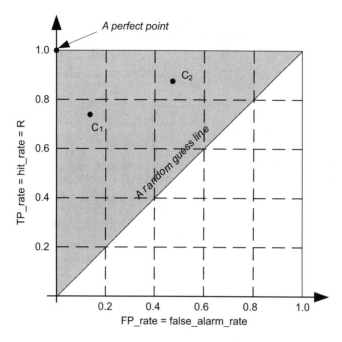

Figure A.4 Characteristic regions of the ROC graph.

produce different points in the ROC space. The points usually fall in the upper triangle of the ROC plane (grayed in Figure A.3), above the random guessing diagonal. However, it can be shown that in a case of a "negated" classifier its response point lies in a lower triangle and can be easily "reflected" to give a meaningful response in the upper triangle of ROC. Comparing classifiers can be carried out by measuring and comparing the distances of their response points from the perfect point. It appears however that a better approach is to generate many different operating points for each of the classifiers and then compare the corresponding areas under the ROC [13].

Generation of the ROC curve requires a scoring parameter for each answer of a classifier. For instance, in the abovementioned simple case of binary classification with the probabilistic Bayes classifier this can be obtained by changing the acceptance parameter κ in Equation (3.77) from 0 to 1. However, for some classifiers acquiring scoring values can be problematic. For instance in the case of the SVM this problem was discussed in Section 4.2.3.2. Efficient methods of ROC generation can be found in the paper by Fawcett [13].

The major advantage of the ROC over the precision/recall measurements is its invariance to the changes of proportions of the positive and negative instances in the test set (so called class skew). However, for some tasks the opposite can hold true. For example in the system for detection of object boundaries Martin *et al.* advocate using precision/recall curves instead of ROC [14]. They formulated the boundary detection problem as a classification problem of discrimination of the non-boundary from boundary pixels. Then the precision-recall is measured and compared with the ground-truth data in human-marked boundaries from the

Berkeley Segmentation Dataset [15]. However, when evaluating boundary detection, the fallout (FPR) parameter of ROC is not meaningful since it depends on image resolution, i.e. the number of pixels. For example, if the resolution of an image is increased by a factor r, then the number of pixels increases as r^2. Nonetheless, objects' boundaries are rather one-dimensional, so the number of true positives progresses in this case as r, whereas the number of true negatives grows as r^2. As a result, the FPR (A.23) will decline by as much as $1/r$. On the other hand, precision P does not pose such a problem since it is normalized by the number of true positives rather than true negative responses.

For further references, an interesting treatment on testing ensembles of classifiers is contained in the last chapter of the book by Rokach [11]. The problem of design and analysis of machine learning experiments is addressed in the book by Alpaydin [16].

A.5.1 Implementation of the Confusion Matrix and Testing Object Detection in Images

The confusion matrix contains just four integer counters of true/false positives/negatives, respectively, as shown in Figure A.3. Thus, it can be easily defined as a support class. An example presents the following C++ code listing.

```
///////////////////////////////////////
// This is called a confusion matrix
// (or a contingency table)
class ConfusionMatrix
{

public:
        long fTP;            // true positive
        long fTN;            // true negative

        long fFP;            // false positive
        long fFN;            // false negative

public:
        ConfusionMatrix( long tp = 0, long tn = 0, long fp = 0, long fn = 0 )
              : fTP( tp ), fTN( tn ), fFP( fp ), fFN( fn ) {}
        ~ConfusionMatrix() {}
        // Copy constructor, operator =, etc.

public:

        // this value is returned if a parameter cannot be computed
        // (e.g. due to a zero denominator, etc.)
        const double kInvalidData;

        double Precision( void ) const;

        // TPR (hit rate, recall R, sensitivity)
        double Recall( void ) const ;
```

```
    // False positive rate FPR (A fallout):
    double Specificity( void ) ;

    // Computes the F-measure
    // alpha denotes a relative cost of a success, in terms of recall R,
    // versus errors that can be tolerated, as described by precision P.
    // By default alpha is set to 0.5.
    double F_Measure( const double alpha = 0.5 );

    // Accuracy (a ratio of good answers to all answers,
    // Percentage of Correct Classifications - PCC)
    double Accuracy( void ) const;

    double G_Measure( void ) const;

    double Jaccard( void ) const;
};
```

Algorithm A.1   Definition of the *ConfusionMatrix* class which defines confusion matrix used in measurements of accuracy of the classifiers.

Since we plan to use the *ConfusionMatrix* as a bare data container it was decided to put all its members to the public section of the class for easier access. However, it is substantial to have good intialization of an object. For this purpose the class default constructor[3] does its job of appropriately setting all data members to zero. Finally, the class contains a number of helper functions which compute all the parameters that specify quality of classification, such as *Precision, Recall*, etc., presented in Table A.1. Such a strategy seems justified and useful since the classification measures can be computed only if a confusion matrix is available. On the other hand, placing them as members of the *ConfusionMatrix* underlines the encapsulation paradigm of the object-oriented methodology, as compared to an alternative in the form of e.g. inline functions taking *ConfusionMatrix* as an input parameter. Such software design issues need to be analyzed in the context of the code use case scenarios.

ConfusionMatrix defined in Algorithm A.1 can be used in all software modules that need to measure true/false positive and negative ratios. A very frequent task is segmentation of an image into two types of area, one pertinent to the sought objects and the second one assigned to all the others (called a background). To measure the quality of an algorithm performing such a type of segmentation, a result of its operation stored in a test image can be compared with a reference image. The latter can be obtained for instance by marking of the object pixels by a human. In effect the confusion matrix with four counters TP, TN, FP, and FN, can be computed. Algorithm A.2 contains a listing of the *CompareImages* function which performs this type of comparison.

---

[3]There are also other construction members, such a copy constructor and an assignment operator, which are not shown here but are present in the attached code [17].

```
///////////////////////////////////////////////////////////
// This function compares the two images.
// It is assumed that there are true positive and negative
// points in the refImage (i.e. an object and the background)
///////////////////////////////////////////////////////////
//
// INPUT:
//                     refImage - the reference image
//                     pretenderImage - an image to compare with the
//                         reference
//                     theParams - output parameters which upon success
//                         contain measured TP, TN, FP, FN values
//                     kBackground - value of a pixel assigned to
//                         the background (false class)
//                     kObject - value of a pixel assigned to
//                         the object (true class)
//
// OUTPUT:
//                     true if success,
//                     false otherwise
//
// REMARKS:
//                     Image dimensions must be the same
//
bool CompareImages( const MonochromeImage & refImage,
                    const MonochromeImage & pretenderImage,
                    ConfusionMatrix & theParams,
                    const MonochromeImage::PixelType kBackground,
                    const MonochromeImage::PixelType kObject )
{
    const int kCols = refImage.GetCol();
    const int kRows = refImage.GetRow();

    if( kCols != pretenderImage.GetCol() || kRows != pretenderImage.GetRow())
        return false;

    MonochromeImage::PixelType theRefPixel = 0;
    MonochromeImage::PixelType theTestPixel = 0;

    // reset the counters
    theParams.fTP = theParams.fTN = theParams.fFP = theParams.fFN = 0;

    register int i = 0, j = 0;

    for( i = 0; i < kRows; ++ i )
    {
        for( j = 0; j < kCols; ++ j )
        {
            theRefPixel = refImage.GetPixel( j, i );
            theTestPixel = pretenderImage.GetPixel( j, i );

            bool agreement = theRefPixel == theTestPixel;
```

```
        if( agreement == true )
        {
            // True answer
            if( theRefPixel == kObject )
            {
                ++ theParams.fTP;    // increase true-positive
            }
            else
            {
                ++ theParams.fTN;    // increase true-negative
            }
        }
        else
        {
            // False answer
            if( theTestPixel == kObject )
            {
                ++ theParams.fFP;    // increase false-positive
            }
            else
            {
                ++ theParams.fFN;    // increase false-negative
            }
        }
    }

    return true;
}
```

Algorithm A.2   A function to compare the reference and test images. Images are assumed to contain pixels with only two different values characteristic of the detected objects and the background.

Objects are assumed to be those pixels with a value *kObject*, whereas background is defined by *kBackground*. Results of computation are returned thanks to the parameter *theParams* which is a reference to the supplied *ConfusionMatrix* object. For the same pixel positions the function checks whether pixel values in the reference and test image are also the same. If so, then the "true value" counters are incremented, i.e. the true-positive TP for objects or true-negative TN for background value pixels. On the other hand, if the test image manifests different pixel values than for the same pixel position in the reference, then the "false" counters are incremented.

A.6   Code Acceleration with OpenMP

The advent of easily available parallel computer architectures, such as FPGA [19], graphic cards [20], or multicore processors, opens new possibilities in the design of parallel algorithms. In this section we provide an introduction to the OpenMP library which can help in software acceleration, adding parallelism that is useful in the case of loops processing massive image

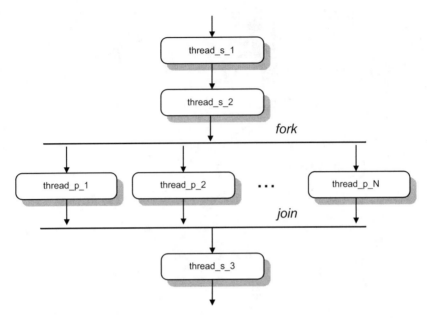

Figure A.5   Activity diagram of a process started with two serial threads followed by a parallel region. A set of threads is forked at the beginning of a parallel region and joined at its end. Joining requires thread synchronizations (a barrier).

data [21, 22]. It was designed for shared-memory parallel programs. An alternative is the Message Passing Interface (MPI) library [23, 24] which facilitates portable programming for distributed-memory architectures in which multiple processes execute independently and communicate by passing messages. However, when considering porting of the existing serial software implementations, OpenMP usually requires many fewer changes to the code and in effect allows faster implementation. This is a very important factor which not only accounts for man-hours spent on the project but also for quality of the resulting code, expressed in a number of potential bugs and time devoted for debugging and maintenance. However, there are systems which can benefit from many programming paradigms and connection of the many parallel libraries.

As already mentioned OpenMP is a parallel programming library for shared memory multithreading execution. It fits best to refactoring of the loops processing significant amount of iterations. This proves most effective if the operations inside a loop are independent, which is often a case in image processing [18, 25].

A key concept is multithreading operating on a shared memory. A thread is a dispatchable unit of work executed concurrently to other threads. A thread contains its own set of local data. However, to exchange information threads need to communicate accessing common data. However, contrary to local data, accessing shared memory needs to be done in a safe way so as not to destroy operations carried out by other threads. This security mechanism is called thread synchronization. Figure A.5 depicts an activity diagram of a process started with two serial threads followed by a parallel region. A set of threads is forked at the beginning of a parallel region and joined at its end. Joining requires thread synchronizations (a barrier).

Table A.2   Definitions of the most common OpenMP combined constructs for work sharing (from [18]).

OpenMP directive	Functionality
#pragma omp parallel for *[clause[[,]* *clause]. . .]*	Fork the for-loop iterations to the parallel threads
#pragma omp parallel sections	Start independent units of work
#pragma omp single	Ensures exactly a single thread to execute a block of code

Table A.2 summarizes definitions of the most common OpenMP constructions for work sharing [21]. In C/C++ code[4] they are entered by the preprocessor pragmas. Recent advances in the standard,[5] as well as available programming platforms, user manuals, and many other useful information can be accessed from the OpenMP web page [22]. Adding OpenMP to a new or existing code is straightforward and can be easily turned on and off with preprocessor directives, as will be discussed.

Directives shown in Table A.2 are so called combined constructs which join the *parallel* keyword with the *for* keyword in the one *#pragma* instead of their separate declarations, which sometimes leads to a slower performance. On the other hand, the OpenMP *parallel for* directive is accompanied by the clauses which set the specific behavior of this parallel construct. These clauses are briefly characterized in Table A.3.

Other parallel constructions in Table A.2, such as *#pragma omp parallel sections* can also be accompanied with some clauses from Table A.3 (with the exception of the *ordered* and *schedule*). Complete definitions of the parallel constructs and clauses, as well as their in-depth treatment with examples are provided in [21, 22].

A.6.1   Recipes for Object-Oriented Code Design with OpenMP

When adding concurrent constructions into the new or existing software components the first issue is to ensure correctness of the new solution. In both cases the aim is to speed up computations with parallel constructs available on the contemporary computer architectures whose use is facilitated by OpenMP.

Frequently the software development process starts with a serial version of an algorithm. Then, after debugging and ensuring that a component executes correctly, it can be ported to its parallel version, e.g. using OpenMP. A piece of practical advice is to keep both versions for further analysis, development, comparison of speed up factor (which sometimes can be a surprise), but most of all for verification of the correctness of a solution. However, in order to obtain the best results, the concurrency should be accounted for at the stage of software architecture design. In this section we provide a set of design recipes for design of object-oriented software components to take full benefit of OpenMP. The discussion is based on the paper [18].

---

[4]OpenMP is also available for the FORTRAN language as a set of code directives.

[5]At the time of writing this book OpenMP committee was about to release OpenMP 4.0 Draft [22].

Table A.3 Definitions of the OpenMP clauses for loop sharing (from [18]).

OpenMP clause	Description
private (*list*)	Declares a variable as private; each thread will contain its own but uninitialized copy (for initialization use firstprivate)
firstprivate (*list*)	Initializes a private variable with a value of the variable with the same name before entering the thread
lastprivate (*list*)	Used to make the last value of a private variable available after termination of a thread
reduction (*operator:list*)	Orders parallel execution of the associative and commutative operations (addition, multiplication, etc.) with aggregation of partial results at the threads barrier
ordered	Forces sequential order of execution within a parallel loop
schedule (*kind[,chunk size]*)	Controls distribution of loop iterations over the threads; the scheduling modes are static, dynamic, or guided
shared (*list*)	Declares data objects which will be shared among the threads' executing regions it is related to (for loop); each data from the list can be freely read and/or written by each of the threads in any moment
nowait	Tells a compiler to skip a synchronization barrier at the end of the parallel loop construction
default (*none|shared*)	Tells how the compiler should treat local variables. If *none*, then each variable needs to be explicitly declared as shared or private. If *shared*, then all variables will be treated as shared
num_threads (*N*)	Specifies a number of threads used in a parallel block
if(*scalar_logic_expr*)	Specifies conditional execution; if *scalar_logic_expr* is true then the parallel region will be executed by a set of threads; otherwise only a single thread is launched

When programming concurrent components with object-oriented methods the first issue is *thread safety* on a level of a class or a component (library). Thread safety means knowing the restrictions (if any) of using a given object in concurrent threads. So whether locally created objects of a given class are totally independent or have anything in common such as static members? Spotting such parts in existing code is sometimes difficult. However, if not recognized and properly synchronized, they lead to very malicious errors which exhibit in random malfunctioning of a parallel version of the software. Thus, the advice to keep a serial version of the code and run cross-validation tests of both.

The next important thing to do is to partition data on local, i.e. the ones which local copies will be created for each of the threads, and shared variables (however, these may not be initialized in an expected way, as will be discussed). The latter group needs to be treated with special care when changing their values. Otherwise race conditions arise which usually lead to wrong values of data. This is a new phenomenon when compared to a serially executed code and wrong categorization of variables is one of the most frequent errors when programming concurrent systems.

When using OpenMP, variables which contain constant values, such as size of an array, can be declared *shared* or *private* with the *firstprivate* clause used for their initialization (see

Table A.3). Sometimes one is preferred over the other (e.g. in the non-uniform memory access (NUMA) architectures). In OpenMP data is shared by default, with some exceptions though. One of which is the iteration variable of the *for* loop(s) inside the parallel region. All other private data can be specified explicitly by the *private* clause. Each thread has its own (and private) set of all private variables. However, there is another trap – one must remember that if not stated specifically they are not initialized. To carry out private data initialization the *firstprivate* clause needs to be used. On the other hand, shared variables are accessible to each of the threads when they execute. No copies are made for shared variables, so for each such variable its value is left by the last thread that wrote to it. There are some cases where we have the option to use either shared or private data. These are for instance constant values or other data which do not change inside the parallel regions, such as loop delimiters, multiplicative factors, iteration steps, etc. These can be made available in the parallel section either as shared or *firstprivate* data. Performance of both is platform dependent. In the following examples the first strategy is undertaken.

Special care needs to be taken for direct access to class members or references. In OpenMP only a variable defined outside a class, or a static data member, can be used in the data-sharing clause shown in Table A.3. Because of this the following code

```
class A
{
        public:
                int f_a;
        public:
                void fun( void );
};

void A::fun( void )
{
        #pragma omp parallel for shared( f_a )
        // for( ... )
}
```

will not compile. We need a way round this, which first requires creation of a local variable, and then its assignment with a data member, as illustrated in the following code fragment

```
void A::fun( void )
{
        int local_shared_f_a = f_a;

        #pragma omp parallel for shared( local_shared_f_a )
        // for( ... )
}
```

Also, temporary objects cannot be used in a direct way. For example, if in some context a function with a temporary object *Real_2D_Point(j, i)* needs to be called as follows

```
// ...
( * coordTransformEngine )( Real_2D_Point( j, i ) );
// ...
```

then this fragment needs to be worked around. First a named local variable can be created and only after this can a function be called with that variable passed as its argument. This idea is shown in the following code snippet

```
// ...
Real_2D_Point tmp_local_point( j, i );
( * coordTransformEngine )( tmp_local_point );
// ...
```

Special care needs to be undertaken when using the C++ exception mechanism. OpenMP places a strict requirement that once *throw* is invoked inside a parallel region, it must cause execution to resume within the same parallel region, and it must be caught (in the *try-catch* block) by the same thread that threw that exception.

A recommended design practice is not to change an existing class at all but use an inheritance mechanism. That is, to port a class to OpenMP, instead of directly changing code of that class, derive from it a new class which carries out multithreading with OpenMP. This leaves the base class unchanged. Such a design strategy brings some benefits. The first is that we still have the option of using either one. Then, we can compare the results of operations to verify the solution, as well as being able to measure execution speed up ratio, etc. This way we can verify whether the multithreading version suffers from race conditions, deadlock, etc. A practical remark is that source code should be separated with the preprocessor directives, such as for example *#if open_mp ... #endif*, into the serial version and the OpenMP part. Also, when refactoring existing code to utilize OpenMP, a handle-body design pattern can be employed.

The best way to learn how to use OpenMP is to try to port an existing function to take full benefit of a multicore system. For this purpose let us transform a function *SumAllImagePixels* which computes the sum of values of all pixels in an image. This is a template function which allows the input image *in* with pixels of any type whose values can be summed up, defined by a template parameter *T*. Thus, *T* is any built-in type, such as *int*, *long*, *double*, etc. or user defined, if objects of such type can be added or can be converted to *double*.

Algorithm A.3 presents source code[6] of the *SumAllImagePixels* function which contains two versions of the algorithm – the concurrent and the serial, respectively. As suggested, compilation of these two versions of code is controlled by the *USE_OPEN_MP* preprocessor

---

[6]These fragments of code were also presented in [18].

```
template< class T >
double SumAllImagePixels( const TImageFor< T > & in )
{
        register double sum = 0.0;
        register T * src_data_ptr = in.GetImageData();
        register int data_num = in.GetElems();
#if USE_OPEN_MP

        register int i = 0;
        #pragma omp parallel for default( none ) \
                        shared( src_data_ptr, data_num ) \
                        private( i ) \
                        reduction( + : sum ) \
                        schedule( static )
        for( i = 0; i < data_num; ++ i )
        {
                sum += (double) src_data_ptr[ i ];
        }        // here we have a common barrier
#else // USE_OPEN_MP

        while( data_num -- != 0 )
                sum += * src_data_ptr ++;

#endif // USE_OPEN_MP
        return sum;
}
```

Algorithm A.3   Implementation of the *SumAllImagePixels* template function in a serial and OpenMP versions controlled by the preprocessor flag *USE_OPEN_MP* (code from [17]).

flag. The serial implementation is placed in the #*else* ... #*endif* section and contains one *while* loop. However, OpenMP does not allow direct parallelization of the *while* loop and therefore needs to be rewritten into an equivalent *for* loop statement. As shown, the *for* loop also accesses all pixels in the input image, however this time in parallel chunks due to the #*pragma omp parallel for* clause. The clause *default(none)* tells a compiler to expect explicit definitions of the data-sharing attributes of all variables for the parallel block. This is a recommended behavior to avoid unexpected errors caused by falsely assumed default data-sharing attributes if assigned without control of a programmer. The *shared(src_data_ptr, data_num)* clause instructs a compiler that the *src_data_ptr* and *data_num* should be treated as shared variables. The next clause *private(i)* specifies *i* as a private variable. As already mentioned, this makes a compiler create private versions of the variable *i* for each of the *N* threads. Obviously, each thread will have its copy of *i* initialized to the beginning of its own chunk of data to process. We don't need to deal with this explicitly.

On the other hand, the *reduction(+ : sum)* introduces the reduction operation which consists of creating *N* local sum variables, which are then summed up at the synchronization barrier executed after the loop terminates. Finally, the *schedule(static)* instructs that iterations should be divided into chunks of the same size. These are then statically assigned to the threads which execute in the round-robin fashion. Since no chunk size is supplied, the iterations are divided into roughly equal chunks.

Nevertheless, in some situations careless application of the OpenMP parallel clause can lead to worse performance when compared to a serial execution (i.e. with only one thread). This happens if thread creation time, as well as the time required for context switching, are not accounted for. Thus, the effectiveness of a parallel section depends on many factors, one of which is the number of data to be processed which sometimes is not known beforehand. This uncertainty can be resolved with the *if* clause, such as in the presented function in Algorithm A.4. This is an OpenMP version of the Algorithm 3.21 (p. 303), which computes a sum of squared differences of its vector arguments (passed as pointers). Constant *kElems*, copied from the *fDim* member, contains a number of components in the two vectors. In this case parallel threads will be launched in the *for* statement only if the number of elements exceeds 1000.

```
virtual double operator() ( double * x, double * y )
{
        // At first compute the squared Euclidean
        // distance between x and y
        register double tmp = 0.0;
        register double sum = 0.0;

#if USE_OPEN_MP

        register int d = 0;
        register const int kElems = fDim;
        #pragma omp parallel for default( none ) \
                    if( kElems > 1000 ) \
                    shared( x, y, kElems ) \
                    private( d, tmp ) \
                    reduction( + : sum ) \
                    schedule( static )
        for( d = 0; d < kElems; ++ d )
        {
                tmp = x[ d ] - y[ d ];
                sum += tmp * tmp;
        }

#else USE_OPEN_MP

        for( register int d = 0; d < fDim; ++ d )
        {
                tmp = * x ++ - * y ++;
                sum += tmp * tmp;
        }

#endif USE_OPEN_MP

        return exp( - fGamma * sum );
}
```

Algorithm A.4 The *if* clause in computation of the kernel function.

Let us observe that the *if* clause in the above code will be executed each time the function is entered, however. This entails a time penalty. A possible way around this problem is application of the strategy design pattern, as discussed in [2]. In Algorithm A.4 let us also point out necessary modifications to the serial code when porting to OpenMP. The most noticeable change is in the access to the components. Access to the components is done with array-like access, e.g. x[*d*], rather than direct manipulation of pointers, such as $* x ++$, which would require their synchronization. Also worth noting is the introduction of the *kElems* local variable to which a data member *fDim* is copied, as already discussed. The *reduction*(+ : *sum*) clause instructs the compiler that the shared variable sum will be used to gather all partial sums of whatever number of threads will be launched to execute the *for* loop. As indicated in Table A.3, this kind of operation is called reduction and can be applied to all associative and commutative operations (addition, subtraction, multiplication, logical operations, etc.) in parallel with aggregation of partial results after the barrier synchronization. The clause *shared*(*x*, *y*, *kElems*) instructs the compiler that *x*, *y*, and *kElems* should be treated as shared variables (a potential necessity of synchronization), while *private*(*d*, *tmp*) tells that *d* and *tmp* that each of the threads will operate on their local copies, thus no synchronization is required in this case. On the other hand, *default*(*none*) instructs the compiler to require each of the variables to be explicitly declared by a programmer as either shared or private. Finally, *schedule*(*static*) tells the compiler to share iterations among the threads in approximately equal chunks. Last but not least, let us observe that the whole directive needs to be written in a single line. Thus, to make code more readable it was split into a number of lines with the backslash \ mark.

A.6.2 Hints on Using and Code Porting to OpenMP

Refactoring serial code to OpenMP is relatively simple, especially when compared with other threading models such as MPI. However, its proper connection to the object-oriented components can sometimes be problematic. Below we provide a number of practical hints which can facilitate this problem.

1. The time necessary to handle threads in the system should be always taken into consideration. Thread creation and scheduling require execution of a large number of machine cycles.
2. Performance of the parallel construct under different settings, amount of data, etc., and comparison with serial performance should be always measured and not guessed.
3. OpenMP does not protect from race conditions or deadlocks among the threads. They need to be checked at the design stage and properly handled with synchronization objects. Race conditions can arise if the following conditions are met:
   o Two or more threads executing the code;
   o Threads access common data (variables);
   o At least one of the threads modifies common data (write).
   As an example of hidden race conditions let us consider the following function *GetImageVarianceValue*, listed in Algorithm A.5. Now we wish to refactor that function to take full advantage of the loop parallelization offered by OpenMP.

```
1        template< class T >
2        double GetImageVarianceValue( const TImageFor< T > & in )
3        {
4              double N = in.GetElems();
5              if( N <= 1  )
6                    throw T_Standard_HIL_Exception( T_Standard_HIL_Exception::
                                                     kResultInfinity );
7
8              double meanVal = GetImageMeanValue( in );
9              double sum = 0.0;
10             double temp;
11
12             register T * src_data_ptr = in.GetImageData();
13             register int data_num = in.GetElems();
14
15             while( data_num -- != 0 )
16             {
17                   temp = (double)* src_data_ptr ++ - meanVal;
18                   sum += temp * temp;
19             }
20
21             return sum / (N - 1);
22       }
```

Algorithm A.5 A sequential version of the function *GetImageVarianceValue*.

The first thing one ought to do is transformation of the *while* loop in the line into an equivalent *for* loop. Then the *#pragma omp parallel for* ... should follow with all necessary accompanying clauses, as discussed (see Table A.3). We might come to such a version:

However, if we run this code a number of times we will notice slightly (but not always obviously) wrong results. So, what is wrong with this code? The problem lurks in line <24> in which a shared variable *src_data_ptr* is updated. This can result in the wrong values being assigned to the *src_data_ptr* due to preemptive multitasking. The way around this problem is to either put a synchronization barrier for line <24> or to use a private variable *i* to index an array of data pointed to by the *src_data_ptr*. The second solution is much better since it allows full utilization of the multithreading environment. A possible realization of this idea is presented in Algorithm A.7.

Now the shared variable *src_data_ptr* in line <26> in Algorithm A.7 is only read by the threads and does not change. Access to the consecutive elements of the array is accomplished thanks to the private variable *i* of which a local private copy is contained in each of the threads executing the code in lines <24–28>. The REQUIRE directive in line <40> is a form of assertion that is active only in a debug version of the code. This very useful method helps in the early detection of some logical problems in the code, as discussed in [2, 25, 26].

4. Data declared *static* should be spotted in the code and controlled. These are shared among objects of a class and therefore should be avoided or synchronized to avoid race conditions.

```
1    template< class T >
2    double GetImageVarianceValue( const TImageFor< T > & in )
3    {
4        double N = in.GetElems();
5        if( N <= 1 )
6            throw T_Standard_HIL_Exception( T_Standard_HIL_Exception::
                                              kResultInfinity );
7
8        double meanVal = GetImageMeanValue( in );
9        double sum = 0.0;
10       double temp;
11
12       register T * src_data_ptr = in.GetImageData();
13       register int data_num = in.GetElems();
14
15       int i = 0;
16
17       #pragma omp parallel for default( none ) \
18           shared( src_data_ptr, data_num, meanVal ) \
19           private( i, temp ) \
20           reduction( + : sum ) \
21           schedule( static )
22       for( i = 0; i < data_num; ++ i )
23       {
24           temp = (double)* src_data_ptr ++ - meanVal;
                                         // WRONG! lacking synchronization
25           sum += temp * temp;
26       }   // here we have a common barrier
27
28       return sum / (N - 1);
29   }
```

do not use this code – it is an example of WRONG porting

Algorithm A.6 An example of a WRONG concurrent implementation of the *GetImageVarianceValue* function.

5. Parallel nested loops should be examined and chosen with care. In many algorithms the best results are obtained with a relatively small number of threads, frequently the same as the number of cores in a processor. However, thread scheduling is not free. Therefore if it is not well justified, a nested parallelism should be rather avoided; nevertheless, different models of parallelism can be implemented and their performance measured in real conditions.

6. In a definition of parallel sections *all* variables and constants of the parallel block should be always checked to see if their sharing modes are properly declared. All should be explicitly expressed since an omission can lead to hazard conditions. Help in this respect comes from the clause *default(none)*. As a rule of thumb, a number of variables used in the parallel block and those declared in the *shared*, *private*, and *lastprivate* clauses should be counted and their amounts should agree.

7. The private loop variable should always be declared as a signed integer type (*int, long*).

```
1    template< class T >
2    double GetImageVarianceValue( const TImageFor< T > & in )
3    {
4        double N = in.GetElems();
5        if( N <= 1.0 )
6            throw T_Standard_HIL_Exception( T_Standard_HIL_Exception::
                                             kResultInfinity );
7
8        double meanVal = GetImageMeanValue( in );
9        double sum = 0.0;
10       double temp = 0.0;
11
12       register T * src_data_ptr = in.GetImageData();
13       register int data_num = in.GetElems();
14
15   #if USE_OPEN_MP
16
17       int i = 0;
18
19       #pragma omp parallel for default( none ) \
20           shared( src_data_ptr, data_num, meanVal ) \
21           private( i, temp ) \
22           reduction( + : sum ) \
23           schedule( static )
24       for( i = 0; i < data_num; ++ i )
25       {
26           temp = (double)src_data_ptr[ i ] - meanVal;
27           sum += temp * temp;
28       }    // here we have a common barrier
29
30   #else // USE_OPEN_MP
31
32       while( data_num -- != 0 )
33       {
34           temp = (double)* src_data_ptr ++ - meanVal;
35           sum += temp * temp;
36       }
37
38   #endif // USE_OPEN_MP
39
40       REQUIRE( N >= 2.0 );
41       return sum / ( N - 1.0 );
42   }
```

Algorithm A.7 An example of a thread safe version of the function *GetImageVarianceValue*. This function allows choice between a single- and multithread solution in the compile time due to the USE_OPEN_MP preprocessor flag (the OpenMP part darker). Now the shared variable *src_data_ptr* is only read by the threads.

8. To avoid errors pronounced variable names should be used. For instance, a variable named *min_val* when used in OpenMP can be named *shared_min_val* or *private_min_val*, etc.
9. The abbreviated parallel construction, shown in Table A.2, is preferred and usually faster. In the case of *for* loops the compiler inserts only *one* barrier, instead of two when two parallel constructs are used.
10. In contrast to the serial implementation, private data are *not* initialized even if initialization code is explicitly written (such as *int a=0*). To properly initialize private data the *firstprivate* clause should be used.

Further suggestions can be found in the literature and most of all in daily coding practice [21]. The paper by Kim and Bond can also be of help, which provides an interesting survey on multicore software technologies [27].

A.6.3   Performance Analysis

All tests presented here were measured on the Dell® Precision M6500 laptop, with the Intel® Core i7 CPU Q820 @ 1.73GHz 4-core dual-thread processor and 8 GB of RAM, running 64-bit Windows® 7 Professional. The second of the compared systems operates with the AMD E-350 processor 1.6 GHz, 4 GB RAM, 32-bit Windows 7 Home Premium edition. For both platforms software was developed with the Visual Studio® .NET 2010 C++ compiler by Microsoft®. Also both versions were created with and without OpenMP being enabled. Time was measured with the time measuring functions of the operating system [18].

Figure A.6 shows a comparison of system performance with and without OpenMP. Performance measurements for the median operation with the 7 × 7 structure element are shown Figure A.6(a). Measurements for computation of the structural tensor are shown in Figure

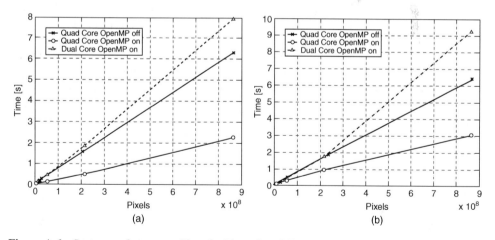

(a)   (b)

Figure A.6   System performance with and without OpenMP, measured for the median operation with the 7×7 structure element (a), and computation of the structural tensor (b). Measurements performed for two microprocessors: Intel® Core i7 CPU Q820 quad core (solid line), dual core AMD E-350 dual core (dashed line).

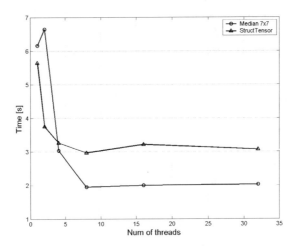

Figure A.7 System performance with respect to the number of threads measured for the median with the 7×7 structure element, and computation of the structural tensor. Measured for Intel® Core i7 CPU Q820 quad core microprocessor.

A.6(b). Tests were performed for the two described microprocessor systems. In the first case a speed up ratio of ~3 can be observed for the quad core system. In the case of the structural tensor the speed up ratio is ~2. A slightly lower result is caused by more steps in the computations, from which only some of them can be parallelized, and the groups of computations need to be synchronized. It is also interesting to note that a system with potentially much lower computational power can be greatly enhanced just by porting to OpenMP (comparing the dashed line in Figure A.6 for a low power machine but with OpenMP enabled, to the quad core system without OpenMP). Another interesting, but expected, observation is that real speed up can be obtained for a large amount of processed data (such as HDTV resolution).

Figure A.7 shows measurements of system performance with respect to the number of threads assigned to a task. In the first case this is the 7×7 median filtering, whereas the second task concerns computation of the structural tensor. This time only the system with the quad core microprocessor was considered. However, the actual number of launched threads can be controlled by the OpenMP function *omp_set_num_threads*.

Based on Figure A.7 it is interesting to note that performance accelerates up to the point of a number of threads approaching the number of available cores on the microprocessor.

A.7 Useful MATLAB® Functions for Matrix and Tensor Processing

In this section we provide basic MATLAB® commands which are useful for checking some of the computations presented in this book, as well as some of the basic information on available packages for tensor processing. However, this is only a very brief overview and more accurate details and examples can be found on the webpage [28], as well as in a number of publications, such as the book by Gilat [29] or the one by Attaway [30]. A practical way to learn MATLAB® programming in the context of pattern recognition is to analyze, then modify and run, well written and published procedures. In this respect a recommended source of MATLAB® procedures for pattern classification is the book by Kuncheva [31], as well as by Theodoridis and Koutroumbas [32]. A number of interesting procedures for nonnegative matrix and tensor decompositions is presented in the book by Cichocki *et al.* [33]. However, one of the most ample sources of pattern recognition procedures in MATLAB® is the PRTOOLS package, available from [34].

Table A.4 presents a very brief overview of MATLAB® functions for eigendecomposition, singular value decomposition of matrices, as well as PCA transformation. The functions are accompanying by the short examples. However, most of the presented functions have many additional options and calling modes, which details can be obtained from the MATLAB® webpage [28].

Computation of the matrix exponent and natural logarithm of the matrices, as well as computation of the distances D_F in (2.28) and also D_{LF} (2.31) were discussed in Section 2.5. These can be computed with the set of MATLAB® functions outlined in Table A.5.

Multiarray elements are very useful in multidimensional data processing, and are listed in the Table A.6. These can also be used to initialized tensor objects.

One of the best MATLAB® packages for tensor decompositions was created by Bader and Kolda[7] [35]. Here, in Table A.7, we present only a small, but hopefully useful, subset of possibilities of this ample package.

It is interesting to note that the two main decomposition functions, *hopm* and *hooi*, return special tensor objects, *cp_tensor* and *tucker_tensor*, respectively, which are complex data structures that convey objects specific to a given decomposition. Apart from the functions shown in Table A.7 the tensor library by Bader and Kolda contains the majority of arithmetic, logical, and comparison functions. Details are available in their paper [35], as well as in the examples attached to the library.

A package for NMF computations is provided in [36]. The other excellent MATLAB® platform for tensor decomposition was created by Cichocki *et al.* [33]. Further details with code listing can be accessed in their book.

A.8   Short Guide to the Attached Software

Procedures and classes discussed in this book can be downloaded from the Internet site [17]. This package builds upon some base classes of the software library attached to our previous book [37]. Both are available for free download for noncommercial use. To keep them up-to-date, details on installation, usage, and further applications of the code are also available from the web page. It contains also a separate manual with a detailed description of the library and its applications.

All sources were written in plain C++ taking special precautions for multiplatform operations. Thus, the sources are available in almost the same form for both Windows® and Linux platforms. Porting to other systems, languages, etc. is also straightforward. Many of the computationally demanding procedures were ported to concurrent versions using OpenMP, as discussed in Appendix A.6. Hence, they take full advantage of multicore and multiprocessor platforms especially. Last but not least, the library is layered to allow easy change of implementation while keeping representation, for example to take advantage of graphic cards.

---

[7]A MATLAB® toolbox from Sandia National Laboratory.

Table A.4 MATLAB® functions eigendecomposition, singular value decomposition of matrices, and PCA.

Function	Description	Example
[E, D] = eig(M)	Computes eigendecomposition of the matrix **M**. The eigenvectors go to the columns of matrix **E**, the eigenvalues to **D**. It holds that: **M*****E**=**E*****D**	>> M = [2, 4; 13, 1]; >> [E, D] = eig(M) E = 0.5110 −0.4597 0.8596 0.8881 D = 8.7284 0 0 −5.7284
[S,V,D] = svd(M)	Computes the singular value decomposition of the matrix **M**. In result: **M** = **S*****V*****D'**, where	>> M = [0 1 15; −2 3 11; 4 4 0]; >> [S, V, D] = svd(M); >> S S = 0.7949 0.0313 −0.6060 0.6065 −0.0124 0.7950 0.0174 −0.9994 −0.0288 >> V V = 18.8224 0 0 0 5.6502 0 0 0 2.4072 >> D D = −0.0608 −0.7032 −0.7084 0.1426 −0.7086 0.6911 0.9879 0.0590 −0.1433 >> >> MM = S ∗ V ∗ D'; >> e = norm(M - MM, 'fro') e = 1.1381e-014

(continued)

Table A.4   (*Continued*)

Function	Description	Example

[C, S] = **princomp(M)**	For the $N \times L$ matrix **M** containing L-dimensional data as its rows, the function does PCA decomposition and returns the $L \times L$ matrix **C** with the principal components in columns, ordered in decreasing variance.
	
	    The optional $N \times L$ matrix **S** contains data transformed to the PCA space
		>> M = rand(100, 3);
		>> [C, S] = princomp(M);
		>> C
		
		C =
		
		    0.6355 −0.4546 0.6241
		    0.4205 0.8817 0.2140
		    0.6476 −0.1265 −0.7514
		
		>> S
		
		S =
		
		    0.1674 0.1804 0.1475
		   −0.0547 0.2291 −0.0332
		    0.3165 −0.7147 −0.0701
		   −0.1634 0.4376 0.0196
		    0.2044 0.3005 0.2314
		    ... ...

Table A.5   MATLAB® functions for computation of the matrix exponent, natural logarithm, and Frobenius norm.

Function	Description	Example
norm(M, 'fro')	Computes the Frobenius norm of a matrix	>> M = [2, 3; −1, 15]; >> norm(M, 'fro'); ans = 15.4596

expm(M)	Computes exponent of the matrix **M**	>> M = [1 2 3; 0 −7 13; 22 −5 4];
		>> expm(M)
		
		ans =
		
		    1.0e+004 *
		
		    2.0613 −0.0209 0.8980
		    3.2612 −0.0330 1.4207
		    4.4112 −0.0447 1.9217

logm(M)	Computes natural logarithm of the matrix M; However, unique computation of the principal logarithm requires nonnegative eigenvalues of M	>> logm(M)
		
		ans =
		
		    1.9014 0.5978 −0.2279
		   −3.9463 2.1458 2.0019
		    0.9594 −1.3771 2.9291

Table A.6   Basics on multidimensional arrays (tensors) in MATLAB®.

Function	Description	Example
construction	Multidimensional arrays are created as a single 2D array, then further dimensions can be added	>> M = [1 2; 3 4]; %% first slice >> M(:, :, 2) = [10, −5; 2 4]; %% 2nd slice, etc. >> M M(:,:,1) = 1 2 3 4 M(:,:,2) = 10 −5 2 4
cat(d, A, B, …)	Creates a multidimensional array concatenating matrices A, B, … along the dimension d	>> M = cat(3, [1 2; 3 4], [10 −5; 2 4]);
size(M)	Returns dimensions of **M**	>> size(M) ans = 2 2 2
ndims(M)	Returns number of dimensions M	>> ndims(M) ans = 3

A.9   Closure

A.9.1   Chapter Summary

This appendix contained a discussion of subjects which are supplementary to the main chapters of this book. These can be split into two groups: mathematical and programming methods. Thus, they concern both theory and practice.

First the morphological scale-space was presented, followed by a discussion on morphological operations on tensors. These methods were used in discussed recognition methods. However, their application abilities are not restricted to those.

A short introduction to the geometry of quadratic forms was aimed at providing some useful equations of this simple concept to facilitate their usage in other methods, such as the mean-shift tracking discussed in Section 3.8.

The next section gathered useful information on testing the quality of classifiers. Specific definitions of the most common performance parameters and measures were collected and explained. ROC plots were also discussed. Implementation of the confusion matrix, as well as the procedure of comparing reference and test patterns followed.

Table A.7 Basic MATLAB® functions for tensor representation and decompositions from the package [35].

Function	Description	Example
tensor(M)	Creating tensor object from multidimensional array **M**	>> M = cat(3, [1 2; 3 4], [10 −5; 2 4]); >> T = tensor(M); >> T T is a tensor of size 2 × 2 × 2 T.data = (:,:,1) = $\quad$ 1 2 $\quad$ 3 4 (:,:,2) = $\quad$ 10 −5 $\quad$ 2 $\quad$ 4
S = ttm(T, M, k)	Computes a product of a tensor T with a matrix M in k-th mode, i.e. **S = T×$_k$ M** (the function has further options for multiplication of series of matrices)	>> T = tensor(cat(3, [−2 1; 0 5], [−7 2; 64 1])); >> M = [1 0; 2 −3]; >> S = ttm(S, M, 1);
S = ttv(T, v, k)	Computes a product of a tensor **T** with a vector v in k-th mode, i.e. **S = T×$_k$ v**	>> T = tensor(cat(3, [−2 1; 0 2], [−1 2; 4 1], $\qquad\qquad\qquad$ [0 4; 1 0])); >> v = [2 3]; >> S = ttv(T, v', 1) S is a tensor of size 2 × 3 S.data = $\quad$ −4 10 3 $\qquad$ 8 $\,$ 7 8
W = ttt(S, T)	Computes the outer product (2.15) of tensors **S** and **T**	>> S = tensor(cat(3, [1 −2; 0 1], [3 0; 7 −2], $\qquad\qquad\qquad$ [0 −8; 11 −11])); >> T = tensor(cat(3, [−2 1; 0 2], [−1 2; 4 1], [0 4; 1 0])); >> W = ttt(S, T); %% 6D tensor
U = ttt(S, T, [1:K], [1:K])	Computes the contracted product (2.16) of tensors **S** and **T** in respect to the indices {1 ... K}	>> S = tensor(cat(3, [1 −2; 0 1], [3 0; 7 −2], $\qquad\qquad\qquad$ [0 −8; 11 −11])); >> T = tensor(cat(3, [−2 1; 0 2], [−1 2; 4 1], [0 4; 1 0])); >> U = ttt(S, T, [1:2], [1:2]) U is a tensor of size 3 × 3 U.data = $\quad$ −2 $\quad$ −4 $\quad$ −8 $\,$ −10 $\quad$ 23 $\quad\;$ 7 $\,$ −30 $\quad$ 17 −21

Table A.7   (*Continued*)

Function	Description	Example
c = **ttt(S, T, [1:N])**	Computes the inner product (2.17) of tensors **S** and **T**	>> S = tensor(cat(3, [1 −2; 0 1], [3 0; 7 −2], [0 −8; 11 −11])); >> T = tensor(cat(3, [−2 1; 0 2], [−1 2; 4 1], [0 4; 1 0])); >> c = double(ttt(S, T, [1:ndims(S)])); >> c c = 0
S = **hopm(T)**	Computes the best rank−1 tensor approximation "CP" (2.232). Result S is a special *cp_tensor* object	>> T = tensor(rand(2, 5, 3)) >> S = hopm(T) S is a CP tensor of size 2 × 5 × 3 S.lambda = 2.9787 S.U{1} = −0.7714 −0.6364 S.U{2} = −0.3833 −0.3698 −0.4114 −0.4603 −0.5790 S.U{3} = 0.5301 0.5333 0.6592
S = **hooi(T, r)**	Computes the best rank−$(R_1, R_2, \dots, R_P)$ approximation "Tucker" (2.251) of tensor T and the specified dimension in vector *r*. The result S is a special *tucker_tensor* object	>> T = tensor(rand(2, 5, 3)) >> S = hooi(T, [2 2 1]) S is a Tucker tensor of size 2 × 5 × 3 S.lambda is a tensor of size 2 × 2 × 1 , S.lambda.data = −2.9784 −0.0000 0.0000 −0.5500 S.U{1} = −0.7708 −0.6371 −0.6371 0.7708 S.U{2} = −0.3821 0.1077 −0.3707 0.6461 −0.4082 −0.1269 −0.4617 −0.7227 −0.5803 0.1806 S.U{3} = −0.5363 −0.5413 −0.6477

The second, implementation part, of this chapter started with a short introduction to the OpenMP library which allowed a straightforward refactoring of serial code into multithreaded. The most common work sharing constructs and their specifying clauses were discussed with examples. Also a number of hints on using OpenMP with object-oriented components was provided. The library allows speed up compared to the serial implementation. However, special attention was paid to avoiding some pitfalls and hazards in concurrent applications, as discussed.

Finally, a set of useful MATLAB® functions for matrix and tensor processing was provided. These require some practice in basic MATLAB® constructions which is not difficult to acquire. The chapter concluded with a brief introduction to the attached software.

A.9.2 Further Reading

Morphological operators are dealt with in the books by Soille [4] and by Ritter and Wilson [38]. Good handbooks on a variety of mathematical methods are Moon and Stirling [9], as well as Press *et al.* [10]. The latter also provides discussion on some important classifiers such as SVM.

Methods of experiment design and testing classifiers are dealt with in the books by Alpaydin [16] and the one by Rokach [11] for ensembles of classifiers. One of the most comprehensive texts on ROC is the paper by Fawcett [13].

An overview of methods, good practice, and some patterns are discussed in the book by Cyganek and Siebert [2]. Description of the OpenCV library can be found in the book by Bradski and Kaehler [39]. Code refactoring is dealt with in the book by Fowler *et al.* [40]. A good guide to OpenMP is the book by Chapman *et al.* [21], as well as manuals on the OpenMP web page [22]. In respect of code acceleration the commonly available graphic cards (GPU) constituted a real breakthrough, for example the ones by nVidia® with their CUDA programming platform [20]. Good reads on the latter are the books by Sanders and Kandrot [41], by Farber [42], as well as numerous manuals and examples on the nVidia® web page [20].

Finally, recommended reading on MATLAB® programming are the books by Gilat [29] and by Attaway [30].

At this point we would like to highly recommend the Python programming language as an alternative tool for scientific computations. This is clear object-oriented language with dozens of useful libraries, and last but not least – all these are free. For computational purposes Python is augmented by the *NumPy* [43] and *SciPy* [44] libraries for numerical and scientifical computations, respectively. Especially impressive is the presence of advanced methods of numerical algebra. For image processing the Python Imaging Library (PIL) [45] can be recommended, as well as the ImageMagick® [46], which can be used to create, edit, make bitmap images, etc. Finally, myriads of amazing scientific plots can be created with help of the *matplotlib* package [47]. Suggested reading on Python in scientific computing is the good book by Langtangen [48], as well as the one by Kiusalaas [49]. Further details on Python are provided in the book by Summerfield [50], as well as on the Python web page [51]. Time spent learning Python will definitely not be wasted.

520   Object Detection and Recognition in Digital Images

Problems and Exercises

1. Write serial implementations of the morphological erosion and dilation discussed in Section A.2. Then write their OpenMP versions. Build the morphological scale-space from some test images. Measure differences in performance on your computer.
2. Convert the morphological median filter discussed in Section 2.6.1 to OpenMP to take benefit from the concurrent run. Measure differences in performance.
3. Write a function for image convolution. Measure its execution time. Port that version to OpenMP and measure performance again.
4. Consider writing exhaustive tests to check if your two versions of convolution are correct.
5. Compile and run three versions of the *GetImageVarianceValue* with code listings in Algorithm A.5 – Algorithm A.7. Compare their results.

References

[1] Jackway P.T., Deriche M.: Scale-Space Properties of the Mutliscale Morphological Dilation-Erosion. IEEE PAMI, Vol. 18, No. 1, pp. 38–51, 1996.
[2] Cyganek B., Siebert J.P.: An Introduction to 3D Computer Vision Techniques and Algorithms, Wiley, 2009.
[3] Mitra S.K., Sicuranza G.L.: Nonlinear Image Processing. Academic Press, 2000.
[4] Soille P.: Morphological Image Analysis. Principles and Applications. Springer, 2003.
[5] Burgeth B., Bruhn A., Papenberg N., Welk M., Weickert J.: Mathematical morphology for matrix fields induced by the Loewner ordering in higher dimensions. Signal Processing, Vol. 87, No. 2, pp. 277–290, 2007.
[6] Golub G.H., Van Loan C.F.: Matrix Computations. The Johns Hopkins University Press, 3rd edition 1996.
[7] Meyer C.D.: Matrix Analysis and Applied Linear Algebra. SIAM, 2000.
[8] Grimmet G., Stirzaker D.: Probability and Random Processes. Third Edition, Oxford University Press, 2001.
[9] Moon T.K., Stirling W.C.: Mathematical Methods and Algorithms for Signal Processing. Prentice-Hall 2000.
[10] Press W.H., Teukolsky S.A., Vetterling W.T., Flannery B.P.: Numerical Recipes in C. The Art of Scientific Computing. Third Edition. Cambridge University Press, 2007.
[11] Rokach L.: Pattern Classification Using Ensemble Methods. Series in Machine Perception and Artificial Intelligence, Vol. 75, World Scientific, 2010.
[12] Tax D.M.J.: One-class classification. PhD thesis, TU Delft University, 2001.
[13] Fawcett T.: An introduction to ROC analysis. Pattern Recognition Letters, Vol. 27, No. 8, pp. 861–874, 2006.
[14] Martin D.R., Fowlkes C.C., Malik J.: Learning to Detect Natural Image Boundaries Using Local Brightness, Color, and Texture Cues. IEEE Transaction on Pattern Analysis and Machine Intelligence, Vol. 26, No. 1, pp. 1–20, 2004.
[15] Martin D., Fowlkes C., Tal D., Malik J.: A Database of Human Segmented Natural Images and its Application to Evaluating Segmentation Algorithms and Measuring Ecological Statistics. Proc. 8th Int'l Conf. Computer Vision, pp. 416–423, 2001.
[16] Alpaydin E.: Introduction to Machine Learning. 2nd edition, MIT Press, 2010.
[17] http://www.wiley.com/go/cyganekobject
[18] Cyganek B.: Adding Parallelism to the Hybrid Image Processing Library in Multi-Threading and Multi-Core Systems. 2nd IEEE International Conference on Networked Embedded Systems for Enterprise Applications (NESEA 2011), Perth, Australia, pp. 103–110, 2011.
[19] http://www.xilinx.com
[20] https://developer.nvidia.com/
[21] Chapman B., Jost G., Van Der Pas A.R.: Using OpenMP. Portable Shared Memory Parallel Programming. MIT Press, 2008.
[22] http://www.openmp.org
[23] http://www.mpi-forum.org
[24] Gropp W., Lusk E., and Thakur R., Using MPI-2. Cambridge, MA: MIT Press, 1999.
[25] Meyer B.: Applying "Design by Contract". IEEE Computer, Vol. 25, No. 10, pp. 40–51, 1992.
[26] McConnell S.: Code Complete. 2nd edition. Microsoft Press, 2004.

[27] Kim H., Bond R.: Multicore Software Technologies. A survey. IEEE Signal Processing Magazine, Vol. 26, No. 6, pp. 80–89, 2009.

[28] http://www.mathworks.com

[29] Gilat A.: MATLAB®: An Introduction with Applications. Wiley; 3rd edition 2007.

[30] Attaway S.: MATLAB®: a Practical Introduction to Programming and Problem Solving. Butterworth-Heinemann, 2009.

[31] Kuncheva L.I.: Combining Pattern Classifiers. Methods and Algorithms. Wiley Interscience, 2005.

[32] Theodoridis S., Koutroumbas K.: Pattern Recognition. Fourth Edition. Academic Press, 2009.

[33] Cichocki A., Zdunek R., Phan A.H., Amari S-I.: Nonnegative Matrix and Tensor Factorizations. Applications to Exploratory Multi-way Data Analysis and Blind Source Separation. Wiley, 2009.

[34] http://www.prtools.org/

[35] Bader W.B., Kolda T.G.: MATLAB® Tensor Classes for Fast Algorithm Prototyping. ACM Transactions on Mathematical Software. Vol. 32, No. 4, pp. 635–653, 2006.

[36] Non-Negative Matrix Factorization Toolbox in MATLAB® (The NMF MATLAB® Toolbox). http://cs.uwindsor.ca/~li11112c/nmf.html, 2012.

[37] http://www.wiley.com/CyganekSiebertBook.html

[38] Ritter G., Wilson J.: Handbook of Computer Vision Algorithms in Image Algebra, CRC Press, 2001.

[39] Bradski G., Kaehler A.: Learning OpenCV. Computer Vision with the OpenCV Library. O'Reilly, 2008.

[40] Fowler M., Beck K., Brant J., Opdyke W., Roberts D.: Refactoring: Improving the Design of Existing Code. Addison-Wesley, 1999.

[41] Sanders J., Kandrot E.: CUDA by Example. An Introduction To General-Purpose GPU Programming. Addison-Wesley, 2011.

[42] Farber R.: CUDA Application Design and Development. Morgan Kaufman, 2011.

[43] http://www.numpy.org/

[44] http://www.scipy.org/

[45] http://www.pythonware.com/products/pil/

[46] http://wiki.python.org/moin/ImageMagick

[47] http://matplotlib.org/

[48] Langtangen H.P.: A Primer on Scientific Programming with Python. Springer, 2009.

[49] Kiusalaas J.: Numerical Methods in Engineering with Python. 2nd edition, Cambridge University Press, 2010.

[50] Summerfield M.: Programming in Python 3: A Complete Introduction to the Python Language. Addison-Wesley, 2009.

[51] http://www.python.org/

Index

Object Detection and Recognition in Digital Images: Theory and Practice, First Edition. Bogusław Cyganek.
© 2013 John Wiley & Sons, Ltd. Published 2013 by John Wiley & Sons, Ltd.

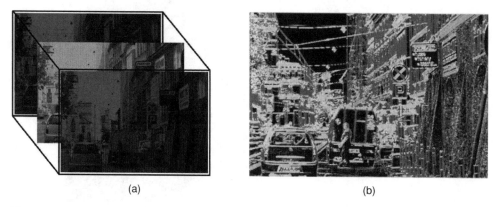

(a)   (b)

Figure 1.3 A color image can be seen as a 3D structure (a). Internal properties of such multidimensional signals can be analyzed with tensors. Local structures can be detected with the structural tensor (b). Here different colors encode orientations of areas with strong signal variations, such as edges. Areas with weak texture are in black. These features can be used to detect pedestrians, cars, road signs and other objects.

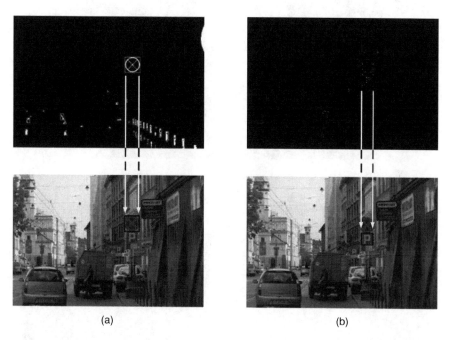

(a)   (b)

Figure 1.5 Circular signs are found by outlining all red objects detected in the scene. Then only those which fulfill the definition and relative size expected for a sign are left (a). Triangular and rectangular shapes are found based on their corner points. The points are checked for all possible rectangles and again only those which comply with fuzzy rules defining sought figures are left (b).

Object Detection and Recognition in Digital Images: Theory and Practice, First Edition. Bogusław Cyganek.
© 2013 John Wiley & Sons, Ltd. Published 2013 by John Wiley & Sons, Ltd.

Figure 2.2   An image and its three color channels as an example of a 3D tensor.

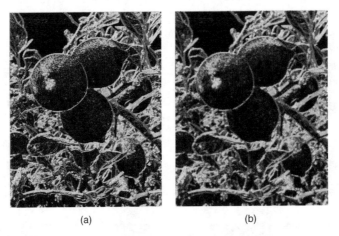

(a)   (b)

Figure 2.7   Color visualization of the structural tensor (2.7) computed from a region in Figure 2.2 in the HSV color encoding (a). Its 5×5 median filtered version (b).

(a)   (b)

Figure 2.10   Exemplary color image (a). Three components of a 2D structural tensor encoded into the HSV color. Phase of the orientation vector **w** represented by the hue channel (H), coherence by saturation (S), and trace with value (V).

(a)   (b)

Figure 2.14 ST of Figure 2.10(a) computed in different scales. Directional filters of $\xi = 5$ and $\rho = 5$ (a), $\xi = 5$ and $\rho = 9$ (b).

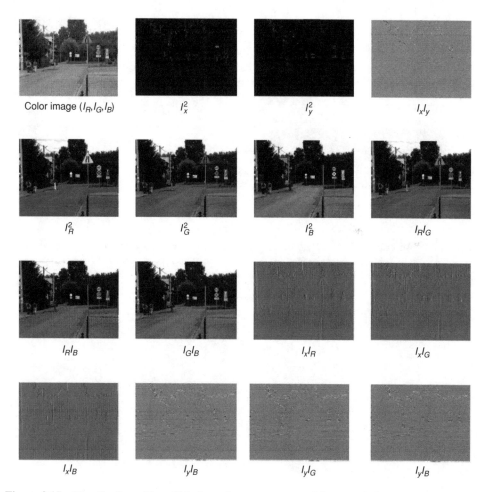

Figure 2.15 Visualization of $k = 15$ independent components of the extended structural tensor computed for the color image shown in the upper left pane.

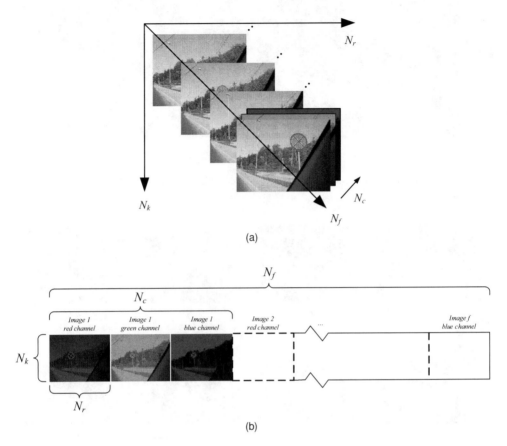

(a)

(b)

Figure 2.24 An example of a 4D tensor representing a traffic video scene with color frames (a). A practical way of tensor flattening – the forward mode – which maps images to the memory in the scanning order (b).

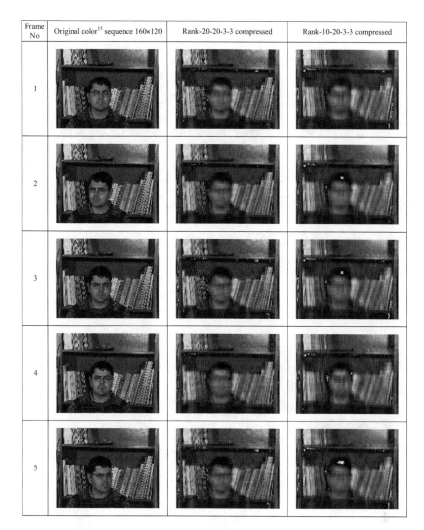

Frame No	Original color[15] sequence 160×120	Rank-20-20-3-3 compressed	Rank-10-20-3-3 compressed
1			
2			
3			
4			
5			

Figure 2.37 Visualization of color sequences of a person from the Georgia Tech face database [141]. Original color sequence (left). Rank -20-20-3-3 compressed version (middle), rank-10-20-3-3 compressed (right).

(a)   (b)   (c)

Figure 3.5 Reconstructed images from the three (a), two (b), and one (c) principal components of the original image from Figure 3.4a. Reconstruction PSNR values: 53.13 dB, 32.36 dB, and 27.08 dB, respectively.

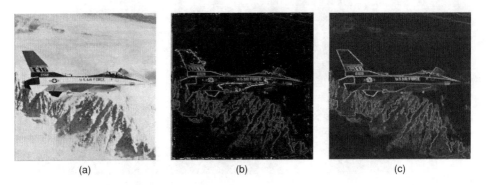

Figure 3.6 A test color image (a) and its morphological gradients computed in each RGB color channel with arbitrary order (b). The same operation but in the PCA decomposed space in order of eigenvalues and after reconstruction (c).

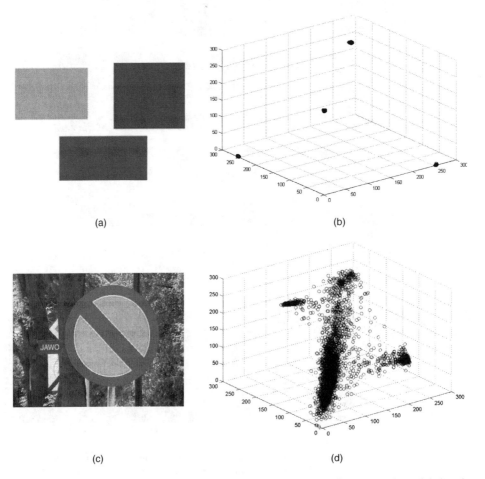

Figure 3.26 Exemplary color images (left column) and the corresponding scatterplots of their colors in the RGB color space (right column).

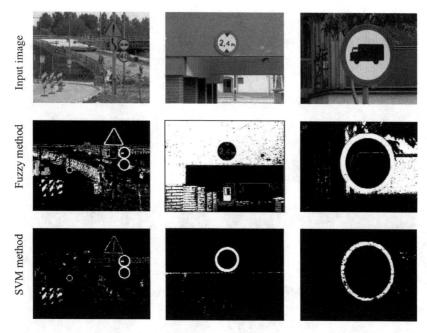

Figure 4.11 Comparison of image segmentation with the fuzzy method (middle row) and the one-class SVM with RBF kernel (lower row) (from [2]).

(a)

(b)

(c)

(d)

Figure 4.14   Red-and-blue color image segmentation with the ensemble of WOC-SVMs trained with the manually selected color samples. An original 640×480 color image of a traffic scene (a). Segmentation results for $M = 1$ (b), $M = 2$ (c), and $M = 5$ (d) (from [26]).

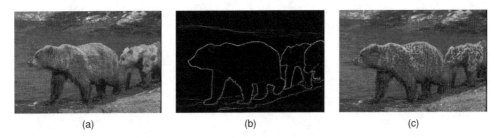

(a)

(b)

(c)

Figure 4.15   A 481×321 test image (a) and manually segmented areas of image from the Berkeley Segmentation Database [27] (b). Manually selected 923 points from which the RGB color values were used for system training and segmentation (c), from [26]. Color versions of the images are available at the book web page [28].

(a)   (b)

Figure 4.22   Detection of ovals in a real image (a). Detected points with the method for the allowable phase change set to $N = 400$ (b).

Figure 4.41   Tracking of a boat in a color video sequence. Fuzzy field initialized once in the first frame (upper left) which defines an object to track. The IJK space provides features to track.

Figure 4.42 Boat tracking with the IJK and structural tensor components. More precise tracking of an object due to more descriptive features.

Figure 4.44 Tracking results of a red car in a traffic sequence with the IJK and ST components. User drawn white rectangle defines an object to track (left image).

Figure 5.39 Steps of recognition of the STOP road sign. Real traffic scene (a). A region of a sign detected in the red color segmented map (b). Blue channel used in the recognition stage (c). Correctly recognized STOP sign outlined in the scene (d).

Figure 5.41 Steps in recognition of the two prohibition signs. A traffic scene (a). Regions of separate red circles detected in the color segmented map (b). Blue channel used in the recognition stage (c). Correctly recognized and outlined signs "No Parking" and "No Left Turn" (d).

(a)　　　　　　　　　　　　　　(b)

(c)　　　　　　　　　　　　　　(d)

Figure 5.42 Recognition of the obligation sign from a road scene (a). A blue object detected from the color segmented map (b). Red channel used in recognition stage (c). Correctly recognized "Turn Right" obligation sign (d).

Printed in the United States
By Bookmasters